Diversity

of

life

▶ **Plants: our source of oxygen.**

▶ **Energy from the sun.**

▶ **Decomposers: the original recyclers.**

▶ **How many humans can the Earth withstand?**

▶ **Ecosystems: complex systems of living things and their surroundings.**

chapter

6

Ecosystem Organization & Energy Flow

Stream-side ecosystem.

learning objectives

- Recognize the relationships that organisms have to one another in an ecosystem.
- Understand that useful energy is lost as energy passes from one trophic level to the next.
- Appreciate that it is difficult to quantify the energy flow through an ecosystem.
- List characteristics of several biomes.
- Understand that humans have converted natural ecosystems to human use.

Ecology and Environment

Today we hear people from all walks of life using the terms *ecology* and *environment*. Students, homemakers, politicians, planners, and union leaders speak of "environmental issues" and "ecological concerns." Many consumer products are advertised as being "environmentally friendly" or "earth friendly." Many "environmental" organizations champion specific causes. States have departments of the environment and the U.S. federal government has an Environmental Protection Agency (EPA). "Ecotourism" is advertised by the tourism offices of many states and countries. It is often difficult to determine the exact meaning of many of these new phrases, but underlying all of them is the understanding that in nature there are many interactions, and changing one aspect of a local ecological system could have profound changes that were not anticipated. It is important that we understand the original meanings of these two popular words, ecology and environment.

Ecology is the branch of biology that studies the interrelationships between organisms and their environment. This is a very simple definition for a very complex branch of science. The underlying idea is that living things interact with their surroundings and are influenced by their surroundings while at the same time changing them. These "surroundings" are often called the "environment." Most ecologists define the word **environment** broadly as anything that affects an organism during its lifetime. These environmental influences can be divided into two categories: other living things

that serve as food, shelter, enemies, or helpers; and physical factors such as temperature, rainfall, sunlight, and pH (figure 6.1). For example, if we consider a fish in a stream, we can identify many environmental factors that are important to its life. The temperature of the water is an important physical factor, but it may be influenced by trees living along the bank that shade the stream and prevent the sun from heating it. The kind and number of food organisms in the stream are also important living portions of the environment. The type of material that makes up the stream bottom and the amount of oxygen dissolved in the water are other important physical factors.

As you can see, describing the environment of an organism is not a simple task; everything seems to be influenced or modified

CONCEPT CONNECTIONS

ecology The branch of biology that studies the relationships between organisms and their environment.

environment Anything that affects an organism during its lifetime.

ecosystem An interacting collection of organisms and the nonliving factors that affect them.

by other factors. A plant is influenced by the types and amounts of minerals in the soil, the amount of sunlight hitting the plant, the animals that eat the plants, and the wind, water, and temperature, among other things. Each item on this list can be further subdivided. For instance, water is important to plants, so rainfall is studied in plant ecology. But even the study of rainfall can be complex. Rain could come during one part of the year and be followed by drought, or the rain could be evenly distributed throughout the year and drought would be uncommon. The rainfall could be a hard, driving rain that would run off the surface, or it could come as gentle showers of long duration that would soak into the soil.

Temperature is also important to the life of a plant. For example, two areas of the world can have the same average daily temperature of 10°C (50°F) but have different plants because of different temperature extremes. In one area, the temperature may be 13°C (55.5°F) during the day and 7°C (44.5°F) at night, for a 10°C average. In another area, the temperature may be 20°C (68°F) during the day and only 0°C (32°F) at night, for the same 10°C average. But plants react to extremes in temperature as well as to the daily average, so the first area may have different plants

(a)

(b)

Figure 6.1 **Living and Physical Environmental Factors** (*a*) The woodpecker feeding its young in the hole in this tree is influenced by several factors. The tree itself is a living factor as is the disease that weakened it, causing conditions that allowed the woodpecker to make a hole in the rotting wood. (*b*) The irregular shape of the trees is the result of wind and snow, both physical factors. Snow driven by the prevailing winds tends to "sandblast" one side of the tree and prevent limb growth.

Photosynthesis and Respiration

Although the process of photosynthesis involves many steps, it can be summarized as follows:

| Sunlight energy captured by leaves | + | Carbon dioxide (CO_2) gas from the atmosphere entering leaves | + | Water (H_2O) from the soil transported to leaves | → | Sugar ($C_6H_{12}O_6$) high in chemical energy | + | Oxygen (O_2) gas released into the atmosphere from leaves |

All organisms obtain energy from organic molecules through the process of respiration, which can be summarized as follows:

| Organic molecules ($C_6H_{12}O_6$) from stored molecules in organisms or from food | + | Oxygen (O_2) gas from the atmosphere | → | Carbon dioxide (CO_2) released to the atmosphere | + | Water (H_2O) released to the atmosphere | + | ATP, a form of chemical energy |

than the second. Furthermore, different parts of a plant may respond differently to temperature. For example, tomato plants will grow at 13°C but require warmer temperatures to develop fruit.

Animals are influenced by physical factors and by other living organisms, which are also influenced by physical factors. If the physical environment is too harsh or does not meet an animal's needs, the animal will die or move on to an area where it can live successfully. If nonliving factors do not favor the growth of plants, animals will find little food and few hiding places. Two areas that support only small numbers of living animals are deserts and polar regions. Near the polar regions, the physical factors of low temperature and a short growing season inhibit plant growth, so there are few species of animals with relatively small numbers of individuals. Deserts receive little rainfall and therefore have poor plant growth and low concentrations of animals. On the other hand, tropical rainforests have high rates of plant growth and large numbers of animals of many kinds.

The Organization of Living Systems

Energy is the ability to do work, the ability to cause something to move. Heat energy causes things to expand; electrical energy causes motors to turn; and the kinetic energy of moving objects causes other objects to move when collisions occur. Our society uses large amounts of energy. Light, heat, motion, and electricity are forms of energy we regularly encounter. We use electrical energy to generate light, run motors, and heat food. We convert the energy of chemical reactions into other forms of energy such as heat and light from a campfire, or motion and heat from an automobile.

All living things require a continuous supply of energy to maintain life. In living systems, light energy from the sun and chemical energy from food allows living things to grow and move as well as carry on other important life functions such as reproduction.

An **ecosystem** is all of the interacting organisms in an area and the physical factors that influence them. Ecologists can take an ecosystem apart and examine it from many

points of view. One view is to study energy flow within an ecosystem. Light energy is trapped by photosynthesis and converted into the chemical energy of complex organic molecules, like sugar. Plants use the light energy to produce organic compounds from inorganic matter; thus they are called **producers.** There is a flow of energy from the sun into the living matter of plants.

You are aware that you need to eat in order to obtain the energy needed to stay alive. Organisms that are deprived of a source of energy die. Energy that plants trap can be used by the plant or it can be transferred to other organisms. The common mechanism by which organisms, including plants, get the energy they need for day-to-day activities is known as *respiration.* Respiration involves converting the chemical bond energy of food molecules into other energy-rich molecules that can be used for movement, growth, and all other energy-requiring processes.

Since all organisms that do not do photosynthesis must obtain energy by consuming organic matter, they are called **consumers.** Consumers cannot capture light energy. They eat plants directly or feed on animals that used plants for energy.

Each time the energy enters a different organism, it is said to enter a new **trophic level,** which is a step, or stage, in the flow of energy through an ecosystem (figure 6.2). The plants (producers) receive their energy directly from the sun and are said to occupy the *first trophic level.* Consumers can be divided into

CONCEPT CONNECTIONS

The word *trophic* comes from the Greek word *trophi* meaning "to feed." An *autotroph* is able to produce its own food (*auto* = self), whereas a *heterotroph* must eat (*hetero* = different).

several categories, depending on how they fit into the flow of energy. Animals such as rabbits, cattle, and caterpillars that feed exclusively on plants are called **herbivores;** they occupy the *second trophic level.* Animals that eat other animals are called **carnivores.** They can be subdivided into different trophic levels depending on what animals they eat. Animals that feed on herbivores occupy the *third trophic level.* Animals that feed on carnivores occupy the *fourth trophic level.* Many animals do not fit neatly into these levels; they may occupy different trophic levels depending on

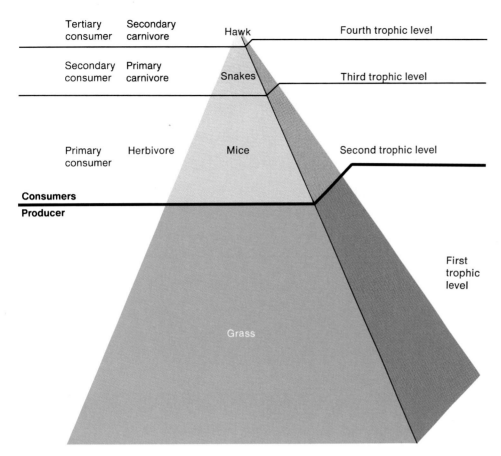

Figure 6.2 The Organization of an Ecosystem Organisms within ecosystems can be divided into several different trophic levels based on how they obtain energy. Several sets of terminology are used to identify these different roles. This illustration shows how the sets of terminology are related to one another.

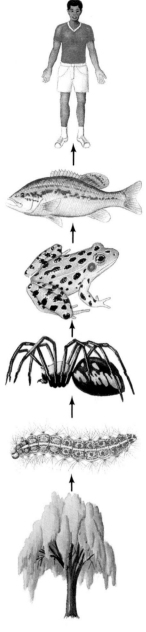

Trophic Levels in a Food Chain As one organism feeds on another organism, energy flows from one trophic level to the next. This illustration shows six trophic levels.

what they are eating at any given time. For example, when you are eating an apple, you are at the second trophic level, and when eating beef you are at the third trophic level. In fact, your level may vary even more!

Some organisms, like humans, rats, robins, and opossums, don't fit neatly into this theoretical scheme. These animals are carnivores at some times and herbivores at others and thus are called **omnivores.** They are classified into different trophic levels depending on what they happen to be eating at the moment.

When an organism dies, its body rots or decomposes. It is quickly populated by numerous organisms that use dead organic matter as a source of food. They use the energy contained within the organic compounds of the dead body for their own purposes. This results in its decomposition and, therefore, these organisms that break down dead material are called **decomposers.** They break down the dead body into carbon dioxide, water, ammonia, and

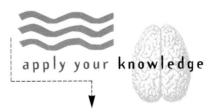

apply your knowledge

Where Are You in the Chain?

A **food chain** is a sequence of organisms feeding upon one another. The following sequence is an example of a food chain: a human eating a fish that ate a frog that ate a spider that ate an insect that consumed plants for food.

At which trophic level would the human be in the above example?

At which trophic level is the spider in this same sequence?

Are humans ever at the first trophic level? What about vegetarians?

What types of organisms other than plants are at the first trophic level?

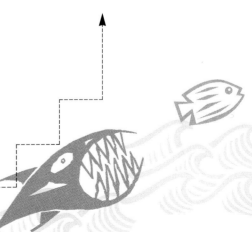

Detritus Food Chains

Although most ecosystems receive energy directly from the sun through the process of photosynthesis, some ecosystems obtain most of their energy from a constant supply of dead organic matter. For example, forest floors and small streams receive a rain of leaves and other bits of material that small animals use as a food source. The small pieces of organic matter, such as broken leaves, feces, and body parts, are known as *detritus*. The insects, slugs, snails, earthworms, and other small animals that use detritus as food are often called *detritivores*.

In the process of consuming leaves, detritivores break the leaves and other organic material into smaller particles that may be used by other organisms for food. The smaller size also allows bacteria and fungi to more effectively colonize the dead organic matter, further decomposing it and making it available to still other organisms as a food source. The bacteria and fungi are in turn eaten by other detritus feeders. Some biologists believe that we greatly underestimate the energy flow through detritus food chains.

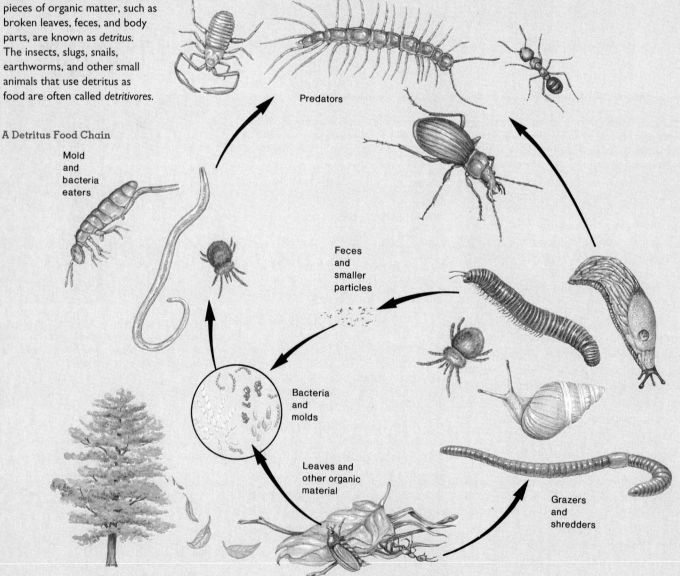

A Detritus Food Chain

Mold and bacteria eaters

Predators

Feces and smaller particles

Bacteria and molds

Leaves and other organic material

Grazers and shredders

table 6.1

Roles in an Ecosystem

Classification	Description	Examples
Producers	Plants that convert simple inorganic compounds into complex organic compounds by photosynthesis.	Trees, flowers, grasses, ferns, mosses, algae
Consumers	Organisms that rely on other organisms as food. Animals that eat plants or other animals.	
Herbivore	Eats plants directly.	Deer, goose, cricket, vegetarian human, many snails
Carnivore	Eats meat.	Wolf, pike, dragonfly
Omnivore	Eats plants and meat.	Rat, most humans
Scavenger	Eats food left by others.	Coyote, skunk, vulture, crayfish
Parasite	Lives in or on another organism, using it for food.	Tick, tapeworm, many insects
Decomposers	Organisms that return organic compounds to inorganic compounds. Important components in recycling.	Bacteria, fungi

other simple inorganic molecules. Common decomposers include bacteria and fungi.

Decomposers efficiently convert nonliving organic matter into simple inorganic molecules that can be used by producers in the process of trapping energy. Decomposers are thus very important components of ecosystems because they recycle materials. As long as the sun supplies the energy, elements are cycled through ecosystems repeatedly. Table 6.1 summarizes the various categories of organisms within an ecosystem.

Figure 6.3 illustrates a forest ecosystem. Can you identify producers, herbivores, carnivores, scavengers, and decomposers? Now that we have a better idea of how ecosystems are organized, we can look more closely at energy flow through ecosystems.

The Great Pyramids: Number, Biomass, and Energy

Certain fundamental rules determine how we interreact with the world around us. One rule involves the production of energy. The food that an animal eats is converted into energy for movement, heat, growth, and reproduction.

Figure 6.3 **A Forest Ecosystem**   This illustration shows many of the organisms in a forest ecosystem. Can you identify the trophic level of each organism within the ecosystem?

Growth and reproduction result in body tissue, which eventually becomes food for other organisms. Not all of the food energy consumed by an organism is converted to its body parts; much of it is lost as motion and heat. Consequently, less energy is available to organisms at higher trophic levels.

When we look at how the various trophic levels of ecosystems are related to one another we see that they form a *pyramid.* You probably are already aware of the pyramidal relationship that exists within ecosystems, though you might not have recognized it. Plants (producers) outnumber animals, and carnivores are always fewer than herbivores. As we go to higher and higher trophic levels, the numbers of organisms get smaller and smaller. The organisms that live at the highest trophic levels are the most rare in an ecosystem. Sharks are more rare than mackerel, wolves are more rare than moose, and snakes are more rare than rodents.

There are several ways to describe how trophic levels in an ecosystem are related to one another. Here we will examine three: the numbers of organisms in each trophic level, the total amount of living material present in each trophic level, and the energy flow through each trophic level.

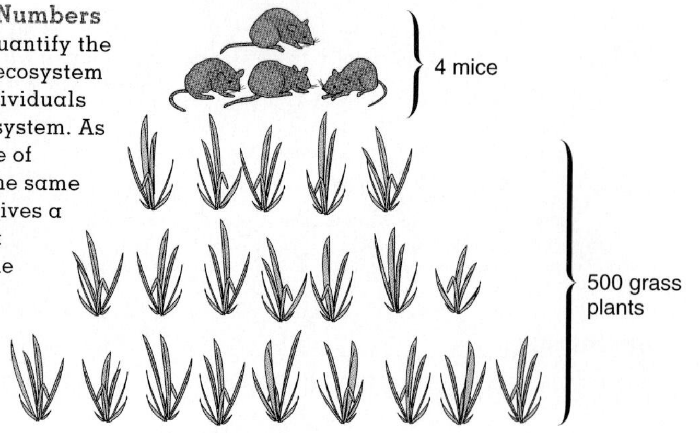

The Pyramid of Numbers

One useful way to examine relationships among trophic levels in an ecosystem is to simply count the number of organisms at each trophic level. This is called a *pyramid of numbers* (figure 6.4). Producers are placed on the bottom, herbivores above them, and carnivores on top. Unfortunately, this is not a good method to use if the organisms at the different trophic levels are of greatly differing size. For example, if you count all the small insects feeding on the leaves of one large tree, you would get an inverted pyramid.

The Pyramid of Biomass

Because of the size-difference problem, many biologists use biomass to measure trophic levels in ecosystems. **Biomass** is usually determined by collecting a sample of the organisms at one trophic level and measuring their dry weight. Weighing the organisms eliminates the size-difference problem. Although a biomass pyramid is better than a pyramid of numbers for measuring some ecosystems, it has shortcomings. For example,

Figure 6.4 A Pyramid of Numbers
One of the easiest ways to quantify the various trophic levels in an ecosystem is to count the number of individuals in a small portion of the ecosystem. As long as all the organisms are of similar size and live about the same length of time, this method gives a good picture of how different trophic levels are related. The relationship between grass and mice is a good example.

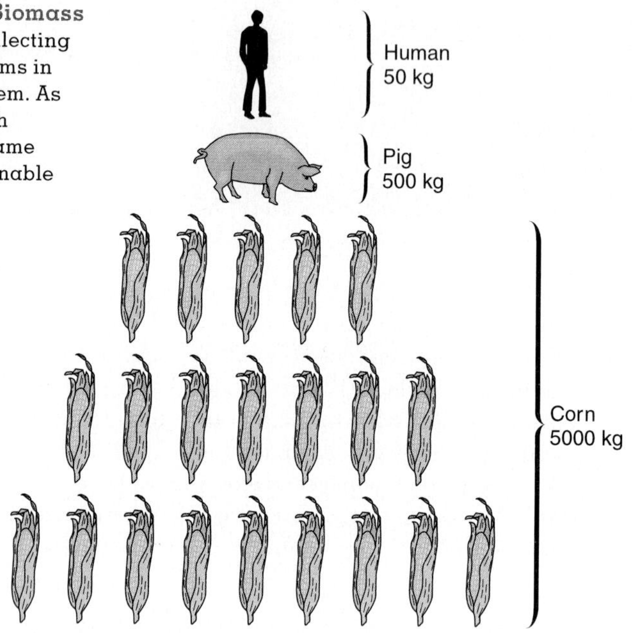

4 mice

500 grass plants

Figure 6.5 A Pyramid of Biomass
Biomass is determined by collecting and weighing all the organisms in a small portion of an ecosystem. As long as the organisms at each trophic level live about the same length of time, this is a reasonable way to estimate the size of different trophic levels.

Human 50 kg

Pig 500 kg

Corn 5000 kg

some organisms accumulate biomass over long periods, while others do not. Many trees live for hundreds of years, while their primary consumers, insects, generally live only one year. Likewise, a whale is a long-lived animal, while the organisms it consumes are relatively short-lived. Figure 6.5 shows a biomass pyramid.

The Pyramid of Energy

A third way to understand how trophic levels in ecosystems are interrelated is a *pyramid of energy.* At the base of this pyramid is the producer trophic level, which contains the most energy of any level. Energy can be measured in several ways. The producer trophic level can be harvested and burned. The calories of heat energy produced by burning equals the energy content of the organic material of the plants. Another way is to measure the rate of photosynthesis and respiration and calculate the amount of energy being trapped in the living material of the plants.

Since only plants are capable of capturing energy from the sun, all other organisms are directly or indirectly dependent on these producers. The second trophic level, herbivores that eat plants, has significantly less energy than does the first trophic level. This loss in energy is primarily due to a fundamental energy principle: Whenever energy is converted from one form to another, some energy is converted to useless heat. Think of any energy-converting machine and how

BIO *feature*

The Importance of Ecosystem Size

Many people interested in songbird populations have documented a significant decrease in the numbers of certain songbird species. Species particularly affected are those that migrate from North America to South America and require relatively large areas of undisturbed forest in both their northern and southern homes. Many of these species are being hurt by human activities that fragment large patches of forest into many smaller patches, creating more edges between different habitat types. Bird and other animal species that thrive in edge habitats replace the songbirds, which require large patches of undisturbed forest. The species most severely affected are those that have both their northern and southern habitats disturbed.

A survey of migrating sharp-shinned hawks indicated that their numbers have also been greatly reduced in recent years. It is thought that since sharp-shinned hawks use small songbirds as their primary source of food, the reduction in hawks is directly related to the reduction in migratory songbirds.

Scarlet Tanager

Indigo Bunting

Wood Thrusher

Human Use of Food Pyramids

Humans are omnivores that can eat both plants and animals as food, so we have a choice in our diet. As the size of the human population increases, however, we cannot afford the 90 percent loss that occurs when plants are fed to animals that are in turn eaten by humans. In much of the less developed world, the primary foods are starchy plants such as rice, corn, potatoes, yams, and cassava, placing people at the herbivore level. Only in the developed countries can people afford to make meat a major part of their diet. This is true from both an energy point of view and a monetary point of view.

Figure A (*a*) shows a pyramid of biomass having a producer base of 100 kilograms of grain. The second trophic level has only 10 kilograms of cattle because of the 90 percent loss typical when energy is transferred from one trophic level to the next. The consumers at the third trophic level, humans in this case, experience a similar 90 percent loss. Therefore, only 1 kilogram of humans can be sustained by the two-step energy transfer.

Figure A Human Biomass Pyramids
Since approximately 90 percent of the energy is lost as energy passes from one trophic level to the next, more people can be supported if they eat producers directly than if they feed on herbivores. Much of the less developed world is in this position today. Rice, corn, wheat, and other producers provide the majority of food for the world's people.

There has been a 99 percent loss in energy: 100 kilograms of grain are necessary to sustain 1 kilogram of humans.

Humans do not need to be carnivores at the third trophic level; we can switch most of our food consumption to the second trophic level. There would then be a 90 percent loss rather than a 99 percent loss, and the 100 kilograms of grain could support 10 kilograms of humans (figure A [*b*]). By eliminating cattle from the human food chain, ten times as much human life can be supported by the same amount of plant material. In parts of the world where food is scarce, people cannot afford the energy loss involved in passing food through the herbivore trophic level. Consequently, most people of the world are consumers at the second trophic level and rely on corn, wheat, rice, and other producers as food. Because much of the world's population is already feeding at the second trophic level, we cannot expect increases in food production that could feed ten times more people than exist today.

Most people cannot fulfill all of their nutritional needs by just eating grains. In addition to calories, people need a certain amount of protein in their diet—and one of the best sources of protein is meat. Although protein is available from plants, the concentration in animal sources is greater. Many people in Africa, Asia, and Latin America have diets that are deficient in both calories and protein. These people have very little food, and what food they do have is mainly from plant sources. In addition, they live in parts of the world where human population growth is most rapid. In other words, these people are poorly nourished, and as the population increases, they will probably experience greater calorie and protein deficiency. Thus, even when people live as consumers at the second trophic level, they may still not get enough food, and if they do, their diets may not have the protein necessary for good health. This results in a general decrease in overall health and mental abilities, making it increasingly difficult for such people to "pull themselves up by their own bootstraps."

100 kilograms of grain

10 kilograms of cow

1 kilogram of people eating steak

100 kilograms of grain

10 kilograms of people eating grain

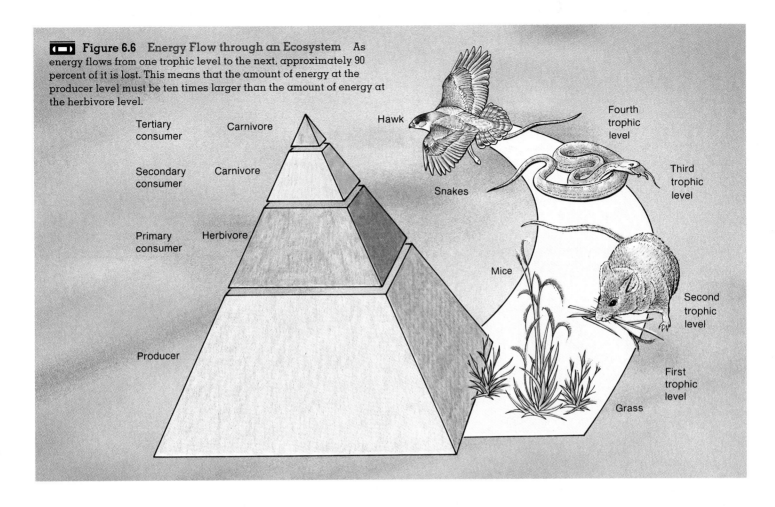

Figure 6.6 Energy Flow through an Ecosystem As energy flows from one trophic level to the next, approximately 90 percent of it is lost. This means that the amount of energy at the producer level must be ten times larger than the amount of energy at the herbivore level.

much heat it gives off: an automobile engine must have a cooling system to get rid of the heat energy produced; a light bulb produces heat you can feel with your hand. Living systems follow the same energy rule. *In general, in every ecosystem there is about a 90 percent loss of energy from one trophic level to the next.* Actual measurements vary from one ecosystem to another, but 90 percent is a good rule of thumb.

In addition to the energy lost as heat, some energy is lost in the capture and processing of food material by herbivores. Although herbivores don't need to chase their food, they do need to travel to where food is available, then gather, chew, digest, and use it for respiration. All these processes require energy. Just as the herbivore trophic level experiences a 90 percent loss in energy content, the higher carnivore trophic levels also experience a reduction in the energy available to

them. Figure 6.6 shows an energy pyramid in which the energy content decreases by 90 percent as we pass from one trophic level to the next.

Ecological Communities

We are all members of communities and we play specific roles within them. In your role as a student you interact with other people in specialized roles. In the same way, organisms in ecological systems are interdependent. We have already discussed how producers, herbivores, carnivores, and decomposers are interrelated. When discussing how organisms interact with one another it is useful to distinguish between an ecosystem and a community. Although the concepts are closely related, the distinctions are important. An ecosystem consists of all the interacting organisms within an area as well as the physical

environment. When we exclude the physical environment and describe only the interacting organisms, we call it a **community.** If we further restrict our description to just the number of individuals of a particular species, we are talking about a **population.** Thus we can look at one organism from several points of view. We can see it as an individual, as part of a population of similar individuals, as part of a community that includes populations of other kinds of organisms, and as part of an ecosystem that includes physical factors as well as living organisms.

As you know from the discussion in the previous section, one of the ways organisms interact is by feeding on one another. A community includes many different food chains. Some organisms may be involved in several food chains at the same time, so the chains become interwoven into a **food web** (figure 6.7). In a community, the interacting food chains

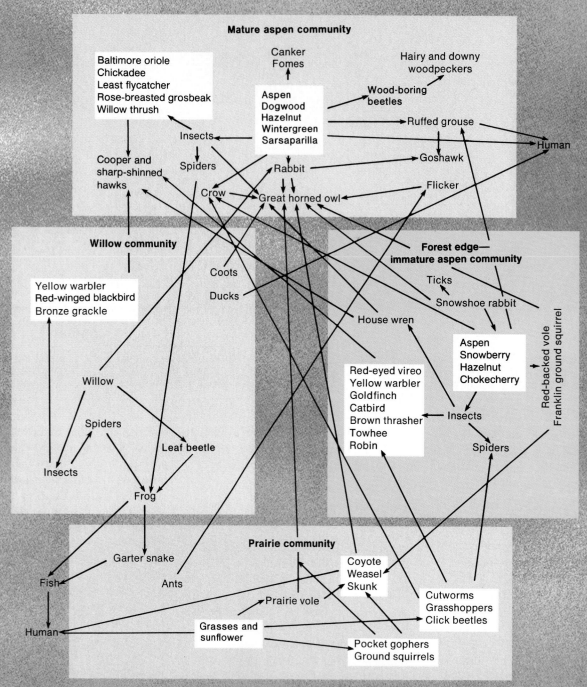

Figure 6.7 A Food Web When many different food chains are interlocked, a food web results. Notice that some organisms—the great horned owl in particular—are part of several food chains. Because of the interlocking nature of the food web, changing conditions may shift the way in which food flows through the system.

usually result in a relatively stable combination of populations. If a particular kind of organism is removed from a community, some adjustment usually occurs in the populations of other organisms in the community. For example, humans have used insecticides to control the populations of many kinds of insects. Reduced insect populations may result in lower numbers of insect-eating birds. Often the indiscriminate overuse of insecticides actually increases the insect problem, because insecticides kill many beneficial predator insects rather than just the one or two target pest species. Plant-eating insects may develop increased populations following insecticide use because there are fewer carnivorous insects to eat them.

You regularly experience changes in your social community. New businesses are established, civic leaders die, new politicians are elected, floods occur, and people move away. Natural biological communities are also dynamic collections of organisms: as one population increases, another decreases. This might occur over several years, or in just one year. This happens because conditions within ecosystems are not constant. During a year there are changes in rainfall, the amount of sunlight, and the average temperature. If a disease kills certain kinds of trees, the organisms that use the trees for food will be negatively affected. Other changes may occur over several years. Periods of drought or the introduction of a new kind of organism may alter a community for many years. We should expect populations to fluctuate as physical factors change. A change in the size of one population will trigger changes in other populations as well. Figure 6.8 shows what happens to the size of a population

BIO *feature*

Zebra Mussels: Invaders from Europe

In the mid 1980s a clamlike organism called the zebra mussel (*Dressenia polymorpha*) was introduced into the waters of the Great Lakes. It probably arrived in the ballast water of a ship from Europe. Ballast water is pumped into empty ships to make them more stable crossing the ocean. Immature stages of the zebra mussel were probably emptied into Lake St. Clair when the ship discharged its ballast water to take on cargo.

This organism has since spread to many areas of the Great Lakes and smaller inland lakes. It has also been discovered in other parts of the United States including the mouth of the Mississippi River. Zebra mussels attach to any hard surface and reproduce rapidly. A square meter (about one square yard) of suitable growing surface in Lake Erie was found to have 20,000 individuals.

These invaders are a concern for two reasons. First, they coat the intake pipes of municipal water plants, boats, and other surfaces not coated with special nonstick paint. Second, they introduce a new organism into the food chain. Zebra mussels efficiently filter small aquatic organisms from the water and may remove food organisms required by native species. There is concern that they will significantly change the ecology of the Great Lakes and other bodies of water where they are found.

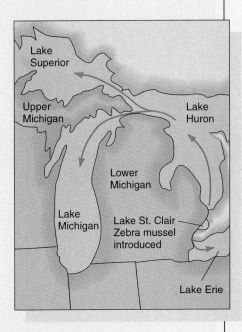

The Spread of the Zebra Mussel

BIO feature

Wolves and Moose on Isle Royale

Isle Royale in Lake Superior has been the site of a long-term study of the relationship between moose and wolf populations. Wolves (probably a single pair) reached Isle Royale earlier in the century by crossing the ice from Canada. By 1958, when studies began on the relationship between the wolves and moose, the population had increased to about twenty. By 1980 the population hit a peak of about fifty individuals, but it has

declined sharply since then. By 1993 the number had fallen to thirteen individuals— two pairs and a pack of nine.

The reproductive success of these wolves is low, and the population will probably eventually die out. Reasons for the decline are unclear, though two possibilities are likely. Inbreeding may have caused a population of individuals that are all genetically similar, and inbred animals often show reduced ability to produce offspring. This would result in fewer surviving pups in each litter. There is also

evidence that a virus found in dogs may have contributed to the decline.

Although wolves do not feed on moose exclusively, there appears to be a relationship between the size of the moose population and the size of the wolf population. When the wolf population is low, the moose population increases. When the wolf population increases, the moose population drops. The current decline in the wolf population has resulted in an increase in moose on Isle Royale.

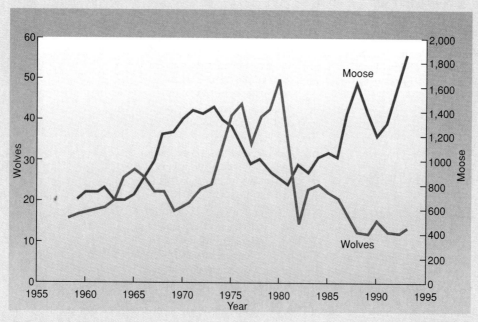

Changes in Wolf and Moose Populations on Isle Royale

of deer as the seasons change. The area can support one hundred deer from January through February, when plant food for deer is least available. It is no accident that deer breed in the fall and give birth in the spring, since during the spring plant populations (producers) are increasing, and the area has more available food to support a large deer population. It is also no accident that wolves and other carnivores that feed on deer give birth in the spring.

The increased food for deer (herbivores) in turn means more energy for the wolves (carnivores) at the next trophic level.

Since ecosystems and communities are complex, it is helpful to set boundaries that allow us to focus our study on a small collection of organisms. An example of a community with easily determined natural boundaries is a small pond (figure 6.9). The water's edge naturally defines the limits of this community. You

would expect to find certain animals and plants living in the pond, such as fish, frogs, snails, insects, algae, pond weeds, bacteria, and fungi. But you might ask, What about the plants and animals that live right at the water's edge? That leads us to think about the animals that spend only part of their lives in the water. That awkward-looking, long-legged bird wading in the shallows and darting its long beak down to spear a fish has its nest atop some tall trees away from the water. Should it be considered part of the pond community? Should we also include the deer that comes to drink at dusk and then wanders away? Small parasites could enter the body of the deer as it drinks. The immature parasites may develop into adults within the deer's body. These same parasites must spend part of their life cycle in the body of a certain snail. Are these parasites part of the pond community? What originally seemed to be a clear-cut example of one community has become less clear. Although the general outlines of a community can be arbitrarily set for the purposes of a study, we must realize that the boundaries of a community or an ecosystem are somewhat artificial.

Types of Communities

Every place on Earth has a particular kind of community that is typical for the region. You easily recognize that there are certain combinations of organisms that fit together. Bison (buffalo), grass, and prairie dogs are typical of North American grasslands; oak trees, squirrels, and leaf-eating insects are typical of deciduous forests; and lizards, cacti, and jackrabbits are typical of deserts.

Types of communities that have a wide geographic distribution are known as **biomes.** Refer to the map of the various biomes in figure 6.10. The nature of terrestrial biomes is primarily determined by two physical factors: the amount of precipitation and the yearly temperature patterns (figure 6.11).

The biome typical of the eastern part of North America is the *temperate deciduous forest.* Biomes

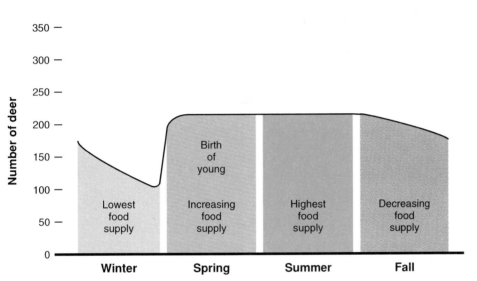

Figure 6.8 Annual Changes in Population Size The number of organisms living in an area varies during the year. The availability of food is the primary factor determining the size of the population of deer in this illustration, but water availability, availability of soil nutrients, and other factors could also be important.

Figure 6.9 A Pond Community Although a pond would seem to be an easy community to characterize, it interacts extensively with the surrounding land-based communities. Some of the organisms associated with a pond community are always present in the water (fish, pond weeds, clams); others occasionally venture from the water to the surrounding land (frogs, dragonflies, turtles, muskrats); still others are occasional or rare visitors (mink, herons, ducks).

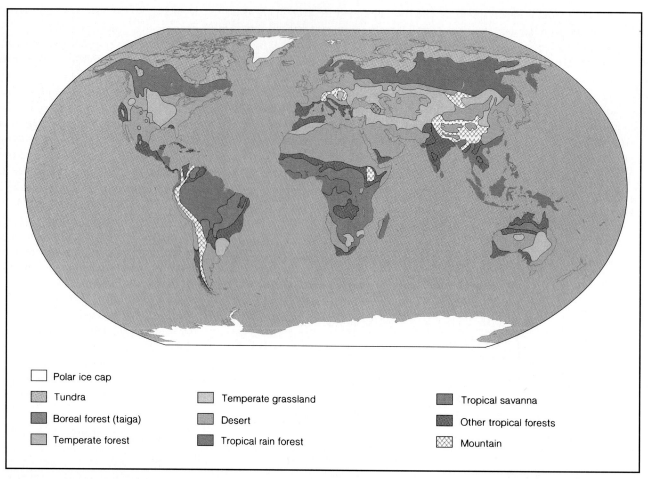

Figure 6.10 **Biomes of the World** Major climate differences determine the kind of vegetation that can live in a region of the world. Associated with specialized groups of plants are particular communities of animals. These regional ecosystems are called biomes.

- ☐ Polar ice cap
- ▨ Tundra
- ▨ Boreal forest (taiga)
- ▨ Temperate forest
- ▨ Temperate grassland
- ▨ Desert
- ▨ Tropical rain forest
- ▨ Tropical savanna
- ▨ Other tropical forests
- ▨ Mountain

Figure 6.11 **Influence of Precipitation and Temperature on Vegetation** Temperature and moisture are two major factors that influence the kind of vegetation that can occur in an area. Areas with low moisture and low temperatures produce tundra; areas with high moisture and freezing temperatures during part of the year produce deciduous or coniferous forests; dry areas produce deserts; moderate amounts of rainfall or seasonal rainfall support grasslands or savannas; and areas with high rainfall and high temperatures support tropical rain forests.

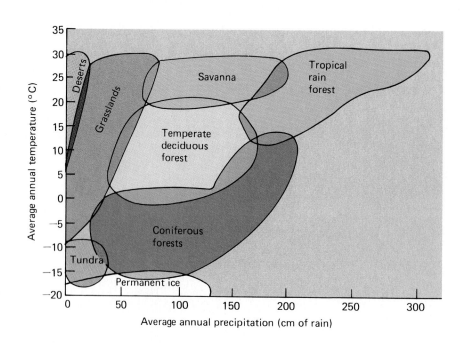

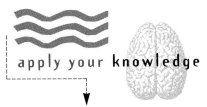

apply your knowledge

Your Local Community

Since most of us live in urban settings, we may need to take a mental trip to recall what the typical community of our region is like. Can you name five plants and five animals that are typical of where you live? Were these plants and animals found in your area before human development?

In your community, what are the producers, consumers, herbivores, carnivores, and decomposers?

are usually named for their major feature, which in this case happens to be the dominant vegetation. The most noticeable plants are large trees that lose their leaves more or less completely during the fall of the year and thus are called *deciduous* trees (figure 6.12).

The predominant climatic features of this biome are a moderate climate with distinct seasons and a fairly high rainfall distributed throughout the year. Deciduous trees require a considerable amount of rainfall to survive. However, the region also has a significant cold and dry part of the year. Snow may be on the ground, but as long as it is frozen it is not available to the trees. The loss of leaves in the fall is thought to be an adaptation to a lack of available moisture during the winter months. The large surfaces of the leaves would tend to lose water rapidly. Dropping the leaves in the fall reduces the amount of water loss.

Altitude and Latitude

The distribution of terrestrial ecosystems is primarily related to precipitation and temperature. The temperature is warmest near the equator and becomes cooler toward the poles. Similarly, as altitude increases, the average temperature decreases. This means that even at the equator it is possible to have cold temperatures on the peaks of tall mountains. As one proceeds from sea level to mountaintops, it is possible to pass through a series of biomes similar to those one would encounter traveling from the equator to the North Pole.

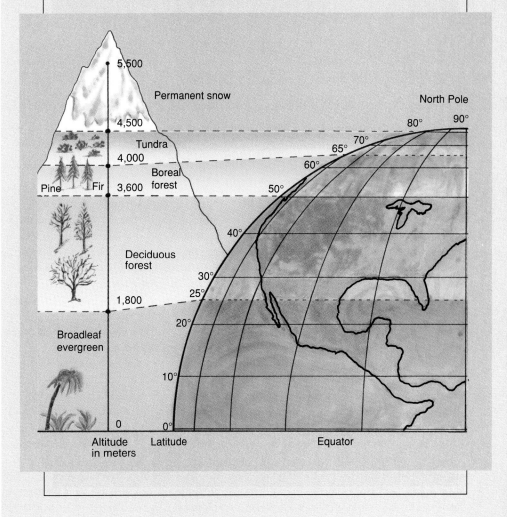

Figure 6.12 **A Temperate Deciduous Forest Biome** This kind of biome is found in parts of the world that have significant rainfall (100 centimeters [40 inches] or more) and cold weather for a significant part of the year when the trees are without leaves.

Using major plant types as a basis for naming biomes works well, since the major type of plant determines what other kinds of plants and animals can live there. Of course, since biomes are so large and have somewhat variable climates, we find differences in the particular species of trees (and other organisms) in various regions within the biome. For instance, in Maryland the tulip tree is one of the common large trees, while in Michigan it is so unusual that people plant it in lawns and parks as a decorative tree. Aspen, birch, cottonwood, oak, hickory, beech, and maple are typical trees found in a deciduous forest. Typical animals are skunks, porcupines, deer, frogs, opossums, owls, mosquitoes, and beetles.

The temperate deciduous forest covers a large area from the Mississippi River to the Atlantic Coast, and from Florida to southern Canada. This type of biome is also found in parts of Europe and Asia. Many local spots within this biome are quite different from one another. Many of them have no trees at all. For example, the tops of some of the mountains along the Appalachian Trail, the sand dunes of Lake Michigan, and the scattered grassy areas in Illinois are natural areas within the biome that lack trees. In much of this region, the natural vegetation has been removed for agriculture, so the original character of the biome is gone except where farming is not practical or the original forest has been preserved.

Our picture of much of the western part of the United States is of a land dominated by grasses. This is the *prairie* biome (figure 6.13). Our image of this biome is colored by the romantic vision we have of Native Americans, cowboys, and cattle barons. Today much of this biome has been extensively modified to raise cattle and crops. This biome is also common in parts of Eurasia, Africa, Australia, and South America. The dominant vegetation is made up of various species of grasses, which are able to live with limited amounts of water. The rainfall in this region is not adequate to support the growth of trees, which are found only along streams, where they can obtain sufficient water. Some grasslands are in areas where the average temperature is high while others are in relatively cool regions of the world. It is the relatively low amount of water available that determines the kind of plants, not the temperature. Prairie fires are common in areas that have not been converted to agriculture, because the grasses often dry out during late summer. Animals found in this biome include the prairie dog, pronghorn antelope, prairie chicken, grasshopper, rattlesnake, and meadowlark. Most of the original grasslands, like the temperate deciduous forest, have been converted to agriculture. Breaking the sod (the thick layer of grass roots) so that wheat, corn, and other grains can be grown exposes the soil to the wind, which may cause excess drying and result in soil erosion that depletes the fertility of the soil.

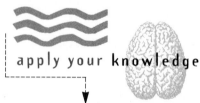

apply your kn**ow**ledge

A Cornfield Ecosystem

Farmers are managers of ecosystems. Consider a cornfield in Iowa. Describe five ways in which the cornfield ecosystem differs from the original prairie it replaced. What trophic level does the farmer fill?

Figure 6.13　A Prairie Biome This typical short-grass prairie of the western United States is associated with an annual rainfall of 25–50 centimeters (10–20 inches). This community contains a unique grouping of plant and animal species.

A *savanna* is a biome similar to a prairie (figure 6.14). Savannas are tropical biomes of central Africa and parts of South America and typically consist of grasses with scattered trees. Such areas generally have a wet season, during which grasses and some scattered trees grow quite rapidly, followed by a dry season in which the grasses dry out and the trees often lose their leaves. This contributes to fires, which are a common feature of this biome. The trees have special adaptations that allow them to survive repeated fires.

Throughout the world, wherever rainfall is low and irregular, a *desert* biome is found. Some deserts are extremely hot, while others can be cool during much of the year. Furthermore, deserts have large daily fluctuations in air temperature. When the sun goes down at night, the land cools off rapidly. There is no insulating blanket of clouds to keep the heat from radiating into space. The distinguishing characteristic of desert biomes is low rainfall, not temperature. A desert biome is characterized by scattered, thorny plants that lack leaves or have reduced leaves

Figure 6.14　A Savanna Biome A savanna is likely to develop in areas that have a rainy season and a dry season. During the dry season, fires are frequent. The fires kill tree seedlings and prevent the establishment of forests.

(figure 6.15). Many of the plants, like cacti, are capable of storing water in their fleshy stems. Although this is a harsh environment, many kinds of flowering plants, insects, reptiles, and mammals live here. The animals usually avoid the hottest part of the day by staying in burrows or other shaded, cool areas.

Through parts of southern Canada, extending southward along the mountains of the United States, and in much of Northern Asia and Europe we find communities that are dominated by evergreen trees. This is the *boreal*, or *coniferous forest*, biome (figure 6.16). The evergreen trees are especially adapted to withstand long, cold winters with abundant snowfall. The needle-shaped leaves of these trees carry on photosynthesis throughout the year but are specially adapted to reduce water loss during the winter months. Most of the trees in the wetter, colder areas are spruces and firs, but some drier, warmer areas have pines. The wetter areas generally have dense stands of small trees intermingled with many other kinds of vegetation and broken up by many small lakes and bogs. Spruces and fir trees are

Figure 6.15   A Desert Biome   The desert gets less than 25 centimeters (10 inches) of precipitation per year, but it teems with life. Cacti, sagebrush, lichens, snakes, small mammals, birds, and insects inhabit the desert. Because daytime temperatures are high, most animals are only active at night, when the air temperature drops significantly.

Figure 6.16   A Boreal Forest Biome   Conifers are the dominant vegetation in most of Canada, in a major part of the former Soviet Union, and at high altitudes in sections of western North America. The boreal forest biome is characterized by cold winters with abundant snowfall.

BIO *feature*

How to Tell a Spruce from a Fir

Spruce trees:

Needles are arranged in compact spirals around their twigs; the needles are square and pointed on the ends; cones usually hang down.

spruce → spiral, square, sharp

Fir trees:

Needles are without stalks; they are generally flattened or grooved on their top surface, not pointed on the end, and flexible; cones point up like candles.

fir → flat, flexible

Spruces and Firs

BIO *feature*

Old-Growth Temperate Rain Forests of the Pacific Northwest

The coastal areas of Northern California, Oregon, Washington, British Columbia, and southern Alaska contain an unusual set of environmental conditions that supports a special kind of forest, a temperate rain forest. The prevailing winds from the west bring moisture-laden air to the coast. As this air meets the coastal mountains and is forced to rise, it cools and the moisture falls as rain or snow. Most of these areas receive 200 centimeters (80 inches) or more precipitation per year. This abundance of water, along with fertile soil and mild temperatures, results in a lush growth of plants.

Sitka spruce, Douglas fir, and western hemlock are typical evergreen coniferous trees. Undisturbed (old-growth) forests of this region have trees as old as 800 years that are almost as tall as the length of a football field. Deciduous trees of various kinds (red alder, big leaf maple, black cottonwood) also exist in places where they can get enough light, and all trees are covered with mosses, ferns, and other plants that grow on the surface of the trees. The dominant color is green, since most surfaces have something photosynthetic growing on them.

When a tree dies and falls to the ground it rots in place and often serves as a new site for the establishment of new trees. This is such a common feature of the forest that the fallen, rotting trees are called nurse trees. The fallen tree also serves as a food source for a variety of insects, which are food for a variety of animals. Some animals,

such as the northern spotted owl, marbled murrelet (a sea bird), and the Roosevelt elk, seem to be dependent on undisturbed forest to be successful.

Because of the rich resource of trees, 90 percent of the original forest has already been logged. What remains may be protected as a remnant of the original forest just as small patches of prairie and eastern woodland have been preserved in other parts of North America.

A Typical Temperate Rain Forest Scene

especially adapted to withstand heavy snowfall. They are cone-shaped and their branches are flexible, allowing the heavy snow to slide off the tree without breaking the branches. In the mountains of the western United States, the pines are often widely scattered and very large, with few branches near the ground. These areas have a parklike appearance because there is very little vegetation on the forest floor. Typical animals in this biome are mice, wolves, squirrels, moose, midges, and flies.

North of the coniferous forest biome is an area known as the *tundra* (figure 6.17). It is characterized by extremely long, severe winters and short, cool summers. The deeper layers of the soil remain permanently frozen and are known as the *permafrost.* Under these conditions, very few kinds of animals and plants can survive. No trees can live in this region. Typical plants and animals of the area are dwarf willow and some other shrubs, reindeer moss, some flowering plants, caribou, wolves, musk oxen, fox, snowy owls, mice, and many kinds of insects. Many kinds of birds are summer residents only. The tundra community is relatively simple, so any changes may have drastic and long-lasting effects. The land is easy to injure and slow to heal; therefore, we must treat it gently. The construction of the Alaskan pipeline has left scars that could still remain a hundred years from now (figure 6.18).

The *tropical rain forest* is at the other end of the climate spectrum from the tundra. Tropical rain forests are found primarily near

the equator in Central and South America, Africa, parts of southern Asia, and some Pacific islands (figure 6.19). The temperature is high, there is no winter, rain falls nearly every day, and there are thousands of species of plants in a small area. Balsa (a very light wood), teak (used in furniture), and ferns the size of trees are examples of plants from the tropical rain forest. Typically, every plant has other plants growing on it. Tree trunks are likely to be covered with orchids, many kinds of vines, and mosses. Tree frogs, bats, lizards, birds, monkeys, and an almost infinite variety of insects inhabit the rain forest.

These forests are very dense, and little sunlight reaches the forest floor. When the forest is opened up (by a hurricane or the death of a large tree) and sunlight reaches the forest floor, the opened area is rapidly overgrown with vegetation. Since plants grow so quickly in these forests, many attempts have been made to bring this land under cultivation. North American agricultural methods require the clearing of large areas and the planting of a single species of crop, such as corn. The constant rain falling on these fields quickly removes the soil's nutrients so that heavy applications of fertilizer are required. Often these soils become hardened when exposed in this way. Although most of these forests are not suitable for agriculture, large expanses of tropical rain forest are being cleared yearly because of the pressure for more farmland in the highly populated tropical countries and the desire for high-quality lumber from many of the forest trees.

Human Use of Ecosystems

Wherever humans go, we alter our surroundings. Our ability to use tools allows us to create conditions favorable to our species. As our numbers have increased from a few million to more than 5 billion, our impact has been considerable. One of our primary concerns is the raising of food. To do this we have modified natural grasslands and forested areas to grow specific food crops. The extent to which humans use an ecosystem depends on many factors. Since plants are the producers, it is their activities that are most important to us. Ecosystems in which conditions are most favorable for plant growth are the most productive. Warm, moist, sunny areas with high levels of nutrients in the soil are ideal. Some areas are less productive because one of the essential

Figure 6.17 **A Tundra Biome** The tundra biome is located in northern parts of North America and Eurasia. It is characterized by short, cool summers and long, extremely cold winters. A layer of soil below the surface remains permanently frozen; consequently, there are no large trees in this biome. Relatively few kinds of plants and animals can survive this harsh environment.

Figure 6.18 **Fragile Biome** Damage done to the tundra of Alaska in order to build the Alaskan pipeline is very slow to heal.

FYI *Biodiversity* is a measure of the number of different kinds of organisms living in an area. In general, ecosystems that are stressed by pollution, severe temperature extremes, drought, and other factors have reduced diversity. The greatest biodiversity is found in the tropical rain forests where the temperature is warm, moisture is abundant, and there are no seasonal changes.

Figure 6.19 **A Tropical Rain Forest Biome** The tropical rain forest is a moist, warm region of the world located near the equator. The growth of vegetation is extremely rapid. There are more kinds of plants and animals in this biome than in any other.

Figure 6.20 Hunter-Gatherer Societies People such as these do not require large amounts of energy to support themselves. They obtain nearly all of their food and other resources from nature. They have few possessions since they must move from place to place in order to obtain food.

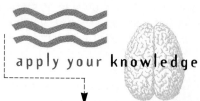

apply your **knowledge**

The High Cost of Eating

The next time you are in the grocery store, write down the unit price of the following items. Be sure that your units are all the same, i.e., price per gram, or price per ounce. What is the ecological basis for these differences in price?

Dry beans

Whole wheat bread

Flour

Sugar

Rice

Hamburger

Frozen fish

Ham

Chicken

factors is missing: deserts have scarce water, arctic areas have low temperatures, and the open ocean has few nutrients. Some communities, such as coral reefs and tropical rain forests, are highly productive. Marshes and estuaries are especially productive because the waters running into them are rich with the nutrients that aquatic plants and algae need for photosynthesis and growth. Furthermore, these aquatic systems are usually shallow so light can penetrate through most of the water column.

Early humans were hunter-gatherers, as are the Kalahari bushmen of Africa and some native tribes of South America and Australia today (figure 6.20). These people are able to make use of productive natural ecosystems by harvesting the food from them. Most of us, however, have lost the knowledge and skills necessary to live this way and thus are dependent on artificially maintained, highly productive agricultural ecosystems. In our dependence, we have destroyed these original ecosystems and replaced them with intensely managed agricultural ecosystems. For example, many Native Americans living in the Great Plains area of the United States used buffalo as a source of food. There was much grass, many buffalo, and few humans. Therefore, in the Native Americans' pyramid of energy, the base was more than ample. However, with the European exploitation and settling of North America, the human population in North America increased at a rapid rate. The top of the pyramid became larger. The food chain (prairie grass—buffalo—human) could no longer supply the food needs of the growing population. As the top of the pyramid grew, it became necessary for the producer base to grow larger. Since wheat and corn yield more biomass for humans than the original prairie grasses could, the settlers' domestic grain and cattle replaced the prairie grass and buffalo. This was fine for the settlers, but devastating for the buffalo and Native Americans.

SUMMARY

Ecology is the study of how organisms interact with their environment. The environment consists of living and physical components that are interrelated in an ecosystem. All ecosystems must have a constant input of energy from the sun. Producer organisms are capable of trapping the sun's energy and converting it into biomass. Herbivores feed on producers and are in turn eaten by carnivores, which may be eaten by other carnivores. Each level in the food chain is known as a trophic level. Other kinds of organisms involved in food chains are omnivores, which eat both plant and animal food, and decomposers, which break down dead organic matter and waste products. All ecosystems have a large producer base with successively smaller amounts of energy at the herbivore and carnivore trophic levels. Each time energy passes from one trophic level to the next, about 90 percent of the energy is lost from the ecosystem.

A community consists of the interacting populations of organisms in an area. The organisms are interrelated in food chains that interlock to create food webs. Because of this interlocking, changes in one part of the community can have effects elsewhere. Major land-based regional ecosystems are known as biomes. The temperate deciduous forest, coniferous forest, tropical rain forest, desert, savanna, and tundra are examples of biomes.

Humans use ecosystems to provide themselves with necessary food and raw materials. As the human population increases, most people live as herbivores at the second trophic level because we cannot afford to lose 90 percent of the energy in plants by first feeding it to animals, which we then eat. Humans have converted most productive ecosystems to agriculture and continue to seek more agricultural land as population increases.

CHAPTER GLOSSARY

biomass (bi′ō-mas) The dry weight of a collection of designated organisms.
biomes (bi′ōmz) Large regional communities.
carnivores (kar′nĭ-vōrz) Animals that eat other animals.
community (ko-miu′nĭ-te) A collection of interacting organisms within an ecosystem.
consumers (kon-soom′urs) Organisms that must obtain energy in the form of organic matter.
decomposers (de-kom-po′zurs) Organisms that use dead organic matter as a source of energy.

ecology (e-kol′o-je) The branch of biology that studies the relationships between organisms and their environment.
ecosystem (e″ko-sis-tum″) An interacting collection of organisms and the nonliving factors that affect them.
energy (e′ner-je) The ability to do work.
environment (en-vi′ron-ment) Anything that affects an organism during its lifetime.
food chain (food chān) A sequence of organisms that feed on one another, resulting in a flow of energy from a producer through a series of consumers.

food web (food web) A system of interlocking food chains.
herbivores (her′bĭ-vōrz) Animals that feed directly on plants.
omnivores (om′nĭ-vōrz) Animals that eat both plants and other animals.
population (pop″u-la′shun) The number of individuals of a specific species in an area.
producers (pro-du′surz) Organisms that produce new organic material from inorganic material with the aid of sunlight.
trophic level (tro′fik lĕ′vel) A step in the flow of energy through an ecosystem.

CONCEPT MAP TERMINOLOGY

Construct a concept map to represent the relationships among the following examples.

carnivore
community
consumer
decomposer
ecosystem
environment

food chain
herbivore
omnivore
producer
trophic level

LABEL • DIAGRAM • EXPLAIN

Choose an ecosystem with which you are familiar. Identify five organisms in at least three different trophic levels and show how the organisms are related to one another.

Multiple Choice Questions

1. In a temperate deciduous forest biome, which of the following populations would have the largest *total biomass?*
   a. red fox
   b. cottontail rabbit
   c. red-tailed hawk
   d. great horned owl
2. Whether an area has a temperate deciduous forest or a grassland is primarily determined by:
   a. the amount of sunlight available to the plants
   b. the amount of rainfall available to the plants
   c. the number of days below freezing
   d. the kinds of animals that live in the region
3. In a food chain of grass being eaten by insects and insects being eaten by birds:
   a. there are three trophic levels
   b. insects will have the most energy
   c. there is no producer level
   d. birds have the largest population
4. Which one of the following would be at the highest trophic level?
   a. grass
   b. cow
   c. decomposer
   d. lion
5. Which one of the following biomes is *not* found in the United States?
   a. temperate deciduous forest
   b. tropical rain forest
   c. desert
   d. prairie
6. A producer will
   a. capture sunlight energy
   b. have a smaller biomass than consumers
   c. be at the highest trophic level
   d. not be affected by decomposer organisms
7. Which of the following biomes have been most modified for agricultural purposes?
   a. grasslands and temperate forests
   b. deserts and tundra
   c. coniferous forest and savannas
   d. tropical rain forests and deserts
8. The major features that determine if an area will be a coniferous forest or a temperate deciduous forest are:
   a. length of winter and amount of snow
   b. the amount of rainfall and length of day
   c. the distance from the equator
   d. the kind of soil present

Questions with Short Answers

1. Why are rainfall and temperature important in an ecosystem?
2. Describe the flow of energy through an ecosystem.
3. What is the difference between the terms *ecosystem* and *environment?*
4. What role does each of the following play in an ecosystem: sunlight, plants, consumers, decomposers, herbivores, carnivores, and omnivores?
5. Give an example of a food chain.
6. What is meant by the term *trophic level?*
7. Why is there usually a larger herbivore biomass than a carnivore biomass?
8. List a dominant plant in each of the following biomes: temperate deciduous forest, coniferous forest, desert, tundra, tropical rain forest, and savanna.
9. Can energy be *recycled* through an ecosystem?
10. What is the difference between an ecosystem and a community?

Community Interactions

Marine community, Red Sea.

Community Interactions

- Understand that organisms interact in a variety of ways within communities.

- Recognize the differences between community, habitat, and niche.

- Describe how water molecules and atoms of carbon and nitrogen are cycled in communities.

- Appreciate that humans alter and interfere with natural ecological processes.

- Recognize that communities proceed through a series of stages to stable climax communities.

Community, Habitat, and Niche

Within ecosystems, organisms influence one another in many ways. Even organisms of the same species affect one another in the course of their normal daily activities. This chapter considers some of the kinds of interactions that occur within ecosystems and describes the various ways in which organisms within communities affect each other in the cycling of matter.

In chapter 6 we examined ecosystems from the point of view of energy flow and recognized that all heterotrophs are dependent on autotrophs for energy. Plants are eaten by herbivores, carnivores need other animals for food, decomposers break down dead organisms and recycle their components, and even plants have needs that are met by animals. For example, the carbon dioxide produced by animal respiration is utilized by plants in the process of photosynthesis, which in turn releases oxygen for the animals. Animals provide many other services for plants: bees pollinate flowers, and many kinds of animals distribute seeds by eating fruits and depositing the seeds in their droppings.

The kind of place an organism lives is known as its **habitat.** The habitat of an organism is where its particular requirements for life can be met; where it is best able to do its job. Today we frequently hear that many species are endangered because their habitats are being altered. Grizzly bears need large patches of land uninhabited by people. Water lilies need shallow freshwater. Pandas need bamboo forests. Many birds must nest in hollow trees, so if all the trees are cut down or the old, diseased trees are removed from the forest, the habitat for the bird is destroyed.

Habitats are usually described in terms of outstanding or particularly significant features in the area where the organism lives. For example, the habitat of a prairie dog is usually described as a grassland, while the habitat of a tuna is described as the open ocean. The habitat of the fiddler crab is sandy ocean shores, and the habitat of various kinds of cacti is the desert. The key thing to keep in mind when you think of habitat is the kind of *place* where a particular kind of organism lives. We don't necessarily need to talk about large spaces when we discuss habitats. It is also possible to describe the habitat of the bacterium *Escherichia coli* as the human gut, or the habitat of a fungus as a rotting log. Organisms that have very specific places in which they live simply have more restricted habitats.

Each species has particular requirements for life and places specific demands on the habitat in which it lives. The specific functional role of an organism is its **niche.** An organism's niche is the way it goes about living its life. Just as the word *place* is the key to understanding the concept of habitat, the word *function* is the key to understanding the concept of a niche. Understanding the niche of an organism involves a detailed understand-

habitat   The place or part of an ecosystem occupied by an organism (its address).

niche   The functional role of an organism (its job).

ing of the factors that affect the organism as well as the impacts an organism has on its surroundings. For example, the niche of an earthworm includes physical factors such as the moisture content, temperature, and particle size of the soil, but it also includes interactions with other living things since it serves as food for birds, moles, and shrews; as bait for anglers; or as a consumer of dead organic

FYI

Many people have the mistaken belief that plants do not need oxygen for respiration. This is not correct. Plants must undergo both photosynthesis and respiration. Photosynthesis allows plants to store sugar, which can then be used later for respiration. Both autotrophs and heterotrophs undergo respiration, but only autotrophs photosynthesize.

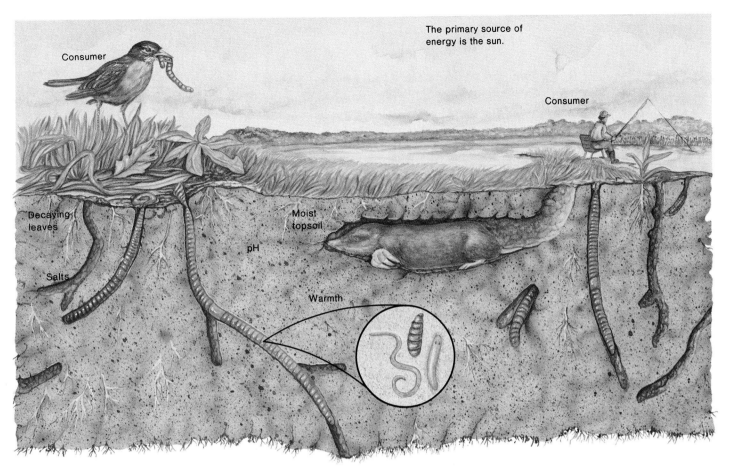

The primary source of energy is the sun.

Consumer

Consumer

Decaying leaves

Moist topsoil

pH

Salts

Warmth

Figure 7.1 The Niche of an Earthworm The niche of an earthworm involves a great many factors. It includes the fact that the earthworm is a consumer of dead organic matter, a source of food for other animals, a host to parasites, and bait for an angler. Furthermore, it includes the fact that the earthworm loosens the soil by its burrowing and "plows" the soil when it deposits materials on the surface. Additionally, the pH, texture, and moisture content of the soil have an impact on the earthworm. Keep in mind that this is but a small part of what the niche of the earthworm includes.

matter (figure 7.1). In addition, an earthworm serves as a host for a variety of parasites, transports minerals and nutrients from deeper soil layers to the surface, and creates burrows that allow air and water to penetrate the soil more easily, thus changing the soil.

Some organisms have rather broad niches; others, with very specialized requirements and limited roles to play, have niches that are quite narrow. The opossum (figure 7.2a) is an animal with a very broad niche. It eats a wide variety of plant and animal foods, can adjust to a wide variety of climates, is used as food by many kinds of carnivores (including humans), and produces large numbers of offspring. Foxes and coyotes are also

very successful. They are able to tolerate humans and often live in suburban areas. By contrast, the koala of Australia (figure 7.2b) has a very narrow niche. It can live only in areas of Australia with specific species of *Eucalyptus* trees, because koalas only eat the leaves of a few kinds of these trees. Furthermore, koalas cannot tolerate low temperatures and do not produce large numbers of offspring. As you might guess, the opossum is expanding its range, and the koala is endangered in much of its range. In North America, the black-footed ferret is an endangered species with a very narrow niche. It is a carnivore that feeds on prairie dogs. Since much of the original prairie has been converted to

agriculture and grazing of cattle, the resulting reduction in prairie dog numbers threatened the ferret with extinction. Protected populations now live in special areas where prairie dogs still exist (figure 7.2c).

The complete description of an organism's niche is a detailed inventory of influences, activities, and impacts. It includes what the organism does and what is done to the organism, how the physical environment influences the organism, and how the organism alters the physical environment. It concerns how one organism helps or hinders another.

Since the niche of an organism is a complex set of items that may not be completely understood, it is often easy to overlook

BIO feature

Habitat Change and the California Condor

The California condor (*Gymnogyps californianus*) is thought to have adapted to feed on the carcasses of large mammals found in North America during the ice age. With the extinction of the large, ice-age mammals their major food source disappeared. By the 1940s, the condor's range had shrunk to a small area near Los Angeles, California. Further fragmentation of their habitat due to human activity, death by shooting, and death by eating animals containing lead shot reduced the total population to about 17 individuals by 1986. This small population of condors produced few offspring, making it difficult for the species to increase in number. California condors do not become sexually mature until six years of age and females typically lay one egg every two years.

In 1987 all of the remaining wild condors were captured to serve as captive breeding populations. The plan was to raise young condors in captivity and ultimately release some of the animals back into the wild. Captive condors bred successfully, and females could be induced to lay a second egg if the first egg was removed from the nest. This increased the number of offspring produced per

female and resulted in a captive population of 51 individuals by 1991. That year two condors were released into the wild and six more were released in 1992. Since they were raised in captivity, are still dependent on food supplied by researchers, and have not reproduced yet, it is still too early to suggest that the California condor will be able to survive on its own and reproduce in the wild.

The California Condor

(a)

(b)

(c)

Figure 7.2 **Broad and Narrow Niches**
(*a*) The opossum has a very broad niche. It eats a variety of foods, is able to live in a variety of habitats, and has a large reproductive capacity. It is generally extending its range in the United States.
(*b*) The koala has a narrow niche. It feeds only on the leaves of the eucalyptus tree, is restricted to relatively warm, forested areas, and is generally endangered in much of its habitat. (*c*) The black-footed ferret has a very specialized niche. It feeds only on prairie dogs. Small populations of black-footed ferrets still can be found in places where prairie dog populations are protected.

important roles played by many organisms. For example, when Europeans introduced cattle into Australia—a continent where there had previously been no large, hoofed mammals—they did not think about the impact of cow manure or the significance of a group of beetles called dung beetles. These beetles rapidly colonize fresh dung, and by using it for food, cause it to be broken down. Since these beetles did not exist in Australia, a significant amount of land became covered with cow manure. The problem was eventually solved by the importation of several species of dung beetles from Africa, where large, hoofed mammals are common. In this newly established community, the cattle were consumers of plant material, the dung beetles made use of what the cattle did not digest, and the plants could make use of the nutrients released by the actions of the dung beetles.

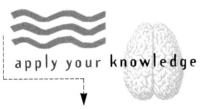

apply your **knowledge**

Heightening Your Awareness

Write a short description of the niche of an organism with which you are familiar. You may want to consider something other than a cat, dog, or other domestic animal. For example, you might describe a squirrel, a rat, grass, or a rose.

Kinds of Organism Interactions

Organisms within communities influence one another in numerous ways. One kind of organism interaction is predation. **Predation** occurs when one animal captures, kills, and eats another animal. The organism that is killed is called the **prey,** and the one that does the killing is called the **predator.** The predator obviously benefits from the relationship, while the prey organism is harmed. Most predators are relatively large compared to their prey and have specific adaptations that aid them in catching prey. Powerful jaws and an ability to run rapidly are important adaptations to the predator niche of lions, tigers, leopards, cheetahs, and wolves (figure 7.3). The keen eyesight and sharp-pointed talons of hawks and owls are important to their success as predators. Although these examples are typical of what we think about when the word *predation* is used, many predators are tiny and are easily overlooked. Most birds prey on insects and many insects prey on other insects. Dragonflies patrol areas where they can capture flying insects, and most fish eat other animals. Many spiders build webs that serve as nets to catch flying insects. The prey are quickly paralyzed by the spider's bite and wrapped in a tangle of silk threads. Other rapidly moving spiders, like wolf spiders and jumping spiders, have large eyes that help them capture prey without the use of a web.

Many kinds of predators are useful to us because they control the populations of organisms that do us harm. For example, snakes eat many kinds of rodents that eat stored grain and other agricultural products. Many birds eat insects that are agricultural pests. It is even possible to think of a predator as having a beneficial effect on the prey species. Certainly the individual organism that is killed is harmed, but the population can benefit. Predators can prevent overpopulation in prey species or reduce the likelihood of epidemic disease by eating sick or diseased individuals. Furthermore, predators shape the future evolution of the prey species. The individuals who fall to them as prey are likely to be less well adapted than the ones that escape predation. Predators usually kill the slow, unwary, sick, or injured individuals. Thus, the remaining population consists of the better-adapted individuals that will pass their characteristics on to their offspring. Because predators eliminate poorly adapted individuals, the population benefits. What is bad for the individual can be good for the population.

Figure 7.3 **The Predator-Prey Relationship** Many predators capture prey by making use of speed. Since strength is needed to kill the prey, the predator is generally larger than the prey. Obviously, predators benefit from the food they obtain to the detriment of their prey. The cheetah can reach speeds of 112 kilometers per hour (70 miles per hour) during a sprint to capture an animal.

Lyme Disease: Parasites, Hosts, and Carriers

Lyme disease is caused by a bacterium but is spread by the bite of a tick. It is currently spreading through the United States, though 97 percent of the cases are centered in three regions: the Northeast (e.g., Connecticut, Rhode Island, New Jersey); the Middle West (e.g., Michigan, Indiana, Ohio); and the West (e.g., California, Oregon, Utah). Once the bacterial parasite, *Borrelia burgdorferi,* has been transferred into a suitable susceptible host (e.g., humans, mice, horses, domestic cats, and dogs), it causes symptoms that have been categorized into three stages. The

first-stage symptoms may appear three to thirty-two days after an individual is bitten by an infected tick (*Ixodes dammini, I. neotomae,* or *I. pacificus*) and include a spreading red rash, headache, nausea, fever, aching joints and muscles, and fatigue. Stage two may not appear for weeks or months after the initial infection and may affect the heart and nervous system. The third stage may appear months or years later and typically appears as severe arthritis. The main reservoir of the disease, that is, the origin of new sources of infection, appears to be the white-footed mouse and dusky-footed woodrat.

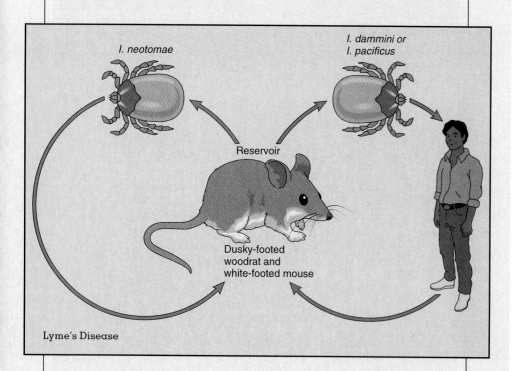

Lyme's Disease

Throughout your life you have encountered many different parasites. Viruses cause colds and flu; bacteria cause skin infections, pneumonia, and strep throat; protozoa may live in your intestines and mouth; worms may have lived in your intestines; or you might have had a fungal infection called athlete's foot. Parasitism is another kind of interaction in which one organism is harmed and the other aided. **Parasitism** specifically involves one organism living in or on another living organism from which it derives nourishment. The **parasite** benefits while harming the **host,** the organism it lives in or on (figure 7.4). Many kinds of fungi live on trees and other kinds of plants, including those that are commercially valuable. Some external parasites live on the surface of their hosts. For example, fleas live on the outside of rats' bodies, where they suck blood and do harm to the rats. Many kinds of worms attach themselves to the

(a)

(b)

Figure 7.4 The Parasite-Host Relationship Parasites benefit from the parasite-host relationship because they obtain nourishment from the host. Tapeworms (a) are internal parasites and lamprey (b) are external parasites. The host may not be killed directly by the relationship, but it is often weakened, thus becoming more vulnerable to predators or diseases. There are more parasites in the world than organisms that are not parasites.

outside of their host and suck body fluids. This is particularly true for host organisms like fish, crabs, and turtles that live in water.

In spite of these examples, we probably most often think of internal parasites, which invade their hosts, as being typical parasites. Dogs, cats, wolves, raccoons, bears, and many other carnivores typically have tapeworms in their intestines. One of the common activities of veterinarians is to monitor and eliminate tapeworms and other worm infections in pet dogs and cats. Another kind of parasite, found in the blood of the rat, is the bacterium *Yersinia pestis*. It does little harm to the rat but can cause a disease known as *plague* or *black death* if it is transmitted to humans. If a flea sucks blood from an infected rat and then bites a human, the bacterium may enter the human bloodstream and cause the disease. During the mid-1300s, when living conditions were poor and rats and fleas were common, epidemics of plague killed millions of people. In some countries in western Europe, 50 percent of the population was killed by this disease.

A relationship in which one organism benefits and the other is neither helped nor harmed is called **commensalism.** Sharks often have another fish, the remora, attached to them. The remora has a sucker on the top side of its head that allows it to attach to the shark and get a free ride (figure 7.5). While the remora benefits from the free ride and by eating leftovers from the shark's meals, the shark does not appear to be troubled by this uninvited guest, nor does it benefit from the presence of the remora.

Figure 7.5 Commensalism In the relationship called commensalism, one organism benefits and the other is not affected. The remora fish shown here hitchhike a ride on the shark. They eat scraps of food left over by the messy eating habits of the shark. The shark does not seem to be hindered in any way.

Problems with Terminology

Predation and parasitism are both relationships in which one member of the pair is helped and the other is harmed. There are several related kinds of interactions that are similar but do not fit our definitions very well. For example, when a cow eats grass, it is certainly harming the grass while deriving benefit from it. We could call cows grass predators, but we usually refer to them as *herbivores.* Likewise, such animals as mosquitoes, biting flies, vampire bats, and ticks take blood meals but don't usually live permanently on the host, nor do they kill it. Are they temporary parasites, or specialized predators? Finally, birds like cowbirds and cuckoos lay their eggs in the nests of other species of birds, who raise these foster young. The adult cowbird and cuckoo or their offspring remove the eggs or the young of the host species, so usually only the cowbird or cuckoo is raised by the foster parents. This kind of relationship has been called *nest parasitism.*

Are humans parasites on dairy cows? What is our relationship to the grass we mow or step on? Are embryos parasites on their mothers? Do we have a mutualistic relationship with pets? Many kinds of interactions between organisms, like these, don't fit neatly into the classification scheme dreamed up by scientists.

(a)

(b)

(c)

Problems with Terminology
(a) Is this mosquito a parasite? (b) Is this cowbird nestling being fed by a red-eyed vireo a parasite? (c) Are bison grass predators?

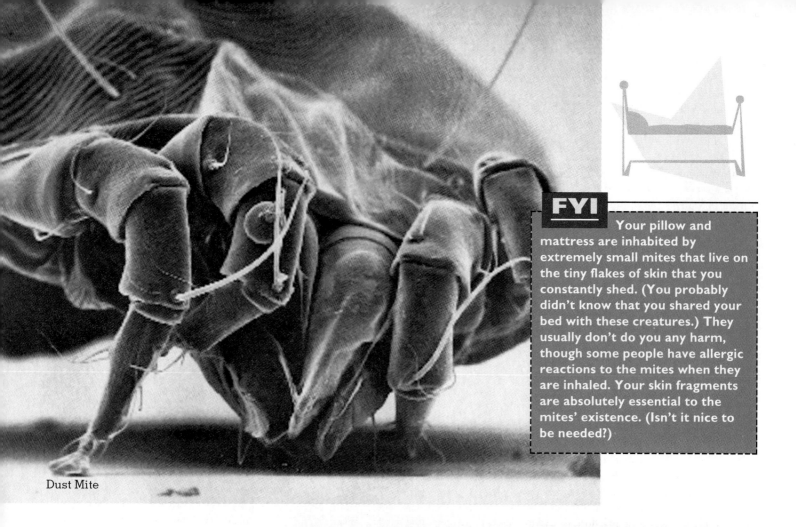

Dust Mite

An example of commensalism in the plant kingdom is the relationship between trees and epiphytic plants. **Epiphytes** are plants that live on the surface of other plants but do not derive nourishment from them. Many kinds of plants (e.g., orchids, Spanish moss, ferns, and mosses) use the surface of trees as a place on which to live (figure 7.6). These kinds of organisms are particularly common in tropical rainforests. Many epiphytes derive benefit from the relationship because they are able to grow at the top of the tree where they receive more sunlight and moisture. The trees derive no benefit from the relationship, nor are they harmed; they simply serve as a support surface for epiphytes.

There are also many situations in which two species live in close association with one another, and both benefit. This is called **mutualism.** One interesting example of mutualism involves digestion in rabbits. Rabbits eat plant material. Plant cell walls contain cellulose, which is made of the sugar glucose. However, rabbits are unable to digest cellulose.

Figure 7.6 Epiphytes Epiphytes are plants that live on the surface of other plants, which they use as a supporting surface without doing any harm.

They manage to get energy out of these cellulose molecules with the help of special bacteria living in their digestive tracts. The bacteria are able to digest cellulose into smaller molecules that the rabbit's digestive system can break down into even smaller glucose molecules. The bacteria benefit because the gut of the rabbit provides them with a moist, warm, nourishing environment in which to live. The rabbit benefits because the bacteria provide them with a source of food. Termites, cattle, buffalo, and antelope also have collections of bacteria and protozoa that live in their digestive tracts and help them to digest cellulose.

Another kind of mutualistic relationship exists between flowering plants and bees. Undoubtedly you have observed bees and other insects visiting flowers to obtain nectar from the blossoms. Usually the flowers are constructed in such a manner that the bees pick up pollen (sperm-containing packages) on their hairy bodies and transfer it to the female part of the next flower they visit (figure 7.7). Because bees normally visit the same species of flower for several minutes while ignoring other kinds of flowers, they can serve as pollen carriers between two flowers of the same species. Plants pollinated in this manner produce less pollen than do plants that rely on the wind to transfer pollen. This saves the plant energy because it doesn't need to produce huge quantities of pollen. It does, however, need to transfer some of its energy savings into the production of nectar and showy flowers that attract the bees to the nectar source. The bees benefit from both the nectar and the pollen: the nectar serves as a source of sugars, while the pollen is a source of protein.

The most common kind of interaction between organisms is **competition.** Competition occurs whenever two organisms both need a vital resource that is in short supply (figure 7.8). The vital resource could be food, shelter, nesting sites, water, mates, or space.

CONCEPT CONNECTIONS

An additional term often used when discussing interactions between organisms is *symbiosis.* **Symbiosis** literally means "living together." Unfortunately, this word is used in several ways, none of which are very precise. It is often used as a synonym for mutualism, but it is also often used to refer to commensalistic relationships and parasitism. The emphasis, however, is on interactions that involve a close physical relationship between two kinds of organisms.

Competition can occur between members of the same species, or it can occur between members of two different species. It can be a snarling tug-of-war between two dogs over a scrap of food, or it can be a silent struggle between plants for access to available light. If you have ever started tomato seeds (or other garden plants) in a garden and failed to eliminate

Figure 7.7 Mutualism Mutualism is an interaction between two organisms in which both benefit. The plant benefits because cross-fertilization (exchange of gametes from a different plant) is more probable; the bee benefits by acquiring nectar for food.

the weeds, you have witnessed competition. If the weeds are not removed, they compete with the garden plants for available sunlight, water, and nutrients, resulting in poor growth of both the garden plants and the weeds. In these examples, competition to some extent results in harm to one of the competing species. The "winner," on the other hand, has had to display its most powerful traits in order to be successful in the competition.

The more similar the requirements of two species, the more intense the competition between them. A basic principle of nature is that no two species of organisms can occupy the same niche at the same time. If two species do occupy the same niche, the competition will be so intense that one will become extinct or be forced to migrate to a different area, or the two species may evolve into slightly different niches. This evolutionary process results in organisms that have minor differences in their structure or behavior, leading to less competition. Their niches are slightly different. For example, many birds catch flying insects as food, but species do not compete directly with each other because some feed at night, some feed high in the air, some feed only near the ground, and still others perch on branches and wait for insects to fly past. As another example, all forest plants need sunlight. The tall trees

Figure 7.8 Competition Whenever a needed resource is in limited supply, organisms compete for it. This competition may be between members of the same species (*intraspecific*), illustrated by the vultures shown in the photograph, or may involve different species (*interspecific*).

have first access to the light. Some small plants of the forest floor grow quickly in the spring when sunlight is available and become inactive during late summer, while others are able to tolerate the shade provided by tall trees.

The Cycling of Materials in Ecosystems

It doesn't appear that the amount of matter on the earth is changing significantly. We might remove some clay from the earth and make bricks to construct a building, but all we have done is move matter from one place to another. We have not created new matter. Similarly, the living world is not accumulating larger and larger amounts of living stuff. As new organisms are created, others die and disintegrate and their components can be used over again.

All matter is made of small units called atoms. There are about 100 different kinds of atoms, such as hydrogen (H), carbon (C), nitrogen (N), phosphorus (P), and oxygen (O), that can be combined with one another to form larger units called molecules. Carbon dioxide (CO_2), water (H_2O), and sugar ($C_6H_{12}O_6$) are examples of common molecules.

Certain kinds of atoms are very common in living things. Carbon atoms are combined with one another in chains or rings to form the large organic molecules typical of living things. **Organic molecules** differ from other molecules in that they consist of long chains of carbon atoms hooked to one another, and they exist in tremendous variety. Examples of organic molecules are carbohydrates, proteins, fats, DNA, and several other special kinds of organic matter. Certain other atoms such as nitrogen, phosphorus, and sulfur are found in specific kinds of organic molecules—nitrogen and sulfur in proteins and nitrogen and phosphorus in DNA. Carbon dioxide, water, table salt, and ammonia are simpler molecules that do not contain carbon atoms in chains and are called **inorganic molecules.**

Although some new atoms are being added to the earth from cosmic dust and meteorites, this amount is not significant in relation to the entire mass of the earth or the biomass of living material. Therefore, living things must reuse the existing atoms again and again. In this recycling process, inorganic molecules are combined to form the organic compounds of living things. If there were no way of recycling this organic matter back into its inorganic forms, organic material would build up as the bodies of dead organisms. Decomposers play a vital role in this recycling process if conditions allow them to operate. If decomposers are kept from destroying organic matter, it builds up as deposits. This is thought to have occurred millions of years ago when the present deposits of coal, oil, and natural gas were formed. Because they are the remains of previous life they are often referred to as fossil fuels.

Living systems contain many kinds of atoms, but some are more common than others. Carbon, nitrogen, oxygen, hydrogen, and phosphorus are found in all living things and must be recycled when an organism dies. Let's look at some examples of this recycling process.

Hydrologic Cycle

Water molecules are essential for life. They are the most common molecule in the body of any organism. The percentage of water varies from one kind of organism to another. In some it may be as low as 60 to 70 percent while in many ocean animals, like jellyfish, it may be closer to 90 percent. All of the metabolic reactions that occur in organisms take place in a watery environment. You cannot live without a constant supply of water, which you consume in the food you eat and in the fluids you drink. In addition, water molecules are involved in the process of photosynthesis as raw material. The hydrogen atoms (H) from water molecules (H_2O) are added to carbon atoms to make carbohydrates and other organic molecules. Furthermore, the oxygen molecules (O_2) released during photosynthesis come from water molecules.

Most of the forces that cause water to be cycled do not involve organisms but are the result of normal physical processes (figure 7.9). Because of the motion of molecules, liquid water evaporates into the atmosphere. Evaporation can occur wherever water is present—from lakes, rivers, soil, or the surface of organisms. Since the oceans contain most of the world's water, an extremely large amount of water enters the atmosphere from the oceans. Plants also transport water from the soil to leaves, where it evaporates. Once the water molecules are in the atmosphere, they are moved by prevailing wind patterns. If warm, moist air encounters cooler temperatures, which often happens over land masses, the water vapor condenses into droplets and falls as rain or snow. When the precipitation falls on land, some of it runs off the surface, some of it evaporates, and some penetrates into the soil. The water in the soil may be taken up by plants and transpired into the atmosphere, or it may become groundwater. Much of the groundwater also eventually makes its way into lakes and streams and ultimately arrives at the ocean from which it originated.

FYI Humans use more water for irrigation than for any other purpose. Since much irrigation water evaporates or is released by transpiration of crop or lawn plants, humans are a significant factor in the water cycle.

Carbon Cycle

Carbon and oxygen combine to form the molecule carbon dioxide (CO_2), which is a gas found in small quantities in the atmosphere. During photosynthesis, carbon dioxide (CO_2) combines with water (H_2O) to form complex organic molecules ($C_6H_{12}O_6$). At the same time, oxygen molecules (O_2) are released into the atmosphere. The organic matter in the bodies of plants may be used by herbivores as food. When an herbivore eats a plant, it breaks down the complex organic molecules into smaller organic building blocks, like simple sugars, amino acids, glycerol, and fatty acids. These building blocks can then be used to construct the large organic molecules that are incorporated into the herbivore's body. Thus, the atoms in the body of the herbivore can be traced back to the plants that were eaten. When herbivores are eaten by carnivores, these same atoms are transferred to them. The waste products of plants and animals and the remains of dead organisms are used by decomposer organisms as sources of carbon, hydrogen, and oxygen atoms.

Finally, all the organisms in this cycle—plants, herbivores, carnivores, and

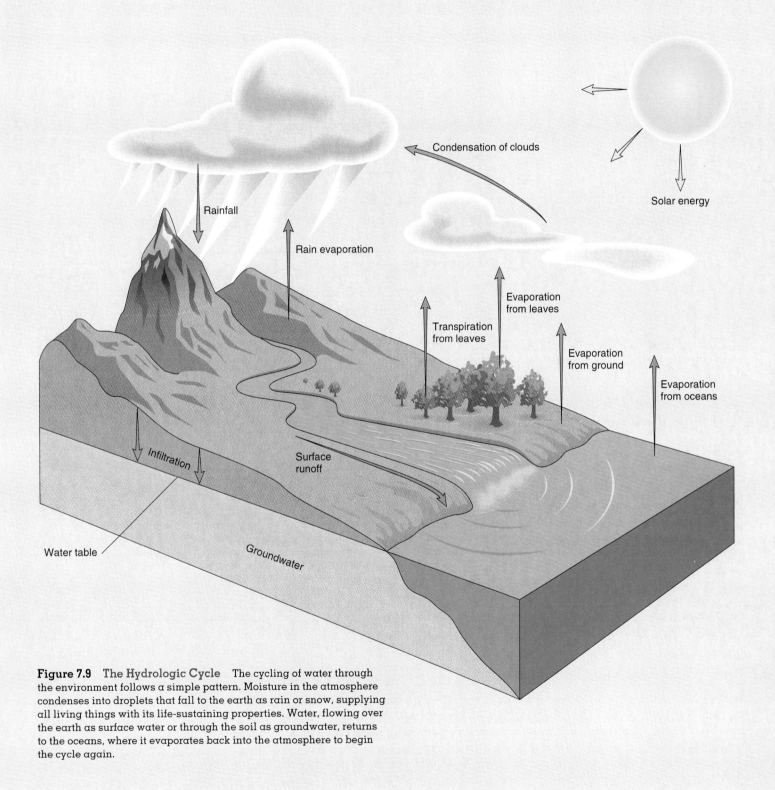

Figure 7.9 The Hydrologic Cycle The cycling of water through the environment follows a simple pattern. Moisture in the atmosphere condenses into droplets that fall to the earth as rain or snow, supplying all living things with its life-sustaining properties. Water, flowing over the earth as surface water or through the soil as groundwater, returns to the oceans, where it evaporates back into the atmosphere to begin the cycle again.

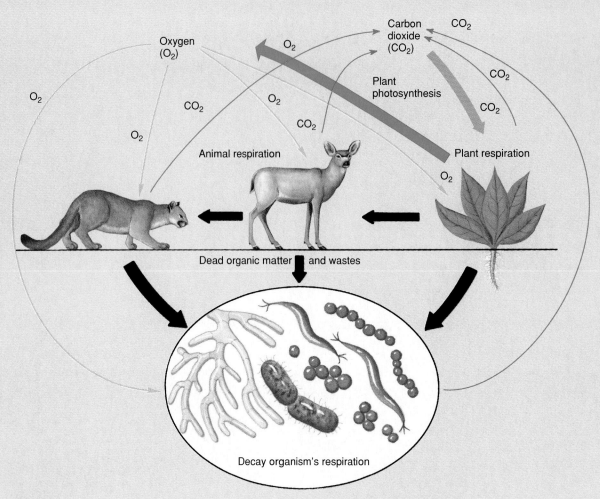

Figure 7.10 **The Carbon Cycle** Carbon atoms are cycled through ecosystems. Carbon dioxide (green arrows) produced by respiration is the source of carbon that plants incorporate into organic molecules when they carry on photosynthesis. These carbon-containing organic molecules (pink arrows) are passed to animals and decomposers when they eat plants and animals. Organic molecules in waste or dead organisms are consumed by decay organisms. All organisms (plants, animals, and decomposers) return carbon atoms to the atmosphere as carbon dioxide when they carry on cellular respiration. Oxygen (blue arrows) is being cycled at the same time that carbon is. The oxygen is released to the atmosphere and into the water during photosynthesis and taken up during cellular respiration.

decomposers—are involved in the process of respiration, in which oxygen (O_2) is used to break down organic compounds into carbon dioxide (CO_2) and water (H_2O). Thus, the carbon atoms that started out as components of carbon dioxide molecules (CO_2) have passed through the bodies of living organisms as parts of organic molecules in their bodies, and returned to the atmosphere as carbon dioxide, ready to be cycled again. Similarly, the oxygen atoms (O) released as oxygen molecules (O_2) during photosynthesis have been used during the process of respiration (figure 7.10).

Nitrogen Cycle

Proper nutrition is important to the life of any organism. Eating the right foods provides the ingredients to construct the molecules each organism needs. Carbon atoms are essential to construct all organic molecules. Nitrogen (N) is essential in the formation of amino acids needed to build proteins and in nitrogenous bases needed to construct DNA and RNA. Nitrogen atoms are passed from one organism to another in the food chain. Nitrogen (N) is found as molecules of nitrogen gas (N_2) in the atmosphere. Although nitrogen gas (N_2) makes up approximately 80 percent of the earth's atmosphere, only a few kinds of bacteria are able to convert it into nitrogen compounds that other organisms can use. Therefore, in most ecosystems the amount of nitrogen available limits the amount of plant biomass that can be produced. Plants are able to obtain nitrogen atoms from several different sources, but the nitrogen atoms are always combined with other atoms such as oxygen (NO_3^-) or hydrogen (NH_3) (figure 7.11).

All animals obtain their nitrogen from the food they eat. Proteins are large molecules constructed of a series of building blocks called amino acids, which contain nitrogen atoms. The proteins eaten by an animal are broken down into their individual amino acids, which can then be reassembled into new proteins needed by the animal. This is similar to taking the individual bricks from a wall and using them to make some other structure such as a sculpture or a brick walkway.

All dead organic matter and waste products of plants and animals are acted upon by decomposer organisms, and the nitrogen is released as ammonia (NH_3), which can be used directly by plants or be acted upon by bacteria to produce different kinds of nitrogen compounds useful to plants.

Finally, other kinds of bacteria are capable of converting certain nitrogen-containing compounds into nitrogen gas (N_2), which is released into the atmosphere. Thus, there is a nitrogen cycle in which nitrogen from the atmosphere is passed through a series of organisms, many of which are bacteria, and ultimately is returned to the atmosphere to be cycled again.

FYI

Question: Why is it necessary to fertilize a crop such as corn but not a forest?

Carbon Dioxide and Global Warming

Humans have significantly altered the carbon cycle. As we burn fossil fuels, the amount of carbon dioxide in the atmosphere continually increases. Many are concerned that increased carbon dioxide levels will lead to a warming of the planet that would cause major changes in climate, leading to the flooding of coastal cities. Although it is uncertain if global warming is occurring, there is no question that the amount of carbon dioxide in the atmosphere has been increasing. In order to reduce carbon dioxide, several countries have planted millions of trees. The thought is that the trees will carry on photosynthesis, grow, and store carbon in their bodies, leading to reduced carbon dioxide levels. At the same time, however, people in other parts of the world continue to destroy forests at an alarming rate. Tree planting does not offset deforestation. In addition, the trees that have been planted will ultimately die and decompose, releasing carbon dioxide back into the atmosphere, so it is not clear that this is an effective means of reducing atmospheric carbon dioxide.

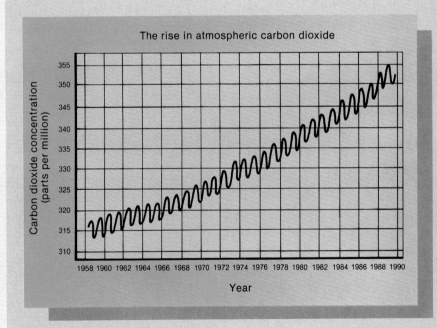

Atmospheric Carbon Dioxide Levels

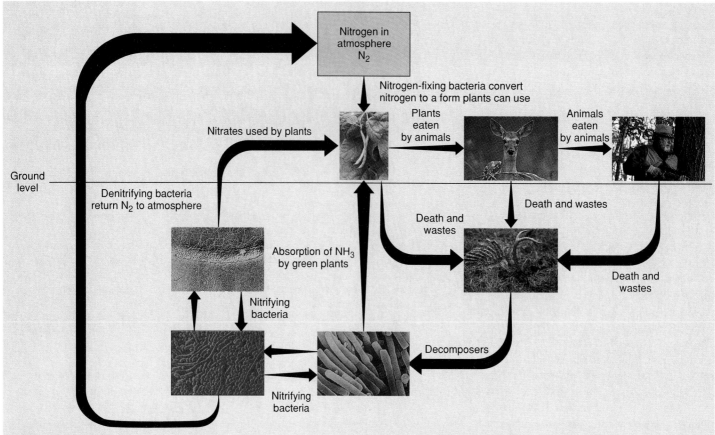

Figure 7.11 **The Nitrogen Cycle** Nitrogen atoms are cycled through ecosystems. Atmospheric nitrogen is converted by nitrogen-fixing bacteria to nitrogen-containing compounds that plants can use to make proteins and other compounds. Proteins are passed to other organisms when one organism is eaten by another. Dead organisms and their waste products are acted upon by decay organisms to form ammonia, which may be reused by plants and converted to other nitrogen compounds by nitrifying bacteria. Denitrifying bacteria return nitrogen as a gas to the atmosphere.

Labels within figure:
- Nitrogen in atmosphere N_2
- Nitrogen-fixing bacteria convert nitrogen to a form plants can use
- Nitrates used by plants
- Plants eaten by animals
- Animals eaten by animals
- Ground level
- Denitrifying bacteria return N_2 to atmosphere
- Absorption of NH_3 by green plants
- Death and wastes
- Death and wastes
- Death and wastes
- Nitrifying bacteria
- Nitrifying bacteria
- Decomposers

The Impact of Human Actions on the Community

As you can see from this discussion and from the discussion of food webs in chapter 6, all organisms are associated in a complex network of relationships. A community consists of all these sets of interrelations. Humans have the capacity to change communities in fundamental ways and do so purposely and accidentally. Agriculture can take place only if the original community is destroyed or significantly modified. Cities and villages eliminate the natural community. Even forests, which we often think of as natural, are modified when trees are cut or wildlife is harvested. Since the relationships within communities can be complex and subtle, it is wise to analyze how the organisms are interrelated when making decisions to manage a community.

Predator Control

During the early years of wildlife management, it was thought that populations of game species could be increased if the populations of their predators were reduced. Consequently, many states passed laws that encouraged the killing of foxes, eagles, hawks, owls, coyotes, cougars, and other predators that use game animals as a source of food. Often bounties were paid to people who killed these predators. In South Dakota, for example, it was decided to increase the pheasant population by reducing the numbers of foxes and coyotes. But when the supposed predator populations were significantly reduced, there was no increase in the pheasant population. There was, however, a rapid increase in the rabbit and mouse populations, and they became serious pests. Evidently the foxes and coyotes were major factors in keeping rabbit and mouse populations under control but had only a minor impact on pheasants.

BIO feature

At the Root of It All

Symbiotic nitrogen-fixing bacteria live in the roots of certain kinds of plants, where they convert nitrogen gas molecules into compounds that the plants can use to make amino acids and nucleic acids. The most common plants that enter into this mutualistic relationship with bacteria are the legumes, such as beans, clover, peas, alfalfa, and locust trees. Some other organisms, such as alder trees, can also participate in this relationship. There are also **free-living nitrogen-fixing bacteria** in the soil that provide nitrogen compounds that can be taken up through the roots of plants, but in this case the bacteria do not live in a close physical union with the plants.

Another way plants get usable nitrogen compounds involves a series of different bacteria. Decomposer bacteria and other bacteria with the ability to convert nitrogen compounds from one form to another convert organic nitrogen-containing compounds into ammonia (NH_3) and other nitrogen-containing molecules such as nitrate (NO_3^-). Many kinds of plants can use this ammonia (NH_3) or nitrate (NO_3^-) from the soil as building blocks for amino acids and nucleic acids.

BIO feature

Agriculture and the Nitrogen Cycle

Since nitrogen is in short supply in most ecosystems, farmers usually find it necessary to supplement the natural nitrogen sources in the soil to obtain maximum plant growth. This can be done in a number of ways. Alternating nitrogen-producing crops with nitrogen-demanding crops helps to maintain high levels of usable nitrogen in the soil. One year, a crop can be planted that has symbiotic nitrogen-fixing bacteria associated with its roots, such as beans or clover. The following year, the farmer can plant a nitrogen-demanding crop, such as corn.

The use of manure is another way of improving nitrogen levels. The waste products of animals are broken down by decomposer bacteria and nitrifying bacteria, resulting in enhanced levels of ammonia and nitrate in the soil. Finally, the farmer can use industrially produced fertilizers containing ammonia or nitrate. These compounds can be used directly by plants or can be converted into other useful forms by nitrifying bacteria.

Fertilizers usually contain more than just nitrogen compounds. The numbers on a fertilizer bag tell you the percentages of nitrogen, phosphorus, and potassium in the fertilizer. For example, a 6-24-24 fertilizer would have 6 percent nitrogen compounds, 24 percent phosphorus compounds, and 24 percent potassium-containing compounds. These other elements (phosphorus and potassium) are also cycled through ecosystems. In natural, nonagricultural ecosystems these elements would be released by decomposers and enter the soil, where they would be available for plant uptake through the roots. However, when crops are removed from fields, these elements are removed with them and must be replaced by adding more fertilizer.

Habitat Destruction

Some communities are quite fragile, whereas others seem to be able to resist major human interference. Communities that have a wide variety of organisms and a high level of interaction are more resistant than those with few organisms and little interaction. In general, the more complex an ecosystem is, the more likely it is to recover after being disturbed. The tundra biome is an example of a community with relatively few organisms and interactions. It is not very resistant to change, and because of its slow rate of repair, damage caused by human activity may persist for hundreds of years.

Some species are more resistant to human activity than others. Rabbits, starlings, skunks, and many kinds of insects and plants are able to maintain high populations despite human activity. Indeed, some may even be encouraged by human activity. By contrast, whales, condors, eagles, and many plant and insect species are not able to resist human interference very well. For most of these endangered species, direct action of humans is not the problem; very few organisms have been driven to extinction by hunting or direct exploitation. Usually, human influence is indirect. Habitat destruction is the main cause of extinction and the endangering of species. As humans convert land to farming, grazing, commercial forestry, and special wildlife management areas, the natural ecosystems are disrupted, and plants and animals with narrow niches tend to be eliminated because they lose critical resources in their environment. Table 7.1 lists several endangered species and the probable causes of their endangerment.

Pesticide Use

Since some insects compete with humans for food and others are involved in the spread of disease, the invention of various kinds of **insecticides** (insect killers) has greatly contributed to the health of the human population. Insecticides, herbicides, and other material used to kill pests are often called **pesticides.** Unfortunately, pesticide use has also presented problems. Sensitive, nonpest animals have been endangered, people worry about pesticide molecules causing health problems, and organisms develop resistance to pesticides.

apply your knowledge

Feeding Starving People

This is a thought puzzle—put it together! Write a paragraph that links all the following bits of information in a way that explains how to solve the problem.

The Problem: How to feed starving people.

Pieces of the Puzzle:

Commercial fertilizer production requires temperatures of 900°C.

Geneticists have developed plants that grow very rapidly and require high amounts of nitrogen to germinate during the normal growing season.

Fossil fuels are stored organic matter.

The rate at which nitrogen can be incorporated into living things depends on the activity of bacteria.

The sun is expected to last for several million years.

Crop rotation is becoming a thing of the past.

The clearing of forests for agriculture changes weather in the area.

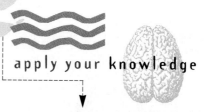

apply your knowledge

Wolves in the Chicken Coop

Should predators be reintroduced to places where they were purposely eliminated in the past? The reintroduction of wolves into Yellowstone National Park in 1995 was very controversial. Park managers felt that since wolves were originally part of the park community, they would help return the park to a more natural condition and control the populations of elk and bison. Farmers and ranchers in the surrounding area are concerned that the wolves may kill livestock. If wolves leave the park and kill livestock, should the farmers and ranchers be reimbursed? Were wolves an important part of the ecosystem previously? Are wolves in the park a danger to tourists?

table 7.1

Endangered and Threatened Species

Species	Reason for Endangerment
Hawaiian crow *Corvis hawaiinis*	Predation by cat and mongoose; disease; habitat destruction
Sonora Chub *Gila ditaenia*	Competition with introduced species
Black-footed ferret *Mustela nigripes*	Poisoning of prairie dogs (their primary food)
Snail kite *Rostrhamus sociabilis*	Specialized eating habits (only eat apple snails); draining of marshes
Grizzly bear *Ursus arctos*	Loss of wilderness areas
California condor *Gymnogyps californianus*	Slow breeding; lead poisoning
Ringed sawback turtle *Graptemys oculifera*	Modification of habitat by construction of reservoir that reduced their primary food source
Scrub mint *Dicerandra frutescens*	Conversion of habitat to citrus groves and housing

CONCEPT CONNECTIONS

Pesticide is the term used to refer to any chemical that kills an unwanted organism. It is a general term for many kinds of *-cides*.

Insecticides are used to kill insects.
Herbicides are used to kill plants (weed killers).
Fungicides are used to kill fungi (molds, etc.).

A number of factors determine how successful you will be in controlling a pest with a pesticide. You must choose a pesticide that will cause the least amount of damage to the harmless or beneficial organisms in the community. The ideal pesticide or insecticide affects only the target pest. Because many of the insects we consider pests are herbivores, you would expect carnivores in the community to use the pest species as prey, and parasites to use the pests as hosts. These predators and parasites would have important roles in controlling the numbers of the pest species. Generally, predators reproduce more slowly than their prey. Because of this, the use of a nonspecific pesticide may actually make matters worse; if such a pesticide is applied to an area, the pest

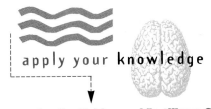

apply your **knowledge**

Are Pesticides and Fertilizers Safe?

Many people will pay more for food that has been "organically grown." They are against the use of both pesticides and chemical fertilizers in the growing of crops. Develop an argument that will convince people that the use of chemical fertilizers is different from the use of pesticides.

is killed but so are its predators. Since the herbivore pest reproduces faster than its predators, the pest population rebounds quickly, unchecked by natural predation. This may necessitate more frequent and more concentrated applications of pesticides. This has actually happened in many cases of pesticide use; the pesticides made the problem worse, and it became increasingly costly to apply them.

Succession

Maintaining lawns, parks, and golf courses involves a tremendous amount of work. The grass grows and is cut, weeds are pulled or killed with weed killer, and grass must be removed from flower beds. All of this work is necessary to prevent the space from changing into some other kind of community. However, a natural piece of forest, desert, or grassland does not seem to change very much. What is the difference between the unstable lawn and an unchanging forest?

Many communities like the biomes we discussed in chapter 6 are relatively stable over long periods of time. A relatively stable, long-lasting community is called a **climax community.** The word *climax* implies the final step in a series of events. That is just what the word means in this context, because communities can go through a series of predictable, temporary stages that eventually result in a long-lasting stable community. The process of changing from one type of community to another is called **succession,** and each intermediate stage leading to the climax community is known as a **successional stage.**

Two different kinds of succession are recognized: **primary succession,** in which a community of plants and animals develops where none existed previously, and **secondary succession,** in which a community of organisms is disturbed by a natural or human-related event (e.g., hurricane, forest fire, harvest) and returned to a previous stage in succession. Primary succession is much more difficult to observe than secondary succession because there are relatively few places on earth that lack communities of organisms. The tops of mountains, newly formed volcanic rock, and rock newly exposed by erosion or glaciers can be said to lack life. Bacteria, algae, fungi, and lichens quickly begin to grow on the bare rock surface, beginning the process of succession. The first organisms to colonize an area are often referred to as **pioneer organisms,** and the community they are a part of is called a **pioneer community.** Lichens are frequently important in pioneer communities. They are unusual organisms that consist of a combination of algal cells and fungal

Bioamplification of DDT in Food Chains

Humans have developed a variety of chemicals to control specific pest organisms. One of these is the insecticide DDT. DDT is an abbreviation for the chemical name dichloro-diphenyl-trichloro-ethane. It is one of a group of organic compounds called *chlorinated hydrocarbons*. Because DDT is a poison used to kill a variety of insects, it is called an insecticide. Although it is no longer used in the United States (its sale and use were banned in the early 1970s), DDT is still manufactured and used in many parts of the world, including Mexico.

When DDT is applied to an area to get rid of insect pests, it is usually dissolved in an oil or fatty compound. It is then sprayed over the area and falls on the insects and plants that the insects use for food. Eventually the DDT enters the insect either directly through the body wall or through the food it is eating and interferes with the normal metabolism of the insect. If small quantities are taken in, the insect can digest and break down the DDT just like any other large organic molecule. Since DDT is soluble in fat or oil, the DDT or its breakdown products are stored in the fat deposits of the insect. Some insects can break down and store all of the DDT they encounter and, therefore, they survive. If an area has been lightly sprayed with DDT, some insects die, some are able to tolerate the DDT, and others break down and store nonlethal quantities. As much as one part DDT per one billion parts of insect tissue can be stored in this manner. This is not much DDT! It is equivalent to one drop of DDT in 100,000 liters (25,000 gallons). However, when an aquatic area is sprayed with a small concentration of DDT, many kinds of organisms in the area may accumulate such tiny quantities in their bodies. Even algae and protozoa found in aquatic ecosystems accumulate pesticides. They may accumulate concentrations in their cells that are 250 times higher than the concentration sprayed on the ecosystem. The algae and protozoa are eaten by insects, which in turn are eaten by frogs, fish, or other carnivores. The concentration in frogs may be 2,000 times what was sprayed. The birds that feed on the frogs and fish may accumulate concentrations that are as much as 80,000 times the original amount. What was originally dilute becomes more concentrated as it accumulates in the food chain because DDT is relatively stable and is stored in the fat deposits of the organisms that take it in.

When DDT was used in the United States, many animals at higher trophic levels died as a result of this accumulation of pesticide in food chains. This process is called **biological amplification.** Even if they were not killed directly by DDT, many birds that feed at higher trophic levels, such as eagles, pelicans, and ospreys, suffered reduced populations because the DDT interfered with the female bird's ability to produce eggshells. Thin eggshells are easily broken, and there were no live young hatched.

What was originally used as an insecticide to control insect pests has been shown to have many harmful consequences for biological communities. Instead of controlling the pest, it selects for resistance and creates populations that can tolerate the poison. Instead of harming only the pest species, it accumulates in food chains to kill nontarget species as well. Furthermore, it is not specific to pest species only, but kills many beneficial species of insects. DDT has not been the gift to humankind that we originally thought it would be.

The Biological Amplification of DDT   All of the numbers shown are in parts per million (ppm). A concentration of one part per million means that in a million equal parts of the organism, one of the parts would be DDT. Notice how the amount of DDT in the bodies of the organisms increases from producers to herbivores to carnivores. Since DDT is persistent, it builds up in the top trophic levels of the food chain.

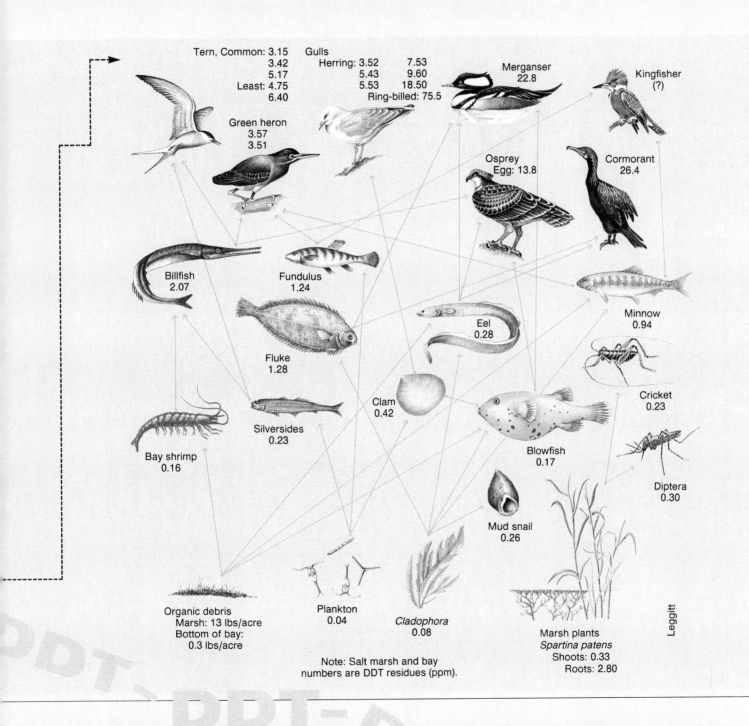

Tern, Common: 3.15
3.42
5.17
Least: 4.75
6.40

Gulls
Herring: 3.52   7.53
5.43   9.60
5.53   18.50
Ring-billed: 75.5

Merganser
22.8

Kingfisher
(?)

Green heron
3.57
3.51

Osprey
Egg: 13.8

Cormorant
26.4

Billfish
2.07

Fundulus
1.24

Minnow
0.94

Fluke
1.28

Eel
0.28

Cricket
0.23

Clam
0.42

Silversides
0.23

Blowfish
0.17

Bay shrimp
0.16

Mud snail
0.26

Diptera
0.30

Organic debris
Marsh: 13 lbs/acre
Bottom of bay:
0.3 lbs/acre

Plankton
0.04

Cladophora
0.08

Marsh plants
Spartina patens
Shoots: 0.33
Roots: 2.80

Leggitt

Note: Salt marsh and bay
numbers are DDT residues (ppm).

BIO feature

Herring Gulls as Indicators of Contamination in the Great Lakes

Herring gulls nest on islands and other protected sites throughout the Great Lakes region. Since they feed primarily on fish, they are near the top of aquatic food chains and tend to accumulate toxic materials from the food they eat. Eggs taken from nests can be analyzed for a variety of contaminants.

Since the early 1970s, the Canadian Wildlife Service has operated a monitoring program to assess trends in the levels of various contaminants in the eggs of herring gulls. In general, the levels of contaminants have declined as both the Canadian and U.S. governments took action to stop new contaminants from entering the Great Lakes. These graphs show the trend in the levels of PCBs in herring gull eggs. PCBs are a group of organic compounds, some of which are much more toxic than others. They were used as fire retardants, lubricants, insulation fluids in electrical transformers, and in some printing inks. Both Canada and the United States have eliminated most uses of PCBs, resulting in the declines in levels found in the eggs; however, research continues to reveal new types of harm caused by members of this complex group of chemicals.

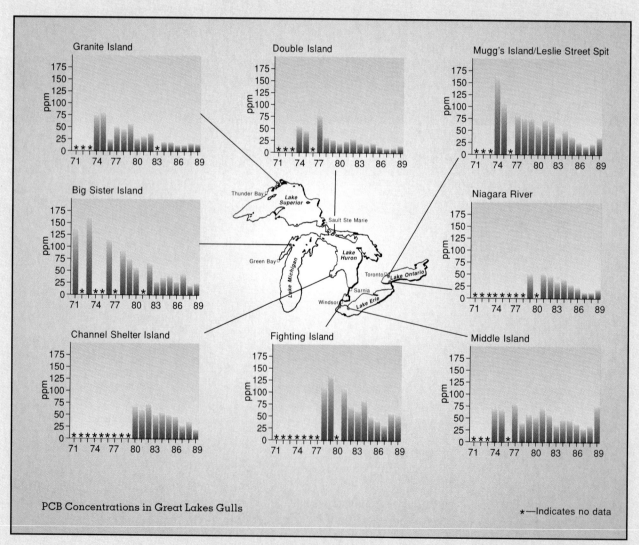

PCB Concentrations in Great Lakes Gulls

★—Indicates no data

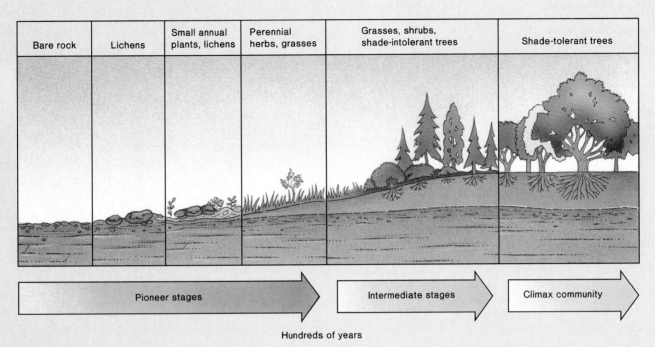

| Bare rock | Lichens | Small annual plants, lichens | Perennial herbs, grasses | Grasses, shrubs, shade-intolerant trees | Shade-tolerant trees |

Pioneer stages → Intermediate stages → Climax community

Hundreds of years

Figure 7.12 Primary Succession   The formation of soil is a major step in primary succession. Until soil is formed, the area is unable to support large amounts of vegetation. The vegetation modifies the harsh environment and increases the amount of organic matter that can build up in the area. The presence of plants eliminates the earlier pioneer stages of succession. If given enough time, a climax community may develop.

cells—a combination that is very hardy and is able to grow on the surface of bare rock (figure 7.12). Since algal cells are present, the lichen is capable of photosynthesis and can form new organic matter. Furthermore, many tiny consumer organisms can make use of the lichens as a source of food and a sheltered place to live. The action of the lichens also tends to break down the rock surface upon which they grow. This fragmentation of rock by lichens is aided by the physical weathering processes of freezing and thawing, dissolving of the rock by water, and wind erosion. It is the first step in the development of a mixture of mineral particles, organic matter, water, air, and organisms called **soil.** Lichens trap dust particles, small rock particles, and the dead remains of lichens and other organisms that live in and on them, resulting in a thin layer of soil.

As more material accumulates, the soil layer becomes thicker and small plants such as mosses may become established, increasing the rate at which energy is trapped by photosynthesis and adding more organic matter to

CONCEPT CONNECTIONS

We commonly use the words *land, earth, soil,* and *dirt* interchangeably, but it would probably be better if usage was restricted as follows:

Land is that part of the earth not covered by water.
The *earth* is the third planet from the sun.
Soil is a complex mixture of air, water, organisms, mineral material, and organic matter that supports the growth of plants.
Dirt is a material that causes something to be unclean.

the soil. Eventually, the soil may be able to support larger plants that are even more efficient at trapping sunlight, and the soil-building process continues at a more rapid pace. Associated with each of the producers in each successional stage is a variety of small animals, fungi, and bacteria. Every change in the community makes it more difficult for the previous group of organisms to maintain itself. Tall plants shade smaller producers; consequently, the smaller organisms become less common, and some may disappear entirely.

Only shade-tolerant species are able to compete successfully. One community has been succeeded by another.

Depending on the climate and other physical environmental factors and the kinds of species locally available to colonize the area, succession can lead to different kinds of climax communities. If the area is dry, it might stop at a grassland stage. If it is cold and wet, a coniferous forest might be the climax community. If it is warm and wet, it may be a tropical rainforest. The rate at which this successional process takes place is variable. In some warm, moist, fertile areas the entire process might take place in less than one hundred years. In harsh environments, like mountaintops or very dry areas, it may take thousands of years.

Another situation that is often called primary succession is the progression from an aquatic community to a terrestrial community. Lakes, ponds, and slow-moving parts of rivers accumulate organic matter. Where the water is shallow, this organic matter supports the development of rooted plants. In deep water, we find only floating plants like water lilies that send

Figure 7.13 Succession from a Pond to a Wet Meadow A shallow pond will slowly fill with organic matter from producers in the pond. Eventually, a floating mat will form over the pond and grasses will become established. In many areas this will be succeeded by a climax forest.

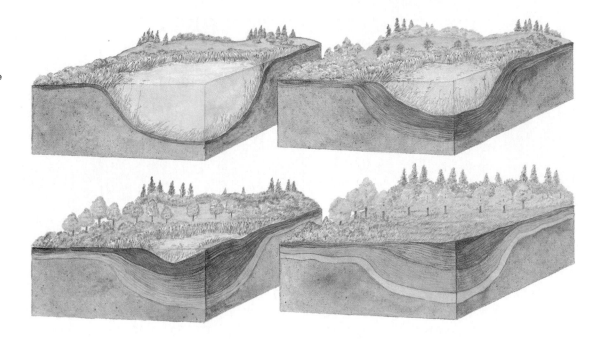

their roots down to the mucky bottom, but in shallower water, upright rooted plants like cattails and rushes develop. The cattail community contributes more organic matter, and the water level becomes more shallow. Eventually, a mat of mosses, grasses, and even small trees may develop on the surface along the edge of the water. If this continues for perhaps one hundred to two hundred years, an entire pond or lake will become filled in. Organic matter accumulates because of the large number of producers and because the depression that was originally filled with water becomes drier. This will usually result in a wet grassland, which in many areas will be replaced by the climax forest community typical of the area (figure 7.13).

Secondary succession occurs when a climax community or one of the successional stages leading to it is changed to an earlier stage. For example, this is what happens when agricultural land is abandoned. One obvious difference between primary succession and secondary succession is that in the latter there is no need to develop a soil layer. If we begin with bare soil the first year, it is likely to be invaded by a pioneer community of weed species that are annual plants. In this situation, a weed is any plant that is a pioneer in an area undergoing secondary succession. Within a year or two, perennial plants like grasses become established. Since most of the pioneer species need bare soil for seed germination, they are replaced by the perennial grasses and other plants that live in association with grasses. The more permanent grassland community is able to

support more insects, small mammals, and birds than the pioneer community could. If rainfall is adequate, several species of shrubs and fast-growing trees that require lots of sunlight (e.g., birch, aspen, juniper, hawthorn, sumac, pine, spruce, and dogwood) will become common. As the trees become larger, the grasses fail to get sufficient sunlight and die out. Eventually, shade-tolerant species of trees (e.g., beech, maple, hickory, oak, hemlock, and cedar) will replace the shade-intolerant species, and a climax community results (figure 7.14).

Most human use of ecosystems involves replacing the natural climax community with an artificial early successional stage. Agriculture involves replacing natural forest or prairie communities with specialized grasses such as wheat, corn, rice, and sorghum. This requires considerable effort on our part because the natural process of succession tends toward the

original climax community. This is certainly true if islands of the original natural community are still available to colonize agricultural land. Small woodlots in agricultural areas of the eastern United States serve this purpose. Much of the work and expense of farming is necessary to prevent succession to the natural climax community. It takes a lot of energy to fight nature.

SUMMARY

Each organism in a community occupies a specific space known as its habitat and has a specific functional role to play, known as its niche. An organism's habitat is usually described in terms of some obvious element of its surroundings. The niche is difficult to describe because it involves so many interactions with the physical environment and other living things.

Interactions between organisms fit into several categories. Predation involves one organism (predator) benefiting at the expense of the organism killed and eaten (prey). Parasitism involves one organism (parasite) benefiting by living in or on another living organism (host) and deriving nourishment from it. Commensal relationships exist when one organism is helped but the other is not affected. Mutualistic relationships benefit both organisms. Symbiosis is any interaction in which two organisms live together in a close physical relationship. Competition may cause harm or benefit to the organisms involved; although an

FYI

Forestry practices often seek to simplify the forest by planting single-species forests of the same age. This certainly makes management and harvest practices easier and more efficient, but these kinds of communities do not contain the variety of plants, animals, fungi, and other organisms typically found in natural climax ecosystems.

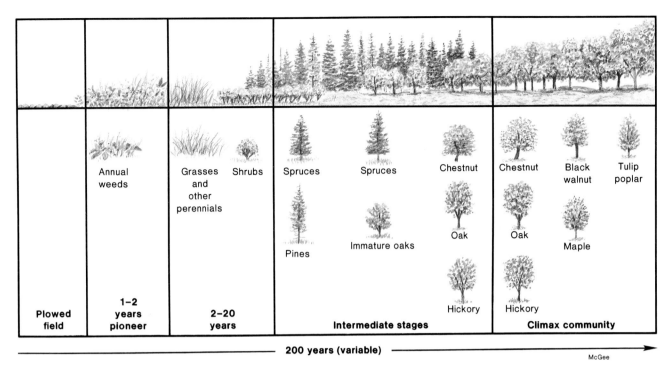

Plowed field	1–2 years pioneer	2–20 years	Intermediate stages	Climax community
	Annual weeds	Grasses and other perennials / Shrubs	Spruces / Pines / Spruces / Immature oaks / Chestnut / Oak / Hickory	Chestnut / Oak / Hickory / Black walnut / Maple / Tulip poplar

200 years (variable)

McGee

Figure 7.14 Secondary Succession on Land A plowed field in the southeastern United States shows a parade of changes over time involving plant and animal associations. The general pattern is for annual weeds to be replaced by grasses and other perennial herbs, which are replaced by shrubs, which are replaced by trees. As the plant species change, so do the animal species.

individual may be harmed, the species may benefit since it may evolve into a different niche or be forced to migrate.

Many atoms are cycled through ecosystems. The carbon atoms of living things come from the carbon dioxide trapped by photosynthesis, are passed from organism to organism as food, and eventually are released to the atmosphere as carbon dioxide by the process of respiration. Water is necessary as a raw material for photosynthesis and as the medium in which all metabolic reactions take place in living things.

Water is cycled by the physical processes of evaporation and condensation. Nitrogen originates in the atmosphere, is trapped by nitrogen-fixing bacteria, passes through a series of organisms, and is ultimately released to the atmosphere by denitrifying bacteria.

Organisms in a community are interrelated to one another in very sensitive ways; thus, changing one part of a community can lead to unexpected consequences. Predator-control practices, habitat destruction, and pesticide use have all caused damage to the ecosystem.

Succession is a series of communities replacing one another. Each community changes the environment to make conditions favorable for a subsequent community and unfavorable for itself. Successional processes may result in a relatively stable stage called the climax community. The stages leading to the climax community are called successional stages. If the process begins with bare rock or water, it is called primary succession. If it begins as a disturbed portion of a community, it is called secondary succession.

CHAPTER GLOSSARY

biological amplification (bi-o-loj´ ĭ-cal am″pli-fĭ-ka´shun) The accumulation of a compound in increasing concentrations in organisms at successively higher trophic levels.

climax community (klī´maks ko-miu´nĭ-te) A relatively stable, long-lasting community.

commensalism (ko-men´sal-izm) A relationship between two organisms in which one organism is helped and the other is not affected.

competition (com-pe-tĭ´shun) A relationship between two organisms in which both organisms are harmed.

epiphyte (ep´e-fīt) A plant that lives on the surface of another plant.

free-living nitrogen-fixing bacteria (ni´tro-jen fik´sing bak-te´re-ah) Soil bacteria that convert nitrogen gas molecules into nitrogen compounds that plants can use.

habitat (hab´ĭ-tat) The place or part of an ecosystem occupied by an organism.

host (hōst) An organism that a parasite lives in or on.

inorganic molecules (in-or-gan´ik mol´uh-kiuls) Molecules that do not contain carbon atoms in rings or chains.

insecticide (in-sek´tĭ-sīd) A poison used to kill insects.

mutualism (miu´chu-al-izm) A relationship between two organisms in which both organisms benefit.

niche (nitch) The functional role of an organism.

organic molecules (or-gan´ik mol´uh-kiuls) Complex molecules whose basic building blocks are carbon atoms in chains or rings.

parasite (pĕr´uh-sīt) An organism that lives in or on another living organism and derives nourishment from it.

parasitism (pĕr′uh-sit-izm) A relationship between two organisms in which one organism lives in or on another organism and derives nourishment from it.

pesticide (pes′tĭ-sīd) A poison used to kill pests. This term is often used interchangeably with insecticide.

pioneer community (pi″o-nēr′ ko-miu′nĭ-te) The first community of organisms in the successional process established in a previously uninhabited area.

pioneer organisms (pi″o-nēr′ or′gun-izms) The first organisms in the successional process.

predation (prĕ-da′shun) A relationship between two organisms that involves the capturing, killing, and eating of one by the other.

predator (pred′uh-tōr) An organism that captures, kills, and eats another animal.

prey (prā) An organism captured, killed, and eaten by a predator.

primary succession (pri′mar-e suk-sĕ′shun) The orderly series of changes that begins in a previously uninhabited area and leads to a climax community.

secondary succession (sek′on-dĕr-e suk-sĕ′shun) The orderly series of changes that begins with the disturbance of an existing community and leads to a climax community.

soil (soyl) A mixture of mineral particles, organic matter, water, air, and organisms.

succession (suk-sĕ′shun) The process of changing one type of community to another.

successional stage (suk-sĕ′shun-al stāj) An intermediate stage in succession.

symbiosis (sim-be-o′sis) A close physical relationship between two kinds of organisms. It usually includes parasitism, commensalism, and mutualism.

symbiotic nitrogen-fixing bacteria (sim-be-ah′tik ni-tro-jen fik′sing bak-tīr′ē-ah) Bacteria that live in the roots of certain kinds of plants, where they convert nitrogen gas molecules into compounds that plants can use.

 EXPLORATIONS **Interactive Software**

Pollution of a Freshwater Lake

This interactive exercise allows students to explore how addition of certain "harmless" chemicals can pollute a lake by allowing algae to grow. Bacteria feeding on dead algae use up all the dissolved oxygen in the lake water, killing the lake. The exercise presents a map of Lake Washington, indicating the location of sewage treatment plants and the chemical composition of their effluent. Students can investigate the nature of pollution by altering the chemicals in the effluent. They will discover that phosphates in the effluent lead to algal growth, which is followed by bacterial growth (they feed on the dead algae) and a precipitous drop in levels of dissolved oxygen. Students can then attempt to "clean up" the lake by altering the nature and amount of effluent. Their success will depend largely upon how early in the pollution process they initiate their recovery efforts.

1. Why does the growth of algae, which are photosynthetic and make oxygen, lead to oxygen depletion of the lake?

2. Why not simply poison the algae in the lake?

3. How long can you wait before cleaning up the lake and still succeed?

4. Can you envision any way to successfully avoid oxygen depletion of the lake without stopping the discharge of treated sewage into the lake?

CONCEPT MAP TERMINOLOGY

Construct a concept map to represent the relationships among the following concepts.

Concept Map 1
climax community
pioneer organisms
soil
succession
symbiotic nitrogen-fixing bacteria

Concept Map 2
commensalism
competition
host
mutualism
niche
parasite
predator
prey

LABEL•DIAGRAM•EXPLAIN

Describe the flow of carbon atoms through an ecosystem. Describe how they enter the food chain, how they pass through it, and how they will eventually leave the food chain.

Multiple Choice Questions

1. Coyotes feed on mice, grouse, and berries. The coyotes have intestinal parasites that they get from the mice they eat. Coyotes are competitors of red foxes and hawks. Coyotes locate their prey by sound and smell.
This is a partial description of:
   a. the niche of coyotes
   b. the biosphere of coyotes
   c. the population of coyotes
   d. the biome of coyotes
2. Which of the following is an example of competition?
   a. a mosquito sucks your blood
   b. a fern lives on the side of a tree
   c. some trees in a forest die because they don't get enough sunlight
   d. all the plants in an area die due to a lack of water
3. With biological amplification, organisms at higher trophic levels:
   a. live longer
   b. accumulate more of a specific compound
   c. grow more rapidly in number
   d. have a high biotic potential
4. If you bought a piece of land that had been used for raising cattle and you left it unused for 50 years, you would expect the land:
   a. to remain unchanged
   b. to become a desert
   c. to become something like what it was before cattle were grazed on it
   d. to have fewer plants growing on it
5. The nitrogen found in carnivores would follow which one of the following paths?
   a. plant → air → carnivore
   b. herbivore → plant → air → carnivore
   c. air → bacteria → plant → herbivore → carnivore
   d. air → plant → bacteria → herbivore → carnivore
6. An interaction between organisms in which both are harmed is called:
   a. predation
   b. symbiosis
   c. parasitism
   d. competition
7. Plants obtain the carbon atoms necessary to manufacture organic matter from:
   a. oxygen in the air
   b. organic matter in the soil
   c. carbon dioxide in the air
   d. manufacturing the carbon from raw materials
8. The practice of agriculture:
   a. destroys natural ecosystems
   b. increases the number of kinds of plants in an area
   c. does not use resources
   d. is not necessary to support the people of the earth

Questions with Short Answers

1. List ten important aspects of your personal ecological niche.
2. What is the difference between a habitat and a niche?
3. What do parasites, commensal organisms, and mutualistic organisms have in common? How are they different?
4. Describe examples of competition between organisms.
5. Trace the flow of carbon atoms through a community that contains plants, herbivores, decomposers, and parasites.
6. Describe four different roles played by bacteria in the nitrogen cycle.
7. Describe the flow of water through the hydrologic cycle.
8. How does primary succession differ from secondary succession?
9. Describe the impact of DDT on communities.
10. How does a climax community differ from a successional community?

chapter

8

Population Ecology

African elephant herd, Kenya.

learning objectives

- Recognize that populations vary in age distribution, sex ratio, size, and density.
- Describe the characteristics of a typical population growth curve.
- Understand why populations grow.
- Recognize the pressures that ultimately limit population size.
- Understand that human populations obey the same rules of growth as populations of other kinds of organisms.

Population Characteristics

Populations of organisms exhibit many characteristics that can vary significantly from one population to another. Recognizing that populations differ from one another is necessary to understanding differences in the way populations grow. While not specifically about growth of the human population, much of the material in this chapter relates to the problems associated with the rapid growth of the human population.

Plants and animals are not evenly distributed around the world but tend to be found in groups. We talk about groves of orange trees, flocks of pigeons, herds of antelope, forests of aspen, packs of wolves, colonies of bacteria, patches of algae, growths of fungi, and nations of people. Sometimes the groups we see are small family units, as in a group of baby geese and their parents walking across a field or park. At other times the groups consist of large numbers of individuals of varying degrees of relatedness, as in the collections of people we call cities. All of these groups are known as *populations*. A **population** is a group of organisms of the *same species* located in the same place at the same time.

In addition to populations, we often see groups of organisms that contain mixtures of species. Forests contain many kinds of trees and other species of plants, birds often migrate in flocks of several species, and many species of insects inhabit one tree. These groups do not represent a population, since a population always consists of individuals of the same species. The terms *species* and *population* are interrelated because a species *is* a population—the largest possible population of a particular kind of organism. Population, however, is often used to refer to subsets of a species living in a particular location at a specific time. For example, the size of the human population in a city changes from hour to hour during the day and varies according to where you set the boundaries of the city. When an individual leaves town that person has left one population and joined another. Since members of a population are of the same species and can reproduce with one another, we expect them to show great similarity in the characteristics they display. For example, all dogs have four legs, a tail, hair, and sharp teeth. However, different populations within one species can show variation in their characteristics. Populations of poodles show many differences from populations of dachshunds.

Another feature of a population is its **age distribution,** which is the number of organisms of each age in the population. You recognize that different populations of humans have different age distributions. School populations have a large number of young individuals and retirement communities have a majority of older residents. Organisms in a population are often grouped into the following categories:

(1) prereproductive juveniles—insect larvae, plant seedlings, or babies
(2) reproductive adults—mature insects, plants producing seeds, or humans in early adulthood

FYI

For many organisms, age distribution changes during the year. Many plants and animals reproduce in the spring or early summer. Early in the spring before reproductive activity has begun, all of the individuals in a population may be reproductive, and there may be no prereproductive individuals. Once young are produced, the number of prereproductive individuals is high for a period of time.

(3) postreproductive adults no longer capable of reproduction—annual plants that have shed their seeds, salmon that have spawned, and many elderly humans

A population is not necessarily divided into equal thirds (figure 8.1). In some situations, a population may be made up of a majority of one age group. If the majority of a human population is prereproductive, then a "baby boom" should be anticipated in the future. If a majority of the population is reproductive, the population should be growing

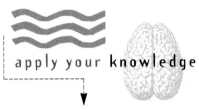

apply your knowledge

Your First Born and the Future Population

1. How many children were in your parents' families? At what age did your grandmothers have their first children?
2. How many children are in your family? At what age did your mother have her first child?
3. How many children do you plan to have? At what age do you plan to have (or did you have) your first child?

Compare the above information for your entire class. Do you notice differences among the generations? If so, how have these changes in trends influenced the rate of population growth? How might the age of first pregnancy and birth influence the rate of population growth in the future? Are these worldwide trends? What factors (social, biological, psychological) might influence your decision on family size?

Graph the data you collect from your class in order to help you see trends.

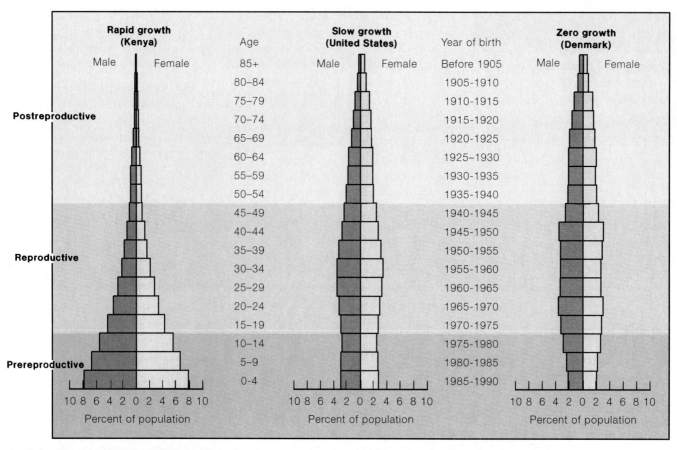

Figure 8.1   Age Distribution in Human Populations   The relative number of individuals in each of the three categories (prereproductive, reproductive, and postreproductive) can give a good clue to the future of the population. Kenya has a large number of young individuals who will become reproducing adults. Therefore, this population will grow rapidly and will double in about nineteen years. The United States has a declining proportion of prereproductive individuals but a relatively large reproductive population. Therefore, it will continue to grow for a time but will probably stabilize in the future. Denmark's population has a large proportion of postreproductive individuals and a small proportion of prereproductive individuals. Its population is stable.

rapidly. If the majority of the human population is postreproductive, a population decline should be anticipated.

Populations can also differ in the ratio of males to females. The **sex ratio** is the number of males in a population compared to the number of females. In bird and mammal species where strong pair-bonding occurs, the sex ratio may be nearly one to one (1:1). Among mammals and birds that do not have strong pair-bonding, sex ratios may show a larger number of females than males. This is particularly true among game species, where more males are shot than females. Since one male can fertilize several females, the population can remain large. If the population of these managed game species becomes large enough to cause a problem, it becomes necessary to harvest some of the females, since their number strongly influences how much reproduction can take place. In addition to these ex-

amples, many species of animals like bison, horses, and elk have mating systems in which one male maintains a harem of females. The sex ratio in these small groups is quite different from the 1:1 ratio typical of the population as a whole (figure 8.2).

There are very few situations in which the number of males exceeds the number of females. In some human and other populations, there may be sex ratios in which the males dominate if female mortality (death

rate) is unusually high or if some special mechanism separates most of one sex from the other.

Regardless of the specific sex ratio in a population, most organisms generate large numbers of offspring, leading to a concentration of organisms in one area. **Population density** is the number of organisms of a species per unit area, for example, the number of dandelions per square meter of lawn, the number of a particular species of bacterium per

FYI

Aphids are insects that have two methods of reproduction. One is the normal method in which males fertilize females. The other involves females producing special eggs that are not fertilized but develop into new females. The sex ratio in these populations is nearly 100 percent female. Usually in the fall of the year some male offspring are produced and normal sexual reproduction occurs.

Figure 8.2 **Sex Ratio in Elk** Some male animals defend a harem of females; therefore, the sex ratio in these groups is several females per male.

Apples

Pigs

Figure 8.3 **Reproductive Capacity** The ability of a population to reproduce greatly exceeds the number necessary to replace those that die. Here are some examples of the excellent reproductive abilities of some species.

square millimeter of skin, or the number of people per square kilometer. Some populations are extremely crowded, while others are scattered. As the population density increases, competition among members of the population for the necessities of life increases. Competition increases the likelihood that some individuals will explore new habitats and migrate to new areas.

Increases in the intensity of competition that cause changes in the environment and lead to dispersal are often referred to as **population pressure.** The dispersal of individuals to new areas can relieve the pressure on the home area and lead to the creation of new populations. Among animals, it is often the juveniles who leave the home population. If dispersal cannot relieve population pres-

sure, there is usually an increase in the rate at which individuals die due to predation, parasitism, starvation, and accidents. In plant populations, dispersal is not very useful for relieving population density; instead, the death of weaker individuals usually results in reduced population density.

Reproductive Capacity

Most organisms produce large numbers of offspring (figure 8.3). We see thousands of seeds from trees, nests full of young birds, and large outbreaks of flies and mosquitos. Sex ratios and age distributions within a population have a direct bearing on the rate of reproduction. Each species has an inherent **reproductive capacity,** which is the theoretical maximum rate of reproduction. Generally, the reproductive capacity is many times larger than the number of offspring needed simply to maintain the population. For example, a female carp may

produce one million to three million eggs in her lifetime. This is her reproductive capacity. However, because so many of her offspring will die, only two or three will ever develop into sexually mature adults. Therefore, her actual rate of reproduction is much smaller than her capacity to reproduce.

A high reproductive capacity is valuable to a species because it provides many opportunities for survival. Since the offspring are not identical, it is likely that some of the individuals will be successful even if the environment changes. From the very beginning of the reproductive process there is a high rate of death and few individuals reach sexual maturity. With most plants and animals, the majority of the eggs produced are never fertilized. An oyster may produce a million eggs a year, but not all of them are fertilized, and most that are fertilized die. An apple tree may display thousands of flowers but produce only a few hundred apples because of unsuccessful pollination. Even after the offspring are produced, the deathrate is usually high among the young. Most apple seeds that fall to the earth do not grow into sexually reproducing trees, and most young animals die as well. But usually enough survive to maintain the population. Organisms that reproduce in this way spend large amounts of energy on the production of potential young, with the probability that only a small number of them will reach reproductive age.

A second way for a species to approach reproduction is to produce relatively few individuals but provide care and protection that ensure a high probability that the young will become reproductive adults. Humans generally produce a single offspring per pregnancy, but nearly all of them live. In effect, the energy has been channeled into the care and protection of the few young produced rather than into the production of incredibly large numbers of young with little chance of survival. Even though fewer young are produced by animals like birds and mammals, their reproductive capacity still greatly exceeds the number required to replace the parents when they die.

The Population Growth Curve

Because most species of organisms have a high reproductive capacity, populations grow rapidly if environmental conditions are favorable. In some environments, there are regular bursts of population growth with the increased reproduction that occurs during the spring and summer. In desert areas this burst occurs following rains.

Suppose a pair of mice were to move into an area with abundant resources, no competition from others of their species, and no predators. (These would be ideal circumstances rarely, if ever, found in nature.) If eight young was the typical litter size for this species, after their first litter there would be eight offspring plus their two parents. Assuming that the original pair were still reproducing and their reproducing offspring were 50 percent

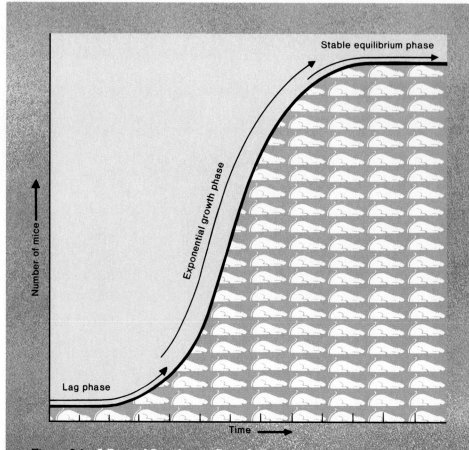

Figure 8.4 **A Typical Population Growth Curve** In this mouse population, the period of time during which there is little growth is known as the lag phase. This is followed by a rapid increase in population as the offspring of the originating population begin to reproduce themselves; this is known as the exponential growth phase. Eventually the population reaches a stable equilibrium phase, during which the birthrate equals the deathrate.

males and 50 percent females, the population could consist of as many as 50 individuals! Should this happen again, the population would be composed of 25 females producing a total of 200 offspring plus the original 50, or 250 mice! Imagine this continuing unchecked! Figure 8.4 shows a graph of change in population size over time known as a **population growth curve.** This kind of curve is typical for the type of situation described above.

How quickly the size of the population changes depends on the rate at which new organisms enter the population compared to the rate at which they leave. We commonly refer to the rate at which individuals enter the population by reproduction as the **birthrate** and the rate at which they leave by death as the **deathrate.**

When a small number of organisms (two mice) first invades an area, there is a period of time before reproduction takes place when the population remains small and relatively constant. This part of the population growth curve is known as the **lag phase.** It will be some time before any new individuals will be born, and if none of the first arrivals die, the population will remain constant. In the case of organisms that take a long time to mature and reproduce such as elephants, deer, and many kinds of plants, the lag phase may be measured in years. With the mice in our example, it will be measured in weeks. With several pairs of mice reproducing, the birthrate increases while the deathrate remains low; therefore, the population begins to grow at an ever-increasing (accelerating) rate. This portion of the population growth curve is known as the **exponential growth phase.**

The number of mice (or any other organism) cannot continue to increase at a faster and faster rate because, eventually, something in the environment will cause an increase in the number of deaths. The number of individuals entering the population will become equal to the number of individuals leaving it by death or migration, and the population size will become stable. Often there is both a decrease in birthrate and an increase in deathrate at this point. This portion of the population growth curve is known as the **stable equilibrium phase.** Reproduction continues and the birthrate is still high, but the deathrate increases and larger numbers of individuals may migrate from the area.

CONCEPT CONNECTIONS

death phase   The portion of some population growth curves in which the size of the population declines.

exponential growth phase   A period of time during population growth when population increases at an accelerating rate.

lag phase   A period of time following colonization when the population remains small or increases slowly.

stable equilibrium phase   A period of time during population growth when the number of individuals entering the population and the number leaving the population are equal, resulting in a stable population.

Ideal Population Growth

Original population size: 2 mice, 1 female and 1 male

↓

First reproductive event

↓

Eight offspring

$1 \times 8 = 8$

Population size: 10 mice (2 parents + 8 offspring),
5 females and 5 males

↓

Second reproductive event

↓

$5 \times 8 = 40$ new offspring

Population size: 50 mice (40 + 10),
25 females and 25 males

↓

Third reproductive event

↓

$25 \times 8 = 200$

Population size: 200 + 50 = 250

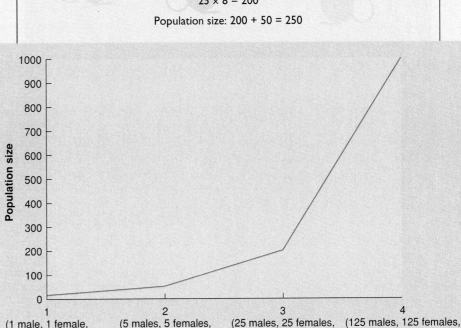

Ideal Population Growth

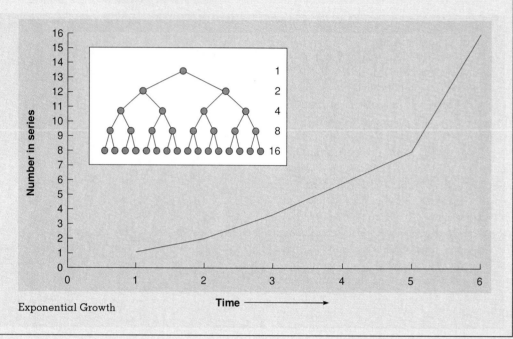

Exponential Growth

Exponential growth can be illustrated with the following series of numbers: 1:2:4:8:16:32:64:128. As we move left to right we double the previous number, but the farther to the right we go the larger the increase in size over the previous number.

Number in series

1
2
4
8
16

Exponential Growth

Time ⟶

Infectious Diseases as Population Problems

Many disease conditions are really population problems. For example, we always have *Streptococcus* bacteria living in our mouth and throat. If for some reason their numbers increase, we may develop strep throat. Similarly, many vaginal and urinary infections are caused by organisms that are normally present but occasionally experience population explosions. Antibiotics are used as limiting factors that decrease the "birthrate" or increase the "deathrate" and return the population to normal. When using antibiotics you should always take the entire quantity prescribed, because if you stop taking the antibiotic too soon, the populations of disease-causing organisms may still be larger than normal and able to quickly increase again to disease-causing levels as soon as the antibiotic is removed.

The George Reserve

In 1930, the George Reserve was established as a wildlife study area. This property, about 1,100 acres in southeastern Michigan, was donated to the University of Michigan to further research on natural populations. It is bounded by a game fence 2.9 meters high. The reserve includes some wooded areas, both deciduous and coniferous, some wet and marshy areas, some permanent and temporary ponds, and some open grassy meadows. Originally, six deer were imported into the reserve: four females (thought to be pregnant) and two males. Natural predators were excluded, so the deer population was expected to increase rapidly.

One goal of the early research was to keep an accurate record of how the deer population grew. Researchers decided to count the number of deer in various age classes so that they could make a cross-check whenever an animal died. Each year, a new census was conducted in the reserve. During the census, a group of individuals (usually graduate students) would line up and walk from one end of the reserve to the other. The people taking the census would ideally be close enough to maintain contact. When a deer passed through the line of census takers, it was counted. Skeletal remains also were counted. At the end of the census, the total deer population was tabulated. The graph shows the actual population changes for almost forty years.

Researchers were surprised by the results of the census. *After only six years, the number of deer had increased from six to over 160.* Why was there such a rapid increase in just six years? Answering this question requires some knowledge of deer biology. Deer are sometimes sexually mature before they are one year old, twins are common,

and triplets can occur. Also, we need to know what phase of population growth the deer population was in in 1930 and again in 1940. What do you think would prevent the number of deer from increasing forever? Note that the curve is not a neat, smooth line but a series of abrupt spikes. What could cause this? What could account for the drastic decline in population in the mid-1930s? Why was there an increase in the next few years?

Recently, one of the research activities at the George Reserve has been an attempt to reduce the population to a particular level by harvesting deer. Contrary to expectations, when the harvest was increased there was not an equivalent steplike decrease in population size.

Can you think of any reason why there is a gradual decline in the population rather than the steplike reduction anticipated?

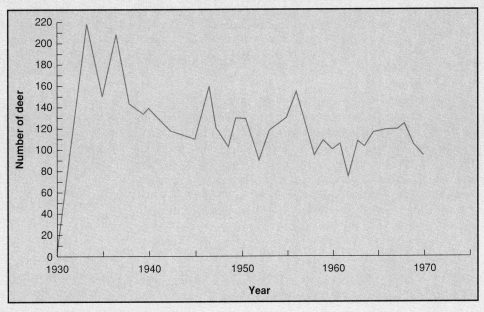

Deer Population at the George Reserve

Population-Size Limitations

While there may be some seasonal changes in the size of a population, we usually are not overrun with any kind of organism. The number of robins, cockroaches, or squirrels remains the same over long time periods. Populations cannot continue to increase indefinitely; eventually, some factor or set of factors acts to limit the size of a population, leading to the development of a stable equilibrium phase or even to a reduction in population size. The specific factors that prevent unlimited population growth are known as **limiting factors.** The number of individuals of a species that can be supported in an ecosystem is known as the **carrying capacity** (figure 8.5). In general, organisms that are small and have short life spans tend to have populations that change a great deal, while large organisms that live a long time tend to reach an optimum population size that can be sustained over an extended period. For example, a forest ecosystem contains populations of many insect species that change by a large amount over time, but the number of specific tree species or large animals such as owls or deer is relatively constant.

CONCEPT CONNECTIONS

carrying capacity The optimum population size an area can support over an extended period of time.

population density The number of organisms of a species per unit area.

The carrying capacity for a species in a certain ecosystem can change. Often such environmental changes as successional modifications, climate fluctuations, disease epidemics, forest fires, or floods can change the capacity of an area to support life. In addition, a change that negatively affects the carrying capacity for one species may increase the carrying capacity for another. For example, the cutting down of mature forests followed by the growth of young trees increases the carrying capacity for deer and rabbits, which use the new growth for food, but decreases the carrying capacity for squirrels, which need mature, fruit-producing trees as a source of food and hollow trees for shelter.

The size of the organisms in a population also affects the carrying capacity. For example, an aquarium of a certain size can support a limited number of fish, but the size of the fish makes a difference. If all the fish are tiny, a large number can be supported; however, the same aquarium may be able to support only one large fish. In other words, the biomass of the population is important (figure 8.6). Similarly, when an area is planted with small trees, the population size is high. But as the trees get larger, competition for nutrients and sunlight becomes more intense, and the number of trees declines, while the biomass increases.

Limiting Factors

You would expect that the amount of food available would limit the size of an animal's population, and you are aware that carnivores limit the population of their prey, but there are many more limiting factors, all of which can be placed in one of four broad categories:

(1) availability of raw materials
(2) availability of energy
(3) production and disposal of waste products
(4) interaction with other organisms

The first category of limiting factors is the availability of raw materials. For plants, magnesium is necessary for the manufacture of chlorophyll, nitrogen is necessary for protein production, and water is necessary for the transport of materials and as a raw material for photosynthesis. If these are not present in the soil, the growth and reproduction of

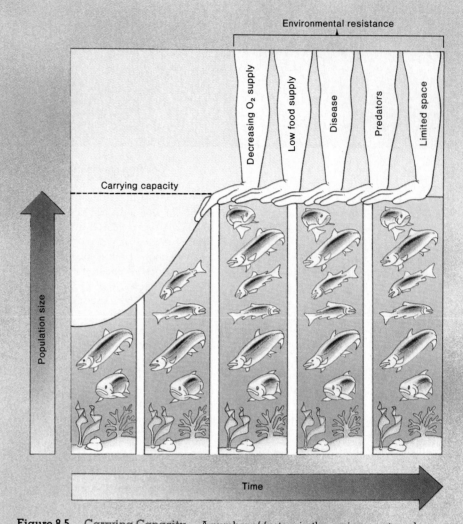

Figure 8.5 Carrying Capacity A number of factors in the environment, such as food, oxygen supply, diseases, predators, and space, determine the number of organisms that can survive in a given area, called the carrying capacity of that area. The environmental factors that limit populations are collectively known as environmental resistance.

plants is inhibited. However, if fertilizer supplies these nutrients, or if irrigation is used to supply water, the effects of these limiting factors can be removed, and some other factor becomes limiting. For animals, the amount of water, minerals, materials for nesting, pH of the environment, suitable burrow sites, or food may be limiting factors. Food for animals really fits into both this category and the next because it supplies both raw materials and energy.

The second major type of limiting factor is the availability of energy. The amount of light available is often a limiting factor for plants, which require light as an energy source for photosynthesis. Since all animals use other living things as sources of energy and raw materials, a major limiting factor for any animal is its food source.

The accumulation of waste products is the third general category of limiting factors. This factor does not usually limit plant populations because plants produce relatively few wastes. However, the buildup of high levels of self-generated waste products is a problem for many bacterial populations and populations of tiny aquatic organisms. As wastes build up, the levels become more and more toxic, and eventually reproduction stops, or the population may even die out. For example, when a few bacteria are introduced into a solution containing a source of food, they go through the kind of population growth curve typical of all organisms. As expected, the number of bacteria begins to increase following a lag phase, increases rapidly during the exponential growth phase, and eventually reaches stability in the stable equilibrium phase. But as waste products accumulate, the bacteria literally drown in their own wastes. When space for disposal is limited, and no other organisms are present to convert the harmful wastes to less harmful products, a population decline known as the **death phase** follows (figure 8.7).

Wine makers deal with this type of situation. When yeasts ferment the sugar in grape juice, they produce ethyl alcohol. When the alcohol concentration reaches a certain level, the yeast population stops growing and eventually declines. Therefore, wine can naturally reach an alcohol concentration of only 12 to 15 percent. To make any drink stronger than that (of a higher alcohol content), water must be removed by distillation or alcohol must be added to fortify the wine. Similarly, in small aquatic pools like

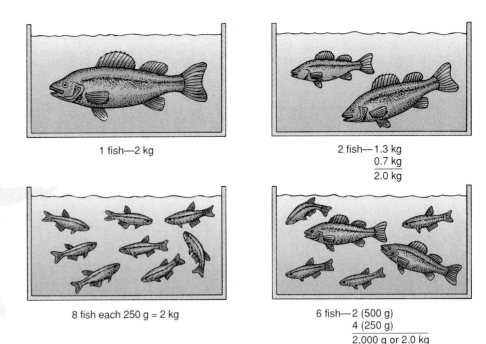

1 fish—2 kg

2 fish—1.3 kg
0.7 kg
————
2.0 kg

8 fish each 250 g = 2 kg

6 fish—2 (500 g)
4 (250 g)
——————
2,000 g or 2.0 kg

Figure 8.6 **The Effect of Biomass on Carrying Capacity** Each aquarium can support a biomass of 2 kilograms of fish. The size of the population is influenced by the body size of the fish in the population.

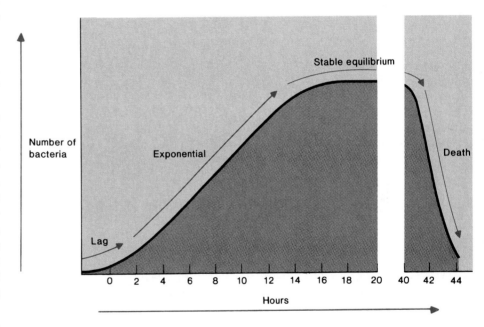

Figure 8.7 **Bacterial Population Growth Curve** The rate of increase in the population of these bacteria is typical of population growth in a favorable environment. As the environmental conditions change as a result of an increase in the amount of waste products, the population first levels off, then begins to decrease. This period of decreasing population size is known as the death phase.

aquariums, it is often difficult to keep populations of organisms healthy because of the buildup of ammonia in the water from the waste products of the animals. This is the primary reason activated charcoal filters and aerators are commonly used in aquariums. The charcoal removes many kinds of toxic compounds and the additional oxygen allows greater breakdown of waste products, thus preventing their buildup.

Forest Management and the Red-Cockaded Woodpecker

Forests managed for lumber production usually have older, diseased trees removed to give healthy trees more room and nutrients for growth. Their removal, however, reduces the number of nesting sites in hollow trees. Sick trees are a resource for birds. The red-cockaded woodpecker, for example, is an endangered species native to the southeastern United States. The birds require old, living pine trees that have a disease known as red-heart in which to build their nests. In order to protect this species, a major lumber company in the area where red-cockaded woodpeckers still exist has agreed not to disturb areas surrounding nest trees. Thus the company is not reducing the nest-tree resource required by this bird.

Red-Cockaded Woodpecker

The fourth set of limiting factors is organism interaction. As we discussed in previous chapters, organisms influence each other in many ways. Some organisms are harmed by these interactions and others benefit. The population size of any organism is negatively affected by parasitism, predation, or competition. Parasitism and predation usually involve interactions between two different species. Cannibalism is rare, but competition among members of the same population is often very intense. Many kinds of organisms perform services for others that have beneficial effects on the population. For example, decomposer organisms destroy toxic waste products, thus benefiting populations of animals. They also recycle materials needed for the growth and development of all organisms. Mutualistic relationships benefit both of the populations involved.

Often, the population sizes of two kinds of organisms are interdependent because each is a primary limiting factor of the other. This is most often seen in parasite-host relationships and predator-prey relationships. A good example is the relationship between the lynx (a predator) and the varying hare (the prey) as it was studied in Canada. The varying hare has a high reproductive capacity that the lynx helps to control by using the varying hare as food. The lynx can capture and kill the weak, the old, the diseased, and the unwary varying hares, leaving stronger, healthier ones to reproduce. Because the lynx is removing unfit individuals, it benefits the varying hare population by reducing the spread of disease and the amount of competition among the hares. While the lynx is helping to limit the varying hare population, the size of the varying hare population determines how many lynx can live in the area, since varying hares are their primary food source. If such events as disease epidemics or unusual environmental conditions cause a decline in the varying hare population, the population of the lynx also falls (figure 8.8).

Density-Dependent and Density-Independent Limiting Factors

Many populations are controlled by limiting factors that become more effective as the size of the population increases. Such

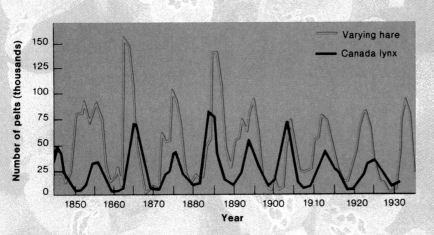

Figure 8.8 **Organism Interaction** The interaction between predator and prey species is complex and often difficult to interpret. These data were collected from the records of the number of pelts purchased by the Hudson Bay Company. It shows that the two populations fluctuate, with changes in the lynx population usually following changes in the varying hare population.
Source: Data from D. A. MacLulich, Fluctuations in the Numbers of the Varying Hare *(Lepus americanus).*

factors are referred to as **density-dependent factors.** Many of the factors we have already discussed are density-dependent. For example, the larger a population becomes, the more likely it is that predators will have a chance to catch some of the individuals. Furthermore, a prolonged period of increasing population allows the size of the predator population to increase. Populations with high densities are more likely to be affected by epidemics of parasites than are populations with widely dispersed individuals, since dense populations allow for the easy spread of parasites from one individual to another. In general, whenever there is competition among members of a population, its intensity increases as the population increases. Large organisms that tend to live a long time and have relatively few young are most likely to be controlled by density-dependent factors.

A second category of population-controlling influences that are not related to the size of the population is known as **density-independent factors.** Density-independent factors are usually accidental or occasional events in nature that happen regardless of the size or density of a population. A sudden rainstorm may drown many small plant seedlings and soil organisms. Many plants and animals are killed by frosts that come late in the spring or early in the fall. A small pond may dry up, resulting in the death of many organisms. The organisms most likely to be controlled by density-independent factors are small, short-lived organisms that can reproduce very rapidly.

> **CONCEPT CONNECTIONS**
>
> **density-independent factors**
> Population-controlling factors whose effects are not related to the size of the population.
>
> **density-dependent factors**
> Population-limiting factors that become more effective as the size of the population increases.

Human Population Growth

Today we hear different stories about the state of the world's population. On one hand we hear that the population is growing very rapidly. In contrast we hear that some countries are afraid that their populations are shrinking. Other countries are concerned about the aging of their populations because the birthrates and deathrates are low. In magazines and on television we see that there are starving people in the world. At the same time we hear discussions about the problem of food surpluses in many countries. It is important to realize that human populations follow the same patterns of growth and are acted upon by the same kinds of limiting factors as populations of other organisms. Look at the curve of human population growth over the past 10,000 years (figure 8.9). Estimates are that the population remained low and constant for thousands of years but increased rapidly in the past few hundred years. Does this mean that humans are the same as or different from other species? Can the human population continue to grow forever?

Although humans have tremendous capacity to alter their environment, the forces that will determine the size of the human population are no different from those that influence other species. There is a carrying capacity to consider, but humans have been able to continuously shift the carrying capacity upward through technological advances and the control of other species. Much of the recent exponential growth phase of the human population can be attributed to the removal of diseases, improvement in agricultural methods, and destruction of natural ecosystems in favor of artificial agricultural ecosystems. But even these achievements have their limits. There must be some limiting factors that will eventually cause a leveling off of our population growth curve. We cannot increase beyond our ability to get raw materials and energy, nor can we ignore the waste products we produce or the other organisms with which we interact.

The first limiting factor we face is our need for *raw materials.* To many of us, raw materials consist simply of the amount of food available, but we should not forget that in a technological society, iron ore, lumber, and irrigation water are also raw materials. However, most people of the world have more basic needs. For the past several decades, large portions of the world's population have not had enough food. Although it is biologically accurate to say that the world can currently produce enough food to feed all the people of the world, there are many reasons why people cannot get food or will not eat it. Many cultures have food taboos or traditions that prevent the use of some available food sources. For example, pork is forbidden in some cultures. Certain groups of people find it almost impossible to digest milk. Some African cultures use a

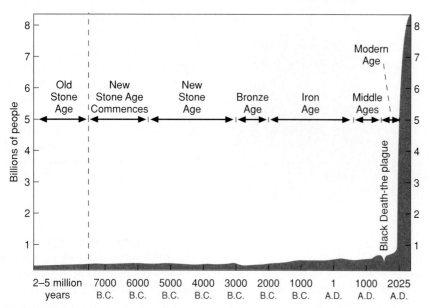

Figure 8.9 **Human Population Growth** The number of humans doubled from A.D. 1850 to 1930 (from one billion to two billion), then doubled again by 1975 (four billion), and could double again (eight billion) by the year 2025. How long can the human population continue to double before the earth's ultimate carrying capacity is reached?

Limiting the Human Population

If you look again at figure 8.9, you will note that it is in some ways similar and in others dissimilar to the population growth curve shown in figure 8.4. What factors have allowed the human population to grow so rapidly? What natural limiting factors will eventually bring this population under control? What is the ultimate carrying capacity of the world? What alternatives to the natural processes of population limitation could bring human population under control?

Consider the following in your answers to these questions: reproduction, death, diseases, food supply, energy, farming practices, food distribution, cultural biases, and anything else you consider to be relevant.

mixture of cow's milk and cow's blood as food, which people of other cultures might be unwilling to eat. But these taboos are much less important than the complex political, economic, and social problems related to the production and distribution of food.

In some cultures, farming is a low-status job, which means that people would rather buy their food from someone else than grow it themselves. This can result in poor use of agricultural resources. War and other political crises disrupt normal life in a country and agricultural production often falls as a result. Food is sometimes used as a political weapon when governments want to control certain groups of people. But probably most important is the fact that transporting food from centers of surplus to centers of need is often very difficult and expensive.

A more fundamental question is whether the world can continue to produce enough food. In 1995 the world population was growing at a rate of 1.5 percent per year. This amounts to nearly three new people being added to the world population every second, and this rate will result in a doubling of the world population in about 45 years. With a continuing increase in the number of mouths to feed, it is unlikely that food production will be able to keep pace with the growth in human population. A primary indicator of the status of the world food situation is the amount of grain produced for each person in the world (per capita grain production).

World per capita grain production peaked in 1984. The poorer nations of the world have had a large increase in population and a decline in per capita grain production because they are less able to afford costly fertilizer, machinery, and the energy necessary to run the machines and irrigate the land to produce their own grain. In addition, the pressure to produce food has led to unwise land use, which has resulted in the loss of topsoil and reduction in the fertility of their agricultural land.

The availability of *energy* is the second broad limiting factor that affects human populations as well as other kinds of organisms. All species on earth—including the human species—ultimately depend on sunlight for energy. Whether one produces electrical power from a hydroelectric dam, burns fossil fuels, or uses a solar cell, the energy is derived from the sun. Energy is needed for transportation, building and maintaining homes, and food production. It is very difficult to develop unbiased, reasonably accurate estimates of global energy "reserves" in the form of petroleum, natural gas, and coal. Therefore, it is difficult to predict how long these "reserves" might last. We do know, however, that the quantities are limited and that the rate of use has been increasing, particularly in the developed and developing countries.

If the less developed countries were to attain a standard of living equal to that of the developed nations, the global energy "reserves" would disappear overnight. Since the United States constitutes about 4.6 percent of the world's population and consumes approximately 25 percent of the world's energy resources, raising the standard of living of the entire world population to that of the United States would result in a 500 percent increase in the rate of consumption of energy and reduce theoretical reserves by an equivalent level. Humans should realize there is a limit to our energy resources; we are living on solar energy that was stored over millions of years, and we are using it at a rate that could deplete it in hundreds of years. Will energy availability be the limiting factor that determines the ultimate carrying capacity for humans, or will problems of waste disposal predominate?

Government Policy and Population Control in China

The actions of government can have a significant impact on the population growth pattern of a nation. Some countries have policies that encourage couples to have children. Canada pays a bonus to couples on the birth of a child. The U.S. tax code indirectly encourages births by providing exemptions for children. Some countries in Europe are concerned about the lack of working-age people in the future and are considering programs to encourage births. Many countries have no official population policy, though the position of most of those that do is to reduce the birthrate.

China has long been the most populated nation of the world. It now contains over 1.2 billion people, more than 21 percent of the world's people. Its history of population control is an interesting study of how government policy affects reproductive activity among its citizens. When the People's Republic of China was established in 1949, the population was about 540 million. The official policy of the government was to encourage births because more Chinese would be able to produce more goods and services, and production was the key to economic prosperity.

Because of its high birthrate and falling deathrate, the population increased to 614 million by 1955. This rapid increase and lack of economic growth led government officials to make changes that would lead to population control. Abortions became legal in 1953, and the first family planning program began in 1955, as a means of improving maternal and child health. Birthrates fell. The failure of government programs during the Great Leap Forward and the establishment of communes to produce food and other products resulted in widespread famine and increased deathrates and low birthrates in the late 1950s and early 1960s. A second family planning program began in 1962 but was not particularly effective.

The present family planning policy began in 1971 with the launching of the *wan xi shao* campaign. Translated, this phrase means "later" (marriages), "longer" (intervals between births), and "fewer" (children). As part of this program the legal ages for marriage were raised. For women and men in rural areas, the ages were raised to twenty-three and twenty-five, respectively; for women and men in urban areas the ages were raised to twenty-five and twenty-eight respectively. These policies resulted in a reduction of birthrates by nearly 50 percent between 1970 and 1979.

An even more restrictive one-child campaign was begun in 1978–1979. The program offered incentives for couples to restrict their family size to one child. Couples enrolled in the program would receive free medical care, cash bonuses for their work, special housing treatment, and extra old-age benefits. Those who broke their pledge were penalized by the loss of these benefits as well as other economic penalties. By the mid-1980s less than 20 percent of the eligible couples were signing up for the program. Rural couples particularly desired more than one child. In fact, in a country where nearly 75 percent of the population is rural, the rural total fertility rate was 2.5 children per woman. In 1988 a second child was sanctioned for rural couples if their first child was a girl, which legalized what had been happening anyway.

The current total fertility rate is 1.9 children per woman. More than 80 percent of couples use contraception; the most commonly used forms of contraception are male and female sterilization and the intrauterine device. Abortion is also an important aspect of this program, with a ratio of more than 600 abortions per 1,000 live births. However, because a large proportion of the population is under 30 years of age, it is still expected to double in about 60 years.

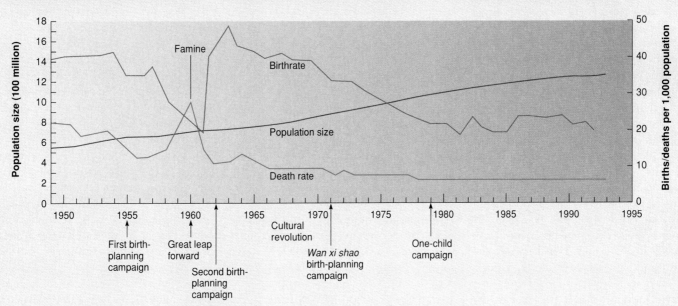

Source: Data from H. Yuan Tien. "China's Demographic Dilemmas" in Population Bulletin 1992, Population Reference Bureau, Inc., Washington, DC and National Family Planning Commission of China.

Population Growth in China

The third category of limiting factors is one of the most talked-about aspects of human activity—the problem of *waste disposal*. Not only do we have biological wastes, which can be dealt with by decomposer organisms, but we generate a variety of technological wastes and by-products that cannot be efficiently degraded by decomposers. Most of what we call pollution results from the waste products of technology. The biological wastes can usually be dealt with fairly efficiently by the building of waste-water treatment plants and other sewage facilities. Certainly these facilities take energy to run, but they rely on decomposers to degrade unwanted organic matter to carbon dioxide and water. Earlier in this chapter we discussed the problem that bacteria and yeasts face when their metabolic waste products accumulate. Are humans in a similar situation on a much larger scale? Are we dumping so much technological waste, much of which is toxic, into the environment that we are being poisoned? Some people believe that disregard for the quality of our environment will be a major factor in decreasing our population growth rate.

The fourth category of limiting factors that determines carrying capacity is *interaction* among organisms. Humans interact with other organisms in as many ways as other species do. We have parasites and occasionally predators. We prey on a variety of animals, both domesticated and wild. Competition is also very important. Insects and rodents compete for the food we raise, and we compete directly with many kinds of animals for the use of ecosystems.

As humans convert more and more land to agricultural and other purposes, many other organisms are displaced. Many of these displaced organisms are not able to compete successfully and must leave the area, have their populations reduced, or become extinct. The American bison (buffalo), African and Asian elephants, the panda, and the grizzly bear are a few species that are much reduced in number because they were not able to compete successfully with the human species. The passenger pigeon, the Carolina parakeet, and the great auk are a few that have become extinct. Animal parks and natural areas throughout the world have become tiny refuges for plants and animals that once occupied vast expanses of the planet. If these refuges are lost, many organisms will become extinct. Even the oceans have been significantly altered by the harvesting of fish for food. Many traditional fishing areas have had the most valuable species removed by

over-fishing, and those who fish commercially have switched to other less desirable fish as the more desirable cod, tuna, and other traditional species have been depleted.

When resources are in short supply, there is competition. Unfortunately, it is usually the young that are least able to compete, and high infant mortality is the result. In many of the less developed countries of the world, where resources are limited and living standards are low, between one and two of every ten children dies before they reach one year of age.

Humans are different from most other organisms in a fundamental way: we are able to predict the outcome of a specific course of action. Current technology and medical knowledge are available to control human population and improve the health and well-being of the people of the world. Why then does the human population continue to grow, resulting in human suffering and stressing the environment in which we live? Since we are social animals that have freedom of choice, we frequently do not do what is considered "best" from an unemotional, unselfish point of view. People make decisions based on historical, social, cultural, ethical, and personal considerations. What is best for the population as a whole may be bad for you as an individual. The biggest problems associated with control of the human population are not biological problems but require the efforts of philosophers, theologians, politicians, sociologists, and others. The knowledge and technology necessary to control the human population are available. What will eventually limit the size of our population? Will it be lack of

resources, lack of energy, accumulated waste products, competition among ourselves, or rational planning of family size?

SUMMARY

A population is a group of organisms of the same species in a particular place at a particular time. Populations differ from one another in individual characteristics, age distribution, sex ratio, and population density. Organisms typically have a reproductive capacity that exceeds what is necessary to replace the parent organisms when they die. This inherent capacity to over-reproduce causes a rapid increase in population size when a new area is colonized. A typical population growth curve consists of a lag phase in which population rises very slowly, followed by an exponential growth phase in which the population increases at an accelerating rate, followed by a leveling-off of the population in a stable equilibrium phase. In some populations, a fourth phase may occur, known as the death phase. This is typical of bacterial and yeast populations.

The carrying capacity is the number of organisms that can be sustained in an area over a long period of time. It is set by a variety of limiting factors. Availability of energy, availability of raw materials, accumulation of wastes, and interactions with other organisms are all categories of limiting factors. Because organisms are interrelated, population changes in one species sometimes affect the size of other populations. This is particularly true when one organism uses another as a source of food. Some limiting factors become more intense as the size of the population increases; these are known as density-dependent factors. Other limiting factors that are more accidental and not related to population size are called density-independent factors.

Humans as a species have the same limits and influences that other organisms do. Our current problems of food production, energy needs, pollution, and habitat destruction are outcomes of uncontrolled population growth. However, humans can reason and predict, thus providing the possibility of population control through conscious population limitation.

CHAPTER GLOSSARY

age distribution (āj dis″tri-biu′shun) The number of organisms of each age in a population.

birthrate (natality) (burth′ rāt) (na-tal′ĭ-te) The number of individuals entering the population by reproduction per thousand individuals in the population.

carrying capacity (ka′re-ing kuh-pas′ĭ-te) The optimum population size an area can support over an extended period of time.

death phase (deth fāz) The portion of some population growth curves in which the size of the population declines.

deathrate (mortality) (deth′ rāt) (mor-tal′ĭ-te) The number of individuals leaving the population by death per thousand individuals in the population.

density-dependent factors (den′sĭ-te de-pen′dent fak′tŏrz) Population-limiting factors that become more effective as the size of the population increases.

density-independent factors (den′sĭ-te in″de-pen′dent fak′tŏrz) Population-controlling factors that are not related to the size of the population.

exponential growth phase (eks-po-nen′shul grōth fāz) A period of time during population growth when the population increases at an accelerating rate.

lag phase (lag fāz) A period of time following colonization when the population remains small or increases slowly.

limiting factors (lim′ĭ-ting fak′tŏrz) Environmental influences that limit population growth.

population (pop″u-la′shun) A group of organisms of the same species located in the same place at the same time.

population density (pop″u-la′shun den′sĭ-te) The number of organisms of a species per unit area.

population growth curve (pop″u-la′shun grōth kurv) A graph of the change in population size over time.

population pressure (pop″u-la′shun presh′yur) Intense competition that leads to changes in the environment and dispersal of organisms.

reproductive capacity (re-pro-duk′tiv kuh-pas′ĭ-te) The theoretical maximum rate of reproduction.

sex ratio (seks ra′sho) The number of males in a population compared to the number of females.

stable equilibrium phase (stā′bul e-kwi-lib′re-um fāz) A period of time during population growth when the number of individuals entering the population and the number leaving the population are equal, resulting in a stable population.

CONCEPT MAP TERMINOLOGY

Construct a concept map to represent the relationships among the following concepts.

birthrate (natality)
carrying capacity
deathrate (mortality)
exponential growth phase
lag phase

limiting factors
population growth curve
reproductive capacity
stable equilibrium phase

LABEL•DIAGRAM•EXPLAIN

Describe what is happening at A, B, and C on the graph and why it happens.

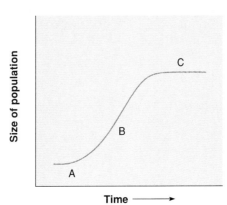

Multiple Choice Questions

1. As the size of a population approaches the carrying capacity:
   a. the number of births and deaths are about equal
   b. the population is in the lag phase
   c. density-dependent limiting factors are not important
   d. the biomass decreases

2. The human population is:
   a. at its carrying capacity
   b. growing rapidly
   c. declining on a worldwide basis
   d. approximately 5 million people

3. Which one of the following populations will have the greatest increase in size?
   a. one that has reached its carrying capacity
   b. one with a high birthrate and a high deathrate
   c. one with a high reproductive capacity and no density-dependent limiting factors
   d. one with a low deathrate and a low birthrate

4. Which one of the following would reduce the birthrate?
   a. increased amounts of food
   b. increased numbers of males in the population
   c. reduced numbers of females in the population
   d. increased carrying capacity

5. Which one of the following human populations would be the least likely to increase over time?
   a. one with 50 percent of the population below 15 years of age
   b. one with 50 percent of the population over 50 years of age
   c. one with a population that is 40 percent males and 60 percent females
   d. a rural population

6. The human population of the world:
   a. must ultimately reach its carrying capacity
   b. cannot be controlled
   c. will stabilize in the next few years
   d. will begin to fall by the year 2020

7. Many kinds of populations grow until they reach a relatively stable size and then do not change much after that. When this stable condition exists:
   a. the number of deaths is greater than the number of births
   b. there is usually a shortage of an important resource
   c. the population will eventually fall to zero
   d. the average age of the population must increase

8. Most populations show a lag phase as the population is first established in a new area. During this time the population does not change very much. This is because:
   a. there are few organisms present
   b. it takes time for reproduction to take place
   c. the environment needs to be modified before organisms can live there
   d. organisms lack the ability to reproduce

Questions with Short Answers

1. Draw the population growth curve of a yeast culture during the wine-making process. Label the lag, exponential growth, stable equilibrium, and death phases.

2. List four ways in which two populations of the same species could be different.

3. Why do populations grow?

4. List four kinds of limiting factors that help to set the carrying capacity for a species.

5. How do the concepts of biomass and population size differ?

6. Differentiate between density-dependent and density-independent limiting factors. Give an example of each.

7. State four measures China has taken to control human population growth.

8. As the human population continues to grow, what should we expect to happen to other species?

9. How does the population growth curve of humans compare with that of other kinds of animals?

10. All organisms over-reproduce. What advantage does this give to the species? What disadvantages?

chapter

9

Behavioral Ecology

Carpet snake constricting a lizard.

Behavioral Ecology

it? Why do you blink when an object rapidly approaches your face? Why do you find it more difficult to communicate with someone on the phone than face to face? Fear of snakes (many are poisonous) and the blinking of eyes can protect us from injury. You may find it more difficult to communicate with someone on the phone because you cannot see their facial expressions. These facial expressions convey part of the message.

Behavior is how any organism acts, what it does, and how it does it. It involves sensing the surroundings and reacting to changes in the environment. The most obvious behaviors involve movement of the animal, although other activities such as blushing or producing odors may also be important. When we think about the behavior of an organism, we should keep in mind that behavior is like any other characteristic displayed. It has value or significance to the organism as it goes about exploiting resources and reproducing more of its own species. Many behaviors are inherited in the same way as are physical characteristics.

Behavior is a very important part of the ecological role of any organism. It allows animals to escape predators, seek out mates, and gain dominance over others of the same species. Plants, for the most part, must rely on structures, physiological changes, or chance to accomplish the same ends. For example, a rabbit can run away from a predator; a plant cannot. But the plant may have developed thorns

learning objectives

- Understand that behavior has evolutionary and ecological significance.

- Distinguish between instinctive and learned behaviors.

- Recognize that there are several kinds of learning.

- Know that animals use sight, sound, and chemicals to communicate for reproductive purposes.

- Appreciate that territoriality and dominance hierarchies allocate resources.

- Know several methods used by animals in navigation.

- Describe why the evolution of social animals is different from that of nonsocial animals.

Understanding Behavior

When you watch a bird or squirrel, its activities appear to have a purpose. Birds search for food, take flight as you approach, and build nests. Likewise, squirrels collect and store nuts and acorns, "scold" you when you get too close, and learn to visit sites where food is available. All of these activities are adaptive. Squirrels collecting food and storing it for the winter, birds fleeing from potential predators, and the complex behavior of building nests are important for the continued survival of the species. Birds that did not take flight at the approach of another animal would be eaten by predators. Squirrels that did not remember the location of sources of food would be less likely to survive, and birds that built obvious nests on the ground would be more likely to lose their young. However, we need to be careful that we not attach too much meaning to what animals do. They may not have the same "thoughts" and "motivations" we do.

Why are most people afraid of snakes? Is this a behavior we are born with or do we learn

FYI

The two most popular exhibits at zoos are primates and reptiles. Our fascination with reptiles is generally based on fear and hate, while interest in primates is based on their similarity to humans. Many of the bizarre behaviors we see in captive animals are not normal. The pacing of many zoo animals does not occur when they are in the wild. Similarly, the begging behavior of bears and elephants is learned.

or toxic compounds in its leaves that discourage animals from eating it. Mate selection in animals often involves elaborate behaviors that assist them in identifying the species and sex of the potential mate. Most plants rely on a much more random method for transferring male sex cells to the female plant. Dominance in plants is often achieved by depriving competitors of essential nutrients or by inhibiting the development of the seeds of other plants. Animals have a variety of behaviors that allow them to exert dominance over members of the same species.

It is not always easy to identify the significance of a behavior without carefully studying the behavior pattern and the impact it has on other organisms. For example, a hungry baby herring gull pecks at a red spot on its parent's bill. What possible value can this behavior have for either the chick or the parent? If we watch, we see that when the chick pecks at the spot, the parent regurgitates food onto the ground, and the chick feeds (figure 9.1). This looks like a simple behavior, but there is more to it than meets the eye. Why did the chick peck to begin with? How did it know to peck at that particular spot? Why did the pecking cause the parent to regurgitate food? These questions are not easy to answer, and many people assume that the actions have the same motivation and direction as similar human behaviors. For example, when a human child points to a piece of candy and makes appropriate noises, it is indicating to its parent that it wants some candy. Is that what the herring gull chick is doing? We really do not know the answer to the question because we have not been able to "crawl inside the brain" of the bird to determine its reason for doing something.

Some people believe that a bird singing on a warm, sunny spring day is making that beautiful sound because it is happy. Students of animal behavior do not accept this idea and have demonstrated that a bird sings to tell others to keep out of its territory. The barbed stinger of a honeybee remains in your skin after you are stung, and the bee tears the stinger out of its body when it flies away. The damage to its body is so great that it dies. Has the bee performed a noble deed of heroism and self-sacrifice? Was it defending its hive from you? We need to know a great deal more about bees to understand the value of their behaviors. The fact that bees are social animals like us makes it particularly tempting to think that they are doing things for the same reasons we are.

FYI

Behaviors You Might Not Think of as Behaviors

Seeds germinating

Leaves falling in autumn

An amoeba moving toward food

A plant bending toward the light

A bacterial cell moving away from a harsh environment

Mushrooms producing spores

A Venus flytrap capturing a fly

CONCEPT CONNECTIONS

- **Anthropomorphism** (*anthropo* = human; *morphi* = shape, or form) is the idea that we can ascribe human feelings, meanings, and emotions to the behavior of animals.

- **Ethology** is the scientific study of the nature of behavior and its ecological and evolutionary significance in natural settings.

The fable of the grasshopper and the ant is another example of crediting animals with human qualities. The ant is pictured as an animal that, despite temptations, works hard from morning until night, storing away food for the winter (figure 9.2). The grasshopper, on the other hand, is represented as a lazy good-for-nothing that fools away the summer when it really ought to be saving up for the tough times ahead. If one is looking for parallels to human behavior, these are good illustrations, but they really are not accurate statements about the lives of the animals from an ecological point of view. Both the ant and the grasshopper are very successful organisms, but each has a different way of satisfying its needs and ensuring that

Figure 9.1 Animal Behavior A baby herring gull causes its parent to regurgitate food onto the ground by pecking at the red spot on its parent's bill. The parent then picks up the food and feeds it to the baby.

Figure 9.2 **The Fable of the Ant and the Grasshopper** In many ways we give human meaning to the actions of animals. The ant is portrayed as an industrious individual who prepares for the future, and the grasshopper as a lazy fellow who sits in the sun and sings all day. This view of animal behavior is anthropomorphic.

FYI An object does not need to be a living organism to respond to a stimulus. Heating and air conditioning systems are triggered to run and stop running in order to maintain a constant temperature, bar codes are "read" at stores, and computers dial your telephone and talk to you.

some of its offspring will be able to produce another generation of organisms. One method of survival is not necessarily better than another, as long as both methods are successful. This is what the study of behavior is all about: looking at the activities of an organism during its life cycle and determining the value of the behavior to the ecological niche of the organism.

Instinct and Learning

Before we go further, we need to discuss how animals generate specific behaviors. The behavior patterns of most organisms involve both instinct and learning. We recognize instinct as behavior that is automatic and inflexible, while learning requires experience and produces behaviors that can be changed. Most animals have a high degree of instinctive behavior and very little learning, while some, like many birds and mammals, are able to demonstrate a great deal of learned behavior.

Instinct

Many animal behaviors are automatic, preprogrammed, and genetically determined. Such behaviors are called **instinctive behaviors** and are found in organisms ranging from simple one-celled protozoans to complex vertebrates. These behaviors are performed correctly the first time without previous experience when the proper stimulus is given. A **stimulus** is some change in the internal or external environment of the organism that causes it to react. The reaction of the organism to the stimulus is called a **response.**

An organism can respond only to stimuli it can recognize. It is difficult for us as humans to appreciate what the world seems like to a bloodhound, for example. The bloodhound is able to identify individuals by smell, whereas we have great difficulty detecting, let alone distinguishing, many odors. Some animals are color-blind and are able to see only shades of gray. Others can see ultraviolet light, which is invisible to us. Many are able to detect the magnetic field of the earth. And some are able to detect infrared radiation (heat) from distant objects.

In our example of the herring gull chick, the red spot on the bill of the adult bird serves as a stimulus to the chick. The chick responds to this spot in a very specific, genetically programmed way. The behavior is innate—it is done correctly the first time without prior experience. (It has been shown that chicks will

peck at red spots on pencils and other objects that do not resemble the adult bird.) The pecking behavior of the chick is in turn the stimulus for the adult bird to regurgitate food, and the regurgitated food is a stimulus for the chick to feed. These behaviors have adaptive value for the gull species. Instinctive behavior has great value because it allows necessary behavior to occur correctly and perfectly without prior experience.

The drawback of instinctive behavior is that it cannot be modified when a new situation presents itself. It can be very effective for the survival of a species if it is involved in fundamental, essential activities that rarely require modification. Instinctive behavior is most common in animals that have short life cycles, simple nervous systems, and little contact with parents. However, some instances of inappropriate behavior may be generated by

FYI

INSTINCT	LEARNING
Animal is born with the behavior. It is hereditary. (Genetic memory.)	Animal is not born with the behavior. Learned behavior is not hereditary but the way in which learning occurs is at least partly hereditary.
Behavior is done correctly the first time. No experience required.	Behavior requires practice. Performance improves with experience.
Behavior cannot be changed.	Behavior can be changed.
Memory is not important.	Memory is important.
Typical of simple animals that have short lives and little contact with parents.	Typical of more complex animals that have long lives and extensive contact with parents.

Figure 9.3 Egg-Rolling Behavior in Geese Geese will use a specific set of head movements to roll any reasonably round object back to the nest. There are several components of this instinctive behavior, including recognition of the object and head-tucking movements. If the egg is removed during the head-tucking movements, the behavior continues as if the egg were still there.

unusual stimuli or by unusual circumstances in which the stimulus occurs. For example, many insects use the sun, moon, or stars as aids to navigation. Over the millions of years of insect evolution, this has been a valuable tool for these animals. However, humans have invented a variety of artificial lights that lead to inappropriate and even destructive behavior, such as when insects collect at a light and batter themselves to death. This mindless, instinctive behavior seems incredibly stupid to us, but even though some individuals die, it is still valuable for the species since the majority do not encounter artificial lights but complete their life cycles normally.

Certain species of geese exhibit a behavior pattern that involves rolling eggs back into the nest. Eggs may roll out of the nest as the parent gets on and off. The developing young in eggs that are not protected from cold or excessive heat will die. Thus, the egg-rolling behavior has the value of protecting future offspring. If, however, the egg is taken from the goose when it is in the middle of egg-rolling behavior, the goose will continue the behavior until it gets back to the nest, even though there is no egg to roll (figure 9.3). This is typical of the inflexible nature of instinctive behaviors. It was also discovered that many other somewhat egg-shaped structures would generate the same behavior. For example, beer cans and baseballs were good triggers for egg-rolling behavior.

Some activities are so complex that it seems impossible for an organism to be born with such abilities. For example, many caterpillars spin cocoons. This is not just a careless jumble of silk threads. A cocoon is so precisely made that you can recognize what species of caterpillar made it. But cocoon spinning is not a learned ability. A caterpillar has no opportunity to learn how to spin a cocoon, since it never observes others doing it. Furthermore, caterpillars do not practice several times before they get a proper, workable cocoon. It is as if a "program" for making a cocoon is in the caterpillar's "computer." If you interrupt its cocoon-making effort in the middle and remove most of the finished part, the caterpillar will go right on making the last half of the cocoon. The caterpillar's "program" does not allow for interruptions and repair (figure 9.4). This inability to adapt as circumstances change is a prominent characteristic of instinctive behavior.

Learned Behavior

The alternative to preprogrammed, instinctive behavior is learned behavior. **Learning** is a change in behavior as a result of experience. Your behavior will be different in some way as a result of reading this chapter. Birds build better nests with practice. Dogs can be taught to do tricks for rewards.

Learning becomes more significant in long-lived animals that care for their young. Animals that live many years are more likely to benefit from an ability to recognize previously encountered situations. They modify their behavior accordingly. Animals that live only a few weeks generally do not exhibit learned behaviors. In addition, young that spend considerable time with their parents have the opportunity to imitate their parents and develop behaviors that are appropriate to local conditions. These behaviors

Half-finished cocoon

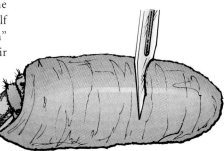

Finished end is removed

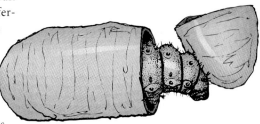

Cocoon "completed"

Figure 9.4 Inflexible Instinctive Behavior If part of the caterpillar's half-complete cocoon is removed while it is still spinning, the animal will finish the job but never repair the damaged end.

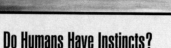

BIO *feature*

Do Humans Have Instincts?

There are certainly human behaviors that are done correctly the first time without prior experience. These are often called *reflexes*. The blink reflex occurs when an object rapidly approaches the face of a newborn. Newborn infants grasp things with their hands and feet. Stroking the side of the face of a newborn causes it to turn toward the side stroked and start sucking activities. It is difficult to see how these behaviors would have been learned. The movements necessary to do the behaviors may have been practiced in the uterus (blinking, grasping, sucking), but it is hard to see how the response to the appropriate stimulus immediately following birth could have been learned. If we look at all of these behaviors from the point of view of the survival of the offspring of our prehistoric ancestors, there appears to be value in having these behaviors be automatic. Blinking would protect the eyes. Grasping would help the infant stay with the mother as she moved about seeking food. Sucking movements are necessary to get the only source of nutrients newborns have: milk.

BIO *feature*

Genes and Human Behavior

It is clear that human behavior is predominantly learned. However, we are finding that even certain fundamental features of behavior such as emotions and motivation may be influenced by genes. Certain kinds of mental illness are clearly genetic. The genes involved probably direct the manufacture of specific chemicals that alter the function of the brain. Several behaviors with suspected genetic involvement are the tendency to develop alcoholism, certain forms of schizophrenia, certain learning disorders, and intelligence. All of these examples should be treated with caution. While genes determine or shape some aspects of our behavior, the environment in which a person functions is also very important. The key idea here is that we should not overlook the probability that some aspects of our behavior are shaped by the genes we inherited from our parents.

take time to develop but have the advantage of adaptability. In order for learning to become dominant over instinct in an animal's life, the animal must also have a large brain to store the new information it is learning. Although learning of some kind has been demonstrated in a wide variety of animals, it is most significant in vertebrates that have relatively large brains.

Kinds of Learning

Learning is not just one kind of activity but can be subdivided into several categories: conditioning, imprinting, and insight.

Conditioning

A Russian physiologist, Ivan Pavlov (1849–1936), was investigating the physiology of digestion when he discovered that dogs can associate an unusual stimulus with a natural stimulus. He was studying the production of saliva by dogs, and he knew that a natural stimulus, such as the presence or smell of food, would cause the dogs to salivate. Pavlov rang a bell just prior to each presentation of food. After a training period, the dogs would begin to salivate when the bell was rung, even though no food was presented. This kind of learning, in which a "neutral" stimulus (the sound of a bell) is associated with a "natural" stimulus (the taste of food), is called **conditioning.** The response produced by the neutral stimulus is called a **conditioned response.**

Figure 9.5 Associative Learning Many animals learn to associate unpleasant experiences with the color or shape of offensive objects and thus avoid them in the future. The blue jay is eating a monarch butterfly. These butterflies contain a chemical that makes blue jays sick. After one or two such experiences, blue jays learn not to eat the monarch or any other butterfly having a similar coloration.

In Pavlov's research the dogs were receiving positive reinforcement, but it is also possible to apply negative reinforcement, in which a response eliminates or prevents the occurrence of a painful or distressing stimulus. For example, if a dog receives an electrical stimulus on the right foot at the same time that a bell is rung, it soon learns to associate the bell with the painful stimulus. It lifts its foot even without an electrical stimulus upon hearing the sound of the bell. Many kinds of animals can be trained to do unusual things by creating artificial situations in which they associate a reward or punishment with particular behaviors. Owners can train their dogs to heel on voice command (walk at the owner's left heel) by firmly saying "heel" and vigorously jerking the dog into the correct position with a leash. (The dog would respond equally well to any sound, not just the word "heel.")

Such associative learning takes place in animals' natural environments as well. Certain kinds of fruits and insects are toxic and have unpleasant tastes. Animals learn to associate the colors and shapes of the offending objects with the bad tastes and avoid these organisms in the future (figure 9.5). Many kinds of birds that live in urban areas have formed an association between getting a meal and automobiles. They can be observed eating squashed insects found on the grills and bumpers of cars and trucks. When a car drives into the birds' territory, they immediately examine the bumper for food. It appears that they have been conditioned to look for food in this unusual place. In some of our national parks, bears have learned to associate people and backpacks with food. These bears are particularly dangerous and are often trapped and moved to other locations. In some cases, attempts have been made to condition the animals to avoid humans. However, some bears continue to associate humans with food and are killed to protect tourists from bear attacks.

The ability to form an association between two events is extremely valuable to an animal. If an association allows the animal to

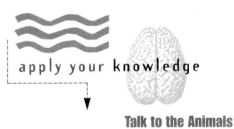

apply your knowledge

Talk to the Animals

If you were going to teach an animal to communicate a message new to that animal, what message would you select? How would you teach the animal to communicate the message at the appropriate time?

BIO *feature*

Conditioning in Humans

Conditioning is a common feature of human behavior. From the time we were very young our parents conditioned us to act in particular ways. In some cases we were rewarded with smiles, pats on the head, or gifts for behaviors our parents wanted to encourage. In other cases we were punished for behaviors our parents wanted to prevent, perhaps with a loud "NO!" when we were about to stick our finger in an electric socket. Even as adults we are constantly subjected to conditioning. The smells of food or even the thought of food causes us to salivate. We respond differently to a person wearing a uniform than we would to the same person in casual dress. We respond almost automatically to the shapes and colors of traffic signs and signals.

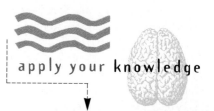

get more food, avoid predators, or protect its young more effectively, it is to the advantage of the species. The association of certain shapes, colors, odors, or sounds with danger is especially valuable.

Imprinting

Imprinting is a special kind of learning in which an animal is genetically primed to learn a specific behavior in a very short period during a particular portion of the animal's life. This type of learning was originally recognized by Konrad Lorenz (1903–1989) in his experiments with geese and ducks. He determined that shortly after hatching, a duckling would follow an object if the object was fairly large, moved, and made noise. The duckling was imprinted on the object.

Lorenz discovered that ducklings will follow only the object on which they were originally imprinted. Under normal conditions, the first large, noisy, moving object newly hatched ducklings see is their mother. Imprinting ensures that the immature birds will follow her and learn appropriate feeding

habits, defensive tactics, and other behaviors by example (figure 9.6). Since they are always near their mother, she can also protect them from enemies or bad weather. If animals imprint on the wrong objects, they are not likely to survive. Since Lorenz's experiments in the

FYI

Characteristics of Imprinting
- **Imprinting occurs at a specific stage in the life of the animal called the *critical period*.**
- **If imprinting does not occur during the critical period, it will never occur.**
- **Only specific kinds of learning can occur.**
- **Imprinting cannot be unlearned once it has occurred.**
- **Imprinting usually involves specific learning critical to the survival of the organism.**

early 1930s, we have discovered that many young animals can be imprinted on several types of stimuli and that their responses include much more than following behaviors.

The way song sparrows learn their song appears to be a kind of imprinting. It has been discovered that the young birds must hear the correct song during a specific part of their youth or they will never be able to perform the song correctly as adults. This is true even if, later in life, they are surrounded by other adult song sparrows singing the correct song. Furthermore, the period of time when they learn the song is prior to the time that they begin singing. Recognizing and performing the correct song is important because it has particular meaning to other song sparrows. For males it conveys the information that a male song sparrow has territory reserved for himself. For females, the male song is an announcement of a possible mate.

Mother sheep and many other kinds of mammals imprint on the odor of their offspring. They are able to identify their offspring among others and will not allow feeding by lambs that are not their own. Shepherds have known for centuries that they can sometimes get a mother that has lost her lambs to accept orphan lambs if they place the skin of the dead lambs over the orphan.

Many fish appear to imprint on odors in the water. Salmon are famous for their ability to return to the freshwater streams where they were hatched. They will jump waterfalls and use specially constructed fish ladders to get around dams. Fish that are raised in artificial fish hatcheries can be imprinted on specific chemicals and induced to return to any river that contains some of that chemical.

Insight

Insight is a special kind of learning in which past experiences are reorganized to solve new problems. When you are faced with a new problem, whether it is a crossword puzzle, a math problem, or any one of a hundred

You Had To Be There

In one of his books, Lorenz described himself squatting on his lawn one day, waddling and quacking, followed by newly hatched ducklings. He was busy being a "mother duck." He was surprised to see a group of tourists on the other side of the fence watching him in amazement. They couldn't see the ducklings hidden by the tall grass. All they could see was this strange performance by a big man with a beard!

Figure 9.6 Imprinting Imprinting is a special kind of irreversible learning that occurs during a specific phase of the life of an animal. These geese have been imprinted on Konrad Lorenz and exhibit a typical "following response."

Do Humans Imprint?

Several situations strongly suggest that imprinting may occur in humans. Many people think a bonding takes place between mother and child immediately after birth and that early contact between mother and child is important in the development of this mother-child bond. It is difficult to prove that this is the case because doing so would require manipulating mother-child interactions, and this is difficult to justify ethically.

Language ability appears to rely on many aspects of imprinting.

Children learn languages relatively easily, and can learn several simultaneously if presented the opportunity. It appears to be much more difficult for adults to learn new languages. Is there a critical period? Many adults who do learn a new language appear to be unable to unlearn previous word patterns and pronunciations and thus speak the new language with a "foreign" accent. Are they unable to unlearn their previous language pattern?

---

other everyday problems, you sort through your past experiences and locate those that apply. You may not even realize that you are doing it, but you put these past experiences together in a new way that may provide the solution to your problem. Because this process is internal and can be demonstrated only through some response, it is very difficult to understand exactly what goes on during it. Behavioral scientists have explored this area for many years, but the study of insight is still in its infancy.

Insight in animals is particularly difficult to study. It is impossible to know for sure whether a novel solution to a problem is the result of "thinking it through" or the result of developing an association between an accidental occurrence and a desirable outcome. For example, a small group of Japanese macaques (monkeys) on an island was studied. They were fed by simply dumping food, such as sweet potatoes or wheat, onto the beach. The food became covered with

sand. Eventually, one of the macaques was observed washing the sand off some sweet potatoes by submerging them in a nearby stream. Later she was also observed sorting wheat and sand by putting the mixture in the water and collecting the wheat that floated on the surface. Are these examples of insight? Or did she accidentally drop the sweet potato

FYI
Cartoonists indicate the internal activities involved in developing insight by drawing a light bulb, and we may even say of ourselves when we understand something that "the light went on."

in the water? Did she associate the accidental washing with sand-free sweet potatoes and then make dropping food in water part of her feeding behavior? We will probably never know, but it is tempting to think so.

Many animals learn by trial and error. They try one approach, then another, then another. Eventually they may accidentally arrive at a solution to a problem. Since they are rewarded when they solve the problem, they may include this behavior in their future activities but not really understand the problem or why the solution works. In your own case, you may have used a computer game without reading the operating instructions. By sitting at the machine using a trial-and-error approach, you may have figured out how to play the game. This does not necessarily mean that you now know how the computer software works well enough to operate another game without going through the trial-and-error process again.

Instinct and Learning in the Same Organism

It is important to recognize that all animals have both instinctive behaviors and various amounts of learned behavior. One basic activity may have instinctive components and learned components. For example, biologists have raised young song sparrows in the

absence of any adult birds, so there was no song for the young birds to imitate. These isolated birds grew up to sing a series of notes similar to the normal song of the species, but not exactly correct. Young birds from the same nest that were not taken from their parents developed a song nearly identical to the parents' song. If bird songs were totally instinctive, there would be no difference between the songs of these two groups of young birds. It appears that the basic melody of the song was inherited by the birds and that the refinements were the result of experience. Therefore, the characteristic song of that species was partly learned behavior (a change in behavior as a result of experience) and partly unlearned (instinctive). It is important to note that many kinds of birds learn most of their song, with very few innate components. A mockingbird is very good at imitating the songs of a wide variety of bird species.

This proportion of learned and instinctive behavior is not the same for all species. Many invertebrate animals rely on instinct for the majority of their behavior patterns, whereas many of the vertebrates (particularly birds and mammals) use a great deal of learning (figure 9.7).

Typically the learned components of an animal's behavior have particular value for the animal's survival. Most of the behavior of a honeybee is instinctive, but it is able to learn new routes to food sources. The style of nest built by a bird is instinctive, but the skill with which it builds improves with experience. The food-searching behavior of birds is probably instinctive, but the ability to modify the behavior to exploit unusual food sources such as bird feeders is learned. On the other hand, honeybees cannot be taught to make products other than beeswax and honey, a robin will not build a nest in a bird house, and insect-eating birds will not learn to visit bird feeders.

Behavior as a Part of an Animal's Niche

Of the examples used so far, some were laboratory studies, some were field studies, and some included aspects of both. Often these studies overlap with the field of psychology. This is

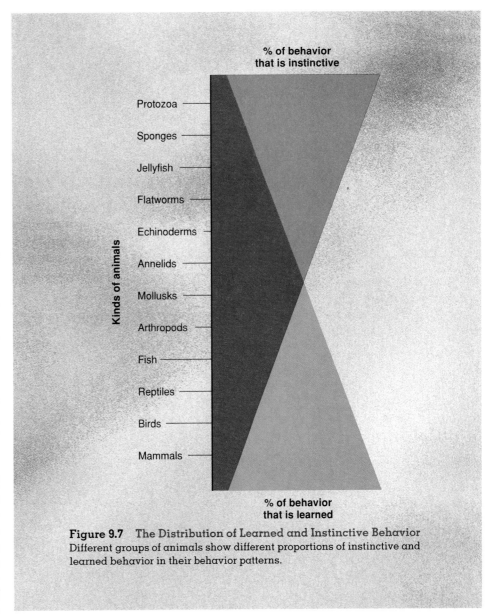

Figure 9.7 **The Distribution of Learned and Instinctive Behavior** Different groups of animals show different proportions of instinctive and learned behavior in their behavior patterns.

particularly true for many of the laboratory studies. You can see that the science of animal behavior is a broad one that draws on information from several fields of study and can be used to explore many different kinds of questions. The topics that follow avoid the field of psychology and concentrate on the significance of behavior from an ecological and evolutionary point of view.

Now that we have some understanding of how organisms generate behavior, we can look at a variety of behaviors in several different kinds of animals and see how they are useful to the animals in their ecological niches.

Reproductive Behavior

Finding Each Other

Obviously, if a species is to survive, it must reproduce. There are many stages in any successful reproductive strategy. First, animals must be able to recognize individuals of the same species that are of the opposite sex. Several techniques are used for this purpose. For instance, frogs of different species produce sounds that are just as distinct as the calls of different species of birds. The call is a code that delivers a very private message, since it is meant only for one species. It is, however, meant for any member of that species near

Figure 9.8 Communication The female moth signals her readiness to mate by releasing a pheromone that attracts males from long distances downwind.

Female

Male

enough to hear. The call produced by male frogs, which both male and female frogs can receive by hearing, results in frogs of both sexes coming together in a small area. Once they gather in a small pond, it is much easier to have the further communication necessary for mating to take place.

Chemicals can also serve to attract animals. **Pheromones** are chemicals produced by animals and released into the environment that trigger behavioral or developmental changes in other animals of the same species. They have the same effect as sound through a different code system. The classic example of a pheromone is the chemical that female moths release into the air. The large, fuzzy antennae of the male

moths can receive the chemical in unbelievably tiny amounts. The male then changes its direction of flight and flies upwind to the source of the pheromone, which is the female (figure 9.8).

The firefly is probably the most familiar organism that uses light signals to bring together males and females. Several different species may live in the same area, but each species flashes its own code. The code is based on the length of the flashes, their frequency, and their overall pattern (figure 9.9). There is also a difference between the signals given by males and females. For the most part, males are attracted to and mate with females of their own species. Once male

Figure 9.9 Firefly Communication The pattern, location, and duration of light flashes all help fireflies identify members of the opposite sex who are of the appropriate species.

Raising the Young

A third element in successful reproduction is providing the young with the resources they need to live to adulthood. Many invertebrate animals spend little energy on the care of the young, leaving them to develop on their own. Usually the young become free-living larvae that eat and grow rapidly. In some species, females make preparations for the young by laying their eggs in suitable sites. Many insects lay their eggs directly on the plant that the larvae will use as food as they develop. Many parasitic species seek out the required host in which to lay their eggs. The eggs of others may be placed in spots that provide safety until the young hatch. Turtles, many fish, and some insects fit into this category. In most of these cases, however, the female lays large numbers of eggs, and most of the young die before reaching adulthood. This is an enormously expensive process: the female invests considerable energy in the production of the eggs but has a low reproductive success rate.

An alternative to this "wasteful" loss of potential young is to produce fewer young but invest large amounts of energy in their care. This is typical of birds and mammals. Parents build nests, share in the feeding and protection of the young, and often assist the young in learning appropriate behavior. Many insects, such as bees, ants, and termites, have elaborate social organizations in which one or a few females produce large numbers of young that are cared for by sterile offspring of the fertile females. Some of the female's offspring will be fertile, reproducing individuals. The activity of caring for young involves many

and female animals have attracted one another's attention, the second stage in successful reproduction takes place.

Assuring Fertilization

The second important activity in reproduction is fertilizing eggs. Many marine organisms simply release their gametes into the sea simultaneously and allow fertilization and further development to take place without any participation by the parents. Sponges, jellyfishes, and many other marine animals fit into this category. Other aquatic animals come together so

that the chances of fertilization are enhanced by the male and female being near one another as the gametes are shed. This is typical of many fish and some amphibians, such as frogs. In most terrestrial organisms, internal fertilization occurs, in which the sperm are introduced into the reproductive tract of the female. Some spiders and other terrestrial animals produce packages of sperm that the female picks up with her reproductive structures. Many of these mating behaviors require elaborate, species-specific communication prior to the mating act. Several examples were given in the previous paragraphs.

complex behavior patterns. It appears that most animals that feed and raise young are able to recognize their own young from those of nearby families and may even kill the young of another family unit. Elaborate greeting ceremonies are usually performed when animals return to the nest or the den. Perhaps this has something to do with being able to identify individual young. Often this behavior is shared among adults as well. This is true for many colonial nesting birds, such as gulls and penguins, and for many carnivorous mammals, such as wolves, dogs, and hyenas.

Allocating Resources

For an animal to be successful, it must receive sufficient resources to live and reproduce. Therefore, we find many kinds of behaviors that divide the available resources so that the species as a whole is benefited, even though some individuals may be harmed.

Territorial Behavior

One kind of behavior pattern that is often tied to successful reproduction is territoriality. **Territoriality** consists of the setting aside of space for the exclusive use of an animal for food, mating, or other purposes. A **territory** is the space an animal defends against others of the same species. When territories are first being established, there is much conflict between individuals. This eventually gives way to the use of a series of signals that define the territory and communicate to others that the territory is occupied.

The male redwing blackbird has red shoulder patches, but the female does not. The male will perch on a high spot, flash his red shoulder patches, and sing to other male redwing blackbirds that happen to venture into his territory. Most other males get the message and leave his territory; those that do not leave are attacked by the occupying male. He will also attack

a stuffed, dead male redwing blackbird in his territory, or even a small piece of red cloth. Clearly, the spot of red is the characteristic that stimulates the male to defend his territory. Once the initial period of conflict is over the birds tend to respect one another's boundaries. All that is required is to frequently announce that the territory is still occupied. This is accomplished by singing from some conspicuous position in the territory. After the territorial boundaries are established, little time is required to prevent other males from venturing close.

Not all male redwing blackbirds are successful in obtaining territories. During the initial period, when fighting is common, some birds regularly win and maintain their territory. Some lose and must choose a less favorable territory or go without. Therefore, territorial behavior is a way to distribute a resource that is in short supply. Since females choose which male's territory they will build their nest in, males that do not have territories are much less likely to fertilize females.

Many carnivorous mammals like foxes, wolves, and coyotes use urine to mark the boundaries of their territories. One of the primary values of the territory for these animals is the food contained in the large space they defend. These territories may include several square kilometers of land.

The possession of a territory is often a requirement for reproductive success. In a way, then, territorial behavior has the effect of allocating breeding space and limiting population size to that which the ecosystem can support. This kind of behavior is widespread in the animal kingdom and can be seen in such diverse groups as insects, spiders, fish, reptiles, birds, and mammals (figure 9.10).

Figure 9.10 Territoriality Colonial nesting seabirds typically have very small nest territories. Each territory is just out of pecking range of the neighbors'.

Dominance Hierarchy

Another way of allocating resources is by the establishment of a **dominance hierarchy.** In this situation a relatively stable, mutually understood order of priority within the group is maintained. A dominance hierarchy is often established in animals that form social groups. One individual in the group dominates all others. A second-ranking individual dominates all but the highest-ranking individual, and so forth, until the lowest-ranking individual must give way to all others within the group. This kind of behavior is seen in barnyard chickens, where it is known as a *pecking order.* Figure 9.11 shows a dominance hierarchy; the lead animal has the highest ranking and the last animal has the lowest ranking.

A dominance hierarchy allows certain individuals to get preferential treatment when resources are scarce. The dominant individual will have first choice of food, mates, shelter, water, and other resources because of its position. Animals low in the hierarchy may fail to mate or may be malnourished in times of scarcity. In some social animals, like wolves, only the dominant males and females reproduce. Poorly adapted animals with low rank may never reproduce. Once a dominance hierarchy is established it results in a more stable social unit with little conflict, except perhaps for an occasional altercation that reinforces the knowledge of which position an animal occupies in the hierarchy. Such a hierarchy frequently results in low-ranking individuals leaving the area. Migrating individuals are often subject to heavy predation. Thus, the dominance hierarchy serves as a population-control mechanism and a way of allocating resources.

Navigation

Since animals move from place to place there is value in being able to return to a nest, water hole, den, or preferred feeding spot. This requires some sort of memory of surroundings (mental map) or some other list of directions that assures safe passage. The activities of honeybees involve communication among the various individuals that are foraging for nectar. The bees are able to communicate information about the direction and distance of the nectar source from the hive. If

Territorial Communication in Gull Colonies

Various kinds of gulls nest in dense concentrations on isolated islands or other inaccessible sites where predators are rare. Within a gull colony, each nest is in a territory of about one square meter. When one gull walks or lands on the territory of another, the defender walks toward the other in the upright threat posture. The head is pointed down with the neck stretched outward and upward. The folded wings are raised slightly as if to be used as clubs. The upright threat posture is one of a number of signals that an animal can use to indicate what it is likely to do in the near future. The bird is communicating an intention to do something, to fight in this case, but it may not follow through.

If the invader shows no sign of retreating, then one or both gulls may start pulling up the grass vigorously with their beaks. This seems to make no sense. The gulls were ready to fight one moment; the next moment they apparently have forgotten about the conflict and are pulling grass. But the struggle has not been forgotten: pulling grass is an example of redirected aggression in which the animal attacks something other than the natural opponent. If the intruding gull doesn't leave at this point, there will be an actual battle. (A person who starts pounding the desk during an argument is showing redirected aggression. Look for examples of this behavior in your neighborhood cats and dogs—maybe even in yourself!)

Redirected Aggression

Figure 9.11 A Dominance Hierarchy Many animals maintain order within their groups by establishing a dominance hierarchy. For example, whenever you see a group of cows or sheep walking in single file, it is likely that the dominant animal is at the head of the line, while the lowest-ranking individual is at the end.

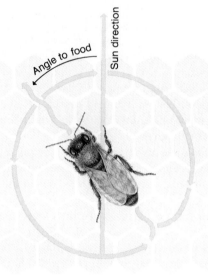

Leggitt

Figure 9.12 Honeybee Communication and Navigation The direction of the straight, tail-wagging part of the honeybee's dance indicates the direction to a source of food. The angle that this straight run makes with the vertical is the same angle at which the bee must fly in relation to the sun to find the food source. The length of the straight run and the duration of each dance cycle indicate the flying time necessary to reach the food source.

the source of nectar is some distance from the hive, the scout bee performs a "wagging dance." The bee walks in a straight line for a short distance, wagging its rear end from side to side. It then circles around back to its starting position and walks the same path as before (figure 9.12). This dance is repeated many times. The direction of the straight-path portion of the dance indicates the direction of the nectar relative to the position of the sun. For instance, if the bee walks straight upward on a vertical surface in the hive, that tells the other bees to fly toward the horizon in the direction of the sun. If the path is thirty degrees to the right of vertical, the source of the nectar is thirty degrees to the right of the sun's position.

The duration of the entire dance and the number of waggles in the straight-path portion of the dance are directly related to the time the bee must fly to get to the nectar. The dance is able to communicate the duration of flight as well as the direction. Since the recruited bees have picked up the scent of the nectar source from the dancer, they also have information about the kind of flower to visit when they arrive at the correct spot.

FYI

Seasonal Migration Animals often use the ability to sense changes in time to prepare for seasonal changes. In areas away from the equator, the length of the day changes as the seasons change. The length of the day is called the **photoperiod**. Many birds prepare for migration and have their migration direction determined by the changing photoperiod. For example, in the fall of the year many birds instinctively change their behavior, store up fat, and begin to migrate from northern areas to areas closer to the equator. This seasonal migration allows them to avoid the harsh winter conditions signaled by the shortening of days. The return migration in the spring is triggered by the lengthening photoperiod. This migration certainly requires a lot of energy, but it allows many birds to exploit temporary food resources in the north during the summer months.

Since the sun is not stationary in the sky, the bees must constantly adjust their angle to the sun. It appears that they do this with some kind of internal clock. Bees that are prevented from going to the source of nectar or from seeing the sun will still fly in the proper direction sometime later, even though the position of the sun is different.

Like honeybees, some birds that migrate during the daylight hours use the sun to guide them. Two instruments are needed to navigate by the sun—an accurate clock and a sextant for measuring the angle between the sun and the horizon. Can a bird perform such measurements without instruments? It is unquestionably true! For nighttime migration, some birds use the stars to help them find their way. In one interesting experiment, warblers, which migrate at night, were placed in a planetarium. The pattern of stars as they appear at any

Food Storage Behavior

Resource allocation becomes most critical during periods of scarcity. In some areas, the dry part of the year is most stressful. In temperate areas, winter reduces many sources of food and forces organisms to adjust.

Animals have several ways of coping with seasonal shortages of food. Some animals simply store food in their bodies as fat and avoid the stress by hibernating. Hibernation is a physiological slowing of all body processes that allows an animal to survive on food it has stored within its body. It is typical of many insects, bats, marmots, and some squirrels. Other animals have instinctive, built-in behavior patterns that cause them to store food during seasons of plenty for periods of scarcity. Squirrels bury nuts, acorns, and other seeds. (They also plant trees because they never find all the seeds they bury.) Chickadees stash seeds in cracks and crevices when seeds are plentiful and spend many hours during the winter exploring similar places for food. Some of the food they find is food they stored. Honeybees store honey, which allows them to live through the winter when nectar is not available. This requires a rather complicated set of behaviors that coordinates the activities of thousands of bees in the hive.

Biological Clocks

Many animals besides birds and bees have a time sense built into their bodies. For instance, you have one. Travelers who fly partway around the world by nonstop jet plane need time to recover from "jet lag." Their digestion, sleep, or both may be upset. Their discomfort is not caused by altitude, water, or food, but by having rapidly crossed several time zones. There is a great difference in the time as measured by the sun or local clocks and that measured by the body; the body's clock is not able to adjust quickly to the change in time.

In the animal world, mating is the most obviously timed event. In the Pacific Ocean off some of the tropical islands lives a marine worm known as the *palolo worm*. Its habit of making a well-timed brief appearance in enormous swarms is a striking example of a biological-clock phenomenon. At mating time, these worms swarm into the shallow waters of the islands and discharge their sperm and eggs. There are so many worms that the sea looks like noodle soup. The people of the islands find this an excellent time to change their diet. They dip up the worms much as North Americans dip up smelt or other small fish that are making a spawning run. The worms appear around the third quarter of the moon in October or November, the time varying somewhat according to local environmental conditions.

season could be projected onto a large domed ceiling. During autumn, when these birds would normally migrate southward, the stars of the autumn sky were shown on the ceiling. The birds responded with much fluttering activity at the south side of the cage, as if they were trying to migrate southward. Then the experimenters tried projecting the stars of the spring sky, even though it was autumn. Now the birds tended to try to fly northward, although there was less unity in their efforts to head north; the birds seemed somewhat confused. Nevertheless, the experiment showed that the birds recognized star patterns and were influenced by them.

There is evidence that some birds navigate by compass direction—that is, they fly as if they had a compass in their heads. They seem to be able to sense magnetic north. Their ability to sense magnetic fields has been proven at the U.S. Navy's test facility in Wisconsin. The weak magnetism radiated from this test site has changed the flight pattern of migrating birds.

Social Behavior

Many species of animals are characterized by interacting groups called **societies,** in which there is division of labor. Societies differ from simple collections of organisms by the greater specialization of the society's individuals. The individuals performing one function cooperate with others having different special abilities. As a result of specialization and cooperation, the society has characteristics not found in any one member of the group: the whole is more than the sum of its parts. But if cooperation and division of labor are to occur, there must be communication among individuals and coordination of effort. Honeybees, for example, have an elaborate communication system and are specialized for specific functions. A few individuals known as *queens* and *drones* specialize in reproduction, while large numbers of *worker* honeybees are involved in collecting food, defending the hive, and caring for the larvae. These roles are rigidly determined by inherited behavior patterns. Each worker honeybee has a specific task, and all tasks must be fulfilled for the group to survive and prosper. As they age, the worker honeybees move through a series of tasks over a period of weeks. When they first emerge from their wax cells, they clean the cells. Several days later, their job is to feed the larvae. Next they build cells. Later they become guards that challenge all insects that land near the entrance to the hive. Finally they become foragers who find and bring back nectar and pollen to feed the other bees in the hive. Foraging is usually the last job before the worker honeybee dies. Although this progression of tasks is the usual order, workers can shift from their main task to others if there is a need. Both the tasks performed and the progression of tasks are instinctively (genetically) determined.

A hive of bees may contain thousands of individuals, but under normal conditions only the queen bee and the male drones are capable of reproduction. None of the thousands of workers who are also females will reproduce. This does not seem to make sense because they appear to be giving up their chance to reproduce and pass their genes on to the next generation. Is this some kind of self-sacrifice on the part of the workers, or is there another explanation? In general, the workers in the hive are the daughters or sisters of the queen and therefore share a large number of her genes. This means they are really helping a portion of their genes get to the next generation by assisting in the raising of their sisters, some of whom will become new queens. This argument has been used to partially explain behaviors in societies that might be bad for the individual but advantageous for the society as a whole.

Animal societies exhibit many levels of complexity and types of social organization that differ from species to species. Some societies show little specialization of individuals other than that determined by sexual differences or differences in physical size and endurance. The African wild dog illustrates such a flexible social organization. These animals are nomadic and hunt in packs. Although an individual wild dog can kill prey about its own size, groups are able to kill fairly large animals if they cooperate in the chase and the kill. When the dogs are young, they do not follow the pack. When adults return from a successful hunt, they regurgitate food if the proper begging signal is presented to them. Thus the young and the adults that remained behind to guard the young are fed by the hunters. The young are the responsibility of the entire pack, which cooperates in their feeding and protection. During the time that the young are at the den site, the pack must give up its nomadic way of life. Therefore, the young are born during the time of year when prey are most abundant. Only one or two of the females in the pack have young each year. If every female had young, the pack could not feed them all. At about two months of age, the young begin traveling with the pack, and the pack can return to its nomadic way of life.

In many ways the honeybee and African wild dog societies are similar. Not all females reproduce, the raising of young is a shared responsibility, and individuals interact in a cooperative manner that is beneficial to the group. Social behavior is found among insects like bees, wasps, ants, and termites and among many species of birds and mammals, such as crows, wolves, and elephants. Some view the social organization of these animals to be different from that of humans. Others feel that there are underlying similarities that shape all societies, whether they are human or nonhuman.

SUMMARY

Behavior is how an organism acts, what it does, and how it does it. The kinds of responses that organisms make to environmental changes (stimuli) may be simple reflexes, very complex instinctive behavior patterns, or learned responses.

Behaviors represent adaptations to the environment. They increase in complexity and variety the more highly specialized and developed the organism is. All organisms have inborn or instinctive behavior, while higher animals also have one or more ways of learning. These include conditioning, imprinting, and insight. Communication for purposes of courtship and mating is accomplished by sounds, visual displays, and chemicals called pheromones. Many animals have special behavior patterns that are useful in the care and raising of young.

Dominance hierarchies and territorial behavior are both involved in the allocation of scarce resources. Migration to avoid seasonal extremes involves a timing sense and some way of determining direction. Animals navigate by means of sound, celestial light cues, and magnetic fields.

Societies consist of groups of animals that specialize and cooperate. Sociobiology attempts to analyze all social behavior in terms of evolutionary principles, ecological principles, and population dynamics.

Sociobiology

The analysis and comparison of animal societies has led to the thought that there may be fundamental processes that shape all societies. The systematic study of all forms of social behavior, both nonhuman and human, is called **sociobiology.**

How did various types of societies develop? What advantage does a member of a social group have? In what ways are social organisms better adapted to their environment than nonsocial organisms? How does social organization affect the way populations grow and change? These are difficult questions to answer.

The ultimate step in this new science is to analyze human societies according to sociobiological principles. Such an analysis is difficult and controversial, however, since humans have a much greater ability to modify behavior than do other animals. Also, human social structure changes very rapidly compared to that of other animals. Sociobiology will continue to explore the basis of social organization and behavior and will continue to be an interesting and controversial area of study.

The Question of Consciousness

One of the most intriguing questions of animal behavior is the possibility that animals may have a self-concept. Do they know what kind of animal they are and who they are as individuals? It seems clear that in many kinds of social animals individuals recognize other members of the group and identify strangers, know their position in a dominance hierarchy, and communicate desires to others in the group. But an animal would not need to be self-aware to accomplish these activities. A memory would allow for individual recognition without self-awareness. Computers can be used to identify a variety of kinds of items without being conscious of their own existence. Similarly, computers regularly communicate to us about their problems (e.g., *unable to read drive A*) without being self-aware. The biggest barrier to this kind of study is the difficulty of communication. We are not able to "talk to the animals." In the few instances where communication with other animals has been attempted, the conclusions have been difficult to interpret. Some studies of chimpanzees using sign language suggest that chimps recognize themselves and communicate thoughts that indicate their personal likes and dislikes. However, for most animals the communication barrier between our species and theirs is currently too great to allow meaningful exploration of this idea.

CHAPTER GLOSSARY

anthropomorphism (an-thro-po-mōr′fizm) The assigning of human feelings, emotions, or meanings to the behavior of animals.

behavior (be-hav′yur) How an organism acts, what it does, and how it does it.

conditioned response (kon-dĭ′shund re-spons′) The behavior displayed when the neutral stimulus is given after association has occurred.

conditioning (kon-dĭ-shun-ing) A kind of learning in which a neutral stimulus is associated with a natural stimulus to produce a particular response.

dominance hierarchy (dom′in-ants hi′ur-ar-ke) A relatively stable, mutually understood order of priority within a group.

ethology (e-thol′uh-je) The scientific study of the nature of behavior and its ecological and evolutionary significance in its natural setting.

imprinting (im′prin-ting) Learning in which a very young animal is genetically primed to learn a specific behavior in a very short period.

insight (in-sīt) Learning in which past experiences are reorganized to solve new problems.

instinctive behavior (in-stink′tiv be-hāv′yur) Automatic, preprogrammed, or genetically determined behavior.

learning (lur′ning) A change in behavior as a result of experience.

pheromone (fĕr-uh-mōn) A chemical produced by an animal and released into the environment to trigger behavioral or developmental processes in some other animal of the same species.

photoperiod (fō″tō-pĭr′ē-ud) The length of the light part of the day.

response (re-spons′) The reaction of an organism to a stimulus.

society (so-si′uh-te) Interacting groups of animals of the same species that show division of labor.

sociobiology (so-sho-bi-ol′o-je) The systematic study of all forms of social behavior, both human and nonhuman.

stimulus (stim′yu-lus) Some change in the internal or external environment of an organism that causes it to react.

territoriality (tĕr″ĭ-tor′e-al′ĭ-te) A behavioral process in which an animal protects space for its exclusive use for food, mating, or other purposes.

territory (tĕr′ĭ-tor-e) A space that an animal defends against others of the same species.

CONCEPT MAP TERMINOLOGY

Construct a concept map to represent the relationships among the following concepts.

behavior
conditioned response
dominance hierarchy
imprinting
insight

instinctive behavior
learning
response
stimulus
territoriality

LABEL • DIAGRAM • EXPLAIN

Some scientists interested in the behavior of Canada geese are trying to establish new populations of geese along the eastern seacoast of the United States. They hand-raised the geese and used ultralight aircraft to lead the geese on training flights. Eventually the geese followed the ultralight several hundred miles from Canada to the eastern U.S. coast. They migrated north in the spring.

In this study, the scientists make several assumptions about the behavior of the geese. How is each of the following kinds of behavior involved in the scientists' study?

1) instinct
2) imprinting
3) associative learning (conditioning)

Multiple Choice Questions

1. Honeybee dances are an example of:
   a. imprinting
   b. instinct
   c. learning
   d. conditioned responses
2. Learned behaviors differ from instinctive behaviors in that:
   a. learned behaviors can be changed
   b. instinctive behaviors can change if the environment changes
   c. learned behaviors are nearly always superior to instinctive behaviors
   d. instinctive behaviors must be practiced
3. A cardinal that was raised in isolation from all other cardinals until adulthood only associates with other cardinals when released. This is evidence of:
   a. learning
   b. a conditioned response
   c. imprinting
   d. instinctive behavior
4. An animal learning a very specific behavior during a certain short period of time is called:
   a. dominance hierarchy
   b. conditioning
   c. insight learning
   d. imprinting
5. The behavior of animals in most cases is appropriate to the environment in which they find themselves (night-flying lightning beetles communicate by light, bees recognize color patterns, birds build nests of specific materials). The most likely explanation for the appropriateness of their behavior is that:
   a. they learned it from their parents
   b. the behavior has evolved and was passed from parent to offspring through genes
   c. each individual makes its own discoveries and adjusts accordingly
   d. the behaviors are randomly determined

6. Which of the following behaviors is an important part of the behavioral patterns of social animals?
   a. individuals participate in greeting displays or behaviors
   b. each individual has its territory
   c. individuals will not give up resources to others in the group
   d. all individuals mate each year
7. Individuals that have territories:
   a. have a greater chance of reproducing than those that do not have a territory
   b. must continuously fight to maintain the territory
   c. will often win a fight when they visit another's territory
   d. are always able to defend their territory

Questions with Short Answers

1. Why do students of animal behavior reject the idea that a singing bird is a happy bird?

2. Briefly describe the behavior of some animal as an example of unlearned behavior. Name the animal.

3. Briefly describe the behavior of some animal as an example of learned behavior. Name the animal.

4. Give an example of a conditioned response. Can you describe one that is not mentioned in this chapter?

5. Name three behaviors typically associated with reproduction.

6. How do territorial behavior and dominance hierarchies help to allocate scarce resources?

7. How do animals use chemicals, light, and sound to communicate?

8. What is sociobiology? Ethology? Anthropomorphism?

9. What is imprinting, and what value does it have to the organism?

10. Describe how honeybees communicate the location of a nectar source.

Organisms have physical, chemical, and behavorial characteristics that allow them to survive in their particular environments. These characteristics are largely determined by the organisms' genes. Environmental conditions change and so must populations in order to survive. Since the origin of the first living cells, populations have experienced genetic change. This genetic change allows populations to survive in changing environments and has been responsible for the evolution of all of the diverse organisms that have ever inhabited the Earth.

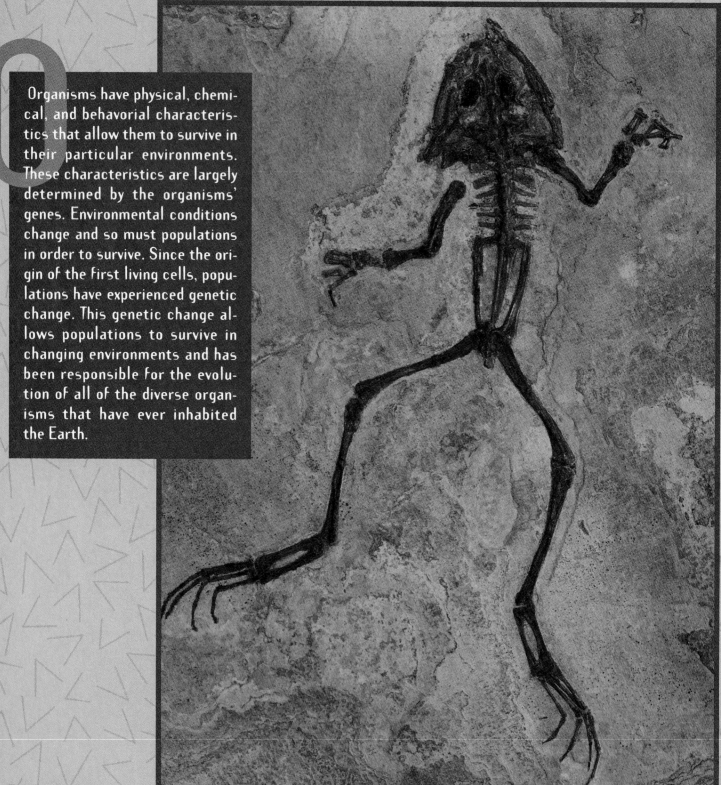

Frog fossil.

The Investigative Process

and the Characteristics of life

Pesticides being sprayed on sugar beet field.

The Investigative Process
and the Characteristics of life

learning objectives

- Be able to distinguish between cause and effect and simple correlation.

- Understand the value of controlled experiments in determining cause-and-effect relationships.

- Distinguish a guess, hypothesis, and theory.

- Recognize the value of clearly stating a question to which you want an answer.

- State the characteristics of a good hypothesis.

- Identify characteristics of living things.

- Identify the experimental group and the control group in a controlled experiment.

- Distinguish between spontaneous generation and biogenesis.

- Understand the difference between science and nonscience.

- Recognize that science has limitations.

- Recognize that pseudoscience appears to be scientific but is really used to mislead.

This introductory chapter will serve two distinct purposes: first, to examine a *scientific process* we can use to identify cause-and-effect relationships; and second, to identify the characteristics of living things and describe how they differ from nonliving things.

The Investigation Process

Before we can examine the scientific process, we need to distinguish between situations that are merely correlated (happen at the same time) and those that show a cause-and-effect relationship. When an event occurs as the result of a known reason, a cause-and-effect relationship exists between the reason and the event. Many events are correlated, but not all correlations show cause-and-effect. For instance, a play-off game might be televised the night before your biology exam. You decide to stop studying and the next day you do poorly on the exam. These events are correlated. The effect is that you received a lower grade than you had hoped for. Did the televised game cause the lower grade? It would certainly be easy for us to place the blame (cause) on the poor timing of the play-off game. The cause of the grade, however, was your lack of sufficient preparation for the exam. The televised game happened at the same time, but it was not the *cause* of the poor grade (figure 1.1).

In this chapter we want to examine how information from a variety of sources can be put to use. We also want to examine the *method* you can use to evaluate facts and events in your life. Our intent is to teach you to recognize when you have applied this method of evaluation. It is also important that you develop the ability to analyze new information using this scientific approach.

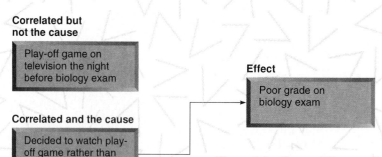

Correlated but not the cause

Play-off game on television the night before biology exam

Correlated and the cause

Decided to watch play-off game rather than study for biology exam

Effect

Poor grade on biology exam

Figure 1.1 Cause, Effect, and Correlation When an event always occurs before a subsequent event and initiates the second event, the events have a cause-and-effect relationship. The two events are also correlated. It is possible to have events that are correlated but do not have a cause-and-effect relationship.

Facts can be **relevant,** meaning they pertain to the matter at hand, or they can be **irrelevant,** meaning they are not related to the situation under discussion.

Since it is frequently difficult to recognize cause-and-effect relationships, we present a story that deals with a person and a significant biological event in her life. Your task is to determine which correlated events in the story show cause and effect. Given all the information, we will use an analytical process to draw conclusions and establish cause-and-effect relationships. As you read this story, you will need to distinguish between relevant and irrelevant facts, assumptions, and opinions. While this event could happen to anyone, the process you will use to determine cause-and-effect relationships is the same one scientists use to draw conclusions about natural events they are studying.

CONCEPT CONNECTIONS

- A **fact** is something true. Information is considered a fact when ample evidence supports it.

- An **assumption** is a speculation, something you think *might* be true.

- An **opinion** is a thought, idea, or belief about something. When you form an opinion, you might use facts and make assumptions.

Figure 1.2 Yard Care Yard care companies apply fertilizer and pesticides to lawns to encourage the growth of grass and to control weeds and other pests.

The Story

Mary Jane, three months pregnant with her first child, was working on the flower bed in her front yard. She began to notice an unusual chemical smell. Looking up from her work, she saw that the *Nifty Green Yard Care Company* was spraying her neighbor's lawn for weeds and insects (figure 1.2). The workers applying the chemical spray were wearing face masks and protective clothing. As they left, they posted a sign warning that children and pets should not play on the grass for two hours, or until it was thoroughly dry. Since Mary Jane was pregnant and particularly sensitive to

FYI

As you read this story, highlight the relevant facts, irrelevant facts, opinions, and assumptions with different colors.

odors, she moved to the other side of the yard and worked on the flower beds there. During the next two weeks Mary Jane had bouts of morning sickness similar to those she had experienced earlier in her pregnancy. The third weekend she had a miscarriage.

Talking the events over with her sister-in-law, Susan, Mary Jane wondered if the chemical spray had anything to do with her miscarriage. Susan said that while waiting in line at the supermarket, she had read a magazine article about women in the Pacific Northwest who claimed that their miscarriages were caused by the spraying of forests with pesticides from helicopters. Susan also recalled a conversation at a party with a woman whose veterinarian told her that her dog's miscarriage might have been related to her house being treated with a termite control pesticide. With this information in mind, Mary Jane decided to contact an attorney she saw advertized on TV and explore the possibility of filing a law-

suit against the lawn care company she assumed had caused her miscarriage. The lawyer agreed to take her case for 50 percent of any damages she might be awarded. They decided to seek $2 million.

Clarifying the Question

Suppose you are a member of the jury deciding this case. You need to determine the key question and what information is relevant. Was the time of Mary Jane's miscarriage merely *correlated* with the chemical spraying of the neighbor's lawn, or is this truly a case of cause and effect? Was the *Nifty Green Yard Care Company* at fault? You might ask yourself, Is there any connection between the reports of miscarriages in the Pacific Northwest and widespread chemical spraying? Is there a correlation between the use of termite control chemicals and miscarriage in a family pet? You must try to distinguish between events that happened at the same time and an event that caused other events to occur. You do not have an easy job! A summary of information presented at the trial follows (figure 1.3).

Outline of the Testimony Presented at the Trial

1. Dr. M. D. Lewis presented medical evidence indicating that Mary Jane had been her patient for prenatal care and that the pregnancy was normal but terminated in a spontaneous abortion.

2. Dr. I. M. Smart reviewed the technical data concerning the animal testing program and the safety of the chemicals used.

3. *Pesticides Unleashed* presented information supporting their claim that home owners use too many artificial methods of controlling pests and that the pesticides used on Mary Jane's neighbor's lawn have been correlated with miscarriages.

4. Victor Fuentes, neighborhood resident, testified that the chemicals were sprayed at Mary Jane's neighbor's address and corroborated the time and date of the application.

5. Meteorologist Jake Chin presented information indicating that the spray could have drifted onto Mary Jane's lawn from next door.

6. *Nifty Green Yard Care Company* representatives testified that their employees follow application regulations established by state and federal government agencies and do not spray neighboring areas.

Figure 1.3   The Court Process   Mary Jane hoped to be compensated for her loss by suing the yard care company.

Gathering the Facts

During the trial, the date and time of the lawn treatment was established. Mary Jane's physician, Dr. M. D. Lewis, indicated that Mary Jane had had a healthy pregnancy and that no restrictions had been placed on her activities. She testified that Mary Jane had miscarried at a local hospital on the date previously established. Dr. Lewis also testified that "fetal development had been insufficiently complete to enable independent survival." Dr. I. M. Smart testified for the *ZAP-UM Chemical Corporation,* the company that manufactured the chemicals used in the spray. He stated that the particular mixture of chemicals used that day is licensed as suitable for neighborhood lawn application. He presented testimony to the effect that, when applied properly, the chemicals are safe. This conclusion was based on years of animal testing required by the *Environmental Protection Agency (EPA).* Dr. Smart presented this information in a very technical way, making it difficult to understand. He discussed several topics, including LD_{50} testing, with which you were unfamiliar. You have never dealt with a live animal testing program and were unsure exactly what was meant by LD_{50}. The juror sitting next to you seemed to understand the technical aspects of Dr. Smart's testimony and seemed to agree with what he was saying.

Pesticides Unleashed (PU), an environmental activist organization, presented data that established increases in production, sale, and use of chemicals to control pests of various types. It was their position that natural plantings around homes could be an alternative to chemical pest control. They cited a tragedy reported in Afghanistan three years previously. An agricultural area was sprayed with one of the same chemicals used by the *Nifty Green Yard Care Company.* After the area was sprayed, 65 percent of the pregnant farm workers experienced miscarriages within three months. *PU* encouraged habitat improvement to control insect pests by natural predators, such as releasing ladybugs into yards to feed on pests.

Several others also testified. Victor Fuentes testified that his neighbor, Mary Jane, was devastated by the loss of her unborn child. He described the care Mary Jane had taken to ensure that the long-awaited baby would be healthy. He also indicated that the *Nifty Green Yard Care Company* sprayed the chemical even though it was very windy on the day in question. Jake Chin, a local meteorologist, testified about the wind speed and direction on the day of the spraying as well as other pertinent aspects of the atmospheric conditions at the time. Representatives from the yard care company explained how they applied the chemical spray and detailed the safety measures used to ensure that their employees did not come into direct contact with the chemicals. They also explained precautions they took to make sure the chemicals were sprayed only on the customer's lawn. In addition, they noted that they placed warning signs that, if followed, would prevent human contact with the chemicals. A great deal of their testimony related to the fact that it was economically unsound as well as environmentally unsafe to put the chemicals anywhere but where the customer paid to have them put. Mary Jane testified that she was devastated by the loss of her baby boy, that she was sure the chemicals caused her miscarriage, and that someone should pay for her grief and pain.

As a member of the jury, you are going to be asked to evaluate all of the testimony. You need to determine what parts of it are relevant so that you and the other jurors can arrive at a sound decision.

Where Do You Go from Here?

One of the first things you might do to distinguish simple correlation from cause and effect in this situation is to organize your information. Start by phrasing the problem as a question. It is important to accurately identify the question you are trying to answer. You must make sure that you are at the right starting point. The question will help you see exactly what you are looking for and will clear away many of the irrelevant issues that could get in the way of your reasoning. Here are some questions about the case you might ask:

- Is the *Nifty Green Yard Care Company* responsible for damage it causes inadvertently?
- Did the chemical used in the spray cause the termination of the pregnancy?
- Was the yard work too strenuous for Mary Jane in her condition?
- Does the chemical company care about where, how, and when its products are used?
- Is the information given by *Pesticides Unleashed* relevant?

Guess, Hypothesis, and Theory

The words *guess, hypothesis,* and *theory* are often used as if they have the same meaning. However, that is not the case.

To **guess** is to form an opinion without supporting evidence. A guess is the least precise of the three. You might guess the time of day or which card is next in the deck. In some cases a guess is simply a random hope, whereas in other circumstances it might be highly predictive. For example, if you have just watched the sun set you could estimate the time, but you probably would not be able to state the exact time.

A **hypothesis** is a logical statement about an event that can be tested for its validity. A **valid** statement is one repeatedly supported by evidence. A hypothesis is a more formal statement than a guess because certain facts are already known and were used to construct the hypothesis

in the first place. For example, you know that in spring sunset is a little later each day. Since sunset the previous day was 7:57 P.M., sunset today *could be* 7:58 P.M. You made use of previous information to form your hypothesis and can test this hypothesis by timing the sunset.

A **theory** is a well-substantiated generalization. It has been supported again and again and usually brings blocks of ideas together. The fact that the earth's axis is at an angle and that during the summer the pole is pointing toward the sun accounts for the changes in day length. The theory that the earth goes round the sun and that the pole is pointing toward the sun in the summer and away from the sun in the winter was a revolution in scientific thinking at one time, but it is unquestioned today.

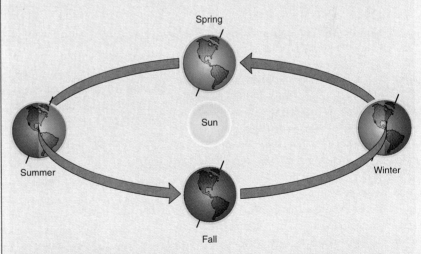

Cause of Seasons Since the earth is tilted on its axis, summer occurs when the pole is pointing toward the sun and winter occurs when the pole is pointing away from the sun.

These and many other questions might arise. The important question for you, the juror, is *did the chemicals sprayed cause the miscarriage?* This question is your starting point.

Cause-and-Effect Relationship: Forming a Hypothesis

After you have established the question to be answered, you are ready to state your hypothesis. Your hypothesis must be consistent with all information you already know and it also must be testable. As a juror you cannot ignore the fact that environmental chemicals can result in miscarriages. Based on the facts in the case, you state the following hypothesis: The chemical mixture sprayed by the company caused Mary Jane's miscarriage.

To find out if your hypothesis is supported by the evidence, the hypothesis should

CONCEPT CONNECTIONS

- A **controlled experiment** is a scientific test that compares two groups. One group is subjected to the suspected cause while the other group is not; all other factors are identical. Differences in the outcomes of the two groups are the result of the presence or absence of the suspected cause.
- The **experimental group** is the group subjected to the suspected cause.
- The **control group** is the group not subjected to the suspected cause and used for comparing outcomes.

be tested. However, in this courtroom situation, the kind of testing required cannot be done. Therefore, you are not able to complete the scientific process of evaluating the evidence. To feel truly confident of the cause of Mary Jane's miscarriage, you would have to perform the following scientific testing procedure. You would purposely expose a large number of pregnant females (the experimental group) to the chemical mixture sprayed by the *Nifty Green Yard Care Company.* You would also expose an equal number of pregnant females to another, harmless spray (the control group). You would then compare the number of miscarriages in both groups.

This testing procedure is an important part of analyzing cause and effect. There are

several aspects that we need to keep in mind. First, we would need to be sure we had a group large enough so that the typical miscarriages due to *other* things would not cloud our information. If you included only one pregnant female in your test and she happened to have a miscarriage due to an infection, you might think her miscarriage was caused by the chemical, when in fact it was not. Therefore, we need to make sure that we start with a lot of pregnant females, so we can feel certain that miscarriages due to other causes will not mislead us.

Second, in order to be able to say that the chemical mixture caused the difference in miscarriage rate in the two groups of pregnant females, you need to be confident that there is no other significant difference between the two groups. Both groups must have similar prenatal care and nutrition and be free of alcohol or drug use. We have arbitrarily placed some females at risk of miscarriage and protected others.

If you actually tried to determine whether a chemical mixture could cause miscarriage in humans, your test would create a moral dilemma. It is not morally defensible to treat pregnant women with such a chemical mixture. Of course you could substitute pregnant female mice for the human subjects. That test would tell you whether the chemical causes miscarriages in mice. Then you could test the chemical mixture on other

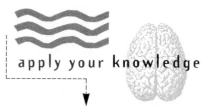

FYI

It's the Law Generally, people think of *law* as referring to civil laws or rules of order such as traffic laws. However, there is a second category that pertains to science: the *scientific laws.* A scientific law is a uniform or constant fact of nature. An example of a biological law is the biogenetic law, which states that all living things come from preexisting living things. You can see from this example that laws are even more general than theories and encompass the answers to even more complex questions. Therefore, there are relatively few scientific laws. Scientific laws are stated only after theories have undergone extensive testing and experimentation. Just because a theory exists does not mean that testing stops. In fact, many scientists see this as a challenge and exert even greater efforts to disprove the theory, and experimentation continues. Should the theory survive this skeptical approach and continue to be supported by experimental evidence, it will become a scientific law. Even after the scientific community has developed a scientific law, it will continue to challenge its validity.

animals such as rabbits, dogs, and chimpanzees. If similar tests are done on other animals, and they all show high numbers of miscarriages in females treated with the mixture, you might feel comfortable saying that the chemical mixture is likely to cause miscarriage in humans.

Drawing a Conclusion

What you have done so far in this analysis is determine a question of cause and effect and suggest an answer that accounts for the information you have. Then you looked at the results of a variety of tests that could tell you whether or not you are dealing with a cause-and-effect relationship. You may find that you need additional information in order to draw a conclusion. If that is the case, you may want to conduct additional tests with different animals or with varying amounts of the chemical mixture.

The final step in this process is to put yourself on the line and say whether you think the chemical mixture caused the miscarriage or not. This conclusion will be based on careful examination of all of the information presented to you. You must make the very best evaluation you can so that you are comfortable with your decision. You must be honest as you look at the information. You cannot prejudge or ignore some pieces of the information and look only at other pieces. This is similar to the process of evaluation good scientists use. The difference between this court case and a scientific investigation is that you are not able to conduct controlled experiments in court to be sure you have reached the correct conclusion.

Evaluating Cause and Effect

1. Make an <u>observation</u> of an event or situation.

2. Clarify the <u>question</u>. Make sure the question can be answered and that it reflects what you want to know.

3. Look for a cause that could account for the effects that have been observed. State a formal <u>hypothesis</u> that links cause and effect. This cause must take into account all of the information available at this time.

4. Set up a controlled <u>experiment</u> to test the hypothesis. Be sure that one group is subjected to the possible cause (experimental group) and the other is not (control group).

5. Gather experimental <u>results</u> and compare the outcome of the two groups in your test situation. Any differences in outcome may be due to the presence or absence of what you thought was the cause.

6. Ask other responsible people to <u>review</u> your thought processes and test situations to make sure you have not overlooked something. They may want to repeat the experiment to see if they get the same outcome.

7. Draw a <u>conclusion</u>. Either this is a cause-and-effect relationship or it is not.

8. If it is not a cause-and-effect relationship, look for another possible cause. <u>Revise</u> the question in step 2.

9. In the future, additional information may cause you to <u>modify</u> your conclusion.

Historical Investigations of Life

In earlier times, no one doubted that life originated from nonliving things. The Greeks, Romans, Chinese, and many other ancient peoples believed that maggots arose from decaying meat; mice developed from wheat stored in dark, damp places; lice formed from sweat; and frogs originated from damp mud. The concept of **spontaneous generation—** the idea that living organisms spontaneously arise from nonliving material—was widely believed until the seventeenth century (figure A). However, some people began to doubt spontaneous generation. They subscribed to **biogenesis,** the concept that life originates only from preexisting life.

the eggs of the flies and not from spontaneous generation in the meat.

Even after Redi's experiment, some people still supported the theory of spontaneous generation. After all, a belief that has been prevalent for more than two thousand years does not die a quick death. In 1748 John T. Needham, an English priest, placed a solution of boiled mutton broth in containers that he sealed with corks. Within several days the broth became cloudy and contained a large population of microorganisms. Needham reasoned that boiling killed all the organisms and that the corks prevented any microorganisms from

Spallanzani's experiment did not completely disprove the theory of spontaneous generation to everyone's satisfaction. The supporters of the old theory attacked Spallanzani by stating that he excluded air, a factor believed necessary for spontaneous generation. Supporters also argued that boiling had destroyed a "vital element." When Joseph Priestly discovered oxygen in 1774, the proponents of spontaneous generation claimed that oxygen was the vital element that Spallanzani had excluded in his sealed containers.

In 1861 the French chemist Louis Pasteur convinced most scientists that spontaneous generation could not occur. He placed a

Figure A Life from Nonlife Many works of art explore the idea that living things could originate from very different types of organisms or even from nonliving matter. M. C. Escher's work entitled "The Reptiles, 1943" shows the life cycle of a little alligator. Amid all kinds of objects, a drawing book lies open at a mosaic drawing of reptilian figures in three contrasting shades. Evidently, one of them is tired of lying flat and rigid among its fellows, so it puts one plastic-looking leg over the edge of the book, wrenches itself free, and launches out into "real" life. It climbs up the back of the zoology book and works its way laboriously up the slippery slope of the set-square to the highest point of its existence. Then, after a quick snort, tired but fulfilled, it goes downhill again, via an ashtray, to the level surface of the flat drawing paper, and meekly rejoins its erstwhile friends, taking up once more its function as one element of surface division.
(© 1995 M.C. Escher/Cordon Art-Baarn-Holland. All Rights Reserved.)

One of the earliest challenges to the theory of spontaneous generation came in 1668 when Francesco Redi, an Italian physician, set up a controlled experiment designed to disprove the theory (figure B). He used two sets of jars that were identical except for one aspect. Both sets of jars contained decaying meat, and both were exposed to the atmosphere, but one set of jars was covered by gauze and the other was uncovered. Redi observed that flies settled on the meat in the open jar, but the gauze blocked their access to the covered jars. When maggots appeared on the meat in the uncovered jars but not on the meat in the covered ones, Redi concluded that the maggots arose from

entering the broth. He concluded that life in the broth was the result of spontaneous generation.

In 1767 another Italian scientist, Lazzaro Spallanzani, challenged Needham's findings. Spallanzani boiled a meat and vegetable broth, placed this medium in clean glass containers, and sealed the openings by melting the glass over a flame. He placed the sealed containers in boiling water to make certain all microorganisms were destroyed. As a control, he set up the same conditions but did not seal the necks of the flasks, allowing air to enter (figure C). Two days later, the open containers had a large population of microorganisms, but there were none in the sealed containers.

fermentable sugar solution and yeast mixture in a flask that had a long swan neck. The mixture and the flask were boiled for a long time. The flask was left open to allow oxygen, the "vital element," to enter, but no organisms developed in the mixture. The organisms that did enter the flask settled on the bottom of the curved portion of the neck and could not reach the sugar-water mixture. As a control, he cut off the swan neck (figure D). This allowed microorganisms from the air to fall into the flask, and within two days the fermentable solution was supporting a population of microorganisms. In his address to the French Academy, Pasteur stated, "Never will the doctrine of spontaneous generation arise from this mortal blow."

Figure B Redi's Experiment The two sets of jars here are identical in every way except one—the gauze covering. The set on the left is called the control group; the set on the right is the experimental group. Any differences seen between the control and experimental groups are the result of a single variable. In this manner, Redi concluded that the presence of maggots in meat was due to flies laying their eggs on the meat and not to spontaneous generation.

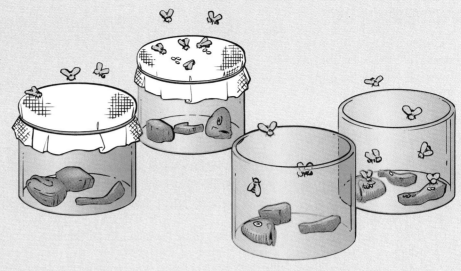

Figure C Spallanzani's Experiment Spallanzani carried the experimental method of Redi one step further. He boiled a meat and vegetable broth and placed this medium in clean flasks. He sealed one and put it in boiling water. As a control, he subjected another flask to the same conditions, except he left it open. Within two days, the open flask had a population of microorganisms. Spallanzani demonstrated that spontaneous generation could not occur unless the broth was exposed to the "germs" in the air.

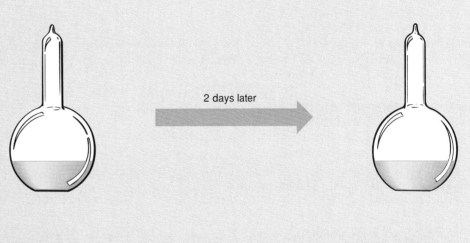

2 days later

2 days later

Figure D Pasteur's Experiment Pasteur used a swan-neck flask that allowed oxygen, but not airborne organisms, to enter. He broke the neck off of another flask. Within two days, there was growth in the second flask. Pasteur demonstrated that germ-free air with oxygen does not cause spontaneous generation.

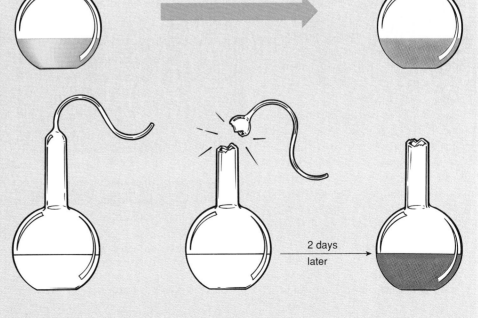

2 days later

Characteristics of Life

Scientists who study living things have utilized the investigative process in order to unravel the mysteries of life. The most fundamental question biologists have attempted to answer is, What is life? Distinguishing life from non-life has at times been difficult: Some objects are clearly alive while others are not. If you watch an insect flying through the air or a tree growing new leaves in the spring or a fish jumping from the water you quickly conclude that each is alive. On the other hand, a mosquito you have just squashed or a rotting log in a forest or a fish carcass lying on the shore are not alive, although they once were. Things that are alive can lose their life. There are also things that were never alive such as erupting volcanoes, falling water, or a bowling ball. We can classify things into three categories: (1) those that are alive, (2) those that were alive but are no longer living, and (3) those that were never

CONCEPT CONNECTIONS

spontaneous generation The theory that living organisms spontaneously arise from nonliving material.

biogenesis The concept that life originates only from preexisting life.

living. What characteristics distinguish living things from things that were never alive, and how do new living things come into existence?

Even with all our knowledge today, determining what is alive is not always easy. Paramedics and doctors are often said to bring a person "back to life." People injured in accidents are sometimes declared "legally dead" though many of their organ systems are still working. Kidney, heart, and liver transplants take place between donors who are legally dead and recipients who are alive, yet the organ transplanted is not dead.

Certain characteristics are typically associated with living things. If something is alive it will show all of these characteristics. It is important to recognize, however, that these characteristics are not necessarily confined only to living things; no-longer-living and never-living things may demonstrate one or more of them.

(1) **Response to a stimulus** is probably the first thing that comes to mind when we think about determining if something is alive. If we saw something and wanted to know if it

were alive we would probably poke it (stimulate it) and look for some response or movement. The responses of organisms are variable and appropriate. Animals, which have nerves, hormones, and muscles, can move in response to a stimulus. Plants lack nerves and muscles, but they can still respond by using hormones to grow toward or away from things. This is why a house plant grows toward the light coming through

Stimulus-Response The actions of the cat suggest that it is responding to the sight and sound (stimuli) of the bird. We cannot tell if the dog has received stimuli from the bird, but it certainly has been stimulated by the cat.

CONCEPT CONNECTIONS

- A **stimulus** is any change that an organism is able to detect.
- A **response** is any change in an organism that results from its detecting a stimulus.
- It is not always clear whether an organism has detected a stimulus and responded.

"Playing Possum" The opossum can often avoid being killed by appearing to be dead. Some people suggest that the animal may be in shock rather than voluntarily becoming limp.

Edward Jenner, the Scientific Method, and the Control of Smallpox

Edward Jenner used the scientific method when he first developed the technique of vaccination in 1795. His vaccine was the result of a twenty-six-year study of two diseases: cowpox and smallpox. Cowpox was known as *vaccinae*. From this word evolved the present-day terms *vaccination* and *vaccine*. Jenner <u>observed</u> that milkmaids developed pocklike sores after milking cows infected with cowpox, but they rarely became sick with smallpox. He asked the <u>question</u>, "Why don't milkmaids get smallpox?" He developed the <u>hypothesis</u> that the mild reaction milkmaids had to cowpox protected them from the often fatal smallpox. This led him to perform an <u>experiment</u> in which he transferred puslike material from cowpox sores to human skin and discovered that people vaccinated this way were protected from smallpox.

When these results became known, public reaction was mixed. Some people thought that vaccination was the work of the devil. Many European rulers, however, supported Jenner by encouraging their subjects to be vaccinated. Napoleon and the Empress of Russia were influential, and in the United States, Thomas Jefferson had some members of his family vaccinated. Many years later, following the development of the germ <u>theory</u> of disease, it was discovered that cowpox and smallpox are caused by viruses that are very similar in structure. Exposure to the cowpox virus allows the body to develop immunity against the cowpox virus and the smallpox virus at the same time. In 1979, almost two hundred years after Jenner developed his vaccination, the Centers for Disease Control and Prevention in the United States and the World Health Organization of the United Nations declared that smallpox was eradicated as a human disease.

Opposition to Vaccination People were suspicious of vaccination against smallpox, and many newspapers published cartoons like this one that reflected the mood of the times.

a window (figure 1.4). An ability to respond to changes in an organism's surroundings requires that the organism detect these changes. In other words, they have sense organs.

(2) **Growth** and (3) **development** are other characteristics that you as a living thing have experienced. Growth results in organisms getting larger over time. This is true for people, grass, snakes, and all other living things. And, as organisms grow, they are likely to develop different forms and abilities. A child gets larger as a result of the growth process and becomes sexually mature as a part of the development process. In many organisms the young may take an entirely different form from the adults, as in maggots that grow and eventually develop into flies (figure 1.5). Plants develop flowers and fruits only after they have reached a certain stage of development.

Normally we associate growth and development with reproduction. (4) **Reproduction** results in the creation of a new copy of the living thing. With some organisms, such as humans, elephants, or trees, it takes many years of growth and development for the offspring to become capable of reproduction. In other cases, for example, many

Figure 1.4   Plants Respond to Light   If a house plant is left in a window, it will grow toward the light.

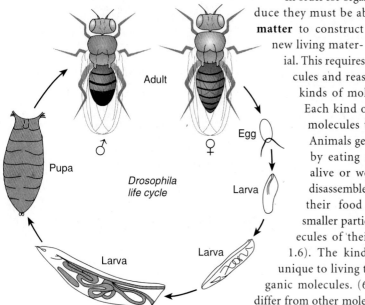

Figure 1.5   Growth and Development The adult fly lays eggs that develop into tiny maggots. As a maggot feeds and grows it eventually develops into a new stage called a pupa, which in turn develops into an adult fly.

single-celled organisms such as bacteria and protozoa, reproduction provides copies that are themselves capable of reproduction in a matter of minutes or days.

In order for organisms to grow or reproduce they must be able to (5) **manipulate matter** to construct new living material. This requires them to capture molecules and reassemble them into the kinds of molecules typical of life. Each kind of organism has many molecules that are unique to it. Animals get the matter they need by eating other things that are alive or were once living. They disassemble the large molecules of their food and reassemble the smaller particles into the large molecules of their own bodies (figure 1.6). The kinds of matter that are unique to living things are made of organic molecules. (6) **Organic molecules** differ from other molecules in that they have large numbers of carbon atoms joined to one another to form long chains or rings (figure 1.7). Carbohydrates, fats, proteins, and DNA are kinds of organic molecules you probably have heard about. In addition to organic

Appendix A

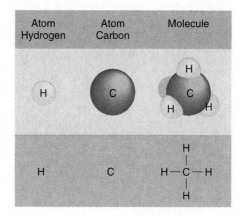

Figure 1.6   Atoms and Molecules   Atoms can be combined to form larger units called molecules.

molecules, living things also contain many kinds of inorganic molecules such as water and salts that do not have chains of carbon atoms (figure 1.7). Plants are able to attach large numbers of carbon atoms together to make organic molecules. In this process, photosynthesis, the inorganic gas carbon dioxide (CO_2) provides the carbon atoms that become joined to form organic molecules.

When organisms manipulate matter, some portions are not usable. These are released as waste products. Atmospheric oxygen could be considered a waste product of photosynthesis. (7) **Waste elimination** is typical of living things.

In the process of reorganizing matter, living things must (8) **manipulate energy.** We get energy in the food we eat and we use this energy in movement and heat production. Plants manipulate sunlight energy and use it to build large organic molecules from smaller inorganic molecules (photosynthesis). The word **metabolism** describes all the chemical and energy changes that take place in living things.

There is an (9) **orderliness** that can be seen in all forms of life. Each kind of living thing has its regular organization. For example, you would recognize any other human as having many similarities to you. Furthermore, we know that the internal machinery of one human is like the internal machinery of others and functions in the same way. The same kind of organizational similarity is found in each kind of living thing. In addition, all organisms are composed of small units, (10) **cells,** which have distinct boundaries. A cell is constructed of organic molecules and regulates the kinds of

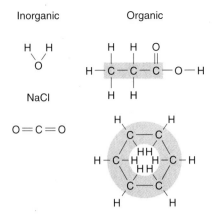

Inorganic | Organic | NaCl

Figure 1.7 Organic and Inorganic Molecules Organic molecules differ from inorganic molecules in that organic molecules contain carbon atoms attached to one another in chains or rings.

materials that enter and leave it. When we look at the cells of different kinds of organisms, we see that there is similarity in their structure and function. Finally, there is an orderliness to the way chemical activities take place within organisms. Chemical reactions occur in a controlled way that prevents damage to the organism, encourages required reactions, and inhibits those that are not necessary. Two major sets of chemical reactions are photosynthesis and respiration.

The orderliness that we see in any one kind of living thing extends to others. All organisms have cells, get energy from molecules in the same way (respiration), make copies of themselves, and contain genetic information within DNA molecules. When we recognize these similarities, it suggests that all living things are related to one another. Offspring share many characteristics with their parents but are somewhat different from individuals in other families and differ in many ways from other kinds of organisms. The fact that all living things share so many fundamental characteristics yet show great differences leads us to consider the concept that all forms of life are related.

We know that the earth is continuously changing. Mountains are formed and worn down, climates change, and ocean levels fluctuate. When conditions change, the kinds of organisms present change as well. We would expect only small changes to occur as organisms reproduce, but if large amounts of time (millions of years) are involved the accumulation of changes may be so great that new kinds of organisms come into being. Furthermore, we know that many kinds of organisms have disappeared from the earth. Mammoths and mastodons are extinct, but their frozen bodies have been found in Siberia. Dinosaurs are extinct, but fossils of their bones can be found throughout the world . Life is capable of change and modification. New forms are generated and some forms are eliminated. All of these ideas are incorporated into the concept of (11) **evolution,** another characteristic of life.

FYI

The words *organism, organ, organize,* and *organic* are all related. Organized objects have parts that fit together in a meaningful way. Organisms are separate living things that are organized. Animals have within their organization organs, and the unique kinds of molecules they contain are called organic. Therefore, organisms consist of organized systems of organs containing organic molecules.

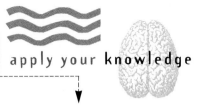

apply your **knowledge**

Alive or Not?

Each of the characteristics of life discussed in the preceding section will be explored in greater detail in later chapters. As you have probably guessed by now, determining if something is alive is not always an easy task. Let's explore it further by considering table 1.1. Look at the items at the top of the table and determine whether they display each characteristic of life. Write yes or no in the boxes. Some boxes have been filled in for you.

In what ways do living things differ from those that are no longer living? In what ways do things that are no longer living differ from things that were never alive?

You may have had difficulty deciding about some of the things listed in the table. You should be able to see that things that were never alive, such as milk and lawn mowers, do not demonstrate all of the characteristics typical of living things. Things that were once alive show some of the characteristics of living things, but they do not reproduce, respond, develop, or evolve. Things that are alive have all the characteristics of life but will not necessarily show all of them at the same time. Some individuals may never show a particular characteristic, but others of its kind will. For example, some people never reproduce, but they are alive.

table 1.1

Alive or Not?

Characteristic of life	Human	Pine tree	Power lawn mower	Lumber	BBQ'd chicken	Pasteurized milk
1. Respond to stimuli		Yes				
2. Grow				No		
3. Undergo development	Yes					
4. Reproduce	Yes					
5. Manipulate matter		Yes				
6. Contain organic molecules					Yes	
7. Eliminate waste					No	
8. Use energy			Yes			
9. Show orderliness			Yes			
10. Contain cells						No
11. Show evolution						No

Pseudoscience

Many people seek to use the appearance of science to convince others to follow a particular course of action. This tactic can be called *pseudoscience*. **Pseudoscience** (*pseudo* = false) takes on the flavor of science but is not supportable as valid or reliable. Often, the purpose of pseudoscience is to confuse or mislead. The area of nutrition, for example, is flooded with pseudoscience. We all know that we must obtain certain nutrients, such as amino acids, vitamins, and minerals, from the food we eat or we may become ill. Many scientific experiments have been performed that reliably demonstrate the validity of this information. However, in most cases, it has not been demonstrated that the nutritional supplements so vigorously advertised are useful or even desirable. Rather, selected bits of scientific information (i.e., amino acids, vitamins, and minerals are essential to good health) have been used to foster a belief that additional amounts of these nutritional supplements are necessary or that they can improve health. In reality, the average person eating a varied diet will obtain all of these nutrients in adequate amounts; nutritional supplements are not required.

In addition, many of these products are labeled "organic" or "natural," implying that they have greater nutritive value because they are organically grown (grown without pesticides or synthetic fertilizers) or because they come from nature. The poisons curare, strychnine, and nicotine are all organic molecules produced in nature by plants that could be grown organically, but we wouldn't want to include them in our diet.

"Nine Out of Ten Doctors Surveyed Recommend Brand X" There are obviously many things wrong with this statement. First, is the person in the white coat really a doctor? Second, if only ten doctors were asked, the sample size is too small. Third, only selected doctors might have been asked to participate. Finally, the question could have been asked in such a way as to obtain the desired answer: "Would you recommend Brand X over Dr. Pete's snake oil?"

Limitations of Science

Science is a way of thinking and seeking and testing the validity of information in order to solve problems. Therefore, the scientific method can be applied only to questions that have a factual basis. Questions concerning morals, value judgments, social issues, and attitudes cannot be answered using the scientific method. What makes a painting great? What is friendship? Should North Americans recycle? These questions are related to values, beliefs, and tastes; therefore, the scientific method cannot be used to answer them.

Many people view science as a powerful tool that will provide answers to the major problems of our time. This is not necessarily true. Most of the problems we face are generated by the behavior and desires of people. Famine, drug abuse, and pollution are human-caused and must be resolved by humans. Science may provide some tools for social planners, politicians, and ethical thinkers, but science does not have, nor does it attempt to provide, all the answers to the problems of the human race. Science is merely one of the tools at our disposal.

BIO *feature*

The Fallibility of Science

Just because scientists say something is true does not necessarily make it true. Everyone makes mistakes, and quite often, as new information is gathered, old laws must be changed or discarded. For example, at one time scientists were sure that the sun revolved around the earth. They observed that the sun rose in the east and traveled across the sky to set in the west. Since scientists could not feel the earth moving, it seemed perfectly logical that the sun traveled around the earth. Once they understood that the earth rotates on its axis, they began to understand that the rising and setting of the sun could be explained in other ways. A completely new concept of the relationship between the sun and the earth developed.

Although this kind of study seems primitive to us today, this change in thinking about the sun and the earth was a crucial step in understanding the universe and how the various parts are related to one another. This information was built upon by many generations of astronomers and space scientists and finally led to space exploration.

In similar fashion, many years passed before people began to recognize that radiation could be dangerous to one's health. When people were exposed repeatedly to small amounts of radiation they did not experience any immediate effects or symptoms of their exposure. As a matter of fact, many people purposely exposed themselves to sources of radiation hoping to relieve symptoms of health problems such as arthritis and skin diseases. Only after years had passed did the effects show up as increased cancer rates in those who were exposed. Because of this information there are now strict regulations governing exposure to radiation.

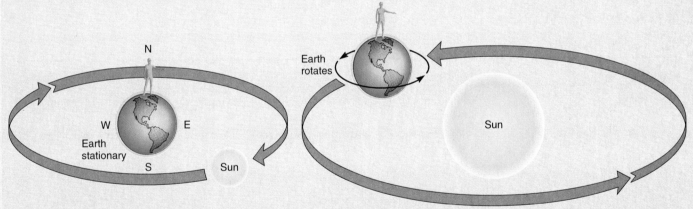

Scientists thought that the sun revolved around the earth.

We now know that the earth rotates on its axis and also revolves around the sun.

Science Must Be Willing to Challenge Previous Beliefs   Science always must be aware that new discoveries may force a reinterpretation of previously held beliefs. Early scientists thought that the sun revolved around the earth in a clockwise direction. This was certainly a reasonable theory at the time. Subsequently, we have learned that the earth revolves around the sun in a counterclockwise direction, at the same time rotating on its axis in a counterclockwise direction. It is this rotation of the earth on its axis that gives us the impression that the sun is moving.

BIO *feature*

Science and Nonscience

The differences between science and nonscience are often based on the assumptions and methods used to gather and organize information and, most importantly, the testing of these assumptions. The difference between a scientist and a nonscientist is that a scientist continually challenges and tests principles and laws, whereas a nonscientist may not feel this is important.

Once you understand the scientific method (table 1.2), you won't have any trouble identifying astronomy, chemistry, physics, and biology as sciences. But what about economics, sociology, anthropology, history, philosophy, and literature? All of these fields may make use of certain laws that are derived in a logical way, but they are also nonscientific in some ways. Some things are beyond science and cannot be approached using the scientific method. Art, literature, theology, and philosophy are rarely thought of as sciences. They are concerned with beauty, human emotion, and speculative thought rather than with facts and verifiable laws. On the other hand, physics, chemistry, geology, and biology are almost always considered sciences. Music is an area of study in a middle ground where scientific approaches may be used to some extent. "Good" music is certainly unrelated to science, but the study of how the human larynx generates the sound of a song is based on scientific principles. Any serious student of music will study the anatomy of the human voice box and how the vocal cords vibrate to generate sound waves. Similarly, economics makes use of mathematical models and established economic laws to make predictions about future economic conditions. Anthropology and sociology also have many aspects that are scientific in nature, but they cannot be considered true sciences because many of the generalizations they have developed cannot be tested by repeated experimentation.

table 1.2

The Scientific Method

Steps	Activity	Example
Observation	Recognize something has happened and that it occurs repeatedly. (Empirical evidence is gained from experience or observation.)	Students in a classroom are stricken with a disease that causes red rashes on their faces. This same situation has been described in several schools in your region. Skin cultures taken from the students indicate that there are some unusual bacteria present.
Question Formulation	Write many different kinds of questions about the observation and keep the ones that will be answerable.	Is the disease psychosomatic (i.e., is this a case of hysteria in which there is nothing organically wrong)? Is the rash caused by a bacterium? Is the disease caused by a virus?
Exploration of Alternative Resources	Go to the library to obtain information about this observation. Also, talk to others who are interested in the same problem.	A search of the medical literature reveals that physicians who used antibiotic X in similar circumstances reported cures, even though they never found a bacterium to be present. Attend scientific meetings where this disease outbreak will be discussed. Contact scientists who are reported to be interested in the same problem.

table 1.2 (continued)

The Scientific Method

Steps	Activity	Example
Hypothesis Formation	Pose a possible answer to your question. Be sure that it is testable and that it accounts for all the known information. Recognize that your hypothesis may be wrong.	Antibiotics do not usually affect viruses. Further, the disease has been reported elsewhere, which tends to rule out psychosomatic disease. Therefore, your hypothesis is that the disease is caused by a bacterium and that antibiotic X can cure the disease by controlling the rate of growth of the bacterial population.
Experimentation	Set up an experiment that will allow you to test your hypothesis using a control group and an experimental group. Be sure to collect and analyze the data carefully.	To test the cause-and-effect relationship between administering antibiotic X and curing the illness, you set up two groups. A control group will be given a placebo (a pill with no active ingredient). The experimental group will receive pills containing antibiotic X. The pills will look identical and will be coded so that neither the person receiving the pill nor the person administering the pill will know which individuals receive the medication and which receive the placebo. After five days you collect the data and find that 90% of those receiving antibiotic X no longer have the rash. By contrast, only 10% of those receiving the placebo have recovered. You conclude that the disease is not psychosomatic and that a bacterium is probably the cause. You publish your results and others in the country report back that they have had similar results.
Theory Formation	Repeat the experiment and share the information with others over a long period of time. Should your information continue to be considered valid and consistent with other closely related research, the scientific community will recognize that a theory has been established or that your information is consistent with existing theories.	Your results support the generally held theory that many kinds of diseases are caused by microorganisms. This generalization is called the *germ theory of disease.*
Law Formation	If your findings are seen to fit with many other major blocks of information that tie together many different kinds of scientific information, it will be recognized by the scientific community as being consistent with current scientific laws. If it is a major new finding, a new law may be formulated.	Your experimental results are consistent with the *biogenetic law* that states that all living things come from previously living things. Your results strongly suggest that the disease was caused by the multiplication of certain bacteria, and that the antibiotic stopped their multiplication.

SUMMARY

In science, our understanding of the natural world and how it operates is usually arrived at by using the scientific method. This process frequently involves observing, questioning, exploring resources, forming hypotheses, and experimenting. The results of the experimentation may allow us to develop a scientific theory or law. This process is an attempt to show cause-and-effect relationships. For example, scientists have used this process to explain the origin of life. When a cause-and-effect relationship is established, it will have predictive value. If, however, a belief concerning a natural event cannot be tested, understanding the event is outside the realm of science. There are many ways of investigating natural events, some scientific and some not. Pseudoscience uses scientific appearances to mislead. The science of biology is the study of living things and how they interact with their surroundings. Living things show the characteristics of (1) response to a stimulus, (2) growth, (3) development, (4) reproduction, (5) manipulation of matter, (6) organic molecules, (7) waste elimination, (8) manipulation of energy, (9) orderliness, (10) cellular structure, and (11) evolution (table 1.3).

table 1.3

A Summary of the Characteristics of Life

Characteristic	Definition
1. Response to a stimulus	Ability of an organism to sense changes and react to them
2. Growth	Increase in size or number of cells
3. Development	Change in form over time
4. Reproduction	Ability of an organism to make copies of itself
5. Manipulate matter	Ability of an organism to rearrange molecular structures
6. Contains organic molecules	Composed of carbon-based molecules
7. Waste elimination	Mechanisms for ridding the organism of unusable or toxic materials produced by the organism
8. Manipulate energy	Ability of an organism to transform and utilize energy to maintain itself
9. Orderliness	Organization of parts into a coordinated whole
10. Cells	Composed of a membrane-bound sac that is able to manipulate matter and energy
11. Evolution	Change in the kinds of living organisms on earth over long periods of time in response to environmental changes

CHAPTER GLOSSARY

assumption (e-sump'shun) A speculation; something you think might be true.

biogenesis (bi-o-jen'uh-sis) The concept that life originates only from preexisting life.

cell (sel) A characteristic of life; the basic structural unit that makes up all living things.

control group (kon-trōl' grūp) The situation used as the basis for comparison in a controlled experiment.

controlled experiment (kon-trōld' ik-sper' e-ment) An experiment that allows for a comparison of two events that are identical in all but one respect.

development (di-ve'lep-ment) A characteristic of life; change in form over time.

evolution (ĕv-o-lu'shun) A characteristic of life; the genetic adaptation of a population of organisms to its environment.

experimental group (ik-sper"e-men'tl grup) The group in a controlled experiment that is identical to the control group in all respects but one.

fact (fakt) Something that is true.

growth (grōwth) A characteristic of life; an increase in size or number of cells.

guess (ges) To form an opinion without supporting evidence.

hypothesis (hy-pa'the-ses) A possible answer to, or explanation of, a question that accounts for all the observed facts and is testable.

irrelevant (i-re'le-vent) Not related to the situation under discussion.

manipulate energy (ma-ni'pu-late e'ner-jee) A characteristic of life; ability of an organism to transform or utilize energy.

manipulate matter (ma-ni'pu-late ma'ter) A characteristic of life; the ability of an organism to rearrange molecular structures.

metabolism 20(me-ta-'bol-izm) The total of all chemical reactions within an organism; for example, nutrient uptake and processing and waste elimination.

opinion (e-pin'yin) A thought, idea, or belief about something.

orderliness (or'der-lee-nes) A characteristic of life; organization of parts into a coordinated whole.

organic molecules (or-gan'ik mol'uh-kiuls) Complex molecules whose basic building blocks are carbon atoms in chains or rings.

pseudoscience (su-dō-si'ens) The use of the appearance of science to mislead. The assertions made are not valid or reliable.

relevant (re′le-vent) Pertaining to the matter at hand.

reproduction (re″pro-duk′shun) A characteristic of life; ability of an organism to make copies of itself.

response (re-spons′) A characteristic of life; the reaction of an organism to a stimulus.

science (si′ens) A process of arriving at a solution to a problem or understanding an event in nature that involves testing possible solutions.

scientific law (si-en-ti′fik law) A uniform or constant fact of nature.

scientific method (si-en-ti′fik meth′ud) A way of gaining information about the world by forming possible solutions to questions, followed by rigorous testing to determine if proposed solutions are valid.

spontaneous generation (spon-ta′ne-us jen-uh-ra′shun) The theory that living organisms arise from nonliving material.

stimulus (stim′yu-lus) Some change in the internal or external environment of an organism that causes it to react.

theory (thee′ah-ree) A plausible, scientifically acceptable generalization supported by several hypotheses and experimental trials.

valid (val′id) A term used to describe meaningful data that fit into the framework of scientific knowledge.

waste elimination (wayst i-li″me-nay′shun) A characteristic of life; mechanism for ridding the organism of unusable or toxic materials produced by the organism.

CONCEPT MAP TERMINOLOGY

At the end of each chapter, you will be asked to construct a *concept map* to show relationships among a list of terms. An example of a concept map of terms from chapter 1 is provided below. In a paragraph, explain what this diagram represents to you.

control group
controlled experiment
experimental group
hypothesis
question
results
theory

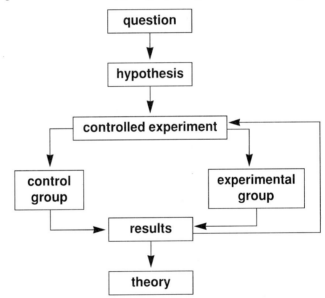

LABEL•DIAGRAM•EXPLAIN

Describe the controlled experiment you would use to determine a cause-and-effect relationship between caffeine and cancer in laboratory rats.

Multiple Choice Questions

1. On a cold winter night, while riding a bus through the city, you observe a man hunched over and shaking in an alley. The passenger next to you says, "The man is cold." This statement expresses a(n):
   a. fact
   b. assumption
   c. theory
   d. observation

2. While driving to school, your car begins to drag and your vehicle is pulled toward the right. You recall a similar experience while riding with your aunt when she had a tire blowout. You also remember learning in drivers education that a blowout of the right front tire will pull the car to the right, while a blowout of the left front tire will pull the car to the left. Based on this, you suspect that your right front tire has just blown and you pull off the road. Your belief that you have a flat is best described as a:
   a. fact
   b. guess
   c. hypothesis
   d. theory

3. A scientifically acceptable generalization supported by several experimental trials is a:
   a. fact
   b. guess
   c. hypothesis
   d. theory

4. In a controlled experiment, the experimental group:
   a. should be identical to the control group
   b. should differ from the control group by one factor
   c. may differ from the control group by several factors
   d. is not necessary if the experiment contains a control group

5. The concept that life originates only from preexisting life is:
   a. biogenesis
   b. spontaneous generation
   c. the germ theory
   d. pseudoscience

6. Which one of the following individuals supported the theory of spontaneous generation?
   a. Francesco Redi
   b. John Needham
   c. Lazzaro Spallanzani
   d. Louis Pasteur

7. The basic structural units of all living things are:
   a. organs
   b. tissues
   c. blood and bones
   d. cells

8. What would be an appropriate first step in studying a cause-and-effect relationship?
   a. conducting an experiment
   b. forming a hypothesis
   c. forming a theory
   d. clarifying the question

9. Pseudoscience:
   a. uses science to solve practical problems
   b. is the purest form of science
   c. has no basis in fact
   d. may alter scientific facts to deceive

10. In an experiment, the group used as a basis of comparison is the:
    a. control group
    b. experimental group
    c. hypothesis group
    d. irrelevant group

Questions with Short Answers

1. List and describe the characteristics of life.

2. List three problems that science cannot solve.

3. Point out pseudoscience aspects of each of the following:
   a. astrology

   b. health food store claims

   c. UFO societies' claims

   d. TV advertisements

4. Describe two ways in which a hypothesis differs from a guess.

5. Distinguish between relevant fact, irrelevant fact, opinion, and assumption.

chapter

2

The Classification of Organisms

West African mandrill.

The Classification of Organisms

learning objectives

- Know the importance of Linnaeus's work.
- Know the categories used in the science of classification.
- Explain the benefits of a system of classification.
- Identify the basic characteristics of a cell.
- Know how prokaryotic cells and eukaryotic cells differ.
- Know the five kingdoms.
- List several characteristics for each kingdom.
- Explain how eukaryotic single-celled organisms could have evolved into multicellular forms.
- Describe the structure and life cycle of a virus.

Classification: A Common Event

Classification, the sorting of objects into groups based on their similarities or differences, is a common human activity. We practice classification every day as we organize objects in our surroundings. For example, most people store materials commonly used in the kitchen (dishes, utensils, spices, food) in their kitchen cabinets, whereas they store personal hygiene items in bathroom cabinets. This seems so logical that we don't even think about it. Yet when you place an item in either your kitchen or bathroom cabinet, you have essentially classified it as a "kitchen thing" or a "bathroom thing." Furthermore, if you store your eating utensils in a drawer separate from other kitchen items, you have created a more specific classification category of "eating utensils." If you separate your spoons from your forks and knives, you are further classifying your eating utensils into groups.

Consider a visit to a shopping mall to purchase the latest compact disc by your favorite music group. The mall will most likely have many shops for you to choose from. There may be shops specializing in women's clothing and shops specializing in men's clothing, a jewelry store, a bookstore, a toy store, and a music store, among others. Inside the music store there may be an area containing tapes and a second area of CDs. Each of these will most likely be further subdivided into sections according to the type of music: rock, rap, country, blues, jazz, gospel, classical. Within each of these groups the music will be arranged according to artist and then by individual recording titles. To locate the CD you want, you will first need to place it in a general category (music recording) and then in a series of more specific categories (recording format, musical style, artist, title). Such systematic arrangement from general groups to more specific groups is called *hierarchical classification.*

Taxonomic Hierarchy

Scientists use classification to show relationships among all types of naturally occurring things, such as the chemical elements that make up our physical world, types of rocks, and kinds of stars. Biologists use classification to show the relationships between living things and to bring order to the over 1.5 million types of organisms that have been identified on earth.

Classification includes two contrasting processes: lumping and splitting. Lumping involves finding similarities among organisms and grouping them together based on these similarities; splitting involves placing organisms into separate categories based on their differences. The science responsible for the lumping and splitting of organisms into logical categories is **taxonomy.**

The modern system of taxonomic classification was developed by Carolus Linnaeus (1707–1778), a Swedish doctor and botanist. Linnaeus recognized a need for placing organisms into groups by a hierarchical classification system (figure 2.1). This system divides all forms of life into **kingdoms,** the largest grouping used in the classification of organisms. Originally there were two kingdoms, Plantae and Animalia, but as an understanding of living things grew, it became apparent that a more complex classification system was needed. Today, many biologists recognize five kingdoms of life: animals, plants, fungi, protists, and bacteria. Several other classification

Figure 2.1 Carolus Linnaeus (1707–1778)
Linnaeus, a Swedish doctor and botanist, originated the modern system of naming organisms. He used binomial (two name) nomenclature that included the genus name and the species name.

Kingdom Animalia

jellyfish   earthworm   dog   house cat   human   sponge   snake   frog   insect   tapeworm   snail   lion   lynx

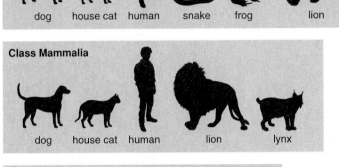

Phylum Chordata

dog   house cat   human   snake   frog   lion   lynx

Class Mammalia

dog   house cat   human   lion   lynx

Order Carnivora

dog   house cat   lion   lynx

Family Felidae

house cat   lion   lynx

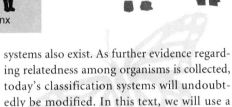

Genus *Felis*

house cat   lynx

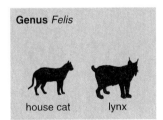

Species *domestica*

house cat

Figure 2.2 Animal phylogeny.

systems also exist. As further evidence regarding relatedness among organisms is collected, today's classification systems will undoubtedly be modified. In this text, we will use a five-kingdom classification.

Each kingdom is divided into smaller and smaller groups until each type of organism is placed in a unique category. The classification categories from most general to most specific are *kingdom, phylum, class, order, family, genus,* and *species* (figure 2.2).

CONCEPT CONNECTIONS

mnemonic devices   A *mnemonic device* is an aid, such as a rhyme or a formula, that is used to trigger the memory. For example, to remember that the classification categories in order from largest to smallest are kingdom, phylum, class, order, family, genus, and species, learn the sentence "<u>K</u>ing <u>P</u>hilip <u>C</u>ame <u>o</u>ver <u>f</u>rom <u>G</u>reat <u>S</u>pain." The first letter in each word of this sentence is the same as the first letter in each of the classification categories. You can create your own mnemonic devices for learning other material found in this text.

Figure 2.3 (a) *Nymphaea odorata,* (b) *Homo sapiens,* and (c) *Felis concolor.*

The taxonomic subdivision after kingdom is usually called a **phylum** (plural *phyla*). In order to determine an organism's place in one of these phyla, the organism is investigated to determine similarities and differences between it and other organisms within the kingdom. An example of a phylum within the animal kingdom is Chordata, which contains all of the animals with backbones. A **class** is a grouping of organisms with similar characteristics within a phylum. For example, within the phylum Chordata there are seven classes: mammals, birds, reptiles, amphibians, and three classes of fishes. An **order** is a grouping within a class. Carnivora is an order of meat-eating animals within the class Mammalia. A **family** consists of a group of closely related organisms within an order. The cat family is a subgrouping of the order Carnivora. Likewise, each family is divided into

smaller groups known as **genera** (sing. *genus*) and each genus is broken down into individual **species.**

A species is a group of organisms of the same type. For example, all humans belong to the same species. Humans are classified into the kingdom Animalia, phylum Chordata, class Mammalia, order Primata (primates), family Hominidae, genus *Homo,* and species *sapiens.*

Linneaus also recognized a need for standardizing the names of organisms. A single species may have several common names worldwide and may even have different names within different regions of a single country. For example, puma, cougar, mountain cat, and mountain lion are all common names for the same species of cat. To eliminate confusion, Linnaeus introduced the **binomial** (two-name) **system of nomenclature.** This system uses two Latin names, genus and species, for

each type of organism. In order to clearly identify the scientific name, binomial names are either *italicized* or underlined. The first letter of the genus name is capitalized. The species name is always written in lowercase. *Felis concolor* is the binomial name for the mountain lion and *Homo sapiens* is the binomial name for humans. When Linnaeus's binomial method was adopted worldwide, it eliminated the confusion that had resulted from using common local names. For example, in the binomial system the white water lily is known as *Nymphaea odorata.* Regardless of which of the 245 common names is used in a botanist's local area, when botanists read *Nymphaea odorata* they know exactly which plant is being referred to (figure 2.3). The binomial name cannot be changed unless there is compelling evidence to justify doing so. The rules that govern the worldwide classification of

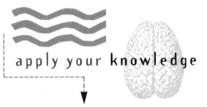

The Modern Classification System

	House Cat	Domestic Dog	Human
Kingdom	Animalia	Animalia	Animalia
Phylum (Division)	Chordata	Chordata	Chordata
Class	Mammalia	Mammalia	Mammalia
Order	Carnivora	Carnivora	Primates
Family	Felidae	Canidae	Hominidae
Genus	Felis	Canis	Homo
Species	domestica	familiaris	sapiens

species are expressed in the International Rules for Botanical Nomenclature, the International Rules for Zoological Nomenclature, and the International Bacteriological Code of Nomenclature.

Tools of Classification

Evolutionary theory states that species existing today developed from species that lived in the past, and that differences between present-day organisms and their ancestors are the result of adaptations to the environment over generations of time. A taxonomic ranking should reflect this evolutionary relationship among the organisms being classified. But just as two cooks may disagree on how a kitchen should be organized, there is no complete agreement as to how organisms are classified or how they are related. People arrive at somewhat different conclusions because they use different kinds of evidence or interpret this evidence differently. Scientists use several lines of evidence to develop evolutionary histories and classify organisms: fossils, comparative anatomy, life-cycle information, and biochemical or molecular evidence.

Fossils

Fossils are physical evidence of previously existing life. They are sometimes preserved intact. For example, mammoths have been found frozen in glaciers, and bacteria and insects have been preserved after becoming embedded in plant resins called amber. Other fossils are only parts of once-living organisms. The outlines or shapes of extinct plant leaves are often found in coal deposits, and individual animal bones are chemically altered (figure 2.4). Animal tracks have also been discovered in the ancient dried mud of river

apply your **knowledge**

Using a Dichotomous Key

A valuable tool for students interested in taxonomy is a dichotomous key. Dichotomous keys help to correctly identify organisms. There are a great variety of dichotomous keys, including keys for fish, plants, reptiles, insects, and all other forms of life.

All dichotomous keys are constructed on the same basis. To identify a species properly, the user is given a series of (usually two) choices. Most libraries have copies of these keys. Pick a group that interests you, such as plants, insects, or fish. Then obtain a specimen that is new to you, follow the directions in the key, and key the organism out to its genus and species.

Use the following key to identify this object.

Key to Common Methods of Travel

1a. No motor, go to 2
1b. Motor, go to 8
2a. Two wheels or fewer, go to 3
2b. Three wheels or more, go to 5
3a. One wheel = unicycle
3b. Two wheels, go to 4
4a. Pedals = bicycle
4b. No pedals = scooter
5a. Three wheels = tricycle
5b. Four wheels, go to 6
6a. Usually 30 cm in length or under
    = roller skate
6b. Usually longer than 30 cm, go to 7
7a. No front handle = skate board
7b. Front handle = wagon
8a. Two wheels = motorcycle
8b. Four wheels = automobile

(a)

(b)

(c)

(d)

Figure 2.4 Fossils Fossils are the remains of organisms, or the evidences that organisms existed in the past. Shown are (a) a fossilized plant leaf, (b) some animal tracks, (c) a whole insect preserved in amber, and (d) some bones of a dinosaur.

A 5,000-year-old Human Fossil

When organisms die, they are rarely preserved as whole, unaltered fossils because such fossils require specific conditions that inhibit decay and physical damage. Whole fossils can be preserved by freezing in ice, dehydration in deserts, acidification in bogs, and entrapment in amber and lava. In 1991, a 5,000-year-old human corpse was discovered by a couple hiking in the Alps along the border of Italy and Austria. The Copper Age male, who had apparently died in a storm, was preserved in a protected pocket of ice that shielded his body from the crushing movement of glaciers. The five-foot, two-inch corpse was almost perfectly preserved with eyes still present and tattoos visible. The only serious damage to the corpse was caused by an overzealous policeman who removed it from the ice. Along with the remains of the man were found scraps of clothing, several tools, and other artifacts that provide clues to his life and culture.

5,000-year-old human fossil

Figure 2.5 Determining the Age of Fossils New layers of sedimentary rock are formed on top of older layers, making it possible to determine the relative ages of fossils found in the various layers. The layers of rock shown here represent hundreds of millions of years of formation. The fossils in the lower layers are millions of years older than the fossils in the upper layers.

beds. It is important to understand that some organisms are more easily fossilized than others. Those that have hard parts, such as skeletons and shells, are more likely to be preserved than are tiny, soft-bodied organisms. Aquatic organisms are much more likely to be buried and fossilized in sediments at the bottom of oceans or lakes than are land animals to be fossilized on land.

Evidence obtained from the discovery and study of fossils allows biologists to place organisms in a time sequence. This can be accomplished by comparing one type of fossil with another. As geological time passes, new layers of sediment are laid down, so older organisms should be found in deeper layers, providing the sequence of layers has not been disturbed. A comparison of the layers indicates the relative age of the fossils found in the rocks. Fossils found in the same layer must have been alive during the same geological period. Even though two fossils may have obvious structural differences, the fact that they are next to one another provides support for their coexistence.

Rocks and fossils can be placed in more specific age ranges by comparing the amounts of certain radioactive materials they contain. By measuring the proportion of carbon-14 in a specimen, fossils less than 60,000 years old can be dated; potassium-40 is used for older specimens. Older sediment layers have less of these radioactive materials than do younger layers (figure 2.5).

Subtle changes in the fossils of organisms over time can be better understood by studying the complete fossil record. For example, the size of the leaf of a certain fossilized plant has been found to change extensively through long geological periods. A comparison of the extremes, the oldest with the newest, would lead to their classification in different categories. However, the fossil links between the extremes clearly show that the younger plant is a descendent of the older (figure 2.6).

Comparative Anatomy

Comparing the anatomy of fossils or currently living organisms can be useful in developing family trees. Since most of the

Figure 2.6 **Fossil Evidences of Changes Over Time** Subtle changes in the nature of the kinds of fossils over time can be better understood by studying the complete fossil record. For example, the size and shape of the leaves of fossil plants has been found to change through long geological periods. The plant, *Ginkgo biloba* (top), was at one time only known from fossil records. Subsequently some specimens of the tree were found living in China. The fossil in the bottom photo is an example of a closely related species, *Ginkgo huttoni,* from the Jurassic period.

BIO *feature*

What Is Carbon Dating?

Carbon is an element that occurs naturally in several forms. The most common form is carbon-12. A second, heavier radioactive form, carbon-14, is constantly being produced in the atmosphere by cosmic rays. Radioactive elements break down into other forms of matter. Hence, radioactive carbon-14 naturally decays. The rate at which carbon-14 is produced is about equal to the rate at which it decays. Therefore, the concentration of carbon-14 on earth stays relatively constant. All living things contain large quantities of the element carbon. Plants take in carbon in the form of carbon dioxide from the atmosphere, and animals obtain carbon from the food they eat. While an organism is alive the proportion of carbon-14 to carbon-12 within its body is equal to its surroundings. When an organism dies, the carbon-14 within its tissues decays, but no new carbon-14 is added. Therefore, the age of plant and animal remains can be determined by their ratio of carbon-14 to carbon-12. The less carbon-14 present, the older the specimen. Radioactive decay rates are measured in half-lives. One half-life is the amount of time it takes for one-half of a radioactive sample to decay.

The half-life of carbon-14 is 5,730 years. Therefore, a bone containing one-half the normal proportion of carbon-14 is 5,730 years old. If the bone contains one-quarter of the normal proportion of carbon-14, it is $2 \times 5,730 = 11,460$ years old, and if it contains one-eighth of the naturally occurring proportion of carbon-14, it is $3 \times 5,730 = 17,190$ years old.

Developmental Biology

Another line of evidence useful in classifying organisms comes from the field of developmental biology. Many organisms have complex life cycles that include several completely different forms or stages. After fertilization, some organisms grow into free-living developmental stages that do not resemble the adults of their species. These are called *larvae* (singular, *larva*). Larval stages often provide clues to the relatedness of organisms. For example, barnacles live attached to rocks and other solid marine objects and look like small, hard cones. Their outward appearance does not suggest that they are related to shrimp; however, the larval stages of barnacles and shrimp are very similar. Detailed anatomical studies of barnacles confirm that they share many structures with shrimp; their outward adult appearance tends to be misleading (figure 2.7). This same kind of evidence is available in the plant kingdom. Many kinds of plants, such as peas, peanuts, and lima beans, produce large, two-parted seeds in pods (you can easily split the seeds into two parts). Even though peas grow as vines, lima beans grow as bushes, and peanuts have their seeds underground, all of these plants are considered to be related (figure 2.8).

DNA Analysis

Like all fields in biology, the science of taxonomy constantly changes as new techniques develop. Recent advances in DNA analysis are being extensively used to determine similarities among species. **DNA** (deoxyribonucleic acid) is the chemical blueprint that determines an organism's characteristics. By comparing DNA, or the chemicals produced from the DNA code, biologists can detect similarities between organisms that are not expressed in physical anatomy. For example, studies of comparative anatomy at one time led researchers to believe that gorillas and chimpanzees were close relatives while humans were more distantly related to the apes. Recent DNA analysis suggests that humans and chimpanzees are actually the more closely related organisms, and gorillas are the distant relative.

These four kinds of evidence (fossils, comparative anatomy, developmental stages, and biochemical evidence) have been used to

(a)

(b)

Figure 2.7 Developmental Biology   The adult barnacle (a) and shrimp (b) are very different from each other, but their early larval stages look alike.

structures of an organism are inherited, organisms having similar structures are thought to be related. Plants, for example, can be divided into two categories: All plants that have flowers are thought to be more closely related to one another than to plants like ferns or evergreens that do not have flowers. In the animal kingdom, all organisms that nurse their young from mammary glands are grouped together, and all animals in the bird category have feathers and beaks and lay eggs with shells. Reptiles also have shelled eggs but differ from birds in that they lack feathers and have scales covering their bodies. The fact that these two groups share this fundamental eggshell characteristic implies that they are more closely related to each other than they are to other groups.

(b)

(c)

(a)

Figure 2.8 **Members of the Bean Family** All of these plants are considered to be related to one another because of the way the seed and pod are constructed. (*a*) Bean, (*b*) peanut, and (*c*) locust.

develop the various taxonomic categories, including kingdoms. Using all these sources, biologists have developed a hypothetical picture of how all organisms are related (figure 2.9). In the present-day science of taxonomy, each organism that has been classified has its own unique binomial name, genus and species. In turn, it is assigned to larger groupings that have a common evolutionary history.

Cells

Taxonomists often rely on the nature of cells to better define how organisms should be categorized. The **cell** is the basic unit of life. All living things are composed of cells and can be distinguished from nonliving things by their cellular structure. Organisms can also be grouped based on their cell structure, how their cells obtain energy, and whether they are *unicellular* (consist of a single cell) or *multicellular* (composed of many cells). Therefore, cells are an essential component in the classification of living things.

Most cells are extremely small and cannot be viewed with the unaided eye. However, if you view a typical cell under a light microscope, you will see a large central structure called the **nucleus.** The nucleus is sometimes called "the control center of the cell" because structures within the nucleus control the cell's activities. The nucleus consists of *chromatin, nucleoli,* and *nucleoplasm* surrounded by an outer boundary called the *nuclear membrane* (figure 2.10).

Outside the nucleus is a watery substance called **cytoplasm.** The cytoplasm contains many dissolved molecules and specific identifiable structures called **organelles** (*elle* = little; *organelle* = little organ). An organelle is to a cell what an organ is to an entire organism. Within your body, you have a heart, lungs, a liver, kidneys, and many other distinguishable organs that perform specific functions. Likewise, a cell contains organelles with unique structures that have particular jobs to accomplish.

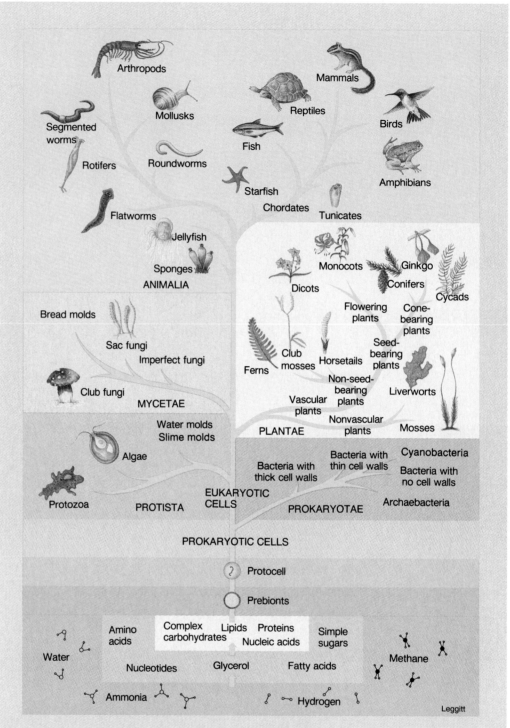

Figure 2.9 Molecules to Organisms The theory of spontaneous generation proposes that the molecules in the early atmosphere and early oceans accumulated to form nonliving structures composed of organic molecules. The nonliving structures are believed to be the forerunners of the first living cells. These first cells probably evolved into prokaryotic cells, on which the kingdom Prokaryotae is based. Some prokaryotic cells probably gave rise to eukaryotic cells. The organisms formed from these early eukaryotic cells were members of the kingdom Protista. Members of this kingdom evolved into the kingdoms Animalia, Plantae, and Mycetae.

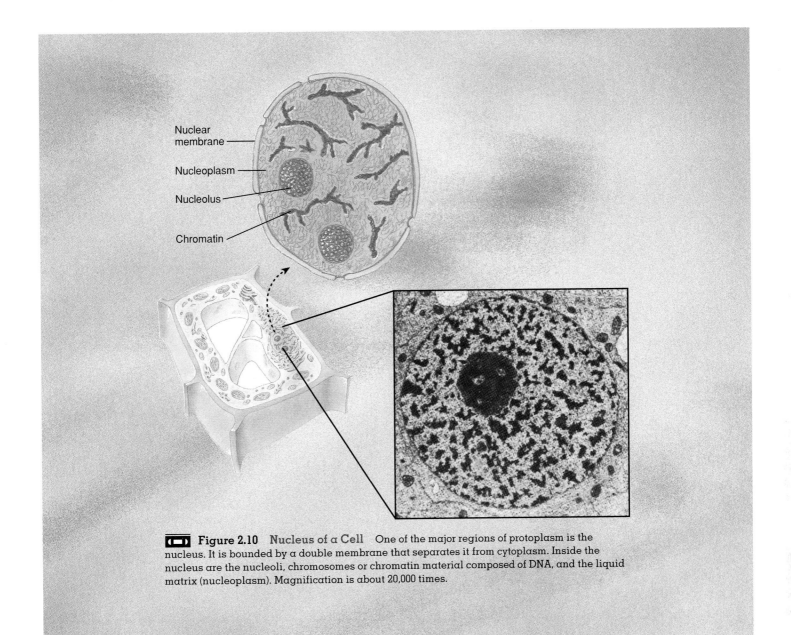

Figure 2.10 Nucleus of a Cell One of the major regions of protoplasm is the nucleus. It is bounded by a double membrane that separates it from cytoplasm. Inside the nucleus are the nucleoli, chromosomes or chromatin material composed of DNA, and the liquid matrix (nucleoplasm). Magnification is about 20,000 times.

A division of labor exists among the organelles of a cell in much the same way it exists in a manufacturing plant. A plant producing computers, for example, needs to bring into the plant the raw materials required for computer production. The computers need to be assembled, packaged, and distributed to merchants. These activities require energy generation and the disposal of waste products. Management oversees production based on decisions made by the board of directors. Although the factory contains many departments, they all work toward the same goal: to produce the best, most cost-effective computer possible. This goal can be met only if all jobs are performed properly.

CONCEPT CONNECTIONS

Components of a Nucleus

- **Chromatin** is made up of DNA and protein molecules. The DNA contains instructions for the growth and maintenance of the cell. During cell division, chromatin coils into dense structures called **chromosomes.** Chromatin and chromosomes are really the same molecules but differ in structural arrangement.

- **Nucleoli** (singular, *nucleolus*) are regions of the chromatin that appear as dark dense areas when the cell is stained. Some cells contain more than one nucleolus (nucleoli). Nucleoli produce molecules that are the building blocks of *ribosomes*. **Ribosomes** that are produced as a result of nucleolar activities in turn produce protein molecules for the cell.

- **Nucleoplasm** is a fluid mixture of water and molecules within the nucleus. The nucleoplasm surrounds the chromatin and provides raw materials necessary for the DNA to function.

- The **nuclear membrane** is composed of two thin sheets that surround the contents of the nucleus. It has large pores that are very selective in allowing materials to pass across the membrane. As a result, a cell maintains a chemical environment inside the nucleus that differs greatly from the chemical environment outside the nucleus.

Table 2.1

Comparison of the Structure and Function of the Cellular Organelles

Organelle	Type of Cell	Structure	Function
Plasma membrane	Prokaryotic Eukaryotic	Typical membrane structure; phospholipid and protein present	Controls passage of some materials to and from the environment of the cell
Granules	Prokaryotic Eukaryotic	Too small to ascribe a specific structure	May have a variety of functions
Chromatin material	Prokaryotic Eukaryotic	Composed of DNA and protein in eukaryotes, but just DNA in prokaryotes	Contains the hereditary information that the cell uses in its day-to-day life and passes it on to the next generation of cells
Ribosome	Prokaryotic Eukaryotic	Protein and RNA structure	Site of protein synthesis
Microtubules	Eukaryotic	Hollow tubes composed of protein	Provide structural support and allow for movement
Microfilament	Eukaryotic	Protein	Provide structural support and allow for movement
Nuclear membrane	Eukaryotic	Typical membrane structure	Separates the nucleus from the cytoplasm
Nucleolus	Eukaryotic	Group of RNA molecules and genes located in the nucleus	Site of ribosome manufacture and storage
Endoplasmic reticulum	Eukaryotic	Folds of membrane forming sheets and canals	Surface for chemical reactions and intracellular transport system
Golgi apparatus	Eukaryotic	Membranous stacks	Associated with the production of secretions and enzyme activation
Vacuoles	Eukaryotic	Membranous sacs	Containers of materials
Lysosome	Eukaryotic	Membranous container	Isolates very strong enzymes from the rest of the cell
Mitochondria	Eukaryotic	Large membrane folded inside of a smaller membrane	Associated with the release of energy from food; site of cellular respiration
Chloroplast	Eukaryotic	Double membranous container of chlorophyll	Site of photosynthesis or food production in green plants
Centriole	Eukaryotic	Microtubular	Associated with cell division
Contractile vacuole	Eukaryotic	Membranous container	Expels excess water
Cilia and flagella	Prokaryotic Eukaryotic	9 + 2 tubulin in eukaryotes; different structure in prokaryotes	Movement

A cell is a factory whose products are the chemicals required to sustain life. All the functions required to produce computers are also necessary to produce the molecules of life. These jobs are accomplished by organelles. Some organelles produce molecules while others package and transport chemicals. Cells have organelles that provide them with energy and structures that dispose of wastes. These processes are managed by molecules called enzymes and are ultimately controlled by DNA. Table 2.1 contains a list of the major cellular organelles and describes their structure and function. Most of these structures are too small to view with a typical light microscope (figure 2.11).

If the computer factory loses its source of electricity, it will immediately shut down. The same is true for cells, which require a constant input of energy. Some organisms are capable of converting simple inorganic molecules into complex energy-containing organic molecules. Most of these organisms, called **autotrophs,** utilize sun energy for this process in a series of reactions called **photosynthesis.** In plants, this energy conversion occurs in organelles called **chloroplasts.** Chloroplasts are composed of thin membrane sheets and contain the green pigment **chlorophyll.**

A second energy-converting organelle is the **mitochondrion.** Mitochondria (plural) also contain thin membrane sheets. Inner membranes of mitochondria form folds and are surrounded by a smooth outer membrane. Mitochondria convert chemical energy into a form of energy that is more readily available to the cell. This process of releasing chemical energy is called **aerobic cellular respiration.** In autotrophs, the chemical energy required for aerobic cellular respiration is produced by photosynthesis. In animals, this energy is provided by food molecules. Organisms like yourself that obtain energy from eating other organisms are called **heterotrophs** (figure 2.12).

All cells are surrounded by a membrane called the **plasma membrane.** This

CONCEPT CONNECTIONS

autotroph: *auto* = self; *troph* = nourisher

heterotroph: *hetero* = other; *troph* = nourisher

Organelles Located in the Cytoplasm

- **Centrioles** are composed of protein tubes called **microtubules** and function in cell division. One curious fact about centrioles is that they are present in most animal cells but not in many plant cells.

- **Chloroplasts** are energy-converting, membranous, saclike organelles in plant cells: the site of photosynthesis. Inside chloroplasts are stacks of membrane sacs that contain chlorophyll and are called **grana.** The space between the grana is called the **stroma.**

- The **endoplasmic reticulum** (ER) is a system of folded membranes and tubes found throughout the cell that provide a large surface upon which chemical activities take place. There are two different types of endoplasmic reticulum—rough ER and smooth ER. The rough ER appears rough because it has **ribosomes** on its surface. Proteins produced by ribosomes enter the endoplasmic reticulum's canal system where they are packaged and transported through the cell. The smooth ER lacks ribosomes but is the site of many other important activities, including fat metabolism and detoxification.

- The **Golgi apparatus** consists of a stack of flattened, smooth, membranous sacs. This organelle is involved in the storage, modification, and packaging of molecules to be secreted from the cell. Mucus, insulin, and digestive enzymes (molecules that aid in the breakdown of other large molecules), for example, are concentrated inside the Golgi and transported to the outside of the cell by tiny vesicles that bud off this organelle.

- **Lysosomes** are specialized organelles that are produced by the Golgi and contain a mixture of digestive enzymes. The enzyme molecules contained within the lysosomes are used by cells to break down worn-out cells or cell parts, to break down digested food particles, and to destroy disease-causing microorganisms.

- **Microfilaments** are long fiberlike structures made of protein , often in close association with the microtubules. Microfilaments are responsible for the flowing movement of the cytoplasm.

- **Microtubules** are small, hollow tubes of protein that function throughout the cytoplasm to provide structural support and enable movement. **Cilia, flagella,** and **centrioles** are composed of microtubules.

- **Mitochondria** are organelles composed of membrane that serve as the site of aerobic cellular respiration. Extensive surface area for chemical reactions is provided by an inner membrane with folded projections called **cristae.**

- **Ribosomes** are small structures that provide a site for the production of protein.

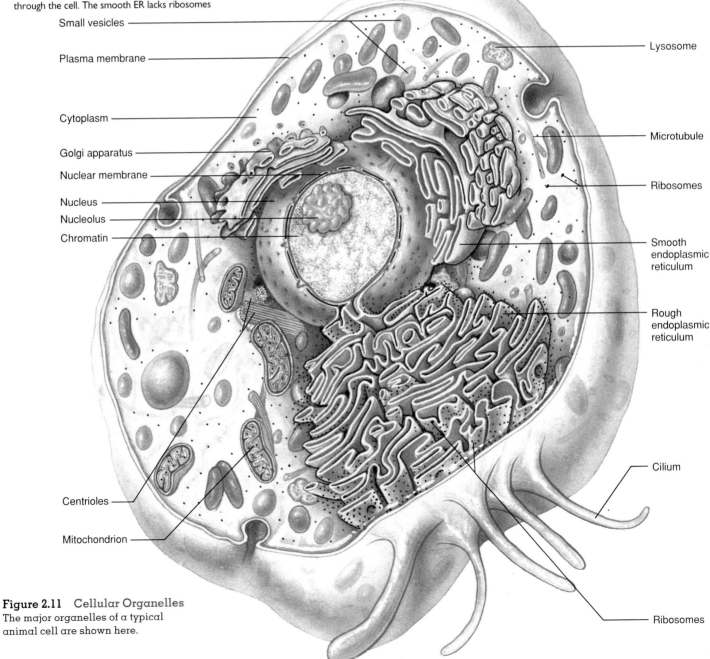

Small vesicles

Plasma membrane

Cytoplasm

Golgi apparatus

Nuclear membrane

Nucleus

Nucleolus

Chromatin

Centrioles

Mitochondrion

Lysosome

Microtubule

Ribosomes

Smooth endoplasmic reticulum

Rough endoplasmic reticulum

Cilium

Ribosomes

Figure 2.11   Cellular Organelles
The major organelles of a typical animal cell are shown here.

Figure 2.12 **Autotrophs and Heterotrophs** The way in which an organism gets its food is important in distinguishing plants and animals. Plants are autotrophs: they can produce their food by photosynthesis in their chloroplasts. Animals are heterotrophs: they must eat other things for their food. Both of these types of organisms release energy from food by aerobic cellular respiration.

structure is sometimes called the "gate-keeper of the cell" because it is the responsibility of the plasma membrane to determine what substances may enter the cell and what substances may exit. It is important to realize that cells are no more isolated from their surroundings than the whole multicellular organism is isolated from its surroundings. In the same way that an organism requires raw materials to function (oxygen, water, and food) and produces waste, so do individual cells. These materials must pass across the plasma membrane. The plasma membrane is selective; it allows only some substances to cross. Oxygen and water may pass through the plasma membrane passively, whereas food molecules like sugar and protein are guided across in a selective manner.

Many cells have tiny, hairlike structures on their surfaces called **cilia** and **flagella.** In general, they are called *flagella* if they are long and few in number, and *cilia* if they are short and numerous. They are similar in structure,

and each functions to move the cell through its environment or to move the environment past the cell. Cilia and flagella are a common means of locomotion for unicellular organisms and small multicellular organisms. In addition, male sex cells (sperm) in animals and many plants utilize flagella for movement. Cilia are commonly found on cells in internal passageways in animals. For exam-

ple, in the air passage known as the trachea, there may be a billion cilia per square centimeter (figure 2.13).

Surrounding the plasma membrane of plants, bacteria, algae, and fungi are thick rigid structures called **cell walls.** Cell walls provide cells with shape and strength while allowing substances like water, air, and dissolved materials to pass through. Several of the strengthening components of cell walls are undigestible to most animals and contribute fiber to the animal's diet.

The Five Kingdoms Based on Cell Structure

Not all of the organelles just described are located in every cell. Some types of cells have combinations of organelles that differ from other types of cells. Some organisms are composed of very simple cells that lack most of the complex organelles described in this chapter. Biologists have classified cells into two major types: *eukaryotic* and *prokaryotic.*

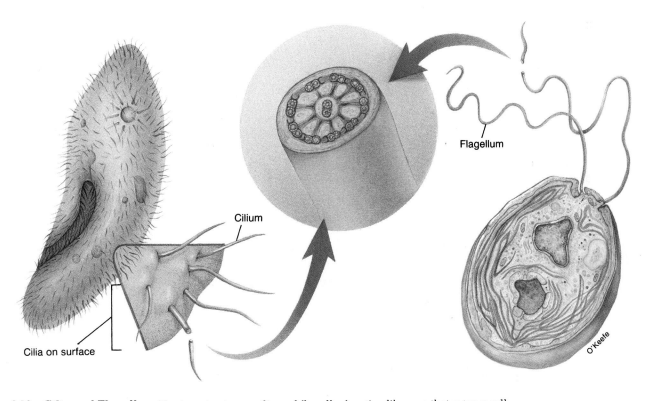

Flagellum

Cilium

Cilia on surface

O'Keefe

Figure 2.13 Cilia and Flagella   The two structures, cilia and flagella, function like oars that move a cell through its environment or move the environment past a cell. They are constructed of groups of microtubules. Flagella are longer than cilia. The cells that line the trachea in humans have many cilia. Foreign material in the air is trapped in mucus and the cilia help to flush it from the airway.

Prokaryotic cells	Eukaryotic cells			
Characterized by few membranous organelles; nuclear material not separated from the cytoplasm by a membrane	Cells larger than prokaryotic cells; nucleus with a membrane separating it from the cytoplasm; many complex organelles composed of many structures including membranes			
Kingdom Prokaryotae	Kingdom Protista	Kingdom Mycetae	Kingdom Plantae	Kingdom Animalia
Unicellular organisms	Unicellular organisms; some in colonies; both photosynthetic and heterotrophic nutrition	Multicellular organisms or loose colonial arrangement of cells; organism is a row or filament of cells; decay fungi and parasites	Multicellular organisms; cells supported by a rigid cell wall of cellulose; some cells have chloroplasts; complex arrangement into tissues	Multicellular organisms with division of labor into complex tissues; no cell wall present; acquire food from the environment
Examples: bacteria and cyanobacteria	Examples: protozoans such as *Amoeba* and *Paramecium* and algae such as *Chlamydomonas* and *Euglena*	Examples: *Penicillium*, morels, button mushrooms, galls, and rusts	Examples: moss, ferns, cone-bearing trees, and flowering plants	Examples: worms, insects, starfish, frogs, reptiles, birds, and mammals

Comparison of Cell Types   The five types of cells illustrated here indicate the major patterns of construction found in all living things. Note the similarities of all five and the subtle differences among them.

Eukaryotic Cell Structure

Eukaryotic cells are complex cells with a true nucleus and organelles composed of membranes. Organisms made up of eukaryotic cells are divided into four kingdoms—animals, plants, fungi, and protists—based on the specific combinations of organelles they contain.

Animals: The Animalia

More than a million species of animals have been classified. These range from microscopic types, like mites or tiny aquatic organisms, to huge animals like elephants or whales. Regardless of their type, all animals have some common traits. They all are composed of eukaryotic cells. All species of animals are heterotrophic, lack cell walls, and are multicellular. Most animals are *motile* (able to move); some, like the sponges and corals, are *sessile* (not able to move). Nearly all animals are capable of sexual reproduction, but a large number can also reproduce *asexually* (without sex) (figure 2.14*a*).

Plants: The Plantae

Members of the plant kingdom are multicellular eukaryotic organisms that contain chlorophyll and produce their own organic compounds. All plant cells have a cell wall.

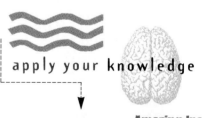

apply your **knowledge**

Amazing Insects

An estimate of the minimum number of species of insects in the world is 750,000. Perhaps, then, it would not surprise you to see a fly with eyes on stalks as long as its wings, a dragonfly with a wingspread of more than 100 centimeters, an insect that can revive after being frozen at −35°C, and a wasp that can push its long, hairlike, egg-laying tool directly into a tree. Of these animals, only the dragonfly is not presently living, but it once was.

What other curious features of this fascinating group can you discover? Have you tried to look at a common beetle under magnification? It will hold still if you chill it.

(a) Kingdom Animalia

Figure 2.14 Representative Organisms of the Five Kingdoms

More than 300,000 species of plants have been classified: 80 percent are flowering plants, 5 percent are mosses and ferns, and the remainder are cone-bearing evergreens (figure 2.14b).

Fungi: The Mycetae

Fungi are nonphotosynthetic eukaryotic organisms with rigid cell walls. The majority are multicellular, but a few, like yeasts, are single-celled. The majority also do not move. All of these organisms are heterotrophs; that is, they must obtain nutrients from organic sources. Most secrete chemicals that digest large molecules into smaller units that are absorbed. Fungi can either be free-living or parasitic. A *parasite* is an organism that receives nourishment by living in or on another organism. The parasite benefits from this relationship and harms the *host,* the organism it lives in or on. Fungi that are free-living, like mushrooms, decompose dead organisms as they absorb nutrients. Fungi that are parasitic are responsible for athlete's foot, vaginal yeast infections, ringworm, and other diseases (figure 2.14c).

apply your **knowledge**

Plant or Fungi?

Some people mistakenly think mushrooms are plants. Why are mushrooms definitely *not* members of the plant kingdom?

(b) Kingdom Plantae

Figure 2.14   Continued

(c)   Kingdom Mycetae

Figure 2.14   Continued

Algae, Protozoans, and Slime Mold: The Protista

Most members of the Protista kingdom are one-celled organisms, although some live in colonies (figure 2.14d). There is a great deal of diversity within the 60,000 known species of algae, protozoa, and slime molds. Many species are free-living freshwater organisms while others are found in marine or terrestrial habitats. Some protists are parasitic, and others form harmless or mutually beneficial associations with other organisms. All species can undergo asexual reproduction. Some species can also reproduce sexually. Many contain chlorophyll in chloroplasts and are autotrophic; others require organic molecules as a source of energy and are heterotrophic. Both autotrophic and heterotrophic protists have mitochondria and can perform aerobic cellular respiration.

Because members of this kingdom are so diverse, most biologists do not feel that the Protista form a valid taxonomic unit. However, it is still a convenient grouping. By placing these organisms together in this group it is possible to gain a useful perspective on how they relate to other kinds of organisms. Three major types of protists exist: plant-like autotrophs (algae); animal-like heterotrophs (protozoa); and the fungus-like heterotrophs (slime molds).

Prokaryotic Cell Structure

Prokaryotic cells are the simplest type of cells. They do not have a nucleus surrounded by a nuclear membrane, nor do they contain mitochondria, chloroplasts, or organelles composed of membranes. However, prokaryotic cells contain DNA and are able to reproduce and engage in chemical reactions. All prokaryotic organisms belong to the bacteria kingdom.

Bacteria: The Prokaryotae

Members of the bacteria kingdom are grouped together because they all have the same cellular structure. They are small, single-celled organisms ranging from 1 to 10 micrometers (μm). Their cell walls typically contain complex organic molecules not found in other kinds of organisms. Some bacteria, such as *Streptococcus pneumoniae*, are disease-causing, but most are not, and many are beneficial to humans. In addition, many are able to photosynthesize.

Bacteria cells reproduce by primitive methods. Some are motile; they move by secreting a slime that glides over the cell's surface, causing it to move through the environment. Others move by means of flagella.

Some bacteria require oxygen and other bacteria can survive in oxygen-depleted environments. There are some autotrophic bacteria, but the majority are heterotrophs. Some heterotrophic organisms obtain energy by decomposing dead organic material; others are parasites that obtain energy and nutrients from living hosts; some live off other organisms without harming them; and still others survive by forming mutually beneficial relationships with other living organisms (figure 2.14e).

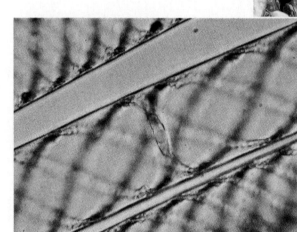

(d)   Kingdom Protista

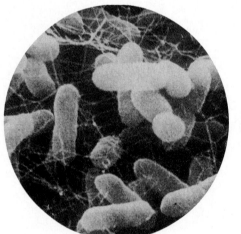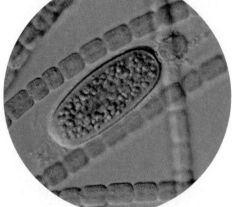

(e)   Kingdom Prokaryotae

Figure 2.14   Continued

Origin of Eukaryotic Cells

Most biologists hypothesize that eukaryotic cells evolved from prokaryotic cells. This hypothesis proposes that structures like mitochondria, chloroplasts, and other membranous organelles originated from small cells that were ingested by larger cells. Once inside, these structures and their functions became integrated with the host cell and ultimately became essential to its survival. This new type of cell was the forerunner of present-day eukaryotic cells. After the origin of unicellular eukaryotic organisms (protists), evolution proceeded along several different pathways. The plant-like autotrophs (algae) probably gave rise to the kingdom Plantae, the animal-like heterotrophs (protozoa) probably gave rise to the kingdom Animalia, and the fungus-like heterotrophs (slime molds) were probably the forerunners of the kingdom Mycetae.

Viruses

A **virus** is hereditary information coated with protein (figure 2.15). All viruses are parasites; they can function only when inside a living cell. Due to their unusual characteristics, viruses are not a member of any kingdom. Biologists do not consider them to be living because they are not cellular and are not capable of living by themselves. Viruses show the characteristics of life only when inside living cells.

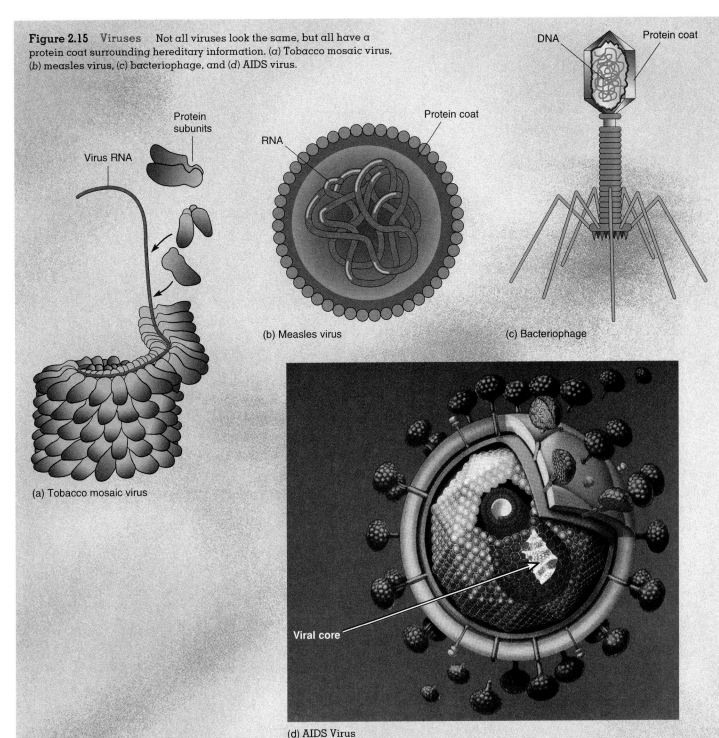

Figure 2.15 Viruses Not all viruses look the same, but all have a protein coat surrounding hereditary information. (a) Tobacco mosaic virus, (b) measles virus, (c) bacteriophage, and (d) AIDS virus.

(a) Tobacco mosaic virus

Protein subunits

Virus RNA

(b) Measles virus

RNA

Protein coat

(c) Bacteriophage

DNA

Protein coat

(d) AIDS Virus

Viral core

Viruses are host-specific: a virus can infect only cells that have the proper sites on their surface to which the virus can attach. For example, the virus responsible for measles attaches to skin cells, hepatitis viruses attach to liver cells, and mumps viruses attach to cells in the salivary glands. Host cells for the HIV virus include some types of human brain cells and several types belonging to the immune system.

Upon entering a host cell, the virus loses its protein coat. Once free in the cell, the hereditary information of the virus may remain free in the cell or it may link with the host's DNA. A virus's hereditary code is able to take command of the host's metabolic pathways and direct it to carry out the work of making new copies of the original virus. The virus makes use of the host's energy and cellular materials for this purpose. When enough new viral components are produced, complete virus particles are assembled and released from the host (figure 2.16). The number of viruses released ranges from ten to thousands. The virus that causes polio releases about 10,000 new virus particles after it has invaded its human host cell.

Viruses vary in size and shape, which helps in classifying them. Some are rod-shaped, others are round, and still others are in

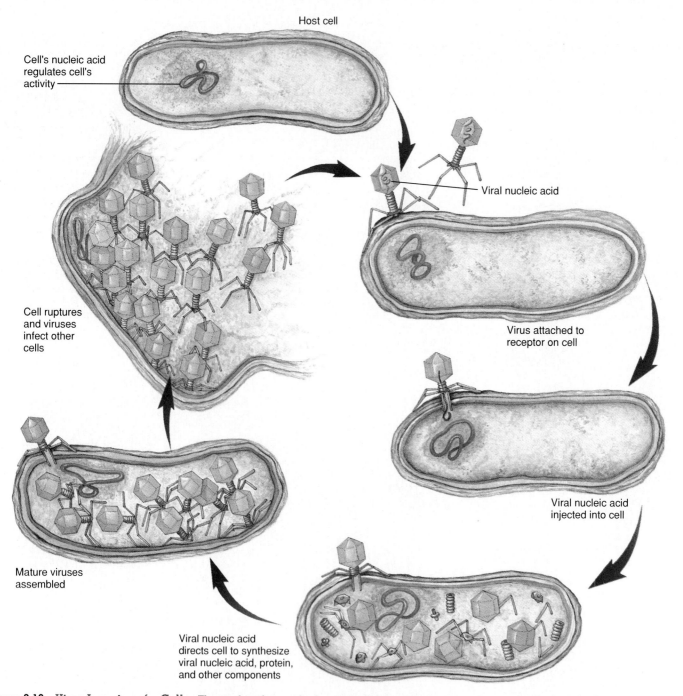

Host cell

Cell's nucleic acid regulates cell's activity

Viral nucleic acid

Cell ruptures and viruses infect other cells

Virus attached to receptor on cell

Mature viruses assembled

Viral nucleic acid injected into cell

Viral nucleic acid directs cell to synthesize viral nucleic acid, protein, and other components

Figure 2.16 Virus Invasion of a Cell   The viral nucleic acid takes control of the activities of the host cell. Because the virus has no functional organelles of its own, it can become metabolically active only when it is within a host cell.

the shape of a coil. Viruses are some of the smallest infectious agents known to humans. Only a few can be seen with a standard laboratory microscope; most require the aid of an electron microscope. A great deal of work is necessary to isolate viruses from the environment and prepare them for observation with an electron microscope. For this reason, most viruses are more quickly identified by their activities in host cells. Almost all of the species in the five kingdoms serve as hosts to some form of virus (table 2.2).

Table 2.2

Viral Diseases

Type of Virus	Disease
Papovaviruses	Warts in humans
Paramyxoviruses	Mumps and measles in humans; distemper in dogs
Adenoviruses	Respiratory infections in most mammals
Poxviruses	Smallpox
Wound-tumor viruses	Diseases in corn and rice
Potexviruses	Potato diseases
Bacteriophage	Infections in many types of bacteria

SUMMARY

The taxonomic ranking of organisms reflects their evolutionary relationships. In addition, fossil evidence, comparative anatomy, developmental stages, and biochemical evidence are employed in the science of taxonomy. To facilitate accurate communication, biologists assign a binomial name to each species that is cataloged. The various species are cataloged into larger groups on the basis of similar traits.

The cell is the basic structure of all living things, and differences in cell structure can be used to group organisms into kingdoms. Complex eukaryotic organisms make up the animal, plant, fungi, and protist kingdoms. The simplest single-celled organisms belong to the bacteria kingdom (Prokaryotae). Although viruses are not considered to be living organisms, they are able to invade many types of cells. Because of their pathogenic effects, viruses are an important factor in the world of living organisms.

CHAPTER GLOSSARY

aerobic cellular respiration (a″ro′bik sel′yu-lar res″pi-ra′shun) The conversion of oxygen and food, such as carbohydrates, to carbon dioxide and water. During this conversion, energy is released.

autotrophs (aw′to-trofs) Organisms able to use light energy to produce organic nutrients from inorganic materials; a self-feeder.

binomial system of nomenclature (bi-no′mi-al sis′tem ov no′men-kla-ture) A naming system that uses two Latin names, genus and species, for each type of organism.

cell (sel) The basic structural unit that makes up living things.

chlorophyll (klo′ro-fil) The green pigment located in the chloroplasts of plant cells that is associated with trapping light energy.

chloroplasts (klo′ro-plasts) Energy-converting organelles in plant cells. Chloroplasts contain the green pigment chlorophyll.

chromatin (kro′mah-tin) Areas or structures within the nucleus of a cell composed of DNA in association with proteins.

chromosomes (kro′mo-somz) Densely coiled chromatin.

class (klas) A group of closely related orders found within a phylum.

classification (kla-se-fe-ka′-shen) The assortment of objects into groups based on their similarities and differences.

cytoplasm (si′to-plazm) The semifluid portion of the cell. Cytoplasm surrounds the nucleus of eukaryotic cells.

DNA (deoxyribonucleic acid) (de-ok″se-ri-bo-nu-kle′ik as′id) A hereditary blueprint located in the nucleus of eukaryotic cells or the cytoplasm of prokaryotic cells.

eukaryotic cells (yu′ka-re-ah″tik sels) One of the two major types of cells; complex cells with a true nucleus and organelles composed of membranes; in plants, fungi, protists, and animals.

family (fam′i-ly) A group of closely related species within an order.

fungus (fun′gus) The common name for the kingdom Mycetae.

genus (je′nus) (pl., *genera*) A group of closely related species within a family.

kingdom (king′dom) The largest grouping used in the classification of organisms.

nuclear membrane (nu′kle-ar mem′bran) The structure surrounding the nucleus that separates the nucleoplasm from the cytoplasm.

nucleoli (nu-kle′o-li) Nuclear structures containing information for ribosome construction.

nucleoplasm (nu′kle-e-pla″zem) The fluid of the nucleus composed of water and the molecules used in the construction of the rest of the nuclear structures.

nucleus (nu′kle-us) The central body that contains the information system for a eukaryotic cell.

order (or′der) A group of closely related organisms within a class.

organelles (or-gan-elz′) Cellular structures that perform specific functions in the cell.

photosynthesis (fo-to-sin′the-sis) The process of combining carbon dioxide and water with the aid of light energy to form sugar and release oxygen.

phylum (fi′lum) A subdivision of a kingdom.

plasma membrane (plaz mah mem brain) The outer boundary membrane of the cell; surrounds all cells.

prokaryotic cells (pro′ka-re-ot″ik sels) One of the two major types of cells. They do not have a typical nucleus bound by a nuclear membrane and lack many of the other membranous cellular organelles; bacteria.

ribosomes (ri′bo-sōmz) Small structures composed of two protein and ribonucleic acid subunits involved in the assembly of proteins from amino acids.

species (spe′shez) The scientific name given to a group of organisms that can potentially interbreed to produce fertile offspring.

taxonomy (tak-son′uh-me) The science of classifying and naming organisms.

virus (vi′rus) A noncellular parasite composed of hereditary material surrounded by a protein coat.

CONCEPT MAP TERMINOLOGY

Construct a concept map to show relationships among the following concepts.

cell structure
classification
comparative anatomy
developmental biology
DNA analysis
fossils
taxonomy

LABEL•DIAGRAM•EXPLAIN

Label each of the following structures on the diagrams of the cells below. Briefly describe the function of each structure.

cell wall
centrioles
chloroplast
chromatin
cilia
cytoplasm
Golgi bodies
lysosomes
microtubules

mitochondria
nuclear membrane
nucleoli
nucleoplasm
plasma membrane
ribosomes
rough endoplasmic reticulum
smooth endoplasmic reticulum

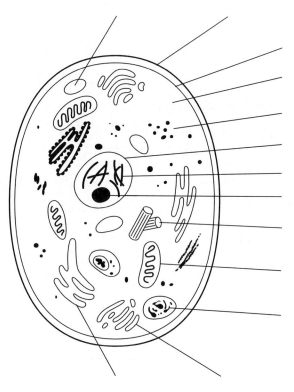

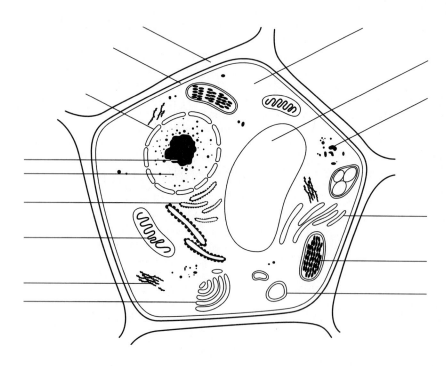

Multiple Choice Questions

1. Which of the following lists the classification categories from largest to smallest?
   a. kingdom, phylum, class, order, family, genus, species
   b. kingdom, phylum, order, class, family, genus, species
   c. kingdom, phylum, order, family, class, genus, species
   d. kingdom, phylum, family, class, order, genus, species

2. Which of the following statements concerning fossils is *false*?
   a. Fossils found in lower sediment layers are generally older than fossils found in upper sediment layers.
   b. Soft-bodied organisms are readily fossilized.
   c. An insect preserved in amber is a fossil.
   d. Aquatic animals are more likely to be fossilized than land animals.

3. The half-life of carbon-14 is 5,630 years. How old is an ancient bone containing one-sixteenth of the carbon-14 typically found in bones today?
   a. 352 years old
   b. 1407.5 years old
   c. 22,520 years old
   d. 90,080 years old

4. Which of the following is true of plants?
   a. Plants are autotrophs.
   b. Photosynthesis occurs in plants.
   c. Aerobic cellular respiration occurs in plants.
   d. All of the above.

5. Which of the following kingdoms contains only heterotrophs?
   a. plants (Plantae)
   b. fungi (Mycetae)
   c. algae, protozoans, and slime molds (Protista)
   d. bacteria (Prokaryotae)

6. The energy conversion organelle(s) is (are) the:
   a. DNA
   b. cilia and flagella
   c. smooth and rough endoplasmic reticula
   d. chloroplasts and mitochondria

7. You will *not* find a cell wall surrounding _____ cells.
   a. plant
   b. fungi
   c. animal
   d. bacteria

8. The structures most frequently associated with movement are:
   a. cilia and flagella
   b. Golgi bodies
   c. mitochondria
   d. ribosomes

9. Viruses are composed of:
   a. prokaryotic cells
   b. eukaryotic cells
   c. hereditary information surrounded by protein
   d. primitive cells containing only DNA, cytoplasm, and a plasma membrane

10. You are a biologist who has just discovered a new life form. This newly described multicellular organism is incapable of photosynthesis and obtains nutrients by absorbing them from the environment. The organism is composed of eukaryotic cells containing cell walls. Into which kingdom will you classify this organism?
    a. animal (Animalia)
    b. plant (Plantae)
    c. fungi (Mycetae)
    d. bacteria (Prokaryotae)

Questions with Short Answers

1. Correctly write the scientific name for the house cat and domestic dog.
   a. house cat:

   b. domestic dog:

2. Name four tools scientists use in classification.
   a.

   b.

   c.

   d.

3. Why are Latin names used for genus and species?

4. What is the value of taxonomy?

5. How do viruses reproduce?

chapter

3

The animal KINGDOM

learning objectives

- Recognize the characteristics of animals.

- Recognize the specializations animals as heterotrophs have for acquiring food.

- Recognize the value of cellular respiration.

- Describe how acids, bases, salts, and the pH scale interrelate.

- Relate ways in which animals are specialized to survive in different habitats.

- Recognize that animals interact with other organisms.

- Describe the ways animals reproduce and care for their developing offspring.

FYI

Slime Molds: Are They Animals? There are several kinds of organisms called slime molds that present problems for taxonomists. These organisms consist of individual eukaryotic cells that lack cell walls. The cells move about engulfing food material and under certain conditions come together in large masses that move about and feed. Sexual reproduction occurs in these masses, and the cells produce spores that do have cell walls. So are they animals? At one time many biologists considered slime molds to be animals. Today most biologists include them with the fungi or place them in a category of their own. Those who place them with the fungi do so because the slime molds have the funguslike characteristics of producing spores and having cell walls.

Characteristics of Animals

Classifying organisms can be tricky business. There always seem to be exceptions to the rules. We need to constantly remind ourselves that classification is a human activity and that the various kinds of organisms evolved to meet the demands of their surroundings, not to fit into a predetermined classification system. So what is an animal? We easily recognize that humans, cats, dogs, birds, snakes, frogs, and fish are animals, but so are insects, worms, jellyfish, clams, sponges, and starfish. Let us look at the characteristics that unite such a wide variety of organisms into a single kingdom.

Animals share the characteristics common to all living things: the ability to manipulate energy and matter; being composed of cells that contain organic molecules; the ability to reproduce, grow, develop, and eliminate wastes; and exhibiting orderliness, evolution, and responsiveness. In addition, animals as a group share several other characteristics. All animals must eat, are able to move during certain stages of their lives, and have bodies made up of several different kinds of nucleus-containing (eukaryotic) cells that lack cell walls. Individually, these characteristics do not completely separate animals from other organisms, but collectively they provide a better feel for what defines an animal. For example, humans eat, move from place to place, and are composed of many different kinds of eukaryotic cells, so we are included in the animal kingdom. So are sharks, grasshoppers, crabs, squids, and alligators.

The Movement of Animals

Running and Walking

Most people associate movement with animals. Humans, for example, move about using their limbs. As a baby you crawled on all fours; as an adult you spend most of your time walking on your hind legs. Since you have practiced doing this for many years, it seems easy. You may even move rapidly, running or jogging. Not only are you using the specialized cells of your muscles and skeleton, but you also are depending on information from your sense organs, your nerves and brain, to get you from place to place.

Limbs or appendages used for locomotion can be found on many types of animals. Primates such as humans and chimpanzees have freedom of movement at the shoulder, hip, and elbow joints, enabling them to easily swing from tree limbs or monkey bars. They can also use their forelimbs to manipulate food, tools, and other items. The limbs of dogs, horses, and other four-legged animals are primarily designed for rapid movement over land and are not able to manipulate objects.

Insects, crabs, lobsters, spiders, and many other similar organisms, classified into the arthropod phylum, have several sets of appendages that allow them to move and manipulate food. Many of the appendages are extremely specialized (figure 3.1). The muscles in these appendages are organized quite

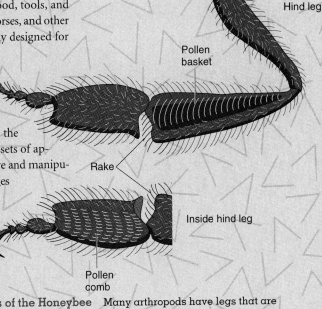

Figure 3.1 Specialized Legs of the Honeybee Many arthropods have legs that are specialized for specific purposes. The three sets of legs of a honeybee demonstrate this extremely well. There are specialized structures for collecting pollen (pollen brush, pollen comb); specialized pollen baskets for carrying pollen; and structures for cleaning the eyes and the antennae.

51

Six Characteristics of Animals

1. During all or part of their lives, animals are able to move from place to place or move one part of their body with respect to other parts.

2. Animals are heterotrophic: they must eat other organisms to obtain molecules that are used as building materials for new cells and for energy to drive life's processes.

3. Animals respond quickly and appropriately to changes in their environment.

4. Sexual reproduction is a characteristic of animals, though many animals reproduce asexually as well.

5. All animal cells have a nucleus but lack cell walls.

6. Multicellular animals contain groups of cells frequently organized into tissues, organs, and organ systems.

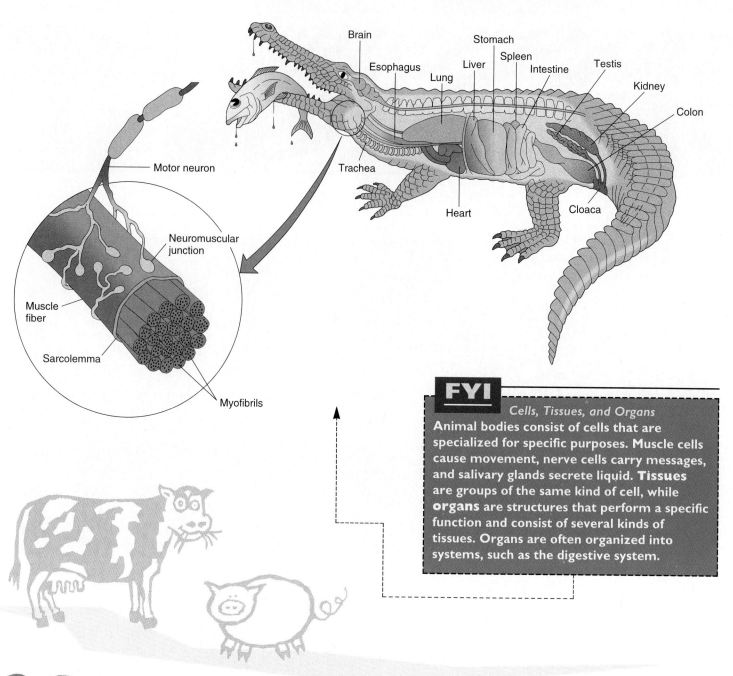

Brain

Stomach

Esophagus   Spleen

Lung   Liver   Intestine   Testis

Kidney

Colon

Motor neuron

Neuromuscular
junction

Trachea

Colon

Muscle
fiber

Sarcolemma

Myofibrils

Heart   Cloaca

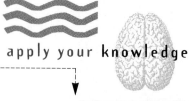

apply your knowledge

Eating Your Way Through

The next time you have a meal of crab legs,
whole crayfish, lobster, or whole shrimp,
examine the object on your plate. What you eat
is the muscle of the animal, which is attached
to projections on the inside of the skeleton. The
animal's skeleton is the hard outer "shell" that
you peel away to get at the meat.

Kinds of Arthropods

Most kinds of arthropods can be distinguished by the number of legs they use for walking.

• Insects have six legs.

• Spiders, mites, and ticks have eight legs.

• Crabs, lobsters, and their relatives typically have ten legs.

• Centipedes have many pairs of legs with one pair per segment.

• Millipedes have many pairs of legs with two pairs per segment.

differently from ours because arthropods do not have bones. Instead, they have a hard, jointed skeleton on the outside of the body. The muscles are attached to the inside of this skeleton. The independent movement of each leg enables arthropods to move rapidly, intricately manipulate their environment in such activities as building structures and handling food, and perform other specialized tasks such as stinging, building webs, and carrying materials.

Flying

Several kinds of animals use appendages to fly through the air. Flight has interested and excited humans for centuries. Long before it was possible to get a jetliner off the ground, humans flirted with the idea of building wings or sails that would allow us to conquer the air. The types of animals that are able to fly include bats, birds, birdlike dinosaurs, and most insects (figure 3.2).

When we compare the wing structure of birds, extinct flying reptiles, and bats we see that they are all based on the same basic plan. There is a set of bones that surround the backbone at the neck region to which the wing bones are attached. You can see this structure when you eat chicken wings. The three parts of the wing include the heavy first bone, similar to your upper arm bone, the two smaller bones in the next section, which resemble your lower arm bones, and the many small bones and cartilage in the third section, which may remind you of your wrist, hand, and finger bones. The bones provide the structural support for a flat surface that makes the wing. In birds, light, hollow feathers fit together to form the wing, while in bats a flap of skin supported by bones serves the same purpose (figure 3.3).

Figure 3.3 Flight Surfaces The flight surface of bats is provided by flaps of skin supported by elongated finger bones (*a*). The flight surface of birds is provided by the overlapping, flat-surfaced feathers of the wings (*b*). Since the feathers are strong, they do not need additional skeletal support.

(a)

Bat

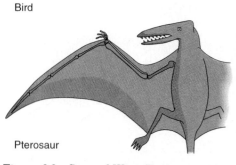

Bird

Pterosaur

Figure 3.2 Several Wing Designs The wings of birds, bats, and extinct flying reptiles are all modifications of the forelimb, with slightly different structures for supporting the wing surface.

(b)

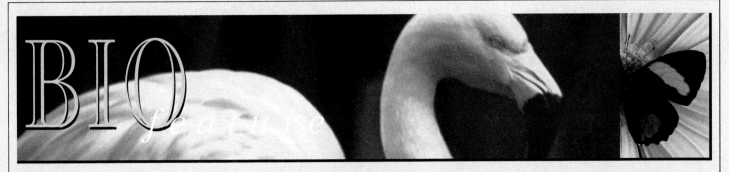

Muscles, Wings, and Flight in Birds

Muscles that move the wings of a bird are located in what we call the breast. There are two different layers of muscle, one that pulls the wings down and one that pulls the wings up. A pulley system consisting of tendons runs through a slot in the shoulder that allows for the wings to be pulled up. The next time you have chicken breasts, look for the two layers of muscle and the tendon (figure A). As the wings move up and down they push against the air, providing forward motion for the bird. Air passing across the surface of the wings provides lift. Some birds, such as eagles, vultures, and hawks, have large wing surfaces and are particularly adept at soaring on air currents. They stay aloft for hours, only occasionally flapping their wings. Most birds with relatively short wings, such as robins, pigeons, hummingbirds, and crows, must continuously beat their wings to remain airborne.

The flight of insects is different from that of bats and birds. Insect wings are modified portions of their external skeleton, so the muscles are attached very differently. The muscles that move the wings are located inside the skeleton (figure B).

Figure A Bird Flight Muscles   There are two layers of flight muscles in birds, one that pulls the wings down and one that pulls the wings up. Both sets of muscles are attached to the enlarged breastbone (sternum), often called a keel bone because it looks like the keel of a sailing ship. A tendon from one set of muscles passes through a groove in the shoulder joint and is attached on the top of the wing bone. Thus the muscle that pulls the wing upward actually can be located below the wing.

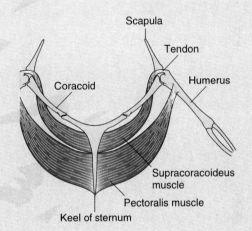

Figure B Insect Flight Muscles   Many kinds of insects have *indirect* flight muscles. In other words, the muscles are not directly attached to the wings. The flight muscles are attached to the inside of the exoskeleton of a region of the insect body known as the thorax. Alternate contractions of the two sets of muscles changes the shape of the thorax and causes the wings to move up and down.

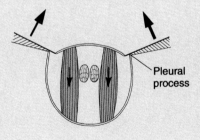

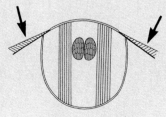

Indirect flight muscles
of flies and midges

The Advantages of Flight

There are several benefits to flight. Animals that fly are able to travel long distances in a short time and use less energy than animals that walk or run. They are able to cross barriers like streams, lakes, oceans, bogs, ravines, or mountains that other animals cannot cross. They can also escape predators by quickly taking flight.

Other Ways of Moving

Many animals do not use legs or wings for locomotion. For example, snakes, worms, clams, and slugs move by waves of muscle contractions down the length of their bodies. Though these animals have no limbs, they can move easily through their environments. An earthworm moves through soil by extending its body to push its head forward. Once inched forward, it expands its front end, creating a larger space in the ground into which the rest of its body moves. Small hairlike projections on its lower surface prevent the worm from slipping backward.

Snakes have powerful muscles that ring and run the length of their bodies. These muscles allow snakes to skillfully move across land as well as climb and swim. To move forward, a snake forms S-shaped curves that proceed from its anterior to its posterior end.

table 3.1
Animal Speeds

Animal	mph	kph
Mosquito	0.9 mph	1.4 kph
Red racer snake	3.6 mph	5.7 kph
Human flea	4.5 mph	7.2 kph
House rat	6.0 mph	9.6 kph
Bumblebee	11.0 mph	17.6 kph
Elephant	24.5 mph	39.2 kph
Rainbow trout	25.0 mph	40.0 kph
Short finned whale	25.3 mph	40.4 kph
Human	27.9 mph	44.6 kph
House cat	30.0 mph	48.0 kph
Horsefly	31.0 mph	50.0 kph
Giraffe	32.0 mph	51.2 kph
Killer whale	34.5 mph	55.0 kph
Dragonfly	36.0 mph	57.6 kph
Red fox	40.0 mph	64.0 kph
Greyhound dog	41.7 mph	66.7 kph
Horse	43.2 mph	69.2 kph
Yellow tuna	46.6 mph	74.5 kph
Turkey	50.0 mph	80.0 kph
Goose	60.0 mph	100.0 kph
Golden eagle	80.0 mph	130.0 kph

As the wave passes along, the animal pushes against objects in its environment and glides along the surface.

The movement of land snails or slugs is very different. These animals lay down a layer of slime or mucus upon which they glide by waves of muscle contractions in their lower flat structure, or foot. Anything that prevents them from contacting the mucus will prevent them from moving. If you are having problems with snails or slugs in your garden, one way to trap them is to set a shallow pan of beer into the ground. The slugs will move into the dish to feed but will not be able to leave because the beer prevents them from making good contact with their slime layer.

Swimming

Many kinds of aquatic animals float or walk on the bottom of the body of water in which they live. For example, many jellyfish float,

and crabs and octopi crawl. Fishes and some other aquatic animals (whales, seals, sea slugs, sea snakes), however, swim through their environment by using rippling or undulating movements. The movement of the tail back and forth or up and down provides most of the forward motion for fishes and aquatic mammals such as dolphins and whales. The limbs of these animals are paddlelike and are used for steering and maintaining an upright position rather than for moving the animals through the water.

Fishes, sharks, and whales all have the same structural arrangement of a collar or girdle to which the limbs are attached. Their limbs are similar, but fishes, sharks, and whales are very different animals. Fishes and whales have a skeleton of bone to which muscles are attached, while sharks have a skeleton of *cartilage*. Furthermore, whales must surface to breathe air into their lungs, while fishes

apply your knowledge

The Movement of Earthworms

When it rains, earthworms emerge from the soil because water has flooded their underground homes. As their tunnels fill, oxygen is reduced to a point where the worms would suffocate should they stay inside. The next time you see one of these "homeless" worms, observe its movement, then pick it up. Carefully draw it through your fingers. Notice the prickly or rough feel of the hairlike projections on its lower body surface. Compare the feel when you stroke the worm from front to back versus back to front. How do your observations help explain how the worm moves through the soil? How do they explain why it is difficult for a bird to pull a worm from its tunnel?

CONCEPT CONNECTIONS

• **Cartilage** is a flexible tissue that resembles soft plastic. In your body, cartilage is the supporting structural material of your external ear and the tip of your nose. It also is in joints such as your knee. When damaged, it is slow to heal.

and sharks are able to separate dissolved oxygen from water. This uptake of dissolved oxygen occurs in gills.

Other animals that live in aquatic environments have unusual methods of locomotion. The jet propulsion of squid, the pulsations of jellyfishes, and the wriggling motion of sea worms are examples of unusual ways in which animals are able to move from place to place.

Stationary Animals

Many aquatic animals are **sessile** or stationary for most of their lives, but all animals at certain times have some cells that are **motile** (able to move). Oysters and mussels are mollusks whose shells are attached to rocks in the ocean, but these organisms produce swimming sperm that fertilize eggs. The eggs develop into an immature or larval stage that swims, eventually settles down on a suitable surface, and grows there the rest of its life as an adult stationary animal. Barnacles (arthropod relatives of shrimp and lobsters) and corals (relatives of jellyfish) have similar life cycles. All of these animals have hard exterior structures that protect them from predators.

Thousands of corals live together in colonies. Each coral constructs cuplike structures into which it can retreat if attacked. The protective structures form massive rocklike ridges known as coral reefs.

Sponges, the simplest of all animals, are also sessile. Like corals, barnacles, oysters, and mussels, sponges have motile sex cells and immature larval stages that swim until they find a suitable place to attach. All sessile animals filter or trap food from the water by creating water currents that bring food to them.

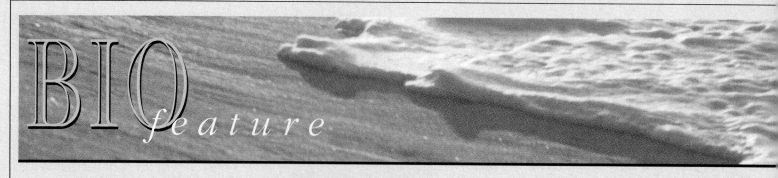

Photo Essay: The Mollusks

Mollusk Variety (a) The bodies of snails are unusually shaped to fit into their coiled shells. (b) Chitons can attach themselves firmly to rock surfaces by their muscular foot. (c) Squids and octopi are predators with well-developed eyes that use jet propulsion to move quickly. (d) Clams create water currents that allow them to filter food from the water.

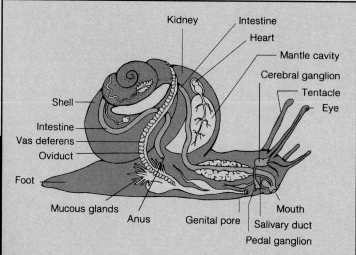

Kidney
Intestine
Heart
Mantle cavity
Cerebral ganglion
Tentacle
Eye
Shell
Intestine
Vas deferens
Oviduct
Foot
Mucous glands
Anus
Genital pore
Salivary duct
Mouth
Pedal ganglion

From Engemann/Hegner, *Invertebrate Zoology*, 3/e, © 1981. Reprinted by permission of Prentice Hall, Upper Saddle River, New Jersey.

(a)

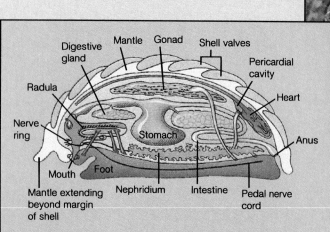

Digestive gland
Mantle
Gonad
Shell valves
Pericardial cavity
Radula
Heart
Nerve ring
Stomach
Anus
Mouth
Foot
Mantle extending beyond margin of shell
Nephridium
Intestine
Pedal nerve cord

(b)

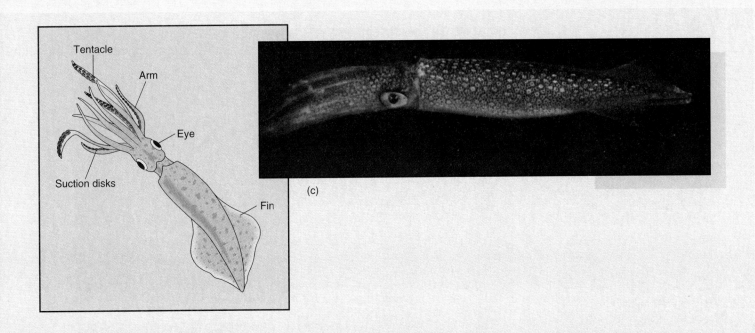

Tentacle

Arm

Eye

Suction disks

Fin

(c)

(d)

Anus

Heart

Kidney

Digestive
gland

Stomach

Gill

Esophagus

Mantle

Brain

Ovary

Mouth

Intestine

Foot

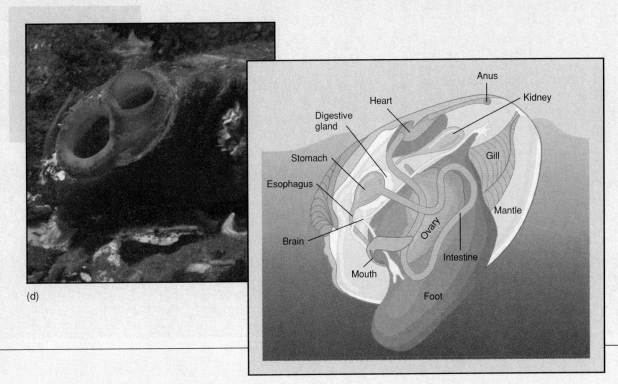

Types of Cnidarians

(a) Jellyfish.
Transparent disks of cells that swim or float through the water. Tentacles around the edge of the disk have surface cells containing a tiny harpoon. These "stinging thread" cells are characteristic of these animals.

(b) Colonial jellyfish.
Large groups of animals that collectively form a giant floating colony with long tentacles and many protective stinging thread cells; an example is the Portuguese man-of-war.

(c) Corals.
Massive colonies of individual animals called polyps. Each individual lives in a tiny tubelike cup composed of a rocklike material. As the colony increases in size, these cuplike structures are placed on top of one another, forming coral reefs. After thousands of years during which the depth of the ocean changes, coral reefs may become thousands of meters (1,000 meters = 1,090 yards) thick. Some islands have been formed by geologic forces thrusting reefs above the ocean surface or by drops in sea level leaving the reefs exposed.

(d) Sea anemones.
Individual polyps that lack the rocklike cups of corals. Many are large and colorful.

(a) The jellyfish *Aurelia;* (b) the Portuguese man-of-war *Physalia;* (c) the coral *Montastrea;* and (d) the sea anemone *Tealia.*

Photo Essay: The Carnivores

Carnivores

(a) Fishing spider

(b) Praying mantis

(c) Black snake

(d) Octopus

Photo Essay: Food Processing in Animals

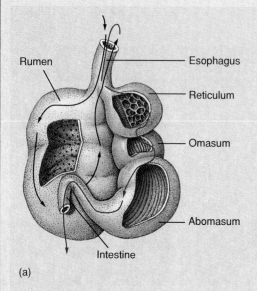

(a)

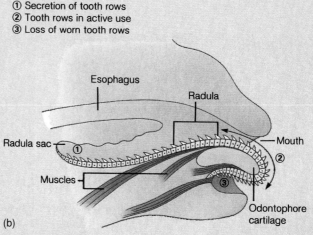

① Secretion of tooth rows
② Tooth rows in active use
③ Loss of worn tooth rows

(b)

From *A Life of Invertebrates* © 1979. W. D. Russell-Hunter. Reprinted by permission.

Different Ways to Grind Food (a) Many kinds of grass-eating mammals have a complex anterior end of the digestive system that assists in grinding food. One portion of this complex is called the rumen; animals like cows, sheep, camels, goats, antelopes, deer, and giraffes are called ruminants. When the food is first consumed it is swallowed and enters the rumen where it is softened by moisture and partially decomposed by microorganisms. Small portions of food, called a cud, are regurgitated and chewed and swallowed a second time. This time the chewed food takes a different route and enters a different part of the "stomach." Eventually it enters the intestine. (b) Many mollusks like snails, chitons, and squids have a radula, a rasping tonguelike structure for reducing the size of food particles. (c) Grasshoppers have several sets of mouthparts that move side to side to cut and grind food. (d) Birds lack teeth to grind food but have a muscular portion of their gut called the gizzard that contains hard materials such as small pieces of rock. The muscular contractions of the gizzard rub the rock particles and food together and the food is ground into smaller pieces.

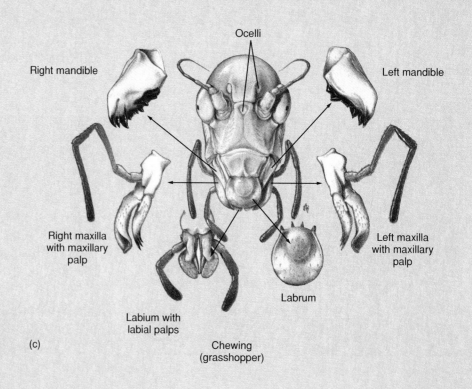

(c)

Chewing (grasshopper)

Animals and Food

Acquiring Food from the Environment

All animals take in food from their environment, alter it while it is within their bodies, utilize parts of the processed food, and then eliminate the leftovers. You will recognize that this is the way in which you are nourished. Using food as an energy source is one of the universal features of all living things; eating food from the environment is characteristic of all animals.

What an animal considers food may vary considerably from species to species. As heterotrophs, animals must rely on their surroundings to provide their nourishment. Snakes eat frogs, sponges eat algae, koalas eat eucalyptus leaves, earthworms eat dead plant matter, and dung beetles eat feces.

A primary activity of any animal is obtaining food. Food can be acquired in a variety of ways. As we have just seen, some aquatic animals filter their food from the water. Clams, oysters, scallops, corals, and sponges strain small organisms and other food particles from the water. The food is concentrated as it enters the animal, where it serves as a source of energy and building materials. Land (terrestrial) animals must move about to obtain their food.

Plant eaters like horses, deer, snails, caterpillars, and parakeets go to their food source. These animals have special features that enable them to process their food. Examples of these specializations are the heavy chewing teeth of horses, the stomach of cows, the rasping tongue of snails, the jaws of caterpillars, and the gizzard of chickens.

Many plant eaters (herbivores) are likely to have specializations such as camouflage coloration, which reduces the likelihood of being attacked while eating. Many large herbivores have colors that resemble their surroundings. Gray, tan, and brown are typical coat colors. Small herbivores, like insects, are likely to be green.

Meat-eating animals (carnivores) that must hunt their food (predators) have specializations related to this lifestyle. They display weapons such as sharp claws and teeth, stingers, or strength that allow them to immobilize prey. Speed, keen senses, mobility, and camouflage are also important. Since the kind of food carnivores eat is different from that of herbivores, their digestive systems are also different. Their teeth or mouthparts are designed for cutting rather than chewing or grinding,

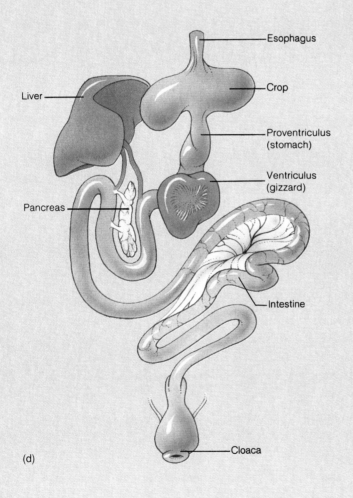

Esophagus

Liver

Crop

Proventriculus
(stomach)

Ventriculus
(gizzard)

Pancreas

Intestine

(d)

Cloaca

and their digestive systems are designed for large, infrequent, highly nutritious meals rather than the continuous, low-quality meals of herbivores.

Processing Food

Many animals, including humans, take in food at one end of a tube specialized for food processing. At various points along the tube the food is ground into smaller and smaller particles, moistened, and mixed with digestive chemicals. When the food has been broken into individual molecules such as sugars, fatty acids, and amino acids, the molecules are removed from the tube and distributed to the cells. When the molecules enter individual cells, some of them are used for energy. Others may be used as construction materials for cells or stored for later use. Meanwhile, the unused particles of food are eliminated at the other end of the tube.

Specialized regions of the food tube may differ from one type of animal to another. For instance, humans use their teeth and stomach as primary grinding organs, whereas birds use small stones in their gizzard to grind up food (figure 3.4). Dogs and cats gulp large chunks of food and rely on their stomach rather than their teeth to break it down.

Getting Energy from Food

The release of energy from food occurs inside cells and, in eukaryotes, involves structures known as mitochondria. Once food molecules have been distributed to the cells and further processed, they enter the mitochondria, where most of the energy is released. The series of reactions involved in releasing energy from food is known as cellular respiration. Large molecules are broken down into small molecules, and energy that held the large molecules together is released and becomes available for the animal to use. This energy is transferred to an energy-carrier molecule called ATP. The energy in ATP is used to power all the activities of life.

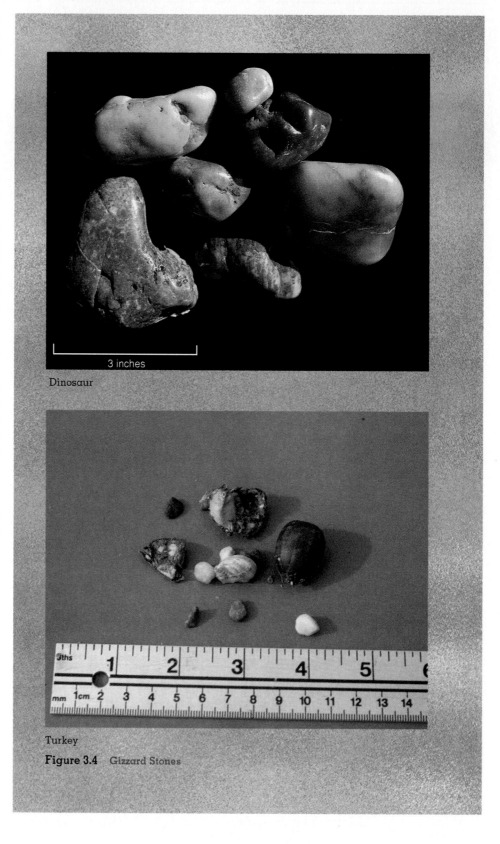

Dinosaur

Turkey

Figure 3.4 Gizzard Stones

BIO *feature*

Parasitism

Many animals live in or on other animals and use them as a source of food. The organism feeding is the *parasite* while the one providing the food is the *host*. Many roundworms, flatworms, segmented worms, and insects are examples of parasites. While some parasites live on the outside of their host, many others live within the body of the organisms on which they are feeding. Parasites have several specializations in order to use a living host for food:

1. They must be able to find a suitable host.

2. They must resist the efforts of the host to rid itself of the parasite.

3. They must have a method of anchoring themselves to the host.

4. They must keep their host alive as long as possible.

We may consider this form of nutrition to be rather unusual, but more animals in the world are parasites than are not.

Parasites

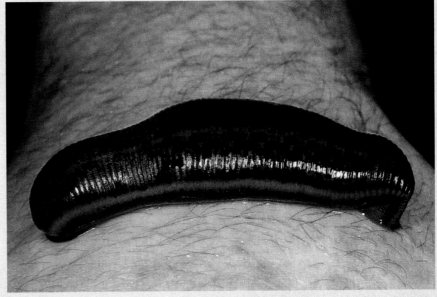

(a) Leech

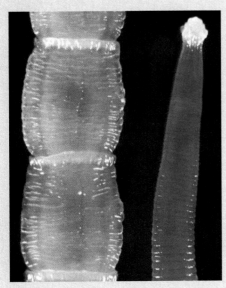

(b) Tapeworm

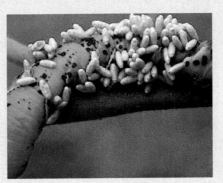

(c) Wasp parasites on a caterpillar

(d) *Ascaris* worms in an intestine

Digestion, Respiration, and Breathing

These words can get pretty confusing! Think about this: All animals digest food *outside of their bodies*. <u>Digestion</u> is the breakdown of large food molecules into smaller ones that are taken into cells. Flies, for example, regurgitate digestive juices onto their food source. The juice digests the food, and the fly sucks up the smaller molecules. Humans swallow food and mix it with digestive juices in the stomach. This might seem to happen "inside" our bodies,

but think about it more carefully. Our digestive system is a tube that runs through our bodies from one end to the other. The inside of the tube is actually outside our bodies! Think of the hole in a donut. When your finger is in the hole, it is actually outside the donut. A hole or a tube is an empty space. Thus, digestion occurs outside the body.

<u>Respiration</u> is a series of chemical reactions that always takes place *inside a*

cell. These reactions result in the extraction of usable energy from digested food.

The term *cellular respiration* may be confused with the respiration that is breathing. <u>Breathing</u> is the exchange of air between an animal's body and the environment. When biologists talk of cellular respiration they mean not breathing but the acquisition of energy via the breakdown of molecules.

Powering the Machinery of the Cell

The letters ATP stand for adenosine triphosphate, a molecule that transfers energy in the cell. When ATP gives up its energy, it is converted into a lower-energy-containing molecule called ADP. ATP is much like a small, rechargeable battery. When charged, the battery contains the

power necessary to run many kinds of machines, such as flashlights and tape recorders. When the energy has been released from the battery, it can be recharged if it is plugged into a recharging unit. The mitochondria act as such recharging units in the cell. ATP can release

its energy to contract muscle fibers, transmit nerve impulses, move nutrients into cells, carry waste out of cells, and even make some animals glow in the dark. In the mitochondria, high-energy ATP is resynthesized from low-energy ADP to be used again.

Cellular Respiration

In aerobic cellular respiration, food and oxygen are chemically combined and rearranged to form carbon dioxide and water. This chemical rearrangement allows the cell to convert the chemical-bond energy from the food molecule into the chemical-bond energy of ATP molecules. These ATP molecules (energy carriers) can then be used to power cellular activity.

In the series of chemical activities known as aerobic cellular respiration, the rate of energy release is controlled to prevent damage to the cell and to maximize ATP production. The chemical equation for the reaction is shown below. Notice that when the food is combined with oxygen, the large food molecule is rearranged into several smaller molecules of carbon dioxide and water. Most importantly, the energy that held the food molecule together is released and is used to charge up the energy carrier molecule, ATP.

$$C_6H_{12}O_6 + 6O_2 \rightarrow 6CO_2 + 6H_2O + 36\ ATP$$

Food / and / oxygen / form / carbon dioxide / and / water / and / usable energy

In eukaryotic cells, the process of energy release from food molecules begins in the cytoplasm and is completed in the mitochondria. There are three major parts of the respiratory process: glycolysis, the Krebs cycle, and the electron transfer system (ETS).

The first major part of the aerobic cellular respiration process is known as *glycolysis* (*glyco* = sugar; *lysis* = breaking), which splits a sugar molecule into two smaller molecules. During glycolysis oxygen is not required and only small amounts of energy are released to form ATP and heat. In the Krebs cycle, the two smaller molecules from glycolysis are taken into the mitochondrion where carbons are removed to form carbon dioxide, a waste product. In addition, during glycolysis and the Krebs cycle, hydrogens are removed. The energy from the hydrogen electrons is transferred to ATP in the electron transfer system. This part of the process requires oxygen (making it aerobic) and produces water. It is during the ETS that most of the energy in food is converted to ATP.

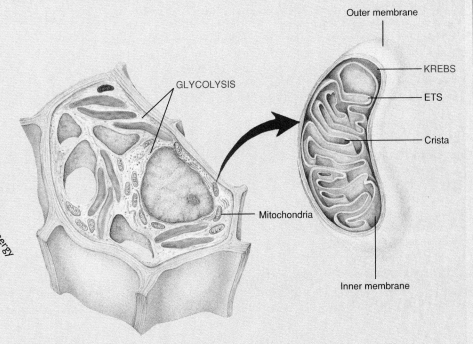

GLYCOLYSIS

Mitochondria

Outer membrane

KREBS

ETS

Crista

Inner membrane

Energy Release in Cells   Glycolysis takes place in the cytoplasm of the cell. The Krebs cycle and electron transfer system take place inside mitochondria.

Acids, Bases, and Salts

Acids and bases are two classes of biologically important compounds that affect the habitats of animals. When acids are dissolved in water, *hydrogen ions* (H^+) are set free. A hydrogen ion is a positively charged hydrogen atom. An **acid**, then, is a compound that releases hydrogen ions when dissolved in a solution. Because the ions have a positive electrical charge, they attract negatively charged particles. A common acid with which you are probably familiar is the sulfuric acid (H_2SO_4) in your automobile battery.

A **base** is the opposite of an acid; it is a compound that releases a group known as a *hydroxyl ion*, or OH^- group. This group is composed of an oxygen atom and a hydrogen atom attached together, but with a negative electrical charge. A base can also be thought of as any substance that is able to attract positively charged particles. A very strong base used in oven cleaners is sodium hydroxide (NaOH).

The degree to which a solution is acidic or basic is represented by a value known as **pH.** The pH scale is a measure of hydrogen ion concentration. A pH of 7 indicates that the solution is neutral; it has an equal number of H^+ ions and OH^- ions to balance each other. As the pH number gets smaller, the number of hydrogen ions in the solution increases. A number higher than 7 indicates that the solution has more OH^- than H^+. As the pH number gets larger, the number of hydroxyl ions increases.

Another group of biologically important ionic compounds is called the *salts*. **Salts** are compounds that do not release either H^+ or OH^-; thus, they are neither acids nor bases. They are generally the result of a reaction between an acid and a base in a solution. For example, when an acid such as H^+Cl^- is mixed with Na^+OH^- in water, the acid and the bases separate into their component parts (H^+, Cl^-, Na^+, and OH^-). The H^+ and the OH^- combine with each other to form water (H_2O). The remaining ions (Na^+ and Cl^-) join to form the salt NaCl (sodium chloride).

$$HCl + NaOH \rightarrow$$
$$(Na^+ + Cl^- + H^+ + OH^-) \rightarrow NaCl + H_2O$$

The chemical process that occurs when acids and bases react is called *neutralization*. The acid no longer acts as an acid (it has been neutralized) and the base no longer acts as a base.

Where Animals Live

Animals inhabit almost every part of the surface of the earth. If we look at various animals we can see structural and behavioral characteristics that allow them to survive in their particular environments or habitats. **Habitat** is the place where an organism lives. It is usually described by a significant feature. Habitats are divided into terrestrial (land) and aquatic; aquatic habitats are either saltwater (marine) or freshwater. Let us look at some of the specializations that enable the animals to live in diverse habitats.

Terrestrial Habitats

Animals that live on land must be able to adjust to widely changing conditions. They are exposed to fluctuations in air temperature, humidity, wind velocity, pH, and soil chemicals. In addition, the "thin" air does not provide the same kind of body support as does the water that buoys aquatic animals. On land, special skeletal structures must provide support and enable movement.

To be successful, terrestrial animals must adjust to changing conditions or avoid them. Animals live in many different microhabitats in order to avoid some of the difficulties of living on land. Microhabitats are small environments such as rodent burrows, worm tunnels, a space under a rock, or the fur of other animals. Many of the difficulties presented by terrestrial habitats arise from animals' need to conserve moisture inside their cells, when the air tends to dry them out. Most terrestrial animals, such as insects, spiders, birds, mammals, and reptiles,

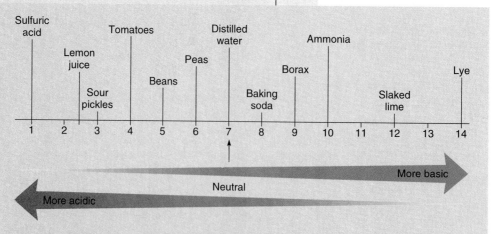

The pH Scale The concentration of acid is greatest when the pH number is lowest. As the pH number increases, the concentration of base increases. At a pH of 7.0, the concentrations of H^+ and OH^- are equal.

Appendix A

have protective outer layers that prevent drying. Other animals, such as toads, frogs, salamanders, and earthworms, move to moist, shady areas under leaves and dead trees or in wetlands to keep from drying out.

In addition to the problems of drying out and body support, many animals must control their body temperature in order to survive. *Ectothermic* animals regulate their body temperature by moving to places where they can be most comfortable. If they become overheated during the hot part of the day, they must move to cooler areas. During the winter, ectothermic animals such as insects, frogs, snakes, and spiders seek sheltered places to avoid being frozen. *Endothermic* animals have internal temperature regulating mechanisms and can maintain a relatively constant body temperature in spite of wide variations in their environment. Many birds migrate to warmer climates to avoid the coldest part of the year; birds that remain in cold climates, as well as resident mammals, survive because they have insulating coats of feathers or fur.

Aquatic Habitats

Aquatic habitats can be divided into freshwater and saltwater (marine). Water provides physical support to animals. You have noticed how water buoys you up when you swim. Certain kinds of physical therapy are done in the water because water helps to support the injured part of the body. Many small aquatic animals such as jellyfish and worms have no skeletal systems; they rely on water to support them.

Aquatic habitats tend to be relatively constant. Temperature does not change rapidly and organisms do not dry out. However, oxygen for respiration must be obtained from the water. Water, or H_2O, is composed of two hydrogen atoms and one oxygen atom, but the oxygen in this molecule is not available for the animal to use in aerobic cellular respiration. Instead, the animal must obtain oxygen (O_2), which is dissolved in the water. The amount of oxygen that can dissolve in water depends on the temperature of the water and the amount of movement of the water. A great deal of oxygen may be dissolved in water that is rushing over rocks in a brook or stream, or in oceans where the waves are crashing against the shore. Water deep in the ocean or in a still lake contains less dissolved oxygen. Some animals, like trout, can live only in water that contains a great deal of dissolved oxygen, while others, like carp, can live in stagnant water where not as much oxygen is available. Certain water insects prefer the highly oxygenated water of a fast-moving stream while others can live in stagnant, temporary ponds. The presence or absence of particular animal species can tell us about the physical and chemical condition of a body of water.

Freshwater and marine environments present slightly different problems for animals. All freshwater animals tend to gain water as a result of osmosis, because their body tissues contain a lower concentration of water than their surroundings. Most marine animals, such as sponges, jellyfish, starfish, and many others, do not have this problem because the water concentration in their bodies is the same as that of the environment in which they are living. Other marine animals, such as fishes, turtles, and birds, actually lose water to their surroundings and must drink water and secrete salt to stay alive. Ocean shore birds secrete excess salt from glands at the base of their bills. A salt buildup can sometimes be seen on their bills.

One of the most interesting groups of inhabitants of saltwater environments is the spiny-skinned animals. Sea stars, starfish, and sand dollars are all examples of this group. These animals are built around a central point and have a radial pattern like spokes on a wheel. If you look closely at a starfish you will notice many bumps or spines on its surface and small, soft tubular extensions on its underside. These small tubes, or tube feet, are controlled by an internal pumping (hydraulic) system. They are used by the animal to hold food and to move about.

CONCEPT CONNECTIONS

- **Poikilotherms** are animals whose body temperature fluctuates with the temperature of their environment. These animals are often called cold-blooded because their body temperature is usually lower than ours.

- **Homeotherms** are animals (birds and mammals) that keep a constant body temperature. These animals are called warm-blooded and use cellular respiration to generate heat. They also have mechanisms for cooling the body if it becomes too warm.

- **Ectothermic** animals can only regulate their body temperature by changing their location to one with a different temperature..

- **Endothermic** animals have internal regulating mechanisms to maintain a relatively constant body temperature in spite of wide temperature variations in their environment.

Responses to the Environment

The degree to which an animal can respond to its environment is determined by its ability to sense what is going on around it, integrate the sensory information from various sources in the nervous system, and send nerve messages to muscles and glands that cause the animal to respond. We are most familiar with the senses we have: touch, smell, hearing, sight, and taste. We can also sense cold and pain, and experience jet lag. Some birds and fishes are able to sense the magnetic poles of the earth, many insects can see ultraviolet light, and many fish can measure electrical changes in the water.

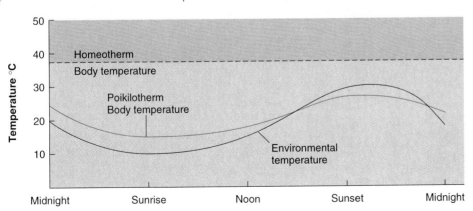

Regulating Body Temperature   The body temperature of a homeotherm remains constant regardless of changes in environmental temperature. The body temperature of a poikilotherm is dependent upon environmental temperature.

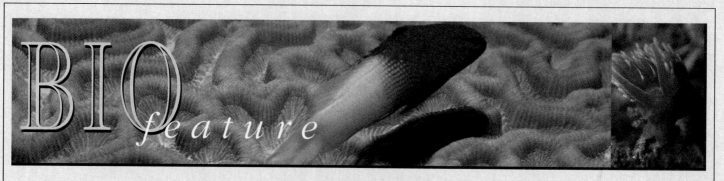

BIO *feature*

Photo Essay: Echinoderms

Echinoderm Variety

(a) Starfish

(b) Brittle star

(c) Sea urchin

(d) Sea cucumber

(e) Sand dollar

(f) Crinoid

Most animals have enlarged portions of the nervous system where integration of sensory information occurs. Often this is called a brain (figure 3.5). A mammal, with its keen sense organs, is constantly receiving information about its surroundings. It may move toward or away from the stimulus; it may respond with aggression or it may hide. The variety of responses mammals can make to a particular stimulus is due to their highly developed brains, keen sense organs, and flexible skeletal and muscular systems.

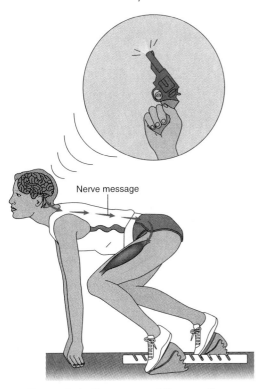

Figure 3.5 Sensing and Responding
All animals are able to sense changes in their surroundings and modify their behavior to respond to those changes. This involves some kind of sensory structure, a nervous system, and specialized cells for movement.

Photo Essay: Animal Sight

Sight

(a) What humans see

(b) What color-blind animals see

(c) What insects that detect ultraviolet light see

Organisms less complex than mammals can also respond to stimuli. Mosquitos can find you (they appear to be able to smell the carbon dioxide in your breath); snails retreat into their shells when touched; earthworms retreat into their burrows when you shine a light on them; and even simple jellyfishes respond to the presence of food or predators. Many of the stimulus and response activities of animals are related to acquiring food, protecting against enemies, finding mates, and producing offspring.

The Cellular Organization of Animals

Animals are multicellular organisms composed of eukaryotic cells that lack cell walls. This can be determined by examining samples of animal tissue with a microscope. All animals have cells that are specialized for particular functions. A group of similar cells that work together to perform a particular function is called a tissue. The tissues in your body include skin (epithelial) tissue, muscle tissue, nervous tissue, and connective (fat, bone, blood, and cartilage) tissue. Every cell in each of these tissues has a nucleus and a surrounding cytoplasm, which contain many kinds of small structures called organelles. Some organelles, like mitochondria, provide energy. Others, like microtubules, provide structure or enable movement. Golgi bodies secrete materials like mucus, digestive juices, or salt to the surface of the cell. While each cell is an independent unit, the cells in a tissue interact with one another to perform a particular service for the entire animal. Even the most uncomplicated sponges have cells organized into tissues.

In complex animals such as fishes, lobsters, insects, and worms, many tissues are organized to form an organ. For example, the heart is composed of muscle tissue, connective tissue, and nervous tissue; the stomach consists of muscle tissue, nervous tissue, epithelial tissue, and connective tissue. If you were to look at thin slices of an organ like the stomach under a microscope you would be able to recognize all the different kinds of tissues (figure 3.6).

In most animals, organs are arranged in groups with each organ contributing to an overall function. These sets of organs are called organ systems.

Reproduction

Most animals reproduce sexually. This involves the joining of two different cells, usually called eggs and sperm. Because each new

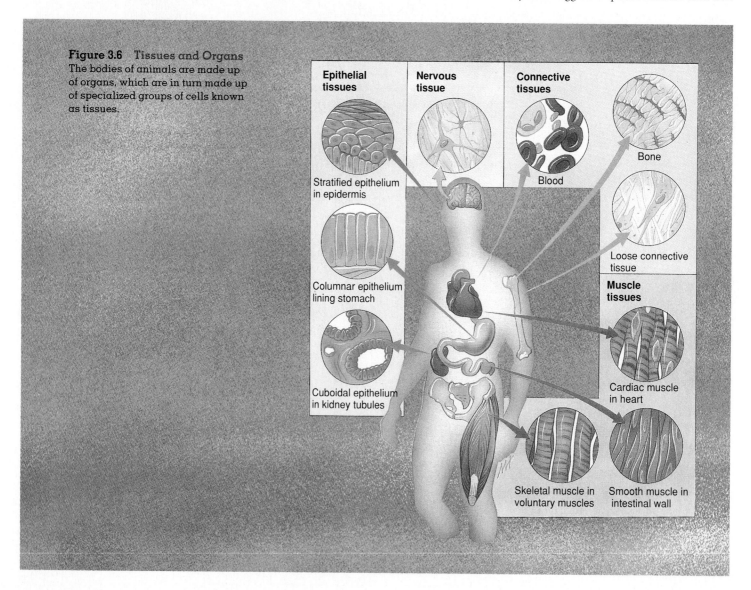

Figure 3.6 Tissues and Organs
The bodies of animals are made up of organs, which are in turn made up of specialized groups of cells known as tissues.

Epithelial tissues

Nervous tissue

Connective tissues

Bone

Stratified epithelium in epidermis

Blood

Columnar epithelium lining stomach

Loose connective tissue

Muscle tissues

Cuboidal epithelium in kidney tubules

Cardiac muscle in heart

Skeletal muscle in voluntary muscles

Smooth muscle in intestinal wall

Organ Systems

Digestive system: Includes mouth, esophagus, stomach, intestines, and associated organs such as the liver, gizzard, gall bladder, and pancreas.

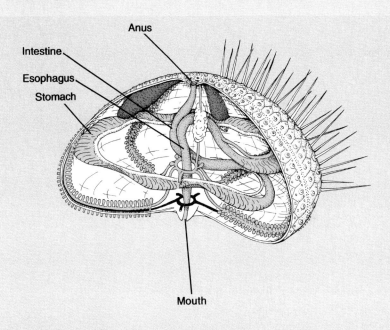

Excretory system: Includes kidneys, urinary bladder, and connecting ducts such as the urethra and ureter.

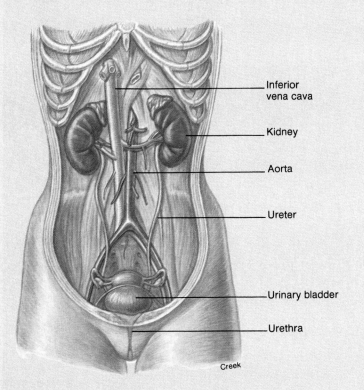

Creek

Continued

Respiratory system: Includes lungs, bronchial tubes, gills, and collar cells.

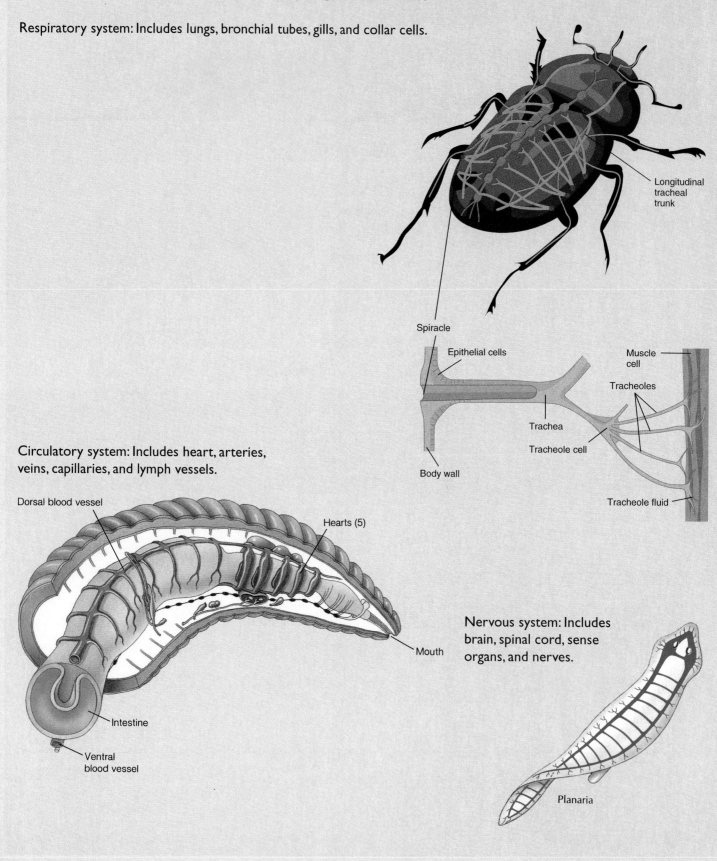

Longitudinal tracheal trunk

Spiracle

Epithelial cells

Muscle cell

Tracheoles

Trachea

Tracheole cell

Body wall

Tracheole fluid

Circulatory system: Includes heart, arteries, veins, capillaries, and lymph vessels.

Dorsal blood vessel

Hearts (5)

Mouth

Intestine

Ventral blood vessel

Nervous system: Includes brain, spinal cord, sense organs, and nerves.

Planaria

Glandular system: Includes mammary, thyroid, testes, ovaries, pituitary, sweat glands, and salivary glands.

From Milton Hildebrand, *Analysis of Vertebrate Structure,* 4th edition. Copyright © 1995 John Wiley & Sons, Inc., New York. Reprinted by permission of John Wiley & Sons, Inc.

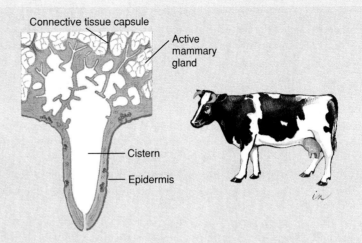

Connective tissue capsule

Active mammary gland

Cistern

Epidermis

Skeletal system: Includes bones and cartilage.

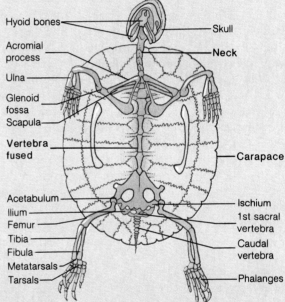

Hyoid bones
Skull
Acromial process
Neck
Ulna
Glenoid fossa
Scapula
Vertebra fused
Carapace
Acetabulum
Ischium
Ilium
1st sacral vertebra
Femur
Tibia
Caudal vertebra
Fibula
Metatarsals
Tarsals
Phalanges

Muscular system: Includes voluntary, involuntary, and heart muscles.

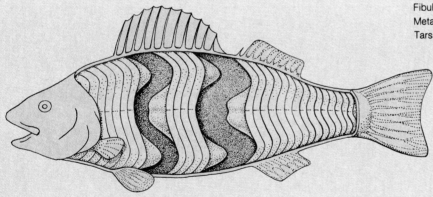

Reproductive system: Includes testes, ovaries, uterus, penis, and vagina.

The excretory and reproductive systems of a female bird.

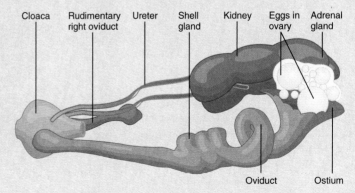

Cloaca
Rudimentary right oviduct
Ureter
Shell gland
Kidney
Eggs in ovary
Adrenal gland
Oviduct
Ostium

animal is the result of input from both parents, it is likely to have a mix of the characteristics of each parent. Each offspring can be different from every other animal in some small way.

In animals, females have the ability to produce many more egg cells than will ever be fertilized, develop, and grow up. Their mates produce even larger numbers of sperm cells. Only one of these sperm cells can fertilize an egg cell, but the production of large numbers of sperm cells is necessary to increase the likelihood that fertilization will occur.

Mammals, birds, reptiles, and insects reproduce by internal fertilization, which improves the chances that sperm cells and egg cells will unite. The eggs and sperm are also protected from drying out. This process of depositing the sperm cells within the reproductive tract of the female is called **copulation**, and since the egg is fertilized inside the reproductive tract of the female it is called *internal fertilization*. Courtship behaviors usually assure that individuals of the same species but of opposite sex engage in copulation.

A disadvantage of sexual reproduction is that the process is costly in terms of energy and resources. After an egg cell is fertilized within the reproductive tract of the female, the female must expend a great deal of time and energy to incubate it. In humans, the female provides nutrients to the developing embryo through the placenta for three quarters of a year. Some mammals are pregnant for longer periods and some for shorter periods; the important point is that the support of this potential offspring is very costly to the female parent.

Pouched mammals, such as the opossum and the kangaroo, incubate their young for much shorter periods of time in their reproductive tracts. Shortly after birth, the young crawl from the birth canal to the pouch where they continue their development. The young are still nourished by the female parent, but the development continues in the pouch rather than in the uterus.

Birds, reptiles, and insects also reproduce by internal fertilization, but the development of the offspring usually continues outside the body of the female. The eggs are large and contain a great deal of stored food material that the young use as they develop inside the protective shell. When birds lay eggs they frequently perform elaborate behavior patterns to ensure that the eggs are protected from predators and from harsh environmental conditions. The parents provide warmth for the developing embryo as they incubate the eggs. Since the newly hatched chicks are unable to care for themselves, the parents provide protection and food for a period of time.

Reptiles like turtles, alligators, lizards, and snakes usually lay eggs in protected nests but do not remain in the area to protect them. There are some exceptions to this general rule: Alligators will protect their nests, and some snakes and lizards retain eggs that hatch inside the body of the mother.

Insects typically lay eggs on the plant or animal that the larva will use as food or in a location where food can be found immediately after the larva emerges. In some cases (bees, ants, wasps, and termites), the young may be fed by other members of the colony.

Many amphibians and fishes release large numbers of egg and sperm into the water rather than having the male deposit sperm within the female. This is called *external fertilization*. Some amphibians use mating songs to attract members of their species. This increases the chances of fertilization by getting males and females near each other. Some male fish build nests to encourage females to deposit eggs where the male can fertilize them. Some species prepare a nest and then protect it; others leave their nest unattended. Still other fishes release eggs and sperm into the water at the same time.

Almost all other animals (sponges, mollusks, jellyfish, and marine worms) reproduce by external fertilization and external incubation. Many are also **hermaphroditic;** that is, they have both sexes in the same body. Examples of hermaphroditic animals are earthworms, snails, and some fishes. Since each individual produces both sperm and eggs, any two individuals are able to exchange sex cells. Some kinds of animals, such as oysters and some kinds of fish, change their sex depending on their age and environmental conditions.

Some animals can also reproduce by asexual means (figure 3.7). A common method of asexual reproduction is fragmentation. When a piece breaks off a parent sponge, floats away, and settles down, it can grow into an adult sponge. Similarly, when starfish are damaged and separated into several pieces, each piece might grow into an adult. Some kinds of jellyfish produce small buds that separate from the parent, move away, and grow into an adult. This budding is a specialized kind of fragmentation. Even many kinds of segmented worms and flatworms reproduce asexually by budding. Asexual reproduction can only form more individuals exactly like the parent.

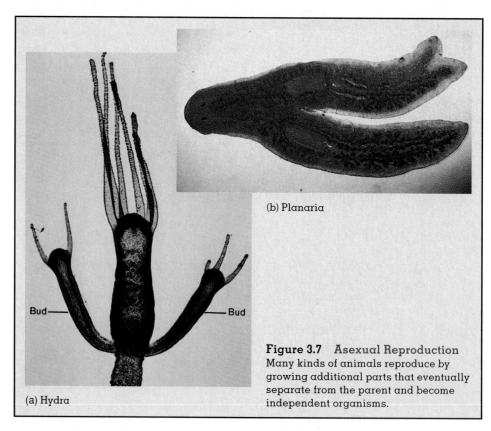

(b) Planaria

(a) Hydra

Bud — — Bud

Figure 3.7 Asexual Reproduction
Many kinds of animals reproduce by growing additional parts that eventually separate from the parent and become independent organisms.

Metamorphosis in Insects and Amphibians

One interesting characteristic of insects is the way in which the young develop and grow. The parent insect deposits eggs on or near the kind of food the hatching egg will require. The fertilized egg cell divides, and the embryo develops certain cells that are specialized for eating and processing food. When the egg hatches it is known as a larva.

At this stage of the insect life cycle the larva could be called an eating machine. As it continues to eat and grow, certain body parts are rearranged and modified and the larval stage gives rise to a pupa stage. During this stage the insect is still sexually immature and likely resembles neither the larva nor the adult that it will become. Finally, this pupa changes into a mature adult insect. The adult is sexually mature and capable of reproduction. When it lays eggs, the cycle is complete.

The process of changing from egg to larva to pupa and finally to adult is known as **metamorphosis.** It is valuable to the insect because it allows different specializations during different stages of the life cycle. Each stage has specific requirements that must be met for the insect to survive.

Some amphibians also undergo metamorphosis during their life cycle. For example, tadpoles develop into frogs. The fertilized frog egg uses stored food in its yolk to grow and divide as it develops into a tadpole. The tadpole feeds on pond plants and grows. Eventually, its tail is absorbed and legs and other structures form.

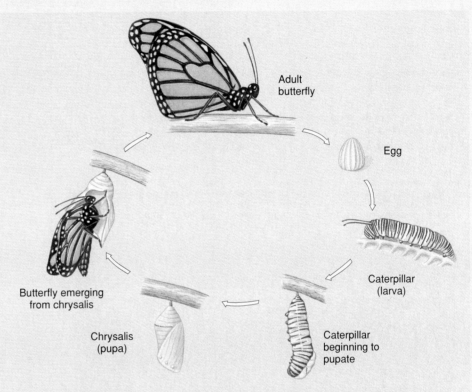

Labels: Adult butterfly · Egg · Caterpillar (larva) · Caterpillar beginning to pupate · Chrysalis (pupa) · Butterfly emerging from chrysalis

Life Cycle of an Insect Many insects go through a series of changes during their lives. The larva leads a very different life from the adult.

Egg	Numerous insect eggs are produced at one time. Because there is no parental care, many of these eggs do not survive. They are deposited on or near the food the developing egg will need as soon as it is hatched.
Larva	Wormlike stage. Specialized for taking in food and processing it.
Pupa	During this stage the internal structures of the organism are modified. Reorganization of the body occurs.
Adult	Specialized for sexual reproduction and dispersal of the population.

BIO *feature*

Parthenogenesis

Parthenogenesis is an unusual method of reproduction used by some insects, crustaceans, and rotifers. In these cases, an "egg" cell is produced but it is not fertilized. These cells develop into exact copies of the female parent. In these species, parthenogenesis may be the primary method of reproduction, but individuals also engage in sexual reproduction at other times. In some species, however, true sexual reproduction is not known to occur. Since the offspring have only one parent, they are genetically identical to that parent. Certain fishes, amphibians, and reptiles also reproduce by parthenogenesis. Populations of certain whiptail lizards are entirely female.

Parthenogenesis   Some species of whiptail lizard reproduce by parthenogenesis.

Animal Kingdom
  Phylum Chordata
    Vertebrates (backbone)
      Mammals (class Mammalia)
        Hair covers body
        Internal fertilization
        Mammary glands to feed young
        Placenta
        Four-chambered heart
          Monotremes
            Egg layers
              Platypus
          Marsupials
            Pouched
              Kangaroo, opossum
          Artiodactyla
            Hoofed, even number of toes
              Sheep, deer, giraffe
          Carnivora
            Meat eaters
              Wolves, weasels
          Cetaceans
            Marine, fishlike
              Whales, dolphins
          Chiroptera
            Flying
              Bats
          Edentata
            No teeth
              Sloth, anteater
          Insectivora
            Insect eaters
              Mole, shrew
          Lagomorphs
            Chisel-like front teeth
              Rabbit, hare
          Perissodactyla
            Hoofed, odd number of toes
              Horse, zebra, rhinoceros
          Primates
            Large brain, eyes front
              Human, lemur, ape
          Proboscidia
            Trunk and tusks
              Elephant
          Rodentia
            Incisor teeth with continual growth
              Squirrel, beaver, mice
          Sirenia
            Aquatic, only forelimbs
              Sea cow, manatee
      Birds (class Aves)
        Feathers cover body
        Internal fertilization
        Eggs with calcified shell
        Forelimbs adapted for flight

Four-chambered heart
  Chicken, eagle, sparrow
Reptiles (class Reptilia)
  Scales cover body
  Internal fertilization
  Membrane-enclosed egg
  Lungs
  Poikilotherm
    Turtle, snake
Amphibians (class Amphibia)
  Moist skin, no scales
  External fertilization typical
  Metamorphosis common
  Lungs in adult form
  Three-chambered heart
    Toad, frog, salamander
Bony fishes (class Osteichthyes)
  Scales
  Gills to acquire oxygen from water
  External fertilization usual
  Limbs are fins
  Two-chambered heart
    Perch, trout, carp
Cartilaginous fishes
(class Chondrichthyes)
  No bone
  Limbs are fins
    Sharks, rays
Spiny-skinned (phylum Echinodermata)
  Spines on surface
  Radial symmetry
  Marine habitats only

Tube-feet and water vascular system
  Starfish, sand dollars
Mollusks (phylum Mollusca)
  Soft body, frequently protected with shell
  Bilateral symmetry
  Anterior head region
  Internal organs in visceral region
  Ventral foot
    Clam, squid, chitin
Arthropods (phylum Arthropoda)
  Insects (class Insecta)
    Head, thorax, and abdomen
    Appendages: antennae, legs,
      mouth parts, wings, etc.
    External skeleton
    Metamorphosis common: from
      egg, larva, pupa, to adult
      Ant, wasp, grasshopper, beetle
  Spiders (class Arachnida)
    Head/thorax and abdomen
    Appendages: four pairs of legs
    External skeleton
      Wolf spider, tick, scorpion, mite
  Millipedes and centipedes
  (classes Diplopoda and Chilopoda)
    Segmented body
    Two or one pair walking legs per
      segment
  Crustaceans (class Crustacea)
    External skeleton
    Walking legs and other appendages

Mostly marine habitat
  Lobster, crab, crayfish,
    isopods, and barnacles
Segmented worms (phylum Annelida)
  Segmented body
  Oxygen uptake through skin
  Circulatory, muscular, digestive, and
    nervous system
    Fanworm, earthworm, leech
Roundworm (phylum Nematoda)
  Cylindrical body
  Not segmented
  Digestive tract
  Many parasitic forms
  Many in soil habitat
    Pinworm, soil nematode
Flatworm (phylum Platyhelminthes)
  Flattened body
  Not segmented
  Sac type digestive system
    Planaria, liver fluke
Cnidaria (phylum Cnidaria)
  Radial symmetry
  Tentacles and stinging cells
  Aquatic habitat
  Some are sessile
    Jellyfish, coral
Sponges (phylum Porifera)
  Pores for water circulation
  Sessile as adults
  Aquatic

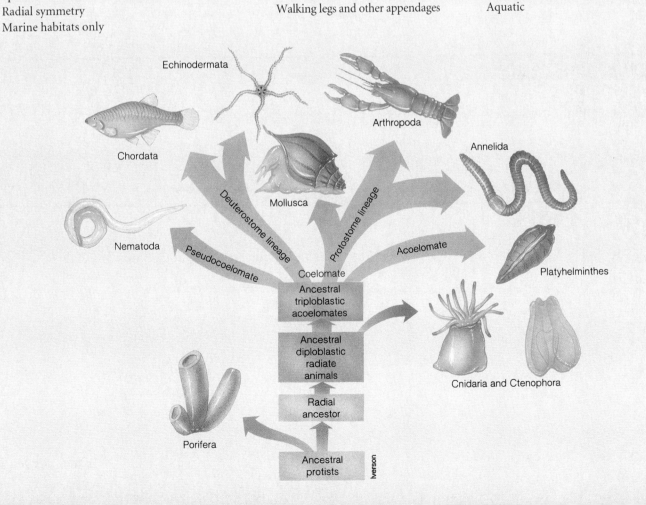

Summary

The 4 million known species of animals inhabit widely diverse environments, and are all multicellular and heterotrophic. A feature of animals is their ability to move through their environments using methods such as running, walking, flying, and swimming. Animals that spend a significant portion of their lives fixed to one spot are sessile. Movement allows animals to acquire food more readily. All animals process their food using aerobic cellular respiration, expressed by the chemical formula $C_6H_{12}O_6 + 6O_2 \rightarrow 6CO_2 + 6H_2O$ + energy. Animals exhibit all the characteristics of life. The characteristic known as responsiveness varies among the animals and involves nerves, muscles, and hormones. Reproduction in animals can be sexual or asexual. In some species, male and female reproductive organs occur in the same animal, while other species have separate sexes. Asexual reproduction includes parthenogenesis, budding, and fragmentation. Certain animals go through developmental states that take on body forms greatly different from the adult form, a process called metamorphosis.

CHAPTER GLOSSARY

acid (ăs′id) Any compound that releases a hydrogen ion in a solution.

base (bas) Any compound that releases a hydroxyl group in a solution.

cartilage (kar′ tl-ij) A flexible, plasticlike supporting tissue in many animals; in humans, it is in the nose, ears, and certain joints.

copulation (kop-yu-la′shun) The mating of male and female; the deposition of the male sex cells, or sperm cells, in the reproductive tract of the female.

ectothermic (ek″te-ther′ mik) Animals whose body temperature is determined by the external environment.

endothermic (en″ de-ther′ mik) Animals that have internal temperature regulating mechanisms.

habitat (hab′ i-tat) The place or part of an ecosystem occupied by an organism.

hermaphroditic (her-ma-fre-di′ tik) Having both sexes in the same body.

homeotherms (ho′ mee-ah-therms″) Animals (birds and mammals) that keep a constant body temperature; they are called warm-blooded.

metamorphosis (me″ te-mor′ fe-ses) The process of changing from one body form to another during the developmental process; for example, from egg to larva to pupa and finally to adult.

motile (mo′ tile) Able to move about freely; not sessile.

organ (or′gun) A structure composed of two or more kinds of tissues.

parthenogenesis (par″ the-no-je′ ne-ses) A form of asexual reproduction in which females produce eggs that are not fertilized and that develop into offspring genetically identical to the parent.

pH A scale used to indicate the strength of an acid or base.

poikilotherms (poy-key′ le-therms″) Animals whose body temperature fluctuates with the temperature of their environment; often called cold-blooded because their body temperature is usually lower than ours.

sessile (se′ sile) Stationary.

tissue (tish′yu) A group of specialized cells that work together to perform a particular function.

CONCEPT MAP TERMINOLOGY

copulation	homeotherms
ectothermic	metamorphosis
endothermic	parthenogenesis
hermaphroditic	poikilotherms

LABEL • DIAGRAM • EXPLAIN

Compare the structures and adaptations of arthropods with those of mammals such as humans in terms of their:

skeletons
sense organs
reproductive processes and potentials
habitats

Multiple Choice Questions

1. Which of the following is *not* true for all members of the animal kingdom?
   a. multicellular
   b. eating
   c. sexual reproduction
   d. responding to the environment
2. Heterotrophy means that animals must:
   a. manufacture their own food
   b. reproduce sexually with two different sexes
   c. eat food
   d. form lasting relationships with members of other groups
3. Cellular respiration is:
   a. a way of releasing energy from food
   b. the way in which a cell breathes
   c. the requirement that all animals must breathe air
   d. a process that produces oxygen
4. Which of the following indicates the greatest amount of acid?
   a. pH of 3
   b. pH of 5
   c. pH of 9
   d. pH of 11
5. Metamorphosis
   a. involves asexual reproduction
   b. allows animals to have specialized stages in the life cycle
   c. reduces the number of eggs produced
   d. is common in birds and mammals
6. Which of the following animals has muscles on the insides of its skeleton?
   a. elephant
   b. jellyfish
   c. lobster
   d. earthworm
7. Which one of the following pairs of items is *not* correctly related?
   a. arthropod–crab
   b. mollusk–squid
   c. cnidarian–coral
   d. echinoderm–centipede
8. Organisms that copulate have:
   a. more offspring than others
   b. a better chance of survival
   c. larger numbers of eggs
   d. internal fertilization

Questions with Short Answers

1. Name four characteristics of an animal.

2. Write the formula for aerobic cellular respiration and explain its significance.

3. Describe three methods used by animals to acquire food.

4. How are animals adapted to survive in an aquatic environment? a terrestrial environment?

5. What advantages does an animal that flies have over an animal that does not fly?

6. How do cells, tissues, organs, and organ systems differ? How are they related to one another?

7. What type of life does a sessile animal have? Give an example of a sessile animal.

8. What happens during the life of an animal that undergoes metamorphosis?

9. Identify different methods of reproduction among animals.

10. Describe how an endothermic animal differs from an ectothermic animal.

The *Life* of *Plants*

Climbing plant on evergreen tree, South Chile.

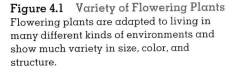

learning objectives

- Recognize the characteristics of plants.
- Realize that plants react to their surroundings.
- Know the basic structures of a plant.
- Describe the value of photosynthesis.
- Understand the difference between sexual and asexual reproduction.
- Realize that plants have varied relationships with other kinds of organisms.
- Understand that mosses, ferns, conifers, and flowering plants are different kinds of plants.

What Is a Plant?

Because plants quietly go about their business of feeding the world, they often are not noticed or appreciated by the casual observer. Yet we are aware enough about plants to associate them with the color green. Grass, garden plants, and trees are all predominantly green, the color associated with photosynthesis. It is the green pigment, chlorophyll, that captures light and allows the process of photosynthesis to store the energy needed by all other organisms. Yet in this quiet sea of green, there is incredible variety and complexity. Plants range in size from tiny floating duckweed the size of your pencil eraser to giant sequoia trees as tall as the length of a football field (figure 4.1). A wide range of colors (e.g., red, yellow, orange, purple, white, pink, violet) stand out against the basic green we associate with plants. Bright spots of color are often flowers and fruit, where the colors serve as attractants for animals.

Plants are adapted to live in just about any environment. They live on the shores of oceans, in shallow freshwater, the bitter cold of the arctic, the dryness of the desert, and the driving rains of tropical forests. There are plants that eat animals, plants that are parasites, plants that don't carry on photosynthesis, and plants that strangle other plants. They show a remarkable variety of form, function, and activity.

If we were asked to decide what makes something a plant, the list might include:

1. being anchored in the soil
2. having hard, woody tissues that support the plant
3. being green and carrying on photosynthesis

Although there are exceptions to these criteria, they are a good starting point to explore what it is to be a plant. We will begin by addressing plants with which you are most familiar: those that produce flowers and fruits. Nearly all flowering plants have the three major body parts called roots, stems, and leaves.

Roots

When you attempt to pull a plant from the ground, you quickly recognize that the underground parts of the plant, the **roots,** anchor it firmly in place. This places serious limitations on plants, since they are unable to move if conditions worsen or new resources become available in different locations. They cannot run away from animals that want to eat them or take evasive action from other dangers. Plants must resist whatever environmental challenges come their way while staying rooted in one place.

Roots have a variety of functions in addition to serving as anchors. Foremost among them is taking up water and nutrients from the soil. The primary nutrients plants obtain from the soil are inorganic molecules, which are incorporated into the organic molecules they build. In order to obtain these molecules, roots must constantly grow so that they are exploring new territory. As a plant becomes larger it needs more root surface to absorb nutrients

(a)

Figure 4.1 **Variety of Flowering Plants** Flowering plants are adapted to living in many different kinds of environments and show much variety in size, color, and structure.

------------ continued

(b)

(f)

(c)

(d)

(e)

BIO *feature*

Carnivorous Plants

Photosynthetic plants make complex molecules from simple inorganic components such as CO_2 and H_2O. One type of organic molecule they must make is protein. In order to do this they must have a source of the element nitrogen. Some plants have evolved unique ways of obtaining nitrogen. Animals have nitrogen in the molecules that make up their muscles and other tissues. (If you wanted a high protein meal you would eat meat.) Some plants obtain nitrogen by capturing and digesting animals.

Venus flytraps have hinged leaves that close on insects that come in contact with sensitive hairs on the leaves. Once the insect is enclosed, the plant digests it by releasing enzymes onto the insect's body. These enzymes break down insect tissue, releasing nitrogen-containing molecules for use by the Venus flytrap.

Pitcher plants have tubular leaves lined with downward pointing hairs. They attract insects because they have reddish-colored veins at the mouth of the tube.

The tubular shape allows the collection of water and nectar at the base of the plant. Insects are attracted by the odor and color pattern and are guided downward by the hairs. Once past the hairs in the plant, the insects slip into a pool of digestive enzymes.

Sundews have sticky hairs on their leaf surfaces that capture insects. Once captured, the insect struggles to escape, but entangles itself even more and is ultimately digested by the plant.

(a) Pitcher plant

(b) Sundew

(c) Venus flytrap

and hold the plant in place. The actively growing portions of the root near the tips have large numbers of small hairlike cells (root hairs) that provide a large surface area for the absorption of nutrients.

We eat many kinds of roots such as carrots, turnips, and radishes. The food value they contain is an indication of another function of roots. Most roots are important storage places for the food produced by the above-ground parts of the plant. Many kinds of plants store food in their roots during the summer months and use this food to stay alive during the winter. The food also provides the materials necessary for growth the following spring. Although we do not eat the roots of plants such as maple trees, rhubarb, or grasses, their roots are as important in food storage as those of carrots, turnips, and radishes (figure 4.2).

Stems

Stems are the above-ground structures of plants that support the light-catching leaves where photosynthesis takes place. Most kinds of perennial plants also have buds on their stems that may grow to produce leaves, flowers, or new branches. Trees have stems that support large numbers of branches; vines have stems that require support; and some

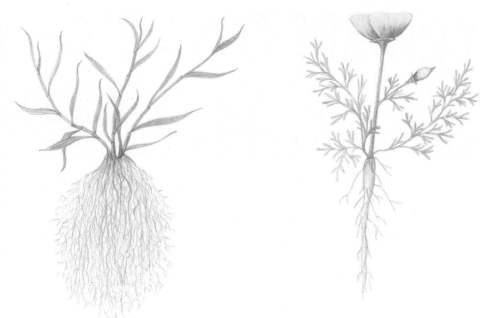

Figure 4.2 Kinds of Roots The roots of grasses are often in the upper layers of the soil and form a dense network that traps water in the dry environment. The roots of trees and many other plants typically extend deep into the soil where they obtain moisture and serve as anchors to hold the large plant upright.

FYI The labels on many plant "foods" can be misleading. The ingredients of the food do not provide the plant with energy, but provide other essential compounds required for plant growth and reproduction.

FYI *When Is a Root Not a Root?* Many foods we eat are not roots, even though they are dug from the soil. Some, like potatoes, sweet potatoes, and yams, are technically called *tubers*. Tubers are underground stems that store food for the plant. A close look at any of these will show "eyes," which are buds on the stem. Onions, leeks, and garlic store food in fleshy buds. When you take an onion apart you see that it consists of layers of "leaves." If you dissect other buds you will see the same arrangement. The little threadlike structures often found growing from the base of an onion are its roots.

A Special Formula to Promote Beautiful Flowers for Season-Long Enjoyment

FLOWER FOOD *Feeds up to 3 times longer!**

contains **POLYON**
Controlled-Release Nitrogen

8-10-8

*As compared to regular 8-10-8 fertilizer without controlled-release nitrogen.

Fertilizer label

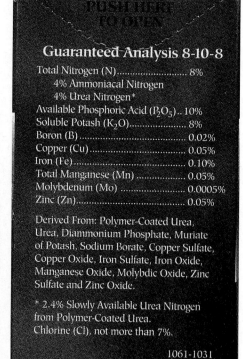

PUSH HERE TO OPEN

Guaranteed Analysis 8-10-8

Total Nitrogen (N).......................... 8%
  4% Ammoniacal Nitrogen
  4% Urea Nitrogen*
Available Phosphoric Acid (P_2O_5).. 10%
Soluble Potash (K_2O)..................... 8%
Boron (B) 0.02%
Copper (Cu) 0.05%
Iron (Fe) 0.10%
Total Manganese (Mn) 0.05%
Molybdenum (Mo) 0.0005%
Zinc (Zn)................................. 0.05%

Derived From: Polymer-Coated Urea, Urea, Diammonium Phosphate, Muriate of Potash, Sodium Borate, Copper Sulfate, Copper Oxide, Iron Sulfate, Iron Oxide, Manganese Oxide, Molybdic Oxide, Zinc Sulfate and Zinc Oxide.

* 2.4% Slowly Available Urea Nitrogen from Polymer-Coated Urea.
Chlorine (Cl), not more than 7%.

1061-1031

BIO *feature*

Successful Transplanting

Transplanting involves uprooting a plant and placing it in a new location. During this process many of the root hairs are broken, so the plant has less ability to absorb water. Therefore, it is very important to water plants well immediately after transplanting.

It also helps to remove many of the leaves, since the larger the leaf surface the more water the plant loses. Removing leaves and providing abundant water allows the plant time to grow new root hairs and repair the damage done during transplanting.

Root Hairs Root hairs increase the ability of plants to obtain water and nutrients from the soil because the roots have more surface area to absorb materials.

BIO *feature*

How Materials Enter Roots: Diffusion, Osmosis, and Active Transport

Materials enter and leave roots by three processes. In nature, all molecules are in constant motion, so when molecules are distributed unequally, they tend to move from an area of high concentration to an area of low concentration. For example, if the cells of a root have little oxygen and the surrounding soil has abundant oxygen, more oxygen molecules will move from the soil to the root cells than from the root cells to the soil. This process, in which there is a net movement of molecules from an area of high concentration to an area of low concentration, is called **diffusion.**

Water diffuses into roots in the same manner. But when water diffuses across a membrane that will not allow other materials to pass (we say the membrane is *semipermeable*), osmosis has taken place. **Osmosis** is a special case of diffusion in which there is a net movement of water molecules from an area of high concentration to an area of low concentration across a semipermeable membrane. Neither diffusion nor osmosis require any effort on the part of the plant. The energy for both processes comes from the random movement of molecules commonly called *molecular motion.*

Some materials are actively moved from the soil to the root. To do this, the plant must use energy and special carrier molecules; thus the process is called **active transport.** Since the cells of the root are expending energy to move the materials from outside the cell to inside the cell, it is possible to use that energy to move materials from an area of *low concentration* to an area of *high concentration.* Therefore, active transport can be used to accumulate and concentrate molecules.

Appendix B

Diffusion The cell membrane allows some materials to pass through it. If there is an unequal distribution of molecules, the net movement of molecules will be from the area of highest concentration to the area of lowest concentration.

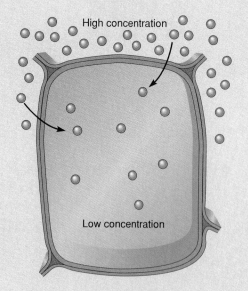

High concentration

Low concentration

FYI
Maple syrup is made from the sugar-rich sap that moves up the tree from the roots in the spring of the year. To make syrup, the sap is boiled to evaporate water. It takes about 40 liters of watery sap to make one liter of thick, sweet syrup.

Collecting sap for maple syrup

BIO *feature*

How We Use Plants

Uses of Plants People use plants for a wide variety of purposes. Food, clothing, shelter, medicines, and fuel are some of the necessities provided by plants.

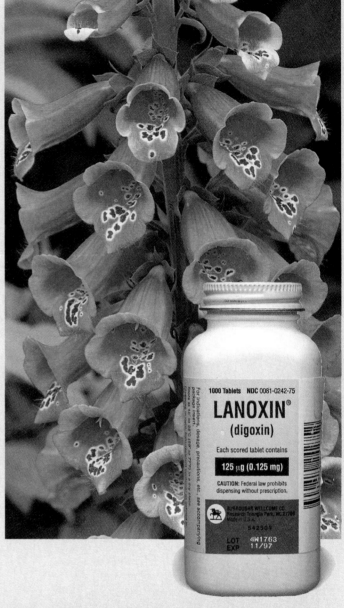

BIO *feature*

Adaptations of the Dandelion

The name dandelion probably comes from the French *dent de lion,* meaning tooth of the lion. These words describe the "toothed" shape of the leaves. The correct scientific name is *Taraxacum officinale.*

Because its leaves are flat on the ground, the dandelion cannot live where there are tall plants. It grows well in disturbed places where the natural vegetation is removed or altered and it can get plenty of sunlight. In many respects lawns are disturbed, because the grass is constantly cut. If lawns were not mowed regularly the grass would grow tall, dandelions would be shaded, and they would not be so successful.

The flower stalks grow rapidly from the short stem at the soil surface and position the seeds with their white, fluffy attachments up high so that they are easily distributed by the wind. The fleshy root at the base of the leaves stores food for the winter and allows the plant to come back year after year.

The creation of well-groomed lawns has provided ideal conditions for growth of this non-native plant, which was probably introduced to this country from Europe. An entire industry focused on destroying these "weeds" has sprung up. The word *weed* has come to mean any plant that

is unwanted or out of place. Yet many people do not see this plant as an ugly duckling in their yard and do not consider it a weed! Some value the dandelion for its beauty, while others use its leaves in salads and ferment its blossoms for dandelion wine.

Taraxacum officinale

plants, like dandelions, have very short stems with all their leaves flat against the ground. Stems have two main functions:

1. to serve as supports for the leaves.
2. to transport raw materials from the roots to the leaves and manufactured food from the leaves to the roots.

CONCEPT CONNECTIONS

- **Annual plants,** like corn and radishes, start from seeds each year, grow during the growing season, produce flowers, release seeds, and die. Their entire life cycle takes place within one year.
- **Biennial plants,** like carrots and parsley, take two years to go through their life cycle. The first year they store food in their roots. The second year they produce flowers and seeds and die.
- **Perennial plants,** like grasses, oak trees, and cacti, live several years (sometimes hundreds) and grow for several years before they are able to produce seeds.

When you chew on celery, carrots, or toothpicks, all of which are stems or made from stems, you recognize that they contain hard, tough materials. These are the cell walls of the plant cells (figure 4.3). All plant cells are surrounded by a cell wall made of a material known as cellulose. Cellulose fibers are interwoven to form a box within which the living portion of the plant cell is contained. Since the cell wall consists of fibers, it can be compared to a layer of fabric made from threads. It has spaces between the fibers through which materials pass relatively easily. However, the cell wall does not stretch very much, and if the cell is full of water and other cellular materials it will become quite rigid. Together, the many cells are able to support large bodies like trees. You might think of a plant body as being similar to the bubble wrap used to protect fragile objects during shipping. Each little bubble contains air and can be easily popped. The entire structure, however, will support considerable weight.

In addition to cellulose, some plants deposit other materials in the cell walls that strengthen them, make them more rigid, and bind them to other neighboring cell walls. Woody plants deposit a material called lignin and grasses deposit silicon dioxide, the material from which sand is made. Stems and roots of

Figure 4.3 Plant Cell Walls A cross section of a portion of stem tissue.

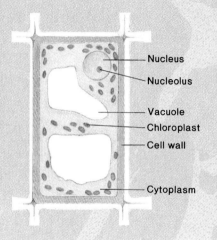

Nucleus
Nucleolus
Vacuole
Chloroplast
Cell wall
Cytoplasm

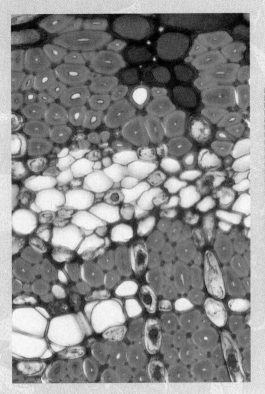

plants tend to have large numbers of cells with strengthened cell walls. This is such an effective support mechanism that large trees are supported against the pull of gravity and can withstand strong winds for centuries without being broken or blown over. Some of the oldest trees on earth have been growing for several thousand years.

When a plant is wounded it usually drips liquid, called sap, from the cut surface. This is because some of the thick-walled cells serve as pipes for transporting liquids throughout the plant. Water and minerals need to be transported to the leaves where photosynthesis takes place, and manufactured food needs to be exported from the leaves to a storage site, like the roots or the stems. Because the function is similar to that of our circulatory system, this tissue is called **vascular tissue.** The vascular tissue consists of many cells connected to one another end to end, thus forming many long tubes similar to a series of pieces of pipe hooked together, only much smaller (figure 4.4). The long strands in celery are good examples of vascular tissue. These strands that get stuck between your teeth are

table 4.1

Molecules Plants Use for Food Storage

Molecules	Examples	How We Use It
Sugar	Sugar beet roots Sugar cane stems Sorghum stems	We extract the sugar from the root or stem and eat the refined product.
Starch	Potatoes (underground stems) Rice seeds Turnip roots	We cook and eat the parts that have stored the starch.
Oil	Seeds of many kinds (peanuts, sunflowers, cotton, corn, coconut) Fruits of olives	Many seeds contain oils that can be squeezed from the seeds. Most cooking oils come from seeds.

bundles of vascular tissue. There are two kinds of vascular tissue: one is used primarily for transporting material from the roots to the leaves and the other is used for transporting material from the leaves to the roots and other parts of the plant.

In addition to the primary functions of support and nutrient transport, stems may also store food. This is true of sugar cane, yams, and potatoes. Many plant stems are green and, therefore, are involved in the process of photosynthesis.

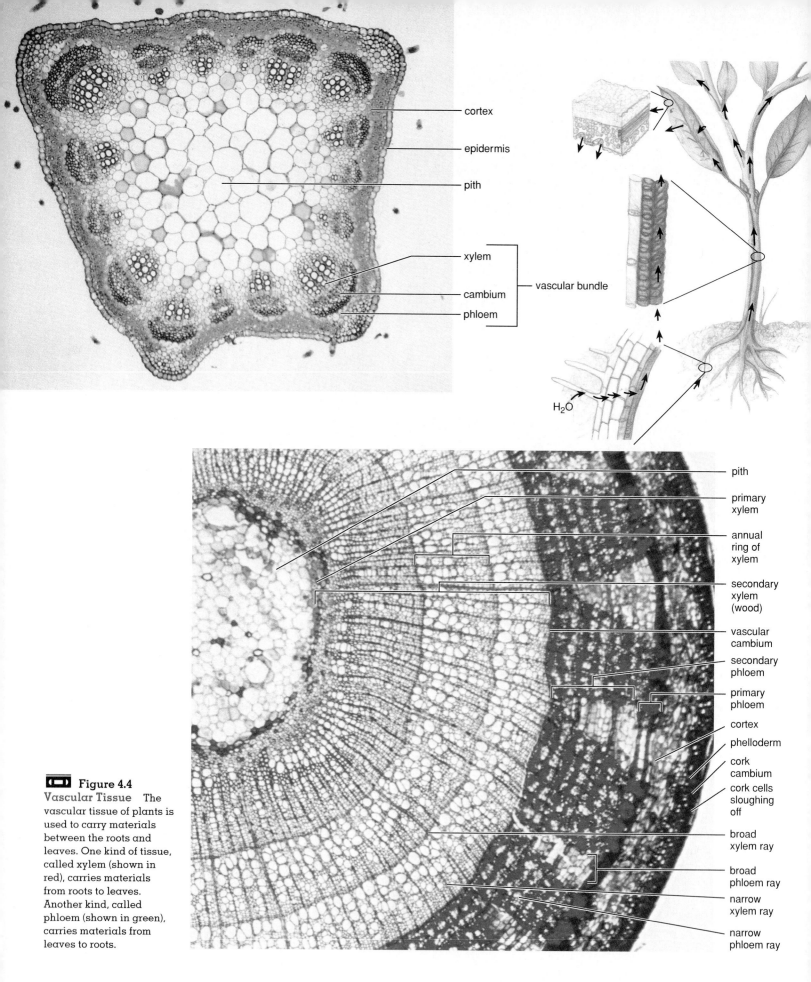

cortex

epidermis

pith

xylem

cambium — vascular bundle

phloem

H_2O

pith

primary xylem

annual ring of xylem

secondary xylem (wood)

vascular cambium

secondary phloem

primary phloem

cortex

phelloderm

cork cambium

cork cells sloughing off

broad xylem ray

broad phloem ray

narrow xylem ray

narrow phloem ray

Figure 4.4
Vascular Tissue The vascular tissue of plants is used to carry materials between the roots and leaves. One kind of tissue, called xylem (shown in red), carries materials from roots to leaves. Another kind, called phloem (shown in green), carries materials from leaves to roots.

BIO *feature*

Meet the Mustard Family

One group of plants belonging to the genus *Brassica* supplies a wide variety of food materials. The group is referred to by biologists as mustards. Several kinds of mustards are grown for their seeds, which are ground up and used as flavorings. Another mustard, called rape, is grown commercially for the seed from which rapeseed oil, a lubricant, is extracted.

Rutabagas and turnips, whose roots are eaten, are also members of this group of plants. But probably the most interesting members are the various cabbagelike varieties of the species, *Brassica oleracea*. When we examine this plant we find that nearly every part of the above-ground portion of the plant is eaten. Cabbage and kale are grown for the leaves, which are

eaten raw or cooked. There are even purple cabbages grown for their colorful leaves. Two kinds of *Brassica oleracea* are grown for their flowers: broccoli and cauliflower are immature flowers that have become large and fleshy. Brussels sprouts are grown for the buds that grow on the stem, and kohlrabi  **Appendix A** is grown for its thickened stems.

(a) Mustard

(b) Rape

(c) Rutabaga or turnip

Continued

Meet the Mustard Family (continued)

(e) Broccoli

(f) Brussels sprouts

(d) Cabbage

(h) Kohlrabi

(g) Cauliflower

Leaves

Green leaves, which trap sunlight, are the site of photosynthesis. **Photosynthesis** involves trapping light energy and converting it into the chemical energy of complex organic molecules, like sugar (figure 4.5).

Although the process of photosynthesis involves many steps, it can be summarized as follows:

| Sunlight energy captured by leaves | + | Carbon dioxide (CO_2) gas from the air (atmosphere) entering leaves | + | Water (H_2O) from the soil transported to leaves | → | Sugar ($C_6H_{12}O_6$) high in chemical energy | + | Oxygen (O_2) gas released into the atmosphere |

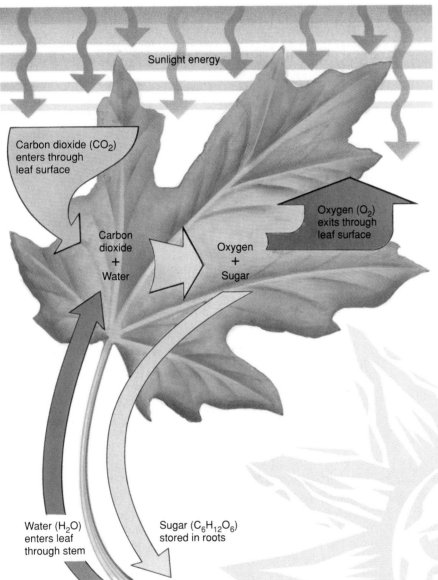

Sunlight energy

Carbon dioxide (CO_2) enters through leaf surface

Oxygen (O_2) exits through leaf surface

Carbon dioxide + Water

Oxygen + Sugar

Water (H_2O) enters leaf through stem

Sugar ($C_6H_{12}O_6$) stored in roots

Green plants and other photosynthetic organisms are, in effect, converting light energy into the energy contained within the chemical bonds of organic compounds. Thus there is a flow of energy from the sun into the organic matter of plants. Light energy is needed to enable the smaller inorganic molecules (water and carbon dioxide) to be combined to make the organic sugar molecules. In the process, oxygen is released.

This process takes place in the green portions of the plant, usually the leaves. Leaves have vascular tissue to allow for the transport of materials, but they also have cells containing chloroplasts. **Chloroplasts** are the cellular structures responsible for photosynthesis. They contain the green pigment **chlorophyll.** The organic molecules produced by plants as a result of photosynthesis can be used by the plant to make the other kinds of molecules needed for its structure and function. In addition, these molecules can satisfy the energy needs of the plant. Organisms that eat plants also use the energy captured by photosynthesis.

Figure 4.5 Photosynthesis The process of photosynthesis takes place in the leaf. Carbon dioxide enters the leaf from the surrounding air. Water is carried to the leaves through the vascular tissue. Light enters the leaf through its surface. The products of photosynthesis are oxygen, which exits the leaf, and sugar, which is transported to other parts of the plant through vascular tissue.

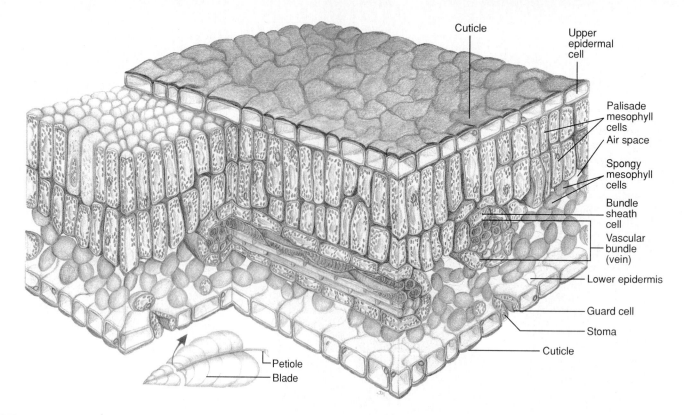

Cuticle

Upper epidermal cell

Palisade mesophyll cells

Air space

Spongy mesophyll cells

Bundle sheath cell

Vascular bundle (vein)

Lower epidermis

Guard cell

Stoma

Cuticle

Petiole

Blade

Figure 4.6 **The Structure of a Leaf** Though a leaf is thin, it consists of several layers. An outer layer has openings called stomatas that allow gases to pass through. The internal layers have many cells with chloroplasts, air spaces, and bundles of vascular tissue.

To carry out photosynthesis, leaves must have certain characteristics. Since it is a solar collector, a leaf should be a flat structure with a large surface area. This assures that the maximum number of cells in the leaf will be exposed to sunlight. A drawback to having thin leaves is an increase in water loss due to evaporation, but thick leaves would not allow penetration of light to the cells on the lower surface. Leaves are generally arranged on plants so they do not shade one another.

Even though leaves are usually thin, they have several layers (figure 4.6). Each layer of cells serves specific functions. One of these layers, the *epidermal* (skin) layer, produces a waxy coat for the outside of the leaf that reduces water loss.

The large surface area of leaves is important for trapping light, but it also leads to rapid evaporation. This loss of water is both an advantage and a disadvantage to a plant. The loss of water helps power the flow of more water and nutrients to the leaves, which the leaves need for photosynthesis, but too much water loss can be deadly. It is necessary that some water be lost and that gases like carbon dioxide and oxygen be able to enter or exit the leaves. Many tiny openings in the epidermis, called **stomates,** regulate these exchanges (figure 4.7). The stomates can close or open to control the rate at which water is lost. Often during periods of drought or during the hottest, driest part of the day the stomates are closed, thus reducing the rate at which the plant loses water.

FYI

There are at least three somewhat different forms of photosynthesis in plants. In addition, the process of photosynthesis is found in other organisms besides plants. Single-celled and more complex forms of algae perform photosynthesis in much the same way plants do. Many kinds of bacteria also perform photosynthesis. Some do not release oxygen as a byproduct. Since it appears that bacteria were on the earth before plants, it is clear that photosynthesis occurred on earth before plants.

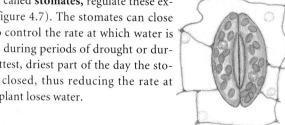

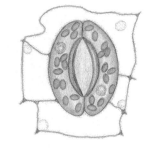

Closed

Open

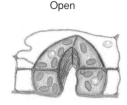

Figure 4.7 **Stomates** The stomates are located in the covering layer on the outside of leaves. When they are in the closed position, leaves do not lose water rapidly nor do they readily exchange oxygen and carbon dioxide. In the open position the leaf loses water but is better able to exchange oxygen and carbon dioxide.

BIO *feature*

Desert Plants

Plants that live in deserts have special adaptations to store water and prevent its loss. They have thick waxy outer layers and reduced numbers of stomates (pores) in their leaves. Many desert plants produce leaves following the infrequent rains but lose them during the driest parts of the year. This reduces the surface available for water evaporation. Some plants, such as the cacti, do not have leaves at all. Photosynthesis takes place in the stem.

Some desert plants have extensive root systems spread just below the surface of the soil that allow them to capture large amounts of water when it rains. Others have deep, penetrating root systems that give them access to sources of water deep in the soil.

Many plants store water. Some, like cacti, store water in their stems. Others store water in thick, fleshy leaves or specialized roots.

(a)

(c)

(b)

Cacti (*a*) Prickly pear cactus. (*b*) A barrel cactus. (*c*) Peyote.

BIO *feature*

Transpiration

Water plays a very important role in the movement of materials in plants. The loss of water from leaves is called **transpiration.** This evaporation lowers the pressure in the leaf, causing additional water to flow into the leaf through the vascular tissue. This is similar to what happens when you suck on a soda straw. You create a low pressure in your mouth, causing liquid to flow up the straw. In plants, transpiration is an important factor in the flow of water and dissolved nutrients upward from the roots.

Plants are continually "sucking" water from the soil and releasing it into the air. In fact, transpiration can significantly cool and humidify the air. Planting trees around homes can lower the air temperature in the house several degrees by providing shade and cooling by evaporation. Trees are natural air conditioners.

> **FYI**
> Bees make honey from a dilute sugar solution called nectar, which is produced in the flowers of many plants. The bees suck the nectar into a sac in their digestive system, carry it back to the hive, regurgitate the nectar into a chamber in the honeycomb, and fan the hive with their wings to evaporate water from the nectar, thus producing the material we know as honey.

Reproduction of Their Kind

Sexual Reproduction

When we "plant" a seed we expect that a plant will grow. The **seed** consists of an embryo plant with some stored food material surrounded by a protective coat (figure 4.8). It is designed to withstand harsh conditions until temperature, sunlight, and moisture are favorable for germination and growth. The seed is the product of many interesting activities in the reproductive life of plants.

> **FYI**
> Conifers like pines, spruces, firs, and cedars produce seeds inside cones instead of flowers.

Some plants have **flowers** that produce seeds. A flower is the reproductive structure. It may contain the male or female reproductive structures or both (figure 4.9). The word *flower* brings to mind the large, colorful parts of a plant. Showy flowers are an important strategy for assuring reproduction. They attract insects that carry pollen from one flower to the next. **Pollen** contains the plant's sperm necessary for fertilizing the plant's eggs, which are located in a part of the flower called the ovary. Even though pollen grains are tiny, plants expend considerable energy and material resources producing them. Producing pollen is an expensive process.

Plants with showy flowers call attention to themselves and attract insects, birds, or mammals for a meal of nectar (sugar water). The animals carry pollen on their bodies as they go from flower to flower and accidently pollinate the flowers. Animals that pollinate include bees, hummingbirds, and bats.

The wind is another way to move pollen from one flower to another. Grasses and many kinds of trees lack showy flowers but produce huge quantities of pollen that are blown by the wind. With large amounts of pollen in the air, it is likely that some pollen grains will land on the right flower and fertilize its eggs. Conifer trees are pollinated with the help of wind. The pollen is produced in small cones and released in such large quantities that clouds of pollen can be seen in the air when sudden gusts of wind shake the branches of the trees (figure 4.10).

> **CONCEPT CONNECTIONS**
> - **Pollination** is the transfer of pollen from one flower to the next.
> - **Fertilization** is the joining of the egg in the ovary with the sperm from the pollen. Fertilization is a separate event that occurs after pollination.

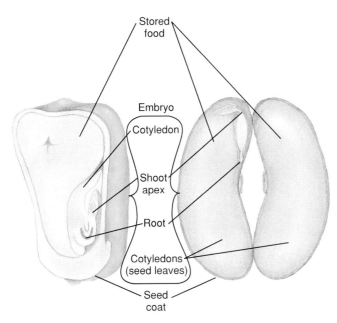

Figure 4.8 Seed Structure   Seeds consist of three important parts: the plant embryo, which will grow into a new plant; stored food, which will provide energy for the early growth of the plant; and a protective seed coat that resists environmental threats.

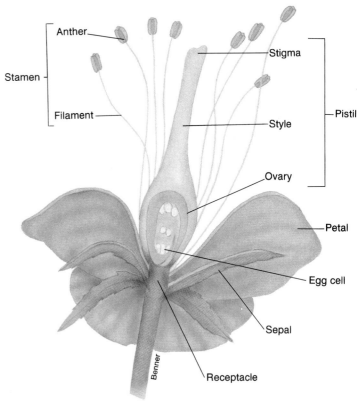

Figure 4.9 Flower Structure   Flowers consist of reproductive structures, including the ovary that contains the eggs and anthers that contain the pollen, and nonreproductive structures such as petals that are associated with the reproductive structures.

(a)

(b)

Figure 4.10 The Difference between Wind-Pollinated and Insect-Pollinated Flowers   The flowers of wind-pollinated plants, such as alder trees, are generally small and inconspicuous (a). They are often present in large numbers on the same stem and in exposed sites where wind can easily carry the pollen away. Insect-pollinated flowers, such as the sitka rose, are typically large and conspicuous (b). Usually they occur in small numbers, often one to a plant.

Hay Fever

Many people suffer from hay fever and as a result develop runny noses, watery eyes, and sneezing, similar to cold symptoms. The symptoms of many kinds of hay fever are caused by the production of histamine in response to the presence of pollen in our respiratory systems. When we breathe in these pollen grains, our bodies recognize the pollen as foreign material and respond by producing histamine. Thus, *anti*histamines are used to treat hay fever symptoms.

People vary in their reactions to pollen. Since each kind of plant produces its flowers and pollen at certain times of the year, hay fever sufferers may be afflicted only during specific periods. The name *hay fever* in itself is interesting. Hay is an animal food made by cutting and drying grasses. Grasses are wind pollinated plants. Although most hay fever sufferers do not develop a fever, the symptoms are like a cold, which is often accompanied by a fever. Hence the name hay fever.

Pollen Pollens of different kinds of plants are distinctly different in size and shape. Wind-pollinated plants typically have small, light-weight grains (a), while insect-pollinated plants typically have larger, heavier grains (b and c). Studies of pollen found in bogs allow scientists to learn how the kinds of plants present changed with time.

(c) (a) (b)

Once the eggs in the flower have been fertilized, they develop into a seed, which is usually enclosed within a structure called a **fruit.** Some fruits, like milkweed and yucca, become dry and brittle and break open to release the seeds. Others, like oaks and walnuts, have hard, woody coverings. Many others, like cherries, oranges, melons, blackberries, strawberries, eggplant, tomatoes, and apples, produce large, conspicuous, edible fruits.

What is the value of a colorful fruit to a plant? In addition to being colorful, most fruits are nutritious. When a fruit is eaten the seeds are usually also consumed. Many kinds of animals eat the fruits and seeds, carry the seeds with them in their guts, and eventually deposit them somewhere else (figure 4.11). The seeds must resist being digested as they travel through the gut of the animal. This method of seed dispersal is a necessary part of

the life cycle of some plants. The next time you are traveling through the countryside, take a close look at the plants growing along fences. There are frequently fruit trees and bushes in large numbers. Were they planted there in the droppings excreted by birds while they sat on the fence?

Plants use animals to distribute their seeds (offspring) in other ways. Fruits like acorns and various kinds of nuts are distributed

Figure 4.11 Fruits Used as Food by Animals Most fruits used as food by animals are large, colorful, and contain a great deal of nutrition. The seeds are spread by the animal as they move about and drop fruit, or when the seeds pass through their digestive system.

by animals that pick them up and store them in holes in the ground. Those that are not retrieved and eaten have been planted in a new place. From the point of view of an oak tree, a squirrel is useful for planting its seeds. Many seeds or fruits stick to animals and are carried on the outside of the animal; they eventually fall off or are purposely removed by the animal (figure 4.12). Some stick because they have a gluelike material, while others have hooks and attach like Velcro.

Many plants use wind as a method of distributing their seeds. A common sight in the spring and early summer is fluffy, white seeds from dandelions and cottonwoods being carried on the wind. Maples, conifers, and many other plants have wings on their seeds that aid in their dispersal.

Asexual Reproduction

Most plants have another method of reproduction and dispersal called *asexual reproduction,* which does not involve pollination or fertilization of eggs (sexual reproduction). New plants produced by asexual reproduction are exactly like the parent. This type of reproduction occurs most commonly by the growth of new above-ground plants from an existing root system. Grasses spread in this manner and rapidly invade adjacent bare patches of soil. Other plants, such as strawberries, send out above-ground runners that take root while still attached to the parent plant.

Many plants with tubers and bulbs also reproduce asexually. They may be fragmented by the action of animals feeding on them, and pieces of the original may be distributed to new areas. Humans do this regularly with tulips and potatoes. A tulip bulb may be separated into parts. When planted, each portion of the original bulb will grow into an adult tulip plant. This also occurs when you try to rid your yard of certain weeds. Cutting the roots into fragments and not removing all the pieces will only result in the asexual reproduction of that plant. You increase your weed problem instead of bringing it under control.

Defense Against Enemies

Plants are vulnerable to being eaten. They contain large amounts of stored food and they cannot run away from enemies. How do plants defend themselves? Many plants have thorns that discourage some animals from eating their twigs and foliage. The thistle, a common weed, has spines on its leaves that at least make it uncomfortable for many animals to feed on them. In pastures where cattle are allowed to graze, it is common to see thistles standing tall and healthy while the grasses near them have been eaten to the ground.

Many plants, such as milkweeds, chili peppers, poison ivy, and the leaves of tomatoes and rhubarb, produce chemicals that may be unpleasant or poisonous to animals. The toxins in members of the chrysanthemum family are actually used as insecticides, and garlic can be used to keep certain insect pests from gardens.

Sticktight

Figure 4.12 Burrs and Stick-tights Many seeds and fruits have special hairs or are sticky and become attached to animals as they move past. Later the seed may be dislodged by its carrier, resulting in the transfer of the seed to a new location distant from its parent.

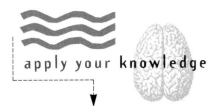

New Plants from Old

Anyone can have a variety of flowering plants at little expense with just a bit of effort. One need only take a sharp knife and cut a piece of the stem of a desirable plant (with your neighbor's permission!). The stem piece should include the tip and about five leaves. Woody stems are apt to be more difficult to root than plants with soft stems. Place some soil in a container, make a hole in the soil with a pencil, and insert the cut stem. The container can be anything, as long as there are a few holes in the bottom for drainage. Cover the container tightly with a transparent plastic bag to prevent water loss through the leaves before the roots develop. Also trim the lower leaves for the same purpose. Water the plant once and keep it out of direct sunlight. Roots may take ten days or longer to develop. If the soil should feel dry to the touch, water the plant again.

Potting soil or vermiculite from a hardware or garden store might be better than soil you dig locally. Another aid is to dip the cut end of the stem into some rooting hormone before planting. However, people have had success for years without these refinements.

Many of these plant materials are unpleasant or toxic to us as well. However, we sometimes use small quantities of these materials to season our food, as with pepper, onion, and horseradish. In other cases the toxic material can be altered to allow the plant to be used for food. For example, cashew nuts are toxic unless roasted. Other substances, such as salicylic acid (aspirin) from willow trees, have been used as pain relievers for generations. Caffeine from the coffee tree and cocaine from the coca plant have a long history of use as stimulants.

Most woody plants and grasses "defend" themselves by having large amounts of hard-to-chew and hard-to-digest material that discourages most animals from feasting on them, although the new growth or the softer tissues may fall prey to hungry insects or other animals.

BIO *feature*

Spices and Flavorings

Think about all the plant materials we use to season our foods. Black pepper comes from the hard, dried berries of a tropical plant, *Piper nigrum.* Cayenne pepper is made from the ground-up fruits of *Capsicum annum;* the hot, spicy chemical in the fruit and seeds is known as capsicin. The seeds of the dill plant, *Anethum graveolens,* are used to flavor pickles and many other foods. The dried or fresh leaves of herbs such as thyme, rosemary, chives, and parsley are also used as flavorings. If you examine your kitchen cabinet you may find the following: cinnamon from the bark of a tree found in India; cloves, which are the dried flower buds of a tropical tree; ginger from the root of a tropical plant of Africa and China; and nutmeg from the seed of a tropical tree of Asia.

Spices like these were so highly prized that fortunes were made in the spice trade. Beginning in the early 1600s, ships from Europe regularly visited the tropical regions of Asia and Africa, returning with cargoes of spices and other rare commodities that could be sold at great profit. Consequently, India has been greatly influenced by Britain, Indonesia has been greatly influenced by the Netherlands, and Britain and France have influenced the development of different portions of Africa.

Relationships with Other Organisms

Plants have a variety of relationships with other kinds of organisms. We have already discussed the relationship between insects and pollination. In fact, relationships between flowering plants and insects can be quite specific. As we noted, many plants entice insects to visit their flowers by providing odors and sugary nectar. Some flowers have shapes or lines on their petals that point certain insects to the source of the nectar.

Other plants may attract insects by smelling like rotting meat or looking like a female insect. In each of these cases the insect is tricked into pollinating the flower as it seeks the meat or mate (figure 4.13). Plant flowers that smell like rotting meat attract flies and other insects that lay their eggs in dead animals. The insect flies around within the flower and is coated with pollen. When it escapes from the bad-smelling flower it is likely to encounter another flower of the same kind and accidentally transfer pollen. Some orchids have flowers that closely resemble the female of certain kinds of wasps. When the male attempts to mate with the flower, pollen-containing packets are attached to his body. When he is seduced by another flower he transfers the pollen.

Figure 4.13 Fly-Pollinated Flowers The carrion flower smells like rotting flesh and is visited by flies that bring about pollination.

Relationships between plants and other organisms may be based on something other than reproductive needs. Some species of woody plants, for example, produce special structures that feed ants. The ants in turn defend their plant by attacking animals that feed on its leaves. The ants will even cut off the leaves of other plants that come in contact with their host. Many plants have bacteria or fungi that live in their roots. The bacteria and fungi receive some benefit from the plant, perhaps some necessary nutrients. In return the bacteria or fungi provide necessary raw materials such as nitrogen or phosphorus to the plant. We still have much to learn about how the roots of plants interact with other organisms in the soil.

Many kinds of plants are parasites on other plants. Mistletoe is a good example. The berries produced by mistletoe are sticky. When birds eat the berries, the seeds become stuck to the beak and may be wiped onto the branch of a tree. When the seeds germinate, they penetrate the tissue of the tree and get nourishment from the tree's sap. Mistletoe is green and carries on photosynthesis, but other plants are totally parasitic, lack chloroplasts, and are white in color. Indian pipe is a good example. This plant lives in the soil as a parasite on the roots of trees. Only when the plant grows a white, above-ground flower stalk do we see that it is there (figure 4.14).

Some plants even grow around other plants and strangle them. The strangler fig tree starts as a young plant growing up the outside of a tall forest tree. Once the fig has reached the bright sunlight, it can carry on photosynthesis more efficiently and grow more rapidly than the other tree can. It sends additional growths down to the soil that take root. Eventually the host tree is completely surrounded by the fig and dies. When the dead tree rots away it leaves behind a fig tree with a hollow space in the middle. A common feature of many tropical forests is the presence of large hollow fig trees (figure 4.15).

Many kinds of plants produce substances that inhibit the germination and growth of their own seeds and the seeds of other kinds of plants. Therefore, seeds that fall near them are prevented from growing. This effectively reduces competition for scarce water and light and helps to ensure the survival of the original plant. Seeds from these chemical-producing plants are able to germinate at a distance, so offspring do not compete with their parents.

(a)

(b)

Figure 4.14 Parasitic Plants Mistletoe (a) and Indian pipe (b) are two flowering plants that are true parasites. They use other living plants as a source of food. The Indian pipe has even lost its ability to carry on photosynthesis.

Figure 4.15 **Strangler Fig Tree** This strangler fig tree began by growing on the surface of another tree. It sent down rootlike structures to the soil and eventually completely surrounded the tree it was growing on. The original tree will die and rot away, leaving a hollow in the middle of the strangler fig.

Figure 4.16 Phototropism The above-ground portions of plants grow toward a source of light.

Response to the Environment

Our casual impression of plants is that they are unchanging objects. However, on closer examination we recognize that plants change over time. They grow new leaves in the spring, produce flowers and fruits at certain times of the year, and grow toward a source of light. Furthermore, they will respond to organisms that harm them, and may even mount an attack against competitors.

One of the first responses studied in plants is their ability to grow toward a source of sunlight. The value of this response is obvious, since plants need light to survive. The mechanism that allows this response involves a **hormone,** a chemical produced in one part of an organism that changes the function of a more distant part. In the case of plants growing toward light, the growing tip of the stem produces a hormone that is transported down the stem. The hormone stimulates cells to divide and grow. If the growing tip of a plant is shaded on one side, the shaded side produces more of the hormone than the lighted side. The larger amount of hormone on the shaded side causes greater growth in that area and the tip of the stem bends toward the light. When all sides of the stem are equally illuminated the stem will grow equally on all sides and will grow straight. If you have house plants in a window it is important to turn them regularly or they will grow more on one side than the other (figure 4.16).

Plants also respond to changes in exposure to daylight. They are able to measure day length and manufacture hormones that cause changes in the growth and development of specific parts of the plant. Some plants produce flowers only when the days are getting longer, some only when the days are getting shorter, and some only after the days have reached a specific length. Other activities are triggered by changing day length. Probably the most obvious is the mechanism that leads to the dropping of leaves in the fall.

BIO *feature*

Fall Color

The autumn color change you see in leaves in certain parts of the world is the result of green chlorophyll breaking down. Other pigments (red, yellow, orange, brown) are present in leaves all summer but are masked by the green chlorophyll pigments. In the fall, a layer of waterproof tissue forms at the base of each leaf and cuts the flow of water and other nutrients to the leaves. The leaf cells die and their chlorophyll disintegrates, revealing the reds, oranges, yellows, and browns that make a trip through the countryside a colorful experience.

Fall colors

Many kinds of climbing vines are able to wrap rapidly growing, stringlike tendrils around sturdy objects in a matter of minutes. As the tendrils grow, they typically wave about. When they encounter an object, they wrap around it and anchor the vine. Once attached, the tendrils change into hard, tough structures that bind the vine to its attachment. Sweet peas, grape vines, and the ivy on old buildings spread in this manner. The ivy can cause great damage to the buildings on which they are growing (figure 4.17).

Plants may even have the ability to communicate with one another. When the leaves of plants are eaten by animals, the new leaves produced to replace those lost often contain higher amounts of toxic materials than the original leaves. An experiment carried out in a greenhouse produced some interesting results. Some of the plants had their leaves mechanically "eaten" by an experimenter while nearby plants were not harmed. Not only did the cut plants produce new leaves with more toxins but the new growth on neighboring, nonmutilated plants had increased toxin levels as well. This raises the possibility that plants communicate in some way, perhaps by the release of molecules that cause changes in the receiving plant.

FYI

Several kinds of weed killers (herbicides) copy the effects of hormones that stimulate growth. The treated plant grows so rapidly that it uses up all its food reserves and dies. An example of such a herbicide is 2,4-D.

Figure 4.17 Clinging Stems Some stems are modified to wrap themselves around objects and give support. The tendrils of this pea plant are a good example.

Evolution Among Plants

When we look at the vast array of plants on earth we see many adaptations that allow plants to live in their environments. Aquatic plants, for example, often contain air spaces in their leaves that allow them to float on the water surface. The stomates that are normally located on the bottom of the leaf surface in land plants are located on the top of aquatic leaves. Some plants are adapted to survive fire and will regrow after a fire has killed the above-ground portions of the plant. Others live in salty soils that would kill most plants. Some plants do not live in soil at all but live suspended from other plants. Orchids, Spanish moss, and other kinds of "air plants" live on the surfaces of other plants. They get all their nutrients from the air and the rain. They need no soil, only a surface to grow on.

All of these adaptations are typically the result of an evolutionary process that has taken millions of years. A look at the fossil record of the plant kingdom reveals that flowering plants are the most recent to have evolved and were preceded by other kinds of plants that are less dominant in the world today than they were in the past. Mosses, ferns, and conifers show up as fossils before flowering plants do. These groups are still present today, but only a few kinds have survived as the flowering plants have increased and replaced them. Although the numbers of mosses, ferns, and conifers is lower today than in the distant past, the species that currently exist continue to evolve and specialize and even thrive in certain situations. The rest of the chapter describes the main characteristics of each group of plants.

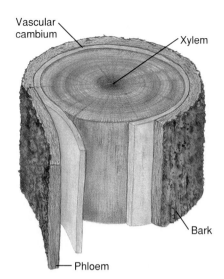

Figure 4.18 Conifer Trees Nearly all cone-bearing plants are large trees with woody stems and needle-shaped leaves. Several are shown here.

Vascular cambium — Xylem — Bark — Phloem

Redwood

Torrey pine

Cedar

CONCEPT CONNECTIONS

- **Angiosperms** are the flowering plants that have their seeds enclosed within a fruit.
- **Gymnosperms** are cone-bearing plants that produce seeds in cones; the seeds are not covered (naked) and are released directly from the cone.

We have already discussed much of the variety of the flowering plants. They are extremely important to humans in many ways. They range in size from small plants that live one year to towering trees that live for centuries. But they all share several characteristics: they have flowers, seeds, and vascular tissue.

The conifers are familiar to us as pine, fir, spruce, cedar, and various other evergreen, needle-leaved trees (figure 4.18). Nearly all are large trees and as such are important as a source of lumber. They resemble flowering plants in that they produce seeds and pollen and have vascular tissue, roots, stems, and leaves. The pollen and eggs are not produced in flowers, however, but are formed in cones. Two different kinds of cones are produced. The more obvious ones that hang on the tree are the female cones. They produce the eggs and store the seeds as they mature. The male cones are smaller and do not stay on the tree for long. They usually release their pollen in the spring, in vast clouds that are carried by the wind. After the pollen is released the male cones usually drop off the tree.

A group more distantly related to the flowering plants is the ferns (figure 4.19). Ferns share fewer characteristics with flowering plants. They have vascular tissue, so some of them can be quite large; tropical regions of the world have tree ferns. They do not reproduce by forming seeds but instead produce structures called spores. These are dispersed by wind as they are released. In addition to leaves, ferns have underground stems and roots. Ferns are used as ornamental plants in houses and yards. Their importance lies in the role they play in stabilizing natural areas such as bogs and wetlands. When we look at the fossil record of 350 million years ago, the ferns were very important in the landscape and are well represented in the coal deposits formed at that time.

Mosses are the most primitive group of plants. They grow as a carpet composed of many parts. Each individual moss plant is composed of a central stalk less than 5 centimeters tall with short, leaflike structures that are the sites of photosynthesis. If you look at the individual cells in the leafy portion of a moss, you can distinguish the cytoplasm, cell wall, and chloroplasts (figure 4.20). Why do botanists consider mosses the lowest step of the evolutionary ladder in the plant kingdom? First, they are considered primitive because they have not developed an efficient way of transporting water throughout their bodies; they must rely on the physical processes of diffusion and osmosis to move materials. The fact that mosses do not have a complex method of moving water limits their size to a few centimeters and their location to moist environments. Another characteristic of mosses points out how closely related they are to their aquatic ancestors: they require water for fertilization.

(a)

(b)

Figure 4.19 Ferns Most ferns are relatively small plants (a), but in tropical areas where there is no frost ferns may be treelike structures several meters tall (b).

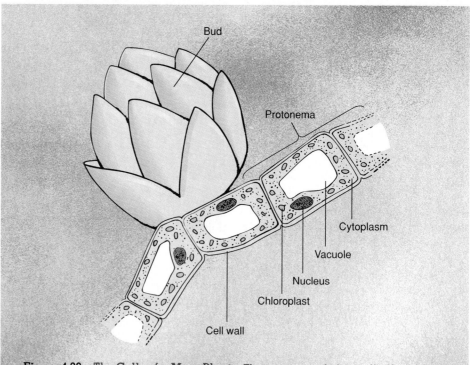

Figure 4.20 The Cells of a Moss Plant These are typical plant cells. Note the large central vacuole and the cell wall. One characteristic that separates mosses from the photosynthetic protists is the presence of the individual chloroplasts.

Figure 4.21 Mosses Since mosses lack vascular tissue they are small and limited to moist places. The structure growing from the top of the green moss plant releases structures called spores that can establish moss plants in new locations.

Sporophytes

Gametophytes

The sperm cells "swim" from the male to the female. Small size, moist habitat, and swimming sperm are considered characteristics of a primitive organism. In a primitive way, mosses have adapted to a terrrestrial niche. They do not have seeds or vascular tissue and are poorly adapted for life on land. They are found in moist conditions and generally live in the shade. Like ferns, they are dispersed by spores carried by the wind (figure 4.21). Some mosses, like sphagnum, are used as a soil conditioner and as the foundation for hanging plants. In some parts of the world, a form of moss known as peat is burned as a fuel.

SUMMARY

Plants are important as organisms that trap light energy. The process of photosynthesis involves manufacturing organic molecules from inorganic molecules in the chloroplasts of plant cells. The light is trapped by the chlorophyll molecule.

Flowering plants are the dominant plants today. They have roots for anchorage and nutrient uptake, stems that support leaves, and leaves for photosynthesis. Vascular tissue allows the flow of materials between the roots and leaves. Sexual reproduction involves the transfer of pollen from one flower to another by animals or wind, fertilization of the egg, and development of the fertilized eggs into seeds. Seed dispersal is assisted by wind, animals, and other means.

Plants have complex interactions with other kinds of organisms. Some are parasites, some are carnivores, and some live on the surface of other plants. Plants intensely compete for nutrients and other resources and respond to injury or changes in their environment. Although flowering plants are the most common forms today, conifers, ferns, and mosses were dominant members of the plant kingdom in the past.

CHAPTER GLOSSARY

active transport (ak'tive trans'port) Use of a carrier molecule to move molecules across a cell membrane often from an area of lower concentration to an area of higher concentration. The carrier requires an input of energy.

angiosperms (an'je-o"spurmz) Plants that produce flowers and fruits.

annual (an'yu-uhl) A plant that completes its life cycle in one year.

biennial (bi-e'ne-al) A plant that requires two years to complete its life cycle.

chlorophyll (klo'ro-fil) The green pigment of plant cells that is responsible for capturing sunlight energy.

chloroplast (klo'ro-plast) A structure found in plant cells that contains chlorophyll and is where photosynthesis takes place.

diffusion (di-fiu'zhun) Net movement of a kind of molecule from a place where that molecule is in higher concentration to a place where that molecule is in lower concentration.

fertilization (fer"ti-li-za'shun) The union of a sperm and egg cell.

flower (flau'er) The plant structure that produces pollen or eggs and usually contains petals.

fruit (frut) A structure that surrounds the seeds in flowering plants.

gymnosperms (jim'no-spurmz) Plants that produce their seeds in cones.

hormone (hor'mon) A chemical produced in one part of an organism that alters the activities of a distant part.

leaf (leef) The thin, flat, green structure of plants where chloroplasts are typically found and photosynthesis takes place.

osmosis (os-mo'sis) The net movement of water molecules through a semipermeable membrane.

perennial (pur-en'e-uhl) A plant that requires many years to complete its life cycle.

photosynthesis (fo-to-sin'the-ses) The process of combining carbon dioxide and water with the aid of sunlight energy to form sugar and release oxygen.

pollen (pah'len) Tiny particles produced by plants that contain the plant equivalent of sperm cells and are carried from one flower to another by wind or animals.

pollination (pol"i-na'shun) The transfer of pollen in gymnosperms and angiosperms.

root (rut) The below-ground portion of a plant responsible for anchoring the plant in the soil and absorbing water and other nutrients.

seed (seed) A structure containing an embryo plant and stored food within a protective coat.

stem (stem) The typically upright portion of the plant that supports the leaves and contains vascular tissue that allows for the flow of material between leaves and roots.

stomate (sto'mate) A tiny opening in the surface of a leaf that allows for the exchange of gases such as carbon dioxide, oxygen, and water vapor.

transpiration (trans"pĭ-ra'shun) The transport of water from the roots to the leaves of the plant and its evaporation from the surface of the leaves.

vascular tissue (vas'kyu-ler tish'yu) Tubes constructed of many cells attached end to end that allow for the transfer of materials from one part of a plant to another.

CONCEPT MAP TERMINOLOGY

Construct a concept map to represent the relationships among the following concepts.

chlorophyll
chloroplast
fruit
leaves
photosynthesis
pollen
root
seed
stem
stomate
vascular tissue

LABEL•DIAGRAM•EXPLAIN

Describe three functional abilities that plants and animals share.

1.
2.
3.

List three functional abilities that only plants have.

1.
2.
3.

Multiple Choice Questions

1. Which of the following is *not* a typical function of a root?
   a. store food
   b. take up water and nutrients
   c. carry on photosynthesis
   d. anchor the plant in the soil

2. During photosynthesis, plants:
   a. create light
   b. manufacture chlorophyll
   c. use up oxygen
   d. use up water

3. Water can only enter the roots of plants if:
   a. the concentration of water is higher in the soil than in the cells of the roots
   b. the cells of the roots are using active transport to pump water into the roots
   c. the cells of the roots already have all the water they can hold
   d. air is able to get to the roots

4. In most plants a major function of the stem is to:
   a. carry on photosynthesis
   b. position the leaves to capture light
   c. transport chlorophyll to the leaves
   d. allow water to enter and leave the plant

5. Which one of the following is *not* a typical ability of plants?
   a. They move various parts of their bodies.
   b. They die during winter or during drought and come back to life when conditions improve.
   c. They reproduce by sexual reproduction.
   d. They change their activities as their environmental surroundings change.

6. Which one of the following is *not* a plant?
   a. pine tree
   b. fern
   c. cactus
   d. mushroom

7. Which one of the following is *not* required for the process of photosynthesis to take place?
   a. oxygen
   b. chloroplasts
   c. water
   d. carbon dioxide

8. One way plants defend themselves from animals is by:
   a. producing poisons
   b. running away
   c. hiding
   d. not reproducing

9. Plants:
   a. interact with other organisms in many ways
   b. seldom hurt other organisms
   c. could not live without animals
   d. have very little ability to adjust to changing surroundings

Questions with Short Answers

1. List three characteristics that distinguish plants from other kinds of organisms.

2. What are the advantages and disadvantages of insect pollination?

3. What are the primary functions of the roots, stems, and leaves?

4. Describe three methods of seed dispersal.

5. What two groups of plants have seeds? What two groups do not have seeds?

6. Why are plant cell walls important to the structure of a plant?

7. What is pollen and why is it important?

8. Describe three ways plants can "fight" animals.

9. List three ways in which plants respond to changes in their surroundings.

10. What is the difference between sexual reproduction and asexual reproduction?

11. How do flowering plants, conifers, ferns, and mosses differ?

c h a p t e r

5

the microorganisms: prokaryotae, protista, and mycetae

The common mold, *Aspergillus nidulans* (color enhanced).

- List the beneficial and harmful effects of bacteria.
- List the beneficial and harmful effects of algae.
- List the beneficial and harmful effects of protozoans.
- Explain how a high reproductive rate and the ability to form spores have enabled microorganisms to survive in changing environments.
- Recognize the importance of algae in aquatic environments.
- Understand the major roles played by microorganisms.
- List some commercial uses of microorganisms.
- List the three groups of Protista and describe their differences.
- List the beneficial and harmful effects of fungi.

Who Are the Microbes?

This chapter gives you a glimpse into the nature of fungi (kingdom Mycetae), protists (kingdom Protista), and bacteria (kingdom Prokaryotae). These three kingdoms share several characteristics that set them apart from the plant and animal kingdoms. They include organisms that reproduce primarily asexually. Because the majority of organisms in these kingdoms are small and cannot be seen without some type of magnification, they are called **microorganisms,** or **microbes.**

There is minimal cooperation among the different cells of microorganisms. Some microbes are free-living, single-celled organisms, while others are collections of cells that cooperate to a limited extent. The latter type are called **colonial.** Some cells within a colony may specialize for reproduction, while others do not. Some colonial microbes coordinate their activities so that the colony functions as a unit.

Microbes are found in aquatic or very moist environments; most lack the specialization required to withstand drying. Since they are small, the habitat does not have to be large. Microbes can maintain huge populations in very small moist places like the skin of your armpits, temporary puddles, and tiled bathroom walls. Others have the special ability to become dormant and survive long periods without water. When moistened, they become actively growing cells again. The simplest of microbes are the bacteria of the kingdom Prokaryotae.

table 5.1

Fungi Characteristics

Division (phylum)	Characteristics
Zygomycota (bread mold)	No cross wall in hypha; mainly saprophytic; form spores asexually; some sexual reproduction
Ascomycota (sac fungi)	Form visible fruiting bodies and spores; sexual reproduction; some one-celled species (yeasts)
Basidiomycota (club fungi)	Spore-forming; sexual reproduction; decomposers; some parasitic on plants; mushrooms
Deuteromycota (imperfect fungi)	Sexual reproduction has never been seen; spore-forming
Mycophycophyta (lichens)	Not a single species, but a symbiotic relationship between alga and fungus

Fungi (Kingdom Mycetae)

Fungus is the common name for members of the kingdom Mycetae. The majority of fungi are nonmotile. They have rigid cell walls. Members of the kingdom Mycetae are non-photosynthetic eukaryotic organisms. The majority are multicellular, but a few, like yeasts, are single-celled (table 5.1). In multicellular fungi the basic structural unit is made up of multicellular filaments. Because all fungi are heterotrophs, they must obtain nutrients from organic sources. Most secrete enzymes that digest large molecules into smaller units that are absorbed.

Even though fungi are stationary, they survive and disperse successfully because of their ability to form spores. An average-sized mushroom can produce more than 20 billion spores; a good-sized puffball can produce as many as 8 trillion spores (figure 5.1). When released, the spores can be transported by wind or water. Because of their small size, they can remain in the air for a long time and travel thousands of kilometers. Fungal spores have been collected as high as 50 kilometers (30 miles) above the earth.

In a favorable environment, a fungus produces dispersal spores, which are short-lived and germinate quickly under suitable conditions. If the environment becomes unfavorable—too cold or hot, or too dry—the fungus produces survival spores. These may live for years before germinating. Fungal spores are almost always present in the air;

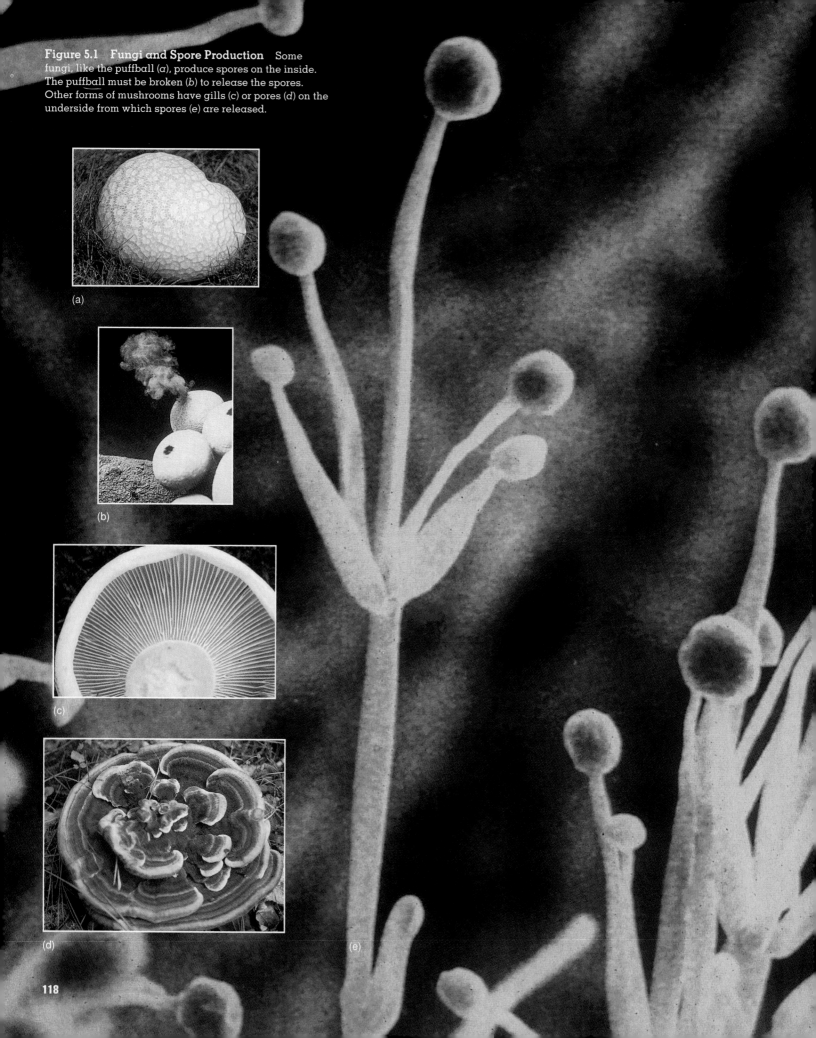

Figure 5.1 Fungi and Spore Production Some fungi, like the puffball (a), produce spores on the inside. The puffball must be broken (b) to release the spores. Other forms of mushrooms have gills (c) or pores (d) on the underside from which spores (e) are released.

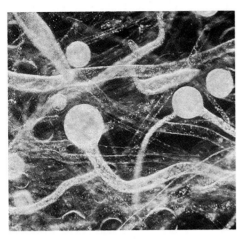

Figure 5.2 Water Mold Rapidly reproducing water molds are fungi that quickly produce a large mass of filaments called hyphae. You often see these masses of hyphae as fuzzy growths on dead fish or other dead material in the water.

as soon as something dies, fungal spores settle on it, and decomposition usually begins (figure 5.2).

Fungi and their by-products have been used as sources of food for centuries. When we think of fungi and food, mushrooms usually come to mind. The common mushroom found in the grocer's vegetable section is grown in seventy countries and has an annual market value in the billions of dollars. But other fungi are used for food. *Shoyu* (soy sauce) was originally made by fermenting a mixture of wheat, soybeans, and a fungus for a year. Most of the soy sauce used today is made by a cheaper method of processing soybeans with an acid, but true gourmets still prefer soy sauce made the original way. Another mold is important to the soft-drink industry. The citric acid that gives a soft drink its sharp taste was originally produced by squeezing juice from lemons and purifying the acid. Today, however, a mold is grown to produce great quantities of citric acid at a low cost.

All fungi are capable of breaking down organic matter to provide themselves with the energy and building materials they need. And since all organisms must have a source of carbon, nitrogen, phosphorus, and other elements that can be incorporated into new carbohydrates, fats, proteins, and other molecules necessary for growth, the fungi, along with bacteria, are the primary recycling agents. Fungi living on a dead log are obtaining energy and nutrients by decomposing the log tissue. Spores are an efficient method

Penicillin: More than Just a Source of Antibiotics

Fungi play a variety of roles. They are used to process food and are vital in recycling. As decomposers, they destroy billions of dollars worth of material each year; as pathogens, they are responsible for certain diseases. Yet they also are beneficial in the production of antibiotics and other chemicals used to treat diseases. *Penicillium chrysogenum* is a mold that produces the antibiotic penicillin, which was the first commercially available antibiotic and is still widely used.

There are more than one hundred species of *Penicillium,* and each characteristically produces spores in a brushlike border; the word *penicillus* means "little brush." Members of this group do more than just produce antibiotics; they are also widely used in processing food. Many people are familiar with the blue, cottony growth that sometimes occurs on citrus fruits. The *P. italicum* growing on the fruit appears to be blue because of the pigment produced in the spores. The blue cheeses, such as the Danish, American, and original Roquefort, all have this color. Each has been aged with *P. roquefortii* to produce their characteristic color, texture, and flavor. Differences in the cheese are determined by the kind of milk used and the conditions under which the aging occurs. Roquefort cheese is made from sheep's milk and aged in Roquefort, France, in particular caves. American blue cheese is made from cow's milk and aged in many places around the United States. The blue color has become an important feature of these cheeses. The same research laboratory that first isolated *P. chrysogenum* also found a mutant species of *P. roquefortii* that would produce spores having no blue color. The cheese made from this mold is "white" blue cheese. The flavor is exactly the same as "blue" blue cheese, but commercially it is worthless: people want the blue color.

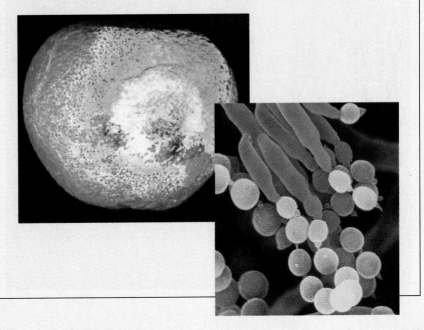

At the Root of It All

Some fungi have a symbiotic relationship with plant roots. **Mycorrhizae** grow inside a plant's root-hair cells, the cells through which plants absorb water and nutrients. The filaments from the fungus grow out of the root-hair cells and greatly increase the amount of absorptive area. Plants with mycorrhizal fungi can absorb as much as ten times more minerals than those without the fungi. Some types of fungi also supply plants with growth hormones, while the plants supply carbohydrates and other organic compounds to the fungi. Mycorrhizal fungi are found in 80 to 90 percent of all plants.

In some situations, mycorrhizae may be essential to the life of a plant. Biologists are investigating a correlation between mycorrhizae and acid-rain damage to trees. Acid-rain conditions can wash out certain necessary plant minerals from the soil, making them less available to plants. The decreased soil pH also makes certain toxic chemicals, such as copper, more accessible to plants. When the roots of trees suspected of being killed by acid-rain are examined, there is often no evidence of the presence of mycorrhizal fungi, while a healthy tree growing next to a dead one has the root fungus.

Mycorrhiza The symbiotic relationship between fungi and the roots of the two plants on the right increases the uptake of water and nutrients into the plant. As a result, these plants show better growth than the control plant on the left, which does not have mycorrhiza.

of dispersal, and when they land in a favorable environment with moist conditions, they germinate and begin the process of decomposition. As decomposers, fungi cause billions of dollars worth of damage each year. Clothing, wood, leather, and all types of food are susceptible to damage by fungi. One of the best ways to protect against such damage is to keep the material dry. Millions of dollars are also spent each year on fungicides to limit damage.

There are also pathogenic fungi that feed on living organisms; those that cause ringworm and athlete's foot are two examples. A number of diseases are caused by fungi that grow on human mucous membranes, such as those of the vagina, lungs, and mouth.

Plants are also susceptible to fungal attacks. Chestnut blight and Dutch elm disease almost caused these two species of trees to become extinct. The fungus that causes Dutch elm disease is a parasite that kills the tree; then it functions as a saprophyte and feeds on the dead tree. Fungi also damage certain domestic crops. Wheat rust gets its common name because infected plants look as if they are covered with rust. Corn smut is also due to a fungal pathogen of plants (figure 5.3).

Figure 5.3 Corn Smut Most people who raise corn have seen corn smut. Besides being unsightly, it decreases the corn yield.

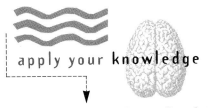

apply your **knowledge**

Declining Mushroom Populations

Throughout much of Europe there has been a severe decline in the mushroom population. On study plots in Holland, data collected since 1912 indicate that the number of different mushroom species has dropped from thirty-seven to twelve per plot in recent years. Along with the reduction in the number of species there is a parallel decline in the number of individual plants; moreover, the surviving plants are smaller.

The phenomenon of the disappearing mushrooms is also evident in England. One study noted that out of sixty fungus species, twenty had declining populations. Biologists are also concerned about a decline in the United States; however, there are no long-term studies, such as those in Europe, to provide evidence for such a decline.

How do you think declines in mushroom populations will impact the environment? Form a panel, review the information in this chapter, and write an opinion with regard to this question.

Fairy Rings

One of the most interesting formations caused by mushroom growth can be seen in lawns, fields, and forests. These formations, known as *fairy rings,* result from the expanding growth of the mushrooms. The inner circle contains normal grass and vegetation. The mushroom population grew out from the center as it used up the soil nutrients necessary for fungal growth. As the microscopic filaments grow outward, they stunt the growth of grass, forming a ring of short grass. Just to the outside of this ring, the grass is healthy because the filaments excrete enzymes that decompose soil material into rich nutrients for growth. The name *fairy ring* comes from an old superstition that such rings were formed by fairies tramping down the grass while dancing in a circle.

Fairy Ring Legends say that fairies danced in a circle in the moonlight and sat upon mushrooms. We understand that the "fairy ring" was formed as the fungus began to grow outward from the center. As organic material was consumed, the fungus grew in an ever widening circle and formed this fairy ring.

A number of fungi produce deadly poisons called **mycotoxins.** There is no easy way to distinguish those that are poisonous from those that are safe to eat. The most deadly mushroom, *Amanita verna,* is known as "the destroying angel" and can be found in woodlands during the summer. Mushroom hunters must learn to recognize this deadly, pure white species. This mushroom is believed to be so dangerous that food accidentally contaminated by its spores can cause illness and possible death.

Algae and Protozoans (Kingdom Protista)

The first protists evolved about 1.5 billion years ago. Like the prokaryotes, most of the protists are one-celled organisms. However, there is a significant difference between the two kingdoms: All of the protists are eukaryotic cells and all of the bacteria are prokaryotic cells. Prokaryotic cells usually have a volume of 1 to 5 cubic micrometers (0.000039 to 0.0002 cubic inches). Most eukaryotic cells have a volume greater than 5,000 cubic micrometers (0.2 cubic inches). This means that eukaryotic cells usually have a volume at least

FYI

A micrometer is a unit of measure equal to one millionth of a meter.
0.000001 meter = 1 micrometer
1 micrometer = 0.000039 inches

table 5.2

Types of Protista

Phylum	Algae Characteristics
Dinoflagellophyta (Dinoflagellates)	Marine; some produce toxins; colonial forms; major photosynthesizers in ocean
Chrysophyta (Golden algae)	Freshwater; form large colonies; no sexual reproduction
Bacillariophyta (Diatoms)	Aquatic; single cells or colonial; two-part silica shell; major aquatic photosynthesizers
Euglenophyta (Euglena)	Freshwater; some both photosynthetic and heterotrophic
Chlorophyta (Green algae)	Marine and freshwater; ancestors of green plants
Phaeophyta (Brown algae)	Marine; large colonial forms; form large seaweed beds
Rhodophyta (Red algae)	Mainly marine, some freshwater; multicellular
Phylum	**Protozoa Characteristics**
Zoomastigina (Flagellates)	Freshwater and marine; some parasitic; one or more flagella
Rhizopoda, Actinopoda, Foraminifera (Amoebae)	Marine and freshwater; some terrestrial; some parasitic
Sporozoa (Nonmotile)	Parasitic; complex life cycle; spore-forming
Ciliophora (Ciliates)	Marine and freshwater; cilia
Phylum	**Funguslike Protist Characteristics**
Myxomycota (Plasmodial slime mold)	Feed by phagocytosis; both haploid and diploid stages
Acrasiomycota (Cellular slime molds)	Freshwater, damp soil; feed by phagocytosis; spore-forming
Oomycota (Water mold)	Parasites or decomposers; filamentous forms; spore-forming

one thousand times greater than prokaryotic cells. The presence of membranous organelles such as the nucleus, ribosomes, mitochondria, and chloroplasts allows protists to be larger than bacteria. These organelles provide a much greater surface area within the cell upon which specialized reactions may occur. This allows for more efficient cell metabolism than is found in bacteria cells.

Because of the great diversity within the more than 60,000 species, it is difficult to separate the kingdom Protista into subgroupings. Usually the species are divided into three groups: **algae,** autotrophic unicellular organisms; **protozoa,** heterotrophic unicellular organisms; and **funguslike** protists (table 5.2).

Plantlike Protists

Algae are protists that have a cellulose cell wall. They contain chlorophyll and can therefore carry on photosynthesis. Single-celled and colonial types occur in a variety of habitats (figure 5.4). There are two major forms of algae in a variety of ocean and freshwater habitats: planktonic and benthic.

FYI

Plankton consists of small floating or weakly swimming organisms. **Benthic** organisms live attached to the bottom or to objects in the water. **Phytoplankton** consists of photosynthetic plankton that forms the basis for most aquatic food chains.

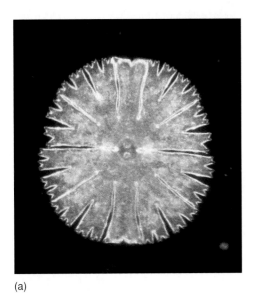

(a)

Figure 5.4 **Algae** Algae is found in a variety of types and colors. (a) A single-celled green alga, *Micrasterias*, and (b) a colonial red alga, *Antithamnium*.

(b)

Figure 5.5 **Zones of the Ocean** Photosynthesis can occur only in the photic zone, the top 100 meters of the oceans. This photosynthetic area contains most of the marine life. The prime source of food in the aphotic zone is the material that filters down from the photic zone. For this reason there is relatively little life in the deep ocean.

The large number of benthic and planktonic algae makes them an important source of atmospheric oxygen (O_2). It is estimated that 30 to 50 percent of atmospheric oxygen is produced by algae.

Since algae require light, phytoplankton is found only near the surface of the water. Even in the clearest water, photosynthesis does not usually occur any deeper than 100 meters (figure 5.5). To remain near the surface, some of the phytoplankton are capable of moving themselves to the top. Others maintain their position by storing food as oil, which is less dense than water and enables the cells to float near the surface.

Three common forms of single-celled algae typically found as phytoplankton are *Euglena*, diatoms, and dinoflagellates. *Euglena* are found mainly in freshwater. They are widely studied because they are easy to grow in the lab. Under certain conditions, these photosynthetic species can ingest food. *Euglena* can be either autotrophic or heterotrophic.

There are over 10,000 species of diatoms. These algae are unique because their cell walls contain silica (silicon dioxide). The algal walls fit together like the lid and bottom of a shoe box; the lid overlaps the bottom (figure 5.6). Because their cell walls contain

Figure 5.6 **Diatoms** Diatoms are single-celled, photosynthetic organisms found in salt water, freshwater, and most soil. Their walls contain silica, and the two parts of the wall fit together like the bottom and lid of a shoe box. Different species show a great variety of shapes.

silicon dioxide, they readily form fossils. The fossil cell walls have large, abrasive surface areas with many tiny holes and can be used in a number of commercial processes. They are used as filters for liquids and as abrasives in specialty soaps, toothpastes, and scouring powders.

Diatoms are commonly found in freshwater and marine environments. They can reproduce both sexually and asexually. When conditions are favorable, asexual reproduction can result in what is called an algal **bloom**—a rapid increase in the population of microorganisms in a body of water. The population can become so large that the water looks murky.

Along with diatoms, dinoflagellates are the most important producers in the ocean. All members of this group of algae have two flagella, which is the reason for their name: *di* = two. Many marine forms are *bioluminescent;* they are responsible for the twinkling lights seen at night in ocean waves or in a boat's wake.

Some species of dinoflagellates have a symbiotic relationship with marine animals such as reef corals; the dinoflagellates provide a source of nutrients for the reef-building corals. Corals that live in the light and contain dinoflagellates grow ten times faster than corals without this symbiont. Thus, in coral reefs, dinoflagellates form the foundation of the food chain.

Multicellular algae, commonly known as *seaweed,* are large colonial forms usually found attached to objects in shallow water. Two types, red algae and brown algae, are mainly marine forms. The green algae are a third kind of seaweed; they are primarily freshwater species.

Red algae live in warm oceans and attach to the ocean floor. They may be found from the shoreline to depths of 100 meters (327 feet). Some red algae become encrusted with calcium carbonate (chalk) and are important in reef building; other species are of commercial importance because they produce agar and carrageenin. *Agar* is widely used as a jelling agent for growth media in microbiology. *Carrageenin* is a gelatinous material used in paints, cosmetics, and baking. It is also used to make gelatin desserts harden faster and to make ice cream smoother. In Asia and Europe some red algae are harvested and used as food.

Brown algae are found in cooler marine environments than the red algae. Colonies of

these algae can reach 100 meters in length (figure 5.7). Brown algae produce *alginates,* which are widely used as stabilizers in frozen desserts, emulsifiers in salad dressings, and as thickeners that give body to foods such as chocolate milk and cream cheeses; they are also used to form gels in such products as fruit jellies.

Green algae are found primarily in freshwater, where they may attach to a variety of objects. Members of this group can also be found growing on trees, in the soil, and even on snowfields in the mountains. Like land plants, green algae have cellulose cell walls and store food as starch. Green algae also have the same types of chlorophyll as do plants. Biologists believe that land plants evolved from the

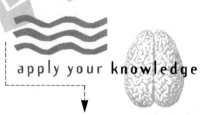

apply your knowledge

Algae In the Kitchen

Algal products are commonly found in many household foods and other items such as cosmetics and toiletries. Take a few minutes and examine the ingredient labels on some of your salad dressings, shampoos, ice cream, gelatin desserts, cookies, and other items. Look for names such as alginate and carrageenin.

Figure 5.7   A Kelp Grove   These multicellular brown algae are attached to the ocean floor by holdfasts. Their blades may reach a length of 100 meters and float upward because they contain bladderlike structures filled with air.

FYI

Seafood Poisoning and Oysters   **Some forms of dinoflagellates produce toxins that can be accumulated by such filter-feeding shellfish as clams, scallops, mussels, and oysters. Filter-feeding shellfish ingest large amounts of the poison, which has no effect on the shellfish but can cause sickness or death in animals that feed on them, such as fish, birds, and humans. Many of the toxin-producing dinoflagellates contain red pigment. Blooms of this kind are responsible for *red tides*. Red tides usually occur in the warm months, during which people should refrain from collecting and eating oysters and other shellfish. The expression "Oysters 'R' in season" comes from the fact that most of the months with an *r* in their spelling are cold weather months, during which oysters are safer to eat. Commercially available shellfish are tested for toxin content; if they are toxic, they should not be marketed.**

green algae. A second major group of organisms in the kingdom Protista, the protozoa, lack all types of chlorophyll.

Animal-like Protists

The word *protozoa* literally means "first animal." The protozoa include all eukaryotic, heterotrophic, single-celled organisms. They are classified according to their method of motion. Most members of the flagellates have flagella and live in freshwater. They have no cell walls and no chloroplasts, and they can be parasitic or free-living.

There is a mutualistic relationship between some flagellates and their termite hosts. Certain protozoa live in termite guts and are capable of digesting cellulose into simple sugars that serve as food for the termite. Of the parasitic protozoa, two different species produce sleeping sickness in humans and domestic cattle. In both cases the protozoan enters the host as the result of an insect bite. The parasite develops in the circulatory system and moves to the spinal fluid surrounding the brain. When this occurs, the infected person develops the "sleeping" condition, which, if untreated, is eventually fatal. Many biologists believe that all other types of protozoa, and even the multicellular animals, evolved from primitive flagellated microbes similar to the flagellates.

Protozoa known as the *Amoeba* have a constantly changing shape (figure 5.8). Amoebae use pseudopods to move about and to engulf food. A *pseudopod* is an extension of the cell that contains moving cytoplasm. Many pseudopods are temporary extensions that form and are retracted as the cell moves. Most amoeboid protists are free-living and feed on bacteria, algae, or even small multicellular organisms.

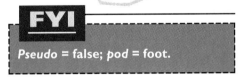

FYI

Pseudo = false; *pod* = foot.

The Bermuda Triangle and the Sargasso Sea

The Sargasso Sea is a large mat of free-floating brown algae between the Bahamas and the Azores. It is thought that this huge mass (as large as the European continent) is the result of brown algae that have become detached from the ocean bottom, carried by ocean currents, and accumulated in this calm region of the Atlantic Ocean. This large mass of floating algae provides a habitat for many marine animals, such as marine turtles, eels, jellyfish, and innumerable crustaceans. Part of the Sargasso Sea makes up the Bermuda Triangle! For centuries sailors have told stories about the dangers of sailing through this region—some true, some not so true.

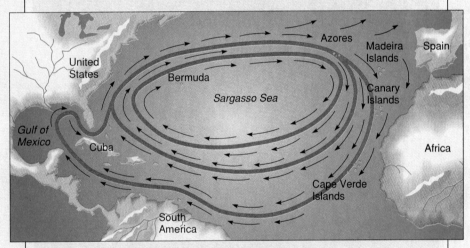

The Bermuda Triangle and the Sargasso Sea

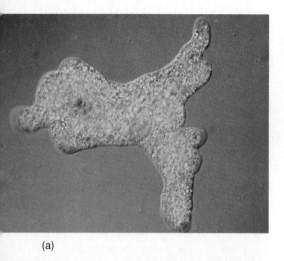

(a)

(b)

Figure 5.8 *Amoeba and Its Relatives*
Several kinds of protozoa use pseudopods for movement and food gathering. (*a*) The pseudopods of *Amoeba* change shape as it moves and feeds. (*b*) Other organisms have long, thin, food-gathering pseudopods extending through a shell that are not used for movement.

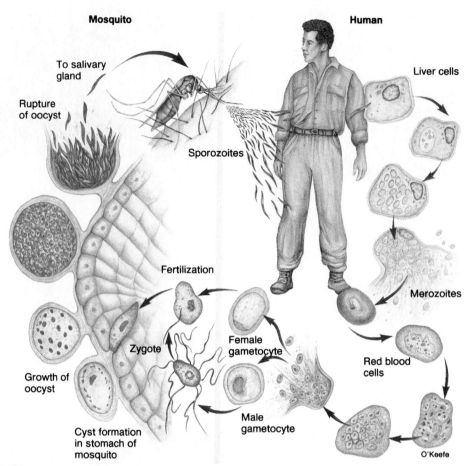

Mosquito

Human

To salivary gland

Rupture of oocyst

Sporozoites

Liver cells

Fertilization

Merozoites

Zygote

Female gametocyte

Growth of oocyst

Red blood cells

Cyst formation in stomach of mosquito

Male gametocyte

O'Keefe

Figure 5.9 The Life Cycle of *Plasmodium vivax*   The life cycle of the malaria parasite requires two hosts, the *Anopheles* mosquito and the human. Humans get malaria when they are bitten by a mosquito carrying the larval stage of *Plasmodium*. The larva undergoes asexual reproduction and releases thousands of individuals that invade red blood cells. Their release causes the chills, fever, and headache associated with malaria. Inside the red blood cell, special male and female cells called gametocytes are formed.

When the mosquito bites a person with malaria, it ingests some gametocytes. Fertilization occurs inside the mosquito and zygotes develop. The resulting larvae are housed in the mosquito's salivary glands. Thus, when the mosquito bites another person, some saliva containing the larvae is released into the person's blood and the cycle begins again.

Some amoebae live in warm oceans and are enclosed in a shell. As the amoebae die, the shells collect on the ocean floor, and their remains form limestone. The cliffs of Dover, England, were formed from such shells. Oil companies are interested in these protozoans because they are often found where oil deposits are located.

Malaria, one of the leading causes of disability and death in the world, is caused by a type of protozoan that is unable to move on its own. Two billion people live in malaria-prone regions of the world. There are an estimated 150 million to 300 million new cases of malaria each year, and the disease kills 2 million to 4 million people annually. The protozoan that causes malaria has a complex life cycle involving a mosquito vector for transmission (figure 5.9).

While in the mosquito vector, the parasite goes through the sexual stages of its life cycle. One of the best ways to control this disease is to eliminate the vector, which usually involves using some sort of pesticide. Many of us are concerned about the harmful effects of pesticides in the environment. However, in parts of the world where malaria is common, the harmful effects of pesticides are of less concern than the harm caused by the disease. Many diseases of domestic and wild animals are also caused by protozoans.

The ciliated protozoans are the most complex. They are commonly known as *ciliates* and derive their name from the fact that they have many short, hairlike structures called *cilia* (figure 5.10). These move in an organized, rhythmic manner and propel the cell through the water. Some types of ciliates, such as *Paramecium*, have nearly 15,000 cilia per cell and move at a rapid speed of one millimeter per second.

CONCEPT CONNECTIONS

• A *vector* is an organism capable of transmitting a parasite from one organism to another.

Most ciliates are free-living cells found in fresh- and salt water, where they feed on bacteria and other small organisms.

Funguslike Protists

There are two kinds of funguslike protists: the slime molds and the water molds. Some slime molds grow on rotting damp logs, leaves, and soil. They look like giant amoebae whose nucleus and other organelles have divided repeatedly within a single large cell (figure 5.11). No cell membranes divide this mass into separate segments. They vary in color from white to bright red or yellow, and may reach relatively large sizes (45 centimeters, or 17.55 inches, in length) when in their best environment.

Other kinds of slime molds exist as large numbers of individual, amoebalike cells. These cells get food by engulfing microbes. When their environment becomes dry or unfavorable, the cells come together. This mass glides along like a garden slug and may flow about for hours before it forms spores. When the mass gets ready to form spores, it forms a stalk with cells that have cell walls. At the top of this specialized structure, cells are modified to become spores. When released, these spores may be carried by the wind and, if they land in a favorable place, develop into new amoebalike cells.

Another group of funguslike protists includes the water molds. This group has reproductive cells with two flagella. A wide variety of

Don't Drink the Water!

Giardia lamblia is a protozoan in streams and lakes throughout the world, including "pure" mountain water in U.S. wilderness areas. More than forty species of animals harbor this organism in their small intestines. Its presence may cause diarrhea, vomiting, cramps, or nausea. *Giardia* can be found even where good human sanitation is practiced. No matter how inviting it may seem to drink directly from that cold mountain stream, *don't*. Deer, beaver, or other animals may have contaminated the water with *Giardia*. Treat the water before drinking. The most effective way to eliminate the spores formed by this protozoan is to use special filters that can filter out particles as small as 1 micrometer; or boil the water for at least five minutes before drinking.

The species called *Entamoeba histolytica* (ent = inside; amoeba = amoeba; histo = tissue; lytica = destroying) is responsible for the diarrheal disease known as dysentery. This is the protozoan most people become infected with when they travel in a foreign country and drink contaminated water. If you plan on such a trip, be sure to see your physician *several weeks before* you go. The infection can be prevented by taking an antiprotozoal antibiotic, but you must start treatment ahead of time.

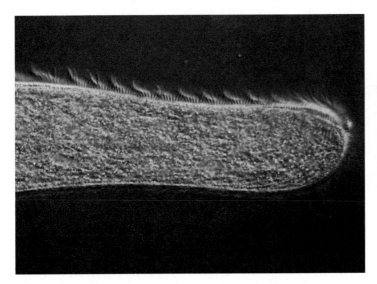

Figure 5.10 Ciliated Protozoa The many hairlike cilia on the surface of this cell are used to propel the protozoan through the water.

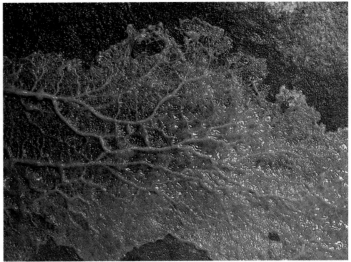

Figure 5.11 Slime Mold Slime molds grow in moist conditions and are important decomposers of dead plants, particularly fallen trees. As the slime mold grows, it forms a large mass with many nuclei. In many kinds of slime molds the mass moves as a unit.

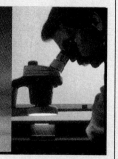

BIO *feature*

Multicellularity in the Protista

The three major types of the kingdom Protista (algae, protozoa, and funguslike protists) include both single-celled and multicellular forms. Biologists believe that evolution in all three of these groups has been similar. In each group, the most primitive organisms are thought to have been single-celled, and we believe these gave rise to the more advanced multicellular forms. Most protozoans, for example, are single-celled, but one protozoan group contains colonial forms. The multicellular forms of funguslike

protists are the slime molds, which have both single-celled and multicellular stages. Perhaps the most widely known example of the evolution from single-celled to multicellular organisms is found in the green algae. A common single-celled green alga is *Chlamydomonas*, which has a cell wall and two flagella. It looks just like the individual cells of the colonial green algae *Volvox*, which can be composed of more than half a million cells. All the flagella of each cell in the colony move, allowing the colony to move in a given direction. Many

of the cells cannot reproduce sexually; other cells assume this function for the colony. In some *Volvox* species, certain cells have even specialized to produce sperm or eggs.

Biologists believe that the division of labor seen in colonial protists represents the beginning of specialization that led to the development of true multicellular organisms with many different kinds of specialized cells. Three types of multicellular organisms—fungi, plants, and animals— eventually developed.

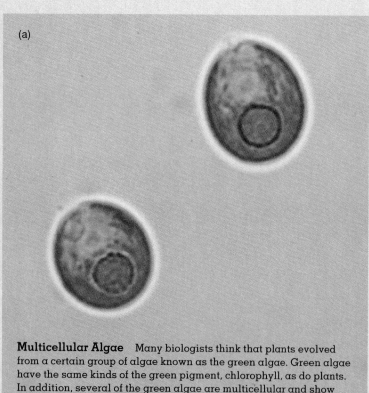

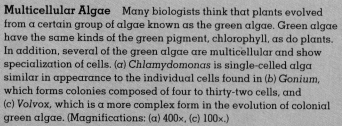

Multicellular Algae   Many biologists think that plants evolved from a certain group of algae known as the green algae. Green algae have the same kinds of the green pigment, chlorophyll, as do plants. In addition, several of the green algae are multicellular and show specialization of cells. (a) *Chlamydomonas* is single-celled alga similar in appearance to the individual cells found in (b) *Gonium*, which forms colonies composed of four to thirty-two cells, and (c) *Volvox*, which is a more complex form in the evolution of colonial green algae. (Magnifications: (a) 400×, (c) 100×.)

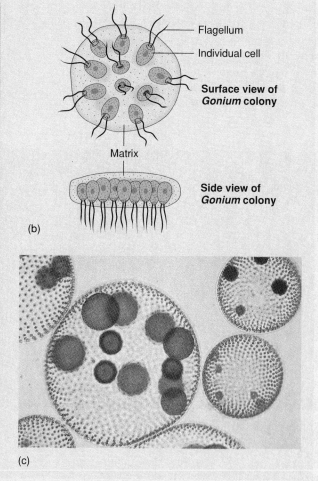

water molds are *saprophytes,* which decompose dead organic matter and usually grow in moist environments. The water molds differ in structure from true fungi in that some filaments have no cross walls, thus allowing the cell contents to flow from cell to cell.

Water molds are important saprophytes and parasites in aquatic environments. They may be seen as fluffy growths on dead fish or other organic matter floating in the water. A parasitic form of this fungus is well known to people who rear tropical fish; it causes a cotton-like growth on the fish. Although water molds are usually found in aquatic environments, they are not limited to this location. Some species cause downy mildew on plants such as grapes. In the 1880s this mildew almost ruined the French wine industry when it spread throughout the vineyards. A copper-based fungicide called Bordeaux mixture—the first chemical used against plant diseases—was used to save the vines.

The Bacteria (Kingdom Prokaryotae)

The kingdom Prokaryotae contains organisms that are commonly referred to as **bacteria.** Other common names for them are *germs* or *microorganisms.* They are single cells that lack an organized nucleus and other complex organelles (figure 5.12). For general purposes, bacteria are divided into the three groups shown in table 5.3.

table 5.3

Types of Bacteria

Type	Characteristics
Eubacteria (true bacteria)	Nitrogen-fixing
	Major pathogens
	Decomposers
Cyanobacteria (blue-green bacteria)	Photosynthetic; release oxygen
Archaebacteria	Tolerate extreme environments
	Anaerobic
	Oxidize inorganic molecules as a source of energy

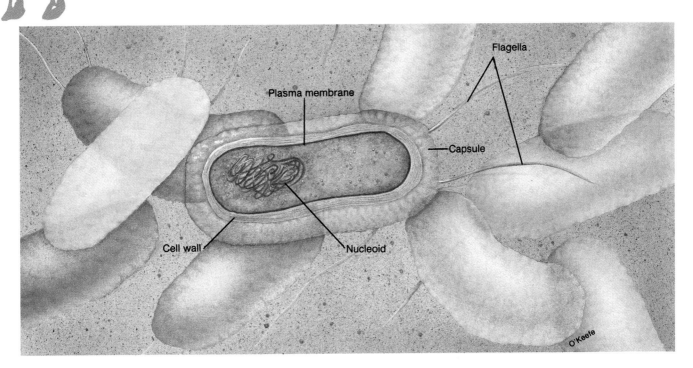

O'Keefe

Figure 5.12 Bacterial Cell Bacteria are simple prokaryotic cells. A plasma membrane regulates the movement of material between the cell and its environment. The genetic material is a single loop of DNA that is not enclosed in a membrane. A rigid cell wall protects and gives shape to the cell. Some bacteria, usually disease-causing organisms, have a capsule that protects them from the immune system of their host. Some bacteria have flagella that allow them to swim through their environment.

BIO *feature*

Beneficial Bacteria

One of the most beneficial bacteria lives right inside your intestines and can be purchased at your drug store. *Lactobacillus acidophilus* lives between the cells that make up the lining of your intestinal tract. It can be helpful to humans in many ways:

1. Produces antibiotics that help to destroy other harmful bacteria that are swallowed.
2. Releases weak acids that enable nutrients to cross into the body more efficiently.
3. Helps prepare your immune system to defend against other invading bacteria.
4. Releases certain vitamins.

Where do these bacteria come from? They can be acquired from the environment and from many uncooked foods such as lettuce, celery, cabbage, and some types of yogurt. You can also buy them at the drug store! In fact, many physicians recommend to their patients with gastrointestinal upset (commonly called diarrhea) that they purchase freeze-dried tablets or granules of these helpful bacteria. Pediatricians also recommend its use for some babies with colic. Certain types of milk also contain this bacteria. Next time you're in the supermarket or drug store, stop by the dairy section or the refrigerator in the pharmacy and check out the various brands that contain *Lactobacillus acidophilus*.

These bacteria establish this mutualistic relationship when they are ingested along with food or drink. When people travel, they consume local bacteria with their meals and may have problems establishing a relationship with these new bacteria. Some people develop traveler's diarrhea as a result.

CONCEPT CONNECTIONS

symbiosis = a long-lasting, close physical relationship between two different species.

mutualism = a relationship between two organisms in which both organisms benefit.

commensalism = a relationship between two organisms in which one organism is helped and the other is not affected.

parasitism = a relationship between two organisms in which one organism lives in or on another organism and derives nourishment from it.

FYI

The Romans knew that bean plants somehow enriched the soil, but it was not until the 1800s that bacteria were recognized as the enriching agents. Certain types of bacteria have a mutualistic relationship with the roots of bean plants. These bacteria are capable of converting atmospheric nitrogen into a form the plants can use.

Many forms of bacteria are beneficial to humans. Some forms of **saprophytic** bacteria decompose dead material, sewage, and other forms of waste into simpler molecules that can be recycled. The food industry uses bacteria to

CONCEPT CONNECTIONS

Saprophytes are organisms that feed on dead or decaying material. Many bacteria are also saprophytic.

produce cheeses, yogurt, sauerkraut, and many other foods. Alcohols, acetones, acids, and other chemicals are produced by bacteria. The pharmaceutical industry employs bacteria to produce antibiotics and vitamins. Some bacteria can even metabolize oil and are used to clean up oil spills.

Bacteria and other organisms may also form mutualistic relationships. In **mutualism,** two different species live together and benefit one another. For example, some intestinal bacteria benefit humans by producing antibiotics that inhibit the development of **pathogenic** bacteria. The helpful bacteria compete with the disease-causing bacteria for nutrients, helping to keep them in check. Intestinal bacteria also aid digestion by releasing small amounts of acid that encourages food molecules to enter the body. They produce and release vitamin K, a vitamin used in the blood-clotting process.

CONCEPT CONNECTIONS

pathogen: *patho* = sad; *gen* = produce.

pathogenic = a microbe that produces sadness (disease).

Animals do not have the ability to digest cellulose, but some bacteria do; they obtain metabolic energy from plant cellulose and release carbon dioxide (CO_2) and methane gas (CH_4). Thus a mutualistic relationship exists, for example, between cows and the bacteria that live in their intestinal tracts. Similar bacteria are found in the human gut and produce intestinal gas. In some regions of the world these same bacteria are used to digest organic waste; the methane is captured and used as a source of fuel.

Early forms of life on Earth consisted of prokaryotic cells. These photosynthetic bacterial cells released oxygen, and the Earth's atmosphere began to change. Such photosynthetic, colonial blue-green bacteria are still present in large numbers and continue to release large amounts of oxygen. Colonies

Common Bacteria Found in or on Your Body

- Skin: *Corynebacterium sp., Staphylococcus sp., Streptococcus sp., E. coli, Mycobacterium sp.*
- Eye: *Corynebacterium sp., Neisseria sp., Bacillus sp., Staphylococcus sp., Streptococcus sp.*
- Ear: *Staphylococcus sp., Streptococcus sp., Corynebacterium sp., Bacillus sp.*
- Mouth: *Streptococcus sp., Staphylococcus sp., Lactobacillus sp., Corynebacterium sp., Fusobacterium sp., Vibrio sp., Haemophilus sp.*
- Nose: *Corynebacterium sp., Staphylococcus sp., Streptococcus sp.*
- Intestinal tract: *Lactobacillus sp., E. coli, Bacillus sp., Clostridium sp., Pseudomonas sp., Bacteroides sp., Streptococcus sp.*
- Genital tract: *Lactobacillus sp., Staphylococcus sp., Streptococcus sp., Clostridium sp., Peptostreptococcus sp., E. coli.*

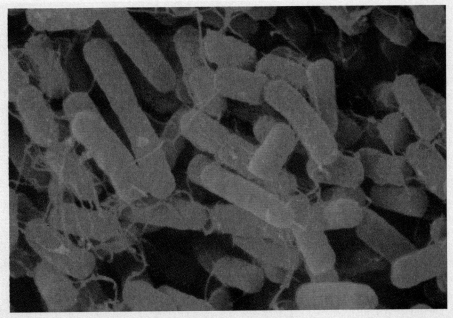

(a) *E. coli*

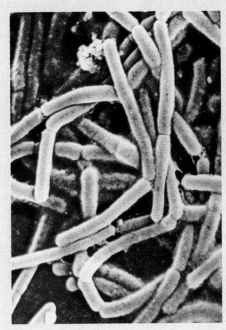

(b) *Lactobacillus*

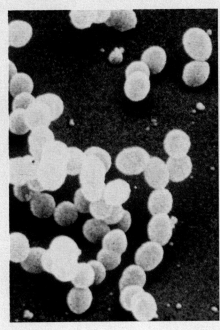

(c) *Streptococcus*

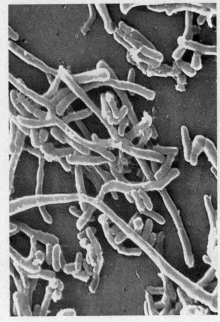

(d) *Corynebacterium*

BIO *feature*

Photosynthetic Bacteria in Drinking Water

Scientists used to call the colonial blue-green bacteria blue-green algae. When they discovered that these organisms are prokaryotic cells, they changed the name and classification.

In 1992 the Centers for Disease Control and Prevention (CDC) in Atlanta, Georgia, reported the first-ever confirmed outbreak (multiple cases) of diarrhea caused by photosynthetic bacteria, the cyanobacteria. Twenty-one people who lived or worked in a hospital dormitory close to Chicago reported having symptoms. The symptoms developed after the building's water tank pump failed. Investigators believe that bacteria lying harmlessly on the bottom of the storage tank were circulated into the drinking water when the tank was refilled following repair. These photosynthetic bacteria caused the outbreak.

FYI

Group A Streptococcus in the News **A very dangerous type of the bacterium *Streptococcus* has recently made headlines. This organism was responsible for the 1990 death of Jim Henson, creator of the Muppets. The Centers for Disease Control and Prevention in Atlanta, Georgia, has estimated that about 15,000 people come down with an infection caused by this form of strep every year. It can be acquired in an open wound, in the throat, or in other locations. The bacteria reproduce rapidly, producing flu-like symptoms of chills and fever that worsen quickly in only a day or two. The pathogen will destroy muscle and fat tissue. If you experience such symptoms that do not get better in a reasonable amount of time, be sure to see a physician for treatment with antibiotics as soon as possible.**

of blue-green bacteria are found in aquatic environments, where they form long, filamentous strands commonly called *pond scum.* Some of the larger cells in the colony are capable of converting atmospheric nitrogen, N_2, to ammonia, NH_3. This provides a form of nitrogen usable to other cells in the colony—an example of division of labor.

The word *bacteria* usually brings to mind visions of tiny things that cause diseases; however, the majority of bacteria are free-living and not harmful. Some diseases are indeed caused by bacteria, but only a minority are pathogens. It is normal for all organisms to have symbiotic relationships with bacteria. Most organisms are lined and covered by bacteria called *normal flora.* In fact, if an organism lacks bacteria it is considered abnormal.

Some potentially harmful bacteria may be associated with an organism yet not cause disease. For example, *Streptococcus pneumoniae* may grow in the throats of healthy people without any pathogenic effects. But if a person's resistance is lowered, as after a bout with viral flu, *Streptococcus pneumoniae* may move into the lungs and reproduce rapidly, causing pneumonia; the relationship has changed from commensalistic to parasitic.

Bacteria may invade a healthy host and cause disease by altering the tissue's normal function. The bacteria release a variety of chemicals that destroy the tissue. The disease ends when the pathogens are killed by the body's defenses or by some outside agent, such as an antibiotic. Examples of such infectious bacterial diseases are strep throat, syphilis, pneumonia, tuberculosis, and leprosy (figure 5.13).

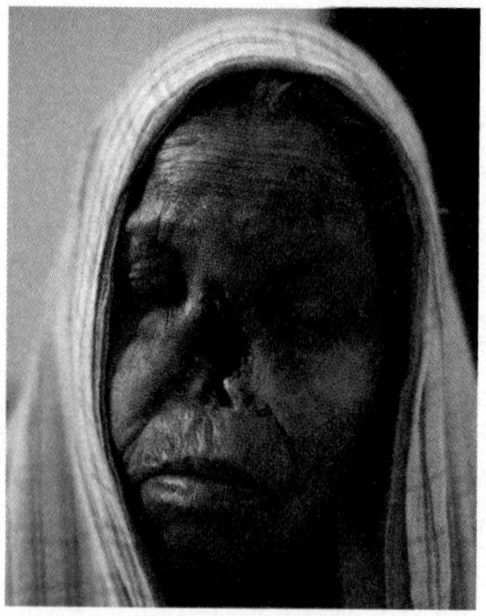

Figure 5.13 Leprosy Over 20 million people worldwide are infected with *Mycobacterium leprae* and have leprosy. The bacterium alters the infected person's physiology, resulting in open sores and loss of tissue.

Many other bacterial illnesses are caused by poisons the bacteria produce. These poisons may be consumed with a meal. In this case, disease can be caused even though the pathogens never enter the host. For example, botulism is an extremely deadly disease that is caused by bacterial toxins in food or drink. Some other bacterial diseases are caused by toxins released from bacteria growing inside the host tissue; tetanus and diphtheria are examples. In general, toxins may cause tissue damage, fever, and aches and pains.

Bacterial pathogens are also important factors in certain plant diseases. Bacteria are responsible for many types of plant blights, wilts, and soft rots. Apple and other fruit trees are susceptible to fire blight, a disease that lowers the fruit yield because it kills the tree's branches. Citrus canker, a disease of

BIO *feature*

The Discovery of Penicillin

The discovery of the antibiotic penicillin is an interesting story. In 1928, Dr. Alexander Fleming was working at St. Mary's Hospital in London. As he sorted through some old petri dishes on his bench, he noticed something unusual. The mold *Penicillium notatum* was growing on some of the petri dishes. Apparently, the mold had found its way through an open window and onto a bacterial culture of *Staphylococcus aureus*. The bacterial colonies that were growing at a distance from the fungus were typical, but there was no growth close to the mold. Fleming isolated the agent responsible for this destruction of the bacteria and named it *penicillin*.

Through Fleming's research and that of several colleagues, the chemical was identified and used for about ten years in microbiological work in the laboratory. Many suspected that penicillin might be used as a drug, but the fungus could not produce enough of the chemical to make it worthwhile. When World War II began and England was being firebombed, there was an urgent need for a drug that would control bacterial infections in burn wounds. Two scientists from England were sent to the United States to find a way to mass produce penicillin.

Their research in isolating new forms of *Penicillium* and purifying the drug were successful. Cultures of the mold now produce one hundred times more drug than did the original mold discovered by Fleming. In addition, the price of the drug has dropped considerably—from $20,000 per kilogram in 1944 to less than $250 today.

The species of *Penicillium* used to produce penicillin today is *P. chrysogenum*, which was first isolated in Peoria, Illinois, from a mixture of molds found growing on a cantaloupe. The species name *chrysogenum* means "golden" and refers to the golden-yellow droplets of antibiotic that the mold produces on the surface of its filaments. The spores of this mold were isolated and irradiated with high dosages of ultraviolet light, which caused mutations to occur in the genes. When some of these mutant spores were germinated, the new filaments were found to produce much greater amounts of the antibiotic.

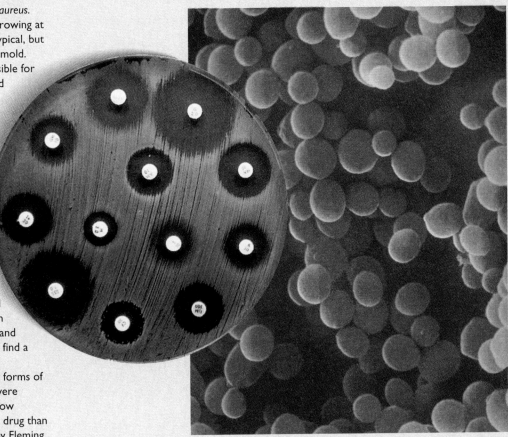

Penicillin   The clear area around the disk on the petri dish is the result of penicillin inhibiting the growth of bacteria. The cloudy nature of the other parts of the petri dish indicates the abundant bacterial growth of *Staphylococcus*.

Figure 5.14 **Bacterial Disease in Plants**
Citrus canker growth on an orange tree
promotes rotting of the infected parts of the tree.

citrus fruits that causes cancerlike growths, can cause widespread damage. In a three-year period, Florida citrus growers lost $2.5 billion because of this disease (figure 5.14).

Despite large investments of time and money, scientists have found it difficult to control bacteria. Two factors operate in favor of the bacteria: their reproductive rate and their ability to form *spores*. Under ideal conditions some bacteria can grow and divide every 20 minutes. If one bacterial cell and all of its offspring were to reproduce at this ideal rate, in 48 hours there would be 2.2×10^{43} cells (figure 5.15). In reality, bacteria cannot achieve such incredibly large populations because they would eventually run out of food and be unable to dispose of their wastes.

	Number of bacteria
Beginning	1
After 1 hour	8
After 2 hours	64
After 3 hours	512
After 4 hours	4,096
After 5 hours	32,768
After 6 hours	262,144
After 7 hours	2,097,152
After 8 hours	16,777,216
After 9 hours	134,217,720
After 10 hours	1,073,741,700
After 11 hours	8,589,933,600
After 12 hours	68,719,468,000
After 13 hours	549,755,720,000
After 14 hours	4,398,045,600,000
After 15 hours	34,184,364,000,000
After 16 hours	281,474,900,000,000
After 17 hours	2,251,799,200,000,000
After 18 hours	18,014,393,000,000,000
After 19 hours	144,115,140,000,000,000
After 20 hours	1,152,921,000,000,000,000
After 21 hours	9,223,368,000,000,000,000
After 22 hours	73,786,948,000,000,000,000
After 23 hours	590,295,560,000,000,000,000
After 24 hours	4,722,364,400,000,000,000,000
After 25 hours	37,778,914,000,000,000,000,000
After 26 hours	302,231,300,000,000,000,000,000
After 27 hours	2,417,850,400,000,000,000,000,000
After 28 hours	19,342,803,000,000,000,000,000,000
After 29 hours	154,742,420,000,000,000,000,000,000
After 30 hours	1,237,939,300,000,000,000,000,000,000
After 31 hours	9,903,514,400,000,000,000,000,000,000
After 32 hours	79,228,112,000,000,000,000,000,000,000
After 33 hours	633,824,880,000,000,000,000,000,000,000
After 34 hours	5,070,598,800,000,000,000,000,000,000,000
After 35 hours	40,564,788,000,000,000,000,000,000,000,000
After 36 hours	324,518,300,000,000,000,000,000,000,000,000
After 37 hours	2,596,146,400,000,000,000,000,000,000,000,000
After 38 hours	20,769,170,000,000,000,000,000,000,000,000,000
After 39 hours	166,153,360,000,000,000,000,000,000,000,000,000
After 40 hours	1,329,226,800,000,000,000,000,000,000,000,000,000
After 41 hours	10,633,814,000,000,000,000,000,000,000,000,000,000
After 42 hours	85,070,512,000,000,000,000,000,000,000,000,000,000
After 43 hours	680,564,080,000,000,000,000,000,000,000,000,000,000
After 44 hours	5,444,512,400,000,000,000,000,000,000,000,000,000,000
After 45 hours	43,556,096,000,000,000,000,000,000,000,000,000,000,000
After 46 hours	348,448,760,000,000,000,000,000,000,000,000,000,000,000
After 47 hours	2,787,590,000,000,000,000,000,000,000,000,000,000,000,000
After 48 hours	22,300,720,000,000,000,000,000,000,000,000,000,000,000,000

Figure 5.15 **Bacterial Growth** If bacteria divided every 20 minutes, the numbers of bacteria would grow astronomically. You can see how uncontrolled disease organisms can increase extremely rapidly, causing acute disease.

Because bacteria reproduce so rapidly, a few antibiotic-resistant cells in a population can increase to dangerous levels in a short time. This requires the use of stronger doses of antibiotics or of new types in order to bring the bacteria under control. Furthermore, these resistant strains can be transferred from one person to another. For example, sulfa drugs and penicillin, once widely used to fight infections, are now ineffective against many strains of pathogenic bacteria. As new antibiotics are developed, resistant bacteria increase. Thus, humans are constantly waging battles against new strains of resistant bacteria.

Another factor that enables some bacteria to survive a hostile environment is their ability to form endospores. An **endospore** is a unique bacterial structure that germinates under favorable conditions to form a new, actively growing cell (figure 5.16). For example, people who preserve food by canning often boil it to kill the bacteria. But not all bacteria are killed by boiling; some of them form endospores. Botulism poison is usually found in foods that are improperly canned. The endospores of *Clostridium botulinum*, the bacterium that causes botulism, can withstand boiling and remain for years in the endospore state. However, if the endospore is in an acid environment, it will not germinate and produce botulism toxin. In that case, the food remains preserved and edible. If conditions become favorable for endospores to germinate, they become actively growing cells and produce toxin. Home canning is the major source of botulism. Using a pressure cooker or pressure canner, and heating the food to temperatures higher than 121°C (249.80°F) for 15 to 20 minutes destroys both botulism toxin and the endospores.

Lichens

Lichens are usually classified with the Mycetae, but they actually represent a very close mutualistic relationship between a fungus and an algal protist or a blue-green bacterium. Algae and blue-green bacteria require a moist environment. Certain species of these photosynthetic organisms grow surrounded by fungus. The fungal covering maintains a moist area, and the photosynthesizers in turn provide nourishment for the fungus. These two species growing together are what we call a **lichen** (figure 5.17). Lichens grow slowly; a patch of lichen may grow only 1 centimeter in diameter per year.

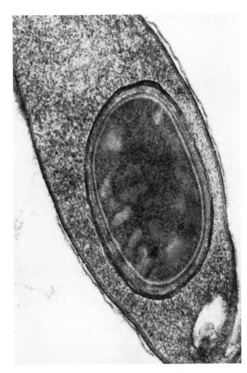

Figure 5.16 Bacterial Endospores
The darker area in the cell is the endospore. It contains the bacterial DNA as well as a concentration of cytoplasmic material that is surrounded and protected by a thick wall (magnification 63,000×).

Since the fungus provides a damp environment and the algae produce the food, lichens require no soil for growth. For this reason, they are commonly found growing on bare rock. Lichens are important in the process of soil formation. They secrete an acid that breaks down the rock and makes minerals available for use by plants. When lichens die, they provide a source of humus—dead organic material—that mixes with the rock particles to form soil.

Lichens are found in a wide variety of environments, ranging from the frigid arctic to the scorching desert. One reason for this success is their ability to withstand drought conditions. Some lichens can survive with only 2 percent water. In this condition they stop photosynthesis and go into a dormant stage, remaining so until water becomes available and photosynthesis begins again.

Another factor in the success of lichens is their ability to absorb minerals. However, because air pollution has increased the amounts of minerals they absorb, many lichens are damaged or killed. For this reason, areas with heavy air pollution are "lichen deserts." Because they can absorb minerals, certain forms of lichens have been used to monitor the amount of various pollutants in the atmosphere, including radioactivity. The absorption of radioactive fallout from Chernobyl by arctic lichens made the meat of the reindeer that fed on them unsafe for human consumption.

(a)   (b)

Figure 5.17 Lichens   Lichens are really two different kinds of organisms growing together: a fungus and an alga. They grow in a variety of habitats. (a) The shrubby lichen is growing on soil; (b) the crustlike lichen is growing on rock. The different coloring is due to the different species of algae or cyanobacteria in the lichens.

SUMMARY

The kingdoms Prokaryotae, Protista, and Mycetae rely mainly on asexual reproduction, and each cell usually satisfies its own nutritional needs. In some species, there is minimal cooperation between cells. The members of the kingdom Prokaryotae are bacteria, which have the genetic ability to function in various environments. Most species of bacteria are beneficial, although some are pathogenic.

Members of the kingdom Protista are one-celled organisms. They differ from the prokaryotes in that they are eukaryotic, not prokaryotic, cells. Protists include algae, autotrophic cells that have a cell wall and carry on photosynthesis; protozoa, which lack cell walls and cannot carry on photosynthesis; and fungus-like protists, whose motile, amoeboid reproductive stage distinguishes them from true fungi.

Some species of Protista developed a primitive type of specialization, and from these evolved the multicellular fungi, plants, and animals.

The kingdom Mycetae consists of nonphotosynthetic, eukaryotic organisms. Most species are multicellular. Fungi are nonmotile organisms that disperse by producing spores. Lichens are a combination of organisms involving a mutualistic relationship between a fungus and an algal protist or blue-green bacteria.

CHAPTER GLOSSARY

algae (al'je) Protists that have cell walls and chlorophyll and can therefore carry on photosynthesis.

bacteria (bak-tīr'ē-ah) Unicellular organisms of the kingdom Prokaryotae that have the genetic ability to function in various environments.

benthic (ben'thik) A term used to describe organisms that live in bodies of water, attached to the bottom or to objects in the water.

bloom (bloom) A rapid increase in the number of microorganisms in a body of water.

colonial (ko-lo'ne-al) A term used to describe a collection of cells that cooperate to a small extent.

commensalism (ko-men'sal-izm) A relationship between two organisms in which one organism is helped and the other is not affected.

endospore (en'do-spōr") A unique bacterial structure with a low metabolic rate that germinates under favorable conditions to grow into a new cell.

fungus (fun'gus) Nonphotosynthetic, multicellular organisms that have cell walls and that reproduce by spores.

lichen (li'kĕn) A mutualistic relation between fungi and algal protists or cyanobacteria.

microorganisms (microbes) (mi"kro-or'guh-niz"mz) Small organisms that cannot be seen without magnification.

mutualism (miu'chu-al-izm) A relationship between two organisms in which both organisms benefit.

mycorrhiza (my"ko-rye'zah) A symbiotic relation between fungi and plant roots.

mycotoxin (mi"ko-tok'sin) A deadly poison produced by fungi.

parasitism (pĕr'uh-sit-izm) A relationship between two organisms in which one organism lives in or on another organism and derives nourishment from it.

pathogen (path'uh-jen) An agent that causes a specific disease.

phytoplankton (fye-tuh-plank'tun) Photosynthetic species that form the basis for most aquatic food chains.

plankton (plank'tun) Small floating or weakly swimming organisms.

protozoa (pro"to-zo'ah) Heterotrophic, unicellular organisms.

saprophyte (sap'ruh-fit) An organism that obtains energy by the decomposition of dead organic material.

symbiosis (sim-be-o'sis) A close physical relationship between two kinds of organisms. It usually includes parasitism, commensalism, and mutualism.

CONCEPT MAP TERMINOLOGY

Construct a concept map to represent the relationships among the following concepts.

antibiotic	infectious
antibody	microorganism
bacteria	pathogen
disease	*Streptococcus*
endospore	toxin

LABEL•DIAGRAM•EXPLAIN

While many people associate the term *microbes* with disease and environmental problems, microorganisms play many beneficial roles in the environment and in our lives. Review the chapter and make a list of the helpful prokaryotes, protistans, and mycetes, and note the nature of their benefits.

Example: Mycetae: fungus—produces the antibiotic penicillin.

Multiple Choice Questions

1. Fire blight and citrus canker are plant diseases caused by:
   a. bacteria
   b. fungi
   c. protozoa
   d. viruses
2. Which of the following is required to destroy bacterial endospores?
   a. boiling water
   b. sunlight
   c. 121°C, for 15–20 minutes under pressure
   d. 100°F for 10 minutes
3. Which of the following is a funguslike protistan?
   a. slime mold
   b. diatom
   c. red algae
   d. athlete's foot
4. Photosynthesis can occur:
   a. in the upper portion of a lake
   b. throughout the water depth
   c. on the bottom only
   d. all of the above
5. Which is *not* a product of algae?
   a. alginate
   b. carrageenin
   c. agar
   d. mycotoxin
6. The vector of malaria is:
   a. a ciliated protozoan
   b. the mosquito, *Anopheles*
   c. a Sporozoan
   d. the common housefly
7. The first chemical used against plant diseases was a copper-based fungicide called:
   a. DDT
   b. penicillin
   c. Bordeaux mixture
   d. silver phosphate
8. Corn smut, fairy rings, and toadstools are all:
   a. fungi
   b. algae
   c. bacteria
   d. protozoa

Questions with Short Answers

1. Why are the protozoa and the algae in different subgroups of the kingdom Protista?

2. What is meant by the term *bloom?*

3. List three microbes that are able to cause disease in humans.

4. Define the terms *symbiosis, mutualism, commensalism,* and *parasitism.*

5. Name two beneficial results of fungal growth and activity.

6. List three microbes that are responsible for the decomposition of dead organisms.

7. Give an example of a microbe that lives in or on another, larger organism.

8. What is a bacterial endospore?

9. What types of microbes live on the surface of lakes, streams, oceans, and other bodies of water?

10. Name three commercial uses of algae.

11. What is the best method to prevent the spread of malaria?

12. What types of spores do fungi produce?

Interrelationships Among Living Things

No organism exists as an isolated individual. All organisms interact with and depend upon the activities of other organisms. Organisms also are influenced by specific physical conditions of their surroundings, such as sunlight, rain, wind, and temperature. Part 2 deals with how plants, animals, fungi, bacteria, and other kinds of organisms interact with their surroundings in networks called ecosystems and includes topics related to population growth, animal behavior, adaptation to surroundings, and energy relationships among kinds of organisms.

Atlantic puffin holding catch of fish.

What is evolution?

Why do populations of organisms differ?

How did life originate?

How did humans evolve?

chapter

10

Evolutionary Change

Galapagos Island land iguana and prickly pear cactus.

- Recognize that evolutionary change is the result of natural selection.
- Understand that genetic variety is essential for natural selection to occur.
- Understand that genes must be expressed to be subjected to natural selection.
- Recognize that reproduction provides excess individuals from which selection can occur.
- Understand that the basic pattern of evolution is divergence, but that several evolutionary patterns can occur.
- Know that the rate of evolutionary change differs with different organisms and at different times.

Evolution

Most people have some sort of conception (or misconception) of evolution. For some people, the word *evolution* may conjure up images of apes and prehistoric humans. In others it may bring about an emotional response stemming from a religious upbringing, or it may bring to mind extinct species of dinosaurs or mammoths. Unfortunately, many perceptions of evolution are far removed from its true meaning. Generally, the word *evolution* means a progressive change over time. We talk about evolving democracies in the former Soviet Union, or the evolution of fashion or musical styles. From a biological perspective, however, the word has a very specific meaning. **Evolution** is the genetic adaptation of a population of organisms to its environment over generations. This adaptation process has led to the development of all of Earth's life-forms.

You have probably noticed that offspring tend to have many of the same characteristics as their parents. This occurs because our traits are determined by information held within our genes. **Genes** are inheritable units that, in part, determine the physical, chemical, and behavioral characteristics of an organism. For example, genes determine that a robin will have wings (physical characteristic), be able to digest earthworms (chemical characteristic),

and be stimulated to fly south in autumn (behavioral characteristic). Half of the genes in each sexually reproducing individual are passed on to each of its offspring through egg or sperm. In this way organisms produce offspring that are, in part, like each parent. You received half of your genes from your mother and half from your father.

Individuals that are best adapted to their environment will obtain more food, find better shelter, and be more protected from predators and disease than those that are not so well adapted. These individuals will, therefore, have a greater chance of successfully producing offspring. The inherited characteristics that made these individuals successful will probably be passed on to the next generation. The various mechanisms that encourage the passage of beneficial genes to future generations and discourage the passage of harmful or less valuable genes are collectively known as **natural selection.** Therefore, evolution (changes in the genes of a population over time) occurs as a result of natural selection.

The Development of Evolutionary Thought

Most scientists accept evolutionary processes as central to an understanding of how various life-forms arose and continue to change today. Changes in bacterial populations are easily observed in the laboratory, while DNA and fossil evidence strongly support the theory of natural selection. This was not always the case, however. For centuries people believed that the various species of plants and animals were fixed and unchanging—that is, they were thought to have remained unchanged from the time of their creation. The process of evolution is generally so slow that people could not see the accumulation of changes during their lifetimes.

In the mid-1700s, Georges-Louis Buffon, a French naturalist, expressed the possibilities of change (evolution) in animals, but he did not suggest any mechanism that would result in evolution. In 1809, Jean Baptiste de Lamarck, a student of Buffon's, suggested a process by which evolution could occur. He proposed that **acquired characteristics,** traits that are gained during an individual's lifetime and not determined by genes, could be transmitted to offspring. For example, he proposed that giraffes originally had short necks. Since

giraffes constantly stretched their necks to obtain food, their necks became slightly longer. This trait of a slightly longer neck could be passed to the offspring, who were themselves stretching their necks, and over time, the necks of giraffes became longer and longer. Although we now know Lamarck's theory was wrong (because acquired characteristics are not inherited), it stimulated further thought as to how evolution could occur. All during this period, from the mid-1700s to the mid-1800s, lively arguments continued about the possibility of evolutionary change. Some, like Lamarck and others, thought that change did take place, while many others said that it was not possible.

CONCEPT CONNECTIONS

- **Evolution** is the change in a population over generations of time. Individuals cannot evolve; only populations evolve.
- **Natural selection** is an explanation of how evolution occurs.
- **Acquired characteristics** are traits gained during an organism's lifetime and cannot be passed on to the next generation.
- **Inherited characteristics** are genetically determined traits that can be passed on to the next generation.

Acquired Characteristics

Many organisms have characteristics that are not genetically determined. These acquired characteristics are gained during the life of the organism and, therefore, cannot be passed on to future generations through sexual reproduction. For example, the breed of dog known as boxers are "supposed" to have short tails. However, the genes for short tails are rare in this breed. Consequently, the tails of these dogs are amputated—a procedure called docking. Similarly, humans frequently alter their physical appearance with hair coloring, tattoos, plastic surgery, or bodybuilding. These acquired characteristics are not passed on to the next generation. Removing the tails of boxers does not remove their genes for long tail production, and dyeing brunette hair to blond will not remove the genes for brunette hair.

It was the thinking of two English scientists that finally provided a mechanism that explained how evolution could occur. In

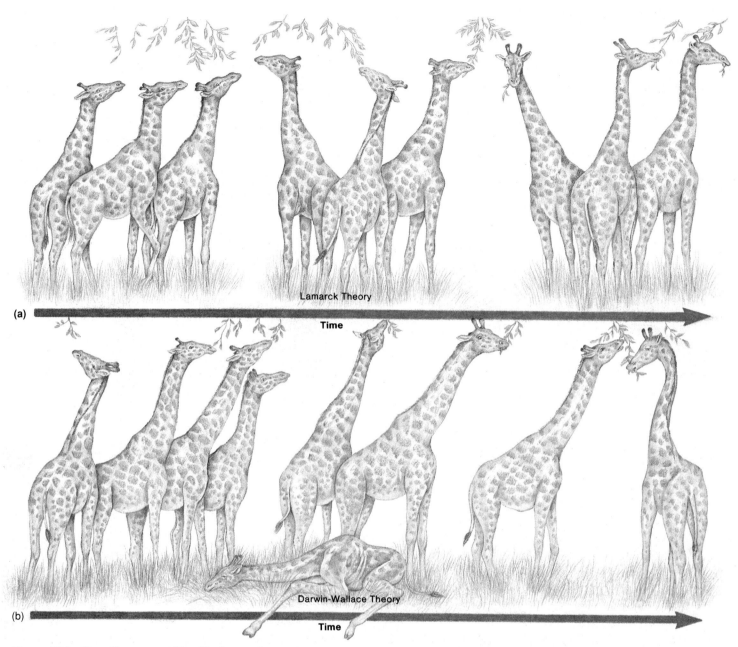

Figure 10.1 **Two Theories of How Evolution Occurs** (a) Lamarck thought that acquired characteristics could be passed on to the next generation. He postulated that as giraffes stretched their necks to get food, their necks got slightly longer. This characteristic was then passed on to the next generation, which in turn stretched their necks. (b) The Darwin-Wallace theory states that there is variation within the population and that those with longer necks are more likely to survive, reproduce, and pass their genes for long necks on to the next generation.

1858, Charles Darwin and Alfred Wallace suggested the theory of natural selection as a way of explaining how evolution occurs. According to the **theory of natural selection,** individuals whose gene combinations favor life in their surroundings will be most likely to survive, reproduce, and pass their genes on to the next generation. The Darwin-Wallace theory of evolution by natural selection explains the development of long necks in giraffes differently than did Lamarck (figure 10.1):

1. In each generation, more giraffes would be born than the food supply could support.
2. In each generation, some giraffes would inherit longer necks, and some would inherit shorter necks.
3. All giraffes would compete for the same food sources.
4. Giraffes with longer necks would obtain more food, have a higher survival rate, and produce more offspring.
5. As a result, more giraffes would have genes for long necks in succeeding generations.

To better understand the logic that led to this explanation, the principles of natural selection will be further addressed.

BIO *feature*

The Voyage of HMS *Beagle*, 1831–1836

Probably the most significant event in Charles Darwin's life was his appointment in 1831, at the age of twenty-two, as naturalist on the British survey ship HMS *Beagle*. Surveys were common at that time, helping to refine maps and chart hazards to shipping.

The voyage of the *Beagle* lasted nearly five years. During the trip, the ship visited South America, the Galápagos Islands, Australia, and many Pacific islands. Darwin suffered greatly from seasickness, and perhaps because of it he made extensive journeys by mule and on foot inland from wherever the *Beagle* happened to be at anchor. His experience was unique for a man so young.

Although many people had seen the places Darwin visited, never before had a student of nature collected volumes of information on these areas. Also, most other people who had visited these exotic places did not recognize the significance of what they saw. Darwin's notebooks included information on plants, animals, rocks, geography, climate, and the native peoples he encountered. The natural history notes he took during the voyage served as a vast storehouse of information, which he used in his writings for the rest of his life. He wrote books on the formation of coral reefs, how volcanos might have been involved in their formation, and finally, *On the Origin of Species*. This last book, written twenty-three years after his return from the voyage, changed biological thinking for all time.

The Voyage of HMS *Beagle*, 1831–1836

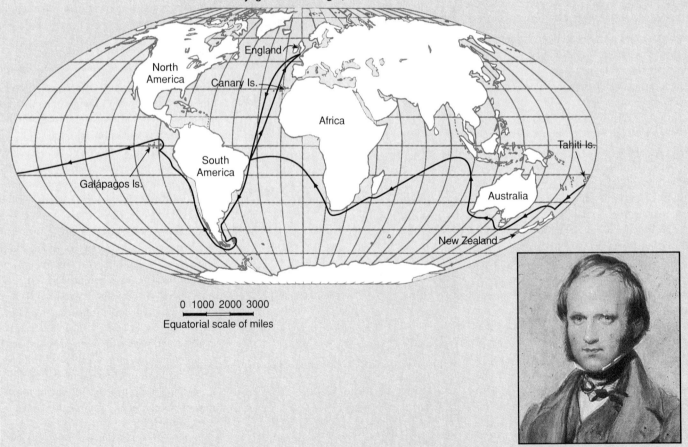

0 1000 2000 3000
Equatorial scale of miles

Charles Darwin

Natural Selection

There are two well-known phrases typically associated with the process of natural selection. These are "survival of the fittest" and "struggle for life." Many people have misconceptions about both of these phrases. For example, the word "survival" in "survival of the fittest" does not refer to an organism's ability to live a long life. Survival is important because those that do not survive will not reproduce, but the more important factor is the number of descendants that an organism leaves. An organism that has survived for hundreds of years but has not reproduced has not contributed any of its genes to the next generation. Nature has selected against (does not favor) such an organism. The key, therefore, is not survival alone but survival and reproduction of the more adapted organisms. *Fitness* in biological terms refers to an organism's ability to reproduce and pass its genes on to the next generation. The fittest organism in a population is not necessarily the strongest or smartest individual, but rather the one who successfully produces the most offspring. Nature selects for (favors) organisms that successfully produce the most offspring.

Similarly, the phrase "struggle for life" does not necessarily refer to open conflict and fighting. The struggle is usually much more subtle than that. When a resource such as nesting material, water, sunlight, or food is in short supply, some individuals survive and reproduce more effectively than others. For example, many kinds of birds require holes in trees as nesting places. If these are in short supply, some birds will be fortunate and find a top-quality nesting site, others will occupy less suitable holes, and some may not find any. There may or may not be fighting for possession of a site. If a site is already occupied, one bird may not necessarily try to drive the other bird out of its nest, but may just continue to search for suitable but less valuable sites. Those that occupy good nesting sites will be much more successful in raising young than will those that must occupy poor sites or those that do not find any (figure 10.2). Plants also struggle for life. With low light levels at the forest floor, some small plants may grow faster and obtain sunlight while shading out those that grow more slowly. The struggle for life in this instance involves a subtle difference in the rate at which the plants grow. But the plants are indeed engaged in a struggle, and a superior growth rate is the weapon for survival.

What Influences Natural Selection?

Now that we have a basic understanding of how natural selection works, we can look at factors that influence it. *Genetic variety* within a species (the fact that not all organisms of a given kind contain the same genes) is brought about by *mutations* (changes to genes) and *genetic recombination* (the mixing of gene combinations that occurs during sexual reproduction). (1) Genetic variety, (2) *gene expression* (the degree to which genes determine a trait), and (3) *excess reproduction* (the ability of most species to reproduce excess offspring) all exert an influence on the process of natural selection.

Genetic Variety

In order for natural selection to occur, there must be genetic differences among the many individual organisms. These differences are known as **genetic variety.** If all individuals were identical genetically, it would not matter which ones reproduce—the same genes would be passed to the next generation and natural selection would not occur. Genetic variety is generated in two different ways: mutations and genetic recombination.

Figure 10.2   Eastern bluebird leaving nest.

Mutations

A **mutation** involves a change in one or more genes. Mutations introduce new genes into a population. All genes for a particular trait originated as a result of mutations sometime in the past and have been maintained within the species as a result of sexual reproduction, passed from one generation to the next. Many mutations are unfavorable since they originate from a random change in a gene or genetic combination that has already proven valuable to the organism. However, in populations of millions of individuals, each of whom has thousands of genes, over thousands of generations it is quite possible that a new beneficial gene could come about as a result of a random mutation. When we look at the various genes that exist in humans or in any other organism, we should remember that every gene originated as a modification of a previously existing gene. For example, the gene for blue eyes in humans may be a mutated brown-eye gene, or blond hair may have originated as a mutated brown-hair gene.

Sexual Reproduction

A second very important process involved in generating genetic variety is sexual reproduction. While sexual reproduction does not generate new genes the way mutation does, it allows for the recombination of genes into combinations that did not occur previously. This combining of genes in new ways in each generation is called **genetic recombination.**

Each individual entering a population by sexual reproduction carries a unique combination of genes. Most sexually reproducing organisms possess two copies of each gene. The sex cells (eggs or sperm) contain only one copy of each gene. When an egg and sperm unite during fertilization, the resulting cell will then contain two copies of each gene. During the formation of sex cells, genes are mixed such that every egg and sperm cell produced by an individual organism is genetically different. When fertilization occurs, one of the millions of possible sperm unites with one of the millions of possible eggs, resulting in a genetically unique individual.

Imagine for example, that you and your friend each have a deck of 26 playing cards composed of two of each card type (two kings,

Figure 10.3 **Chance** In card games, chance determines which cards are dealt to each player.

one heart and one spade; two fives, one heart and one spade) (figure 10.3). You and your friend will play a game in which the object is to select only one of each kind (one king, one five, etc.) of these cards (genes) and place them in a pile. This will represent genes contributed to the offspring you and your friend will create. Whether the king of hearts or the king of spades is dealt to the offspring does not matter and does not influence whether the five of hearts or five of spades is dealt to the offspring. The male and female's contributions (sperm and egg) to the offspring will each be 13 cards. The combining of the two sets would represent fertilization, and the newly created hand of 26 cards (the offspring) would contain a new combination of cards (genes) different from either player (parent). Furthermore, each time this game was played, new card (gene) combinations would be created in the resulting offspring.

There are many kinds of organisms that reproduce primarily *asexually* (without sex) and, therefore, do not benefit from genetic recombination. However, many asexual organisms have the ability to reproduce sexually at certain times. Organisms that reproduce exclusively by asexual methods are not able to generate new gene combinations by sexual reproduction but generate variety through mutations.

Gene Expression

Gene expression is the degree to which a gene shows itself. In order for organisms with certain genes to be selected for or against, the genes must be expressed. Even though an organism contains a certain gene, the gene may not express itself for a number of different reasons. Some genes express themselves only during specific periods in the life of an organism. If the organism dies before the gene has had a chance to express itself, or if the organism completes its reproductive cycle before the gene expresses itself, the gene never had the opportunity to contribute to the reproductive fitness of the organism. A disease, for example, that expresses itself late in life will not influence the fitness of the organism, while a disease that expresses itself before or during the reproductive period may reduce fitness. There are several forms of heart disease that generally develop after the reproductive

years and have no impact on genetic fitness. Yet other heart conditions may reduce an individual's fitness or ability to pass genes on to future generations. *Congenital heart defects* (heart defects that exist at or before birth), for example, if left untreated, will greatly reduce an individual's chance of survival to reproductive age.

Many genes require an environmental trigger to initiate their expression. If the trigger is not encountered, the gene never expresses itself. It is becoming clear that many kinds of human cancers are caused by the presence of genes that require an environmental trigger. Therefore, we seek to identify the triggers and prevent these negative genes from being turned on and causing disease. This will not remove the genes from the population but only prevent their expression.

The expression of many genes is inhibited because an alternative form of the same gene is also present in the organism. Recall that most organisms contain two copies of each gene. For example, the gene for albinism is not expressed in individuals containing one copy of the skin pigment gene for normal pigmentation and one copy of the skin pigment gene for albinism (figure 10.4).

Natural selection evaluates all of an organism's traits. Just because an individual organism has a favorable gene does not guarantee the gene will be passed on. The organism may

Figure 10.4 **Gene Expression** Genes must be expressed to allow the environment to select for or against them. The recessive gene c for albinism shows itself only in individuals who have two genes for the recessive characteristic. The man in this photo is an albino who has the genotype cc.

Figure 10.5 Successful Combinations of Genes Some individuals in a population have valuable combinations of genes and therefore are much more likely to survive and reproduce. If all flies had good eyesight and resistance to pesticides but some had normal wings while others were unable to fly, those with wings would be more likely to avoid predators or flyswatters.

also have unfavorable genes in combination with the favorable gene. All individuals produced by sexual reproduction probably have certain genes that are extremely valuable for survival and others that are less valuable or harmful. Therefore, it is the combination of characteristics that is evaluated—not each characteristic individually. For example, fruit flies may show resistance to insecticides or lack of resistance, well-formed or shriveled wings, and normal vision or blindness. An individual with the combination of insecticide resistance, shriveled wings, and normal vision has two good characteristics and one negative one, but it would not be as successful as an individual with insecticide resistance, normal wings, and normal vision (figure 10.5).

Excess Reproduction

Successful organisms generally reproduce in excess (figure 10.6). For example, geese have a life span of about ten years, and on average a single pair can raise a brood of about eight young each year. In eight reproductive years, a single pair of geese can produce 64 offspring! However, goose populations and most other populations do not grow unchecked. Changes in number may occur, but a high death rate frequently offsets the high reproductive rate

and prevents uncontrolled growth. Extravagant reproduction provides the large surplus of genetically different individuals that allows natural selection to take place. In fact, to maintain itself in an ever-changing environment, each species must change in ways that enhance its ability to adapt to its new environment. For this to occur, members of the population must be eliminated in a non-random manner. Individuals that survive are, for the most part, better suited to the environment than others. They reproduce more of their kind and transmit more of their genes to the next generation than do individuals with genes that do not allow them to be well adapted to the environment in which they live.

How Natural Selection Works

Several mechanisms allow selection of certain individuals for successful reproduction. The specific environmental factors that favor certain characteristics are called **selecting agents.** All selecting agents influence the likelihood that certain characteristics will be passed to future generations.

Differential Survival

As stated previously, the phrase "survival of the fittest" is often associated with the theory of natural selection. Although this is recognized as an oversimplification of the concept,

Figure 10.6 Reproductive Potential The ability of a population to reproduce greatly exceeds the number necessary to replace those who die.

survival is an important factor influencing the passage of genes to following generations. If a population consists of a large number of genetically different individuals, and some of them possess characteristics that make their survival difficult, they are likely to die earlier in life and not have an opportunity to pass their genes on to the next generation. This process is called *differential survival.*

The English peppered moth provides a classic example. Two color types are found in the species: One form is light-colored and the other is dark-colored. These moths normally rest on the bark of trees during the day, where they may be spotted and eaten by birds. Two hundred years ago, the light-colored moths were most common. However, with the advance of the Industrial Revolution in England, which involved an increase in the use of coal, air pollution increased. The soot in the air settled on the trees, changing the bark to a darker color. Because the light moths were more easily seen against a dark background, the birds ate them (figure 10.7). The darker ones were less conspicuous, so they were eaten less frequently and were more likely to reproduce successfully. The light-colored moth, which was originally the more common type, became much less common. This change in gene frequency occurred within the short span of fifty years. Scientists who have studied this situation have estimated that the dark-colored moths had a 20 percent better chance of reproducing than did the light-colored moths. This study is continuing today. As England is solving some of its air-pollution problems, the tree bark is becoming lighter in color, and the light-colored form of the moth is increasing in frequency.

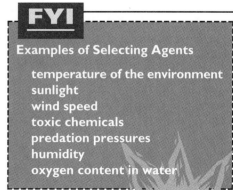

FYI

Examples of Selecting Agents

temperature of the environment
sunlight
wind speed
toxic chemicals
predation pressures
humidity
oxygen content in water

Figure 10.7 The Peppered Moth This photo of the two variations of the peppered moth shows that the light-colored moth is much more conspicuous against the dark tree trunk. (The two dark moths are indicated by arrows.) The trees are dark because of an accumulation of pollutants from the burning of coal. The more conspicuous light-colored moths are more likely to be eaten by bird predators, thus the genes for light color should become more rare in the population.

Differential Reproductive Rates

Survival does not always ensure reproductive success. For a variety of reasons, some organisms may be better able to utilize available resources to produce offspring. If one individual leaves 100 offspring and another leaves only two, the first organism has passed more copies of its genetic information to the next generation than has the second. For example, biologists have studied the frequencies of genes for the height of clover plants (figure 10.8). Two identical fields of clover were planted and cows were allowed to graze in only one of them. Cows acted as a selecting agent by eating the taller plants first. These tall plants rarely got a chance to reproduce. Only the shorter plants flowered and produced seeds. After some time, seeds were collected from both the grazed and ungrazed fields and grown in a greenhouse under identical conditions. The average height of the plants from the ungrazed field was compared to that of the plants from the grazed field. The seeds from the ungrazed field produced

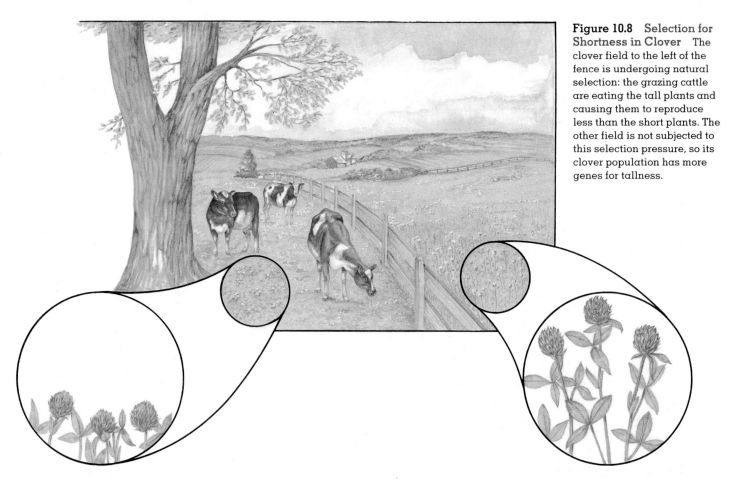

Figure 10.8 Selection for Shortness in Clover The clover field to the left of the fence is undergoing natural selection: the grazing cattle are eating the tall plants and causing them to reproduce less than the short plants. The other field is not subjected to this selection pressure, so its clover population has more genes for tallness.

Natural Selection and Insecticide Resistance

As an example of how differential survival can lead to changed gene frequencies, consider what has happened to many insect populations as we have subjected them to a variety of insecticides. Since there is genetic variety within all species of insects, an insecticide that is used for the first time on a particular species kills all those that are genetically susceptible. However, a few individuals with slightly different genetic compositions may not be killed by the insecticide.

Suppose that in a population of a particular species of insect 5 percent of the individuals have genes that make them resistant to a specific insecticide. The first application of the insecticide could, therefore, kill 95 percent of the population. However, tolerant individuals would then constitute the majority of the breeding population that survived. This would mean that many insects in the second generation would be tolerant. The second use of the insecticide on this population would not be as effective as the first. Many species of insects produce a new generation each month. With continued use of the same insecticide, each

generation would become more tolerant. Ninety-nine percent of the population could become resistant to the insecticide in just five years. As a result, the insecticide would no longer be useful in controlling the species. As a new selecting agent (the insecticide) was introduced into the environment of the insect, natural selection resulted in a population that was tolerant of the insecticide. The graph indicates that in 1990 about 500 species of insects had populations that were resistant to many kinds of insecticides.

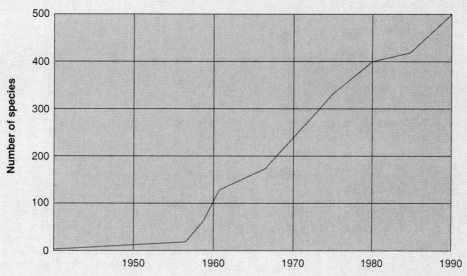

Pest species resistant to insecticides

Resistance to Insecticides   The continued use of insecticides has constantly selected for genes that provide resistance to insecticides. As a result, many species of insects and other arthropods are now resistant to many kinds of insecticides, and the number continues to increase.

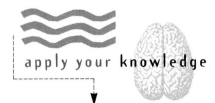

Natural Selection: Look About You

You can demonstrate variety in a population in the following manner. As members of the class enter the room, ask them how tall they are. Perhaps you could have them record their heights on the front board. On a piece of graph paper, plot height against the number of people at each height. Plot men and women separately. Calculate the average height for males and females. What are the extremes for both sexes? Discuss other variations you can observe in natural populations.

some tall, some short, but mostly medium-sized plants. The seeds from the grazed field produced many more shorter plants than medium or tall ones. The cows had selectively eaten the plants that had the genes for tallness. Since the flowers are at the tip of the plant, tall plants were less likely to successfully reproduce, even though they might have been able to survive grazing by cows.

Evolutionary Patterns

Higher levels of evolutionary change are the result of differences accumulated over time due to many selective factors that lead to greater and greater diversity. This diversity leads to the evolution of new species. The basic pattern in evolution is one of **divergent evolution** in which the development of each individual species causes successive branches in the evolution of a group of organisms. This basic pattern is well illustrated by the fossil record of the horse (figure 10.9). The modern horse, with its large size, single toe on each foot, and teeth designed for grinding grasses, is thought to be the result of accumulated changes, beginning from a small, dog-sized

**Figure 10.9 Divergent Evolution In the evolution of the horse, many speciation events have followed one after another. What began as a small, leaf-eating, four-toed animal of the forest has evolved into a large, grass-eating, single-toed animal of the plains. There are many related animals alive today, but early ancestral types are extinct.

BIO *feature*

Natural Selection in a Nutshell

1. An organism's capacity to over-reproduce results in surplus organisms.
2. Because of mutation, new genes enter the gene pool. Because of sexual reproduction, new combinations of genes are present in every generation. These processes result in each individual in a sexually reproducing population being genetically unique and expressing different traits than other individuals.
3. Resources such as food, water, mates, and nesting materials are in short supply, so some individuals have to do without. Other environmental factors, such as disease organisms,

predators, and defense mechanisms, affect survival. These are called selecting agents.
4. Individuals with the best combination of genes are more likely to survive and reproduce, passing more of their genes on to the next generation. An organism can be selected against if it has fewer offspring than better-adapted species members have. It does not need to die to be selected against.
5. Therefore, genes or gene combinations that produce characteristics favorable to survival become more common, and the species changes and becomes better adapted to its environment.

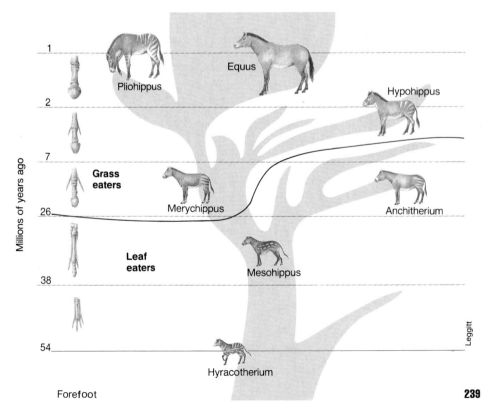

Forefoot

BIO *feature*

Resistance of the Tuberculosis Bacterium to Antibiotics

Tuberculosis (TB) is caused by the bacterium *Mycobacterium tuberculosis*. Many people thought the disease had been eliminated from the United States and the rest of the economically developed world. However, it is rebounding to the point that some U.S. cities now have higher rates of TB than those of developing African nations. The incidence of TB has steadily become more frequent since 1985.

What has caused this recurrence of an old enemy? Several factors are involved. Poverty is important because it creates ideal conditions for rapid spread of this airborne disease. People with lack of access to medical care often live in crowded housing, allowing for easy spread of the bacterium. AIDS sufferers have an impaired immune system; thus, they are susceptible to

the disease if they encounter it. Perhaps the most important contributor to the problem is the evolution of drug-resistant strains of *Mycobacterium tuberculosis*. After years of being subjected to the same kinds of antibiotics, selection has taken place and certain populations of the bacterium are now able to live and reproduce in the presence of antibiotics that once controlled their spread. When the environment of *Mycobacterium tuberculosis* was changed by the addition of antibiotics (a selecting agent), some individual tuberculosis organisms had genes that allowed them to resist certain drugs. Years of selection have given rise to strains that are resistant to many kinds of drugs, resulting in disease outbreaks that are very difficult to control.

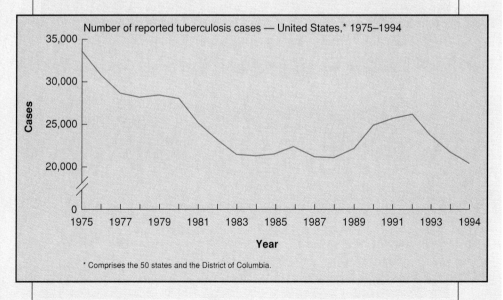

Number of reported tuberculosis cases — United States,* 1975–1994

* Comprises the 50 states and the District of Columbia.

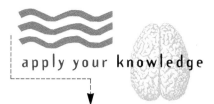

Gonorrhea and Natural Selection

Penicillin was first used to control bacterial infections in the early 1940s. Since that time, it has been found to be effective against the bacteria that cause gonorrhea, a sexually transmitted disease. The drug acts on dividing bacterial cells by preventing the formation of a new protective cell wall. Without the wall, the bacteria can be killed by normal body defenses. Penicillin was routinely and successfully used to treat gonorrhea until a new strain of this disease-causing bacterium was discovered. This particular bacterial strain produces an enzyme that destroys penicillin. How can gonorrhea be controlled now that this organism is resistant to the antibiotic? How did a resistant strain develop?

animal with four toes on its front feet, three toes on its hind feet, and teeth designed for chewing leaves and small twigs. Though we know much about the evolution of the horse, there are still many gaps to fill before we have a complete evolutionary history.

Another basic pattern in the evolution of organisms is extinction. Notice in figure 10.9 that most of the species that developed during the evolution of the modern horse are extinct. Overall estimates of extinction are around 99 percent; that is, 99 percent or more of all the species of organisms that ever existed are now extinct. Extinct species possessed specializations that made them at one time successful in their environments. However, the environment does not remain constant and often changes in such a way that the species that previously thrived are unable to adapt to the new set of conditions. The early ancestors of the modern horse were well adapted to a moist tropical environment, but when the climate became drier, most were no longer able to survive. Only some kinds had the genes necessary for survival and reproduction in the changing environment that led to the development of modern horses.

It is important to recognize that many extinct species were very successful organisms for millions of years. They were not failures in

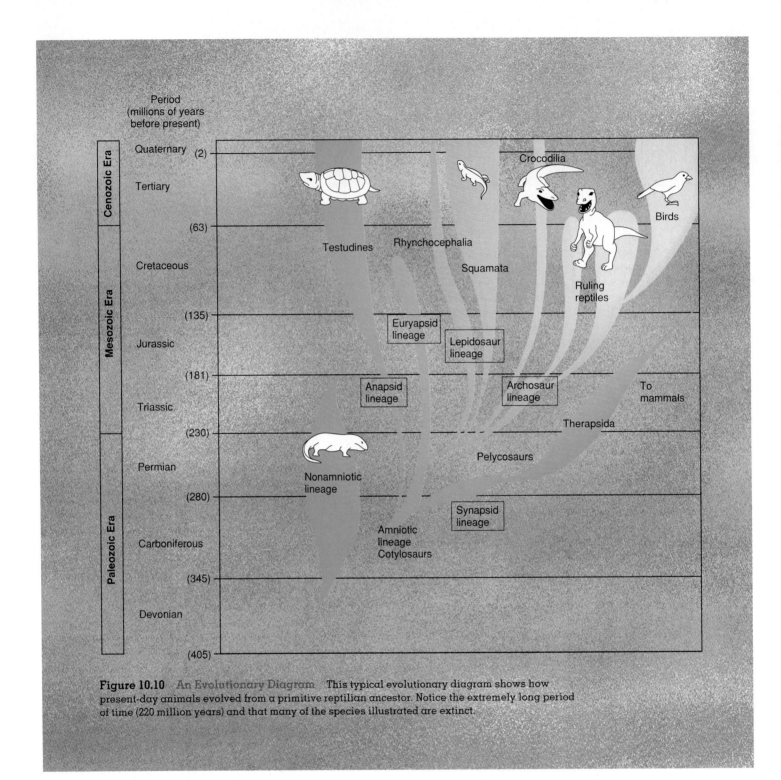

Figure 10.10 *An Evolutionary Diagram* This typical evolutionary diagram shows how present-day animals evolved from a primitive reptilian ancestor. Notice the extremely long period of time (220 million years) and that many of the species illustrated are extinct.

their time; rather, they simply did not survive to the present. It is also important to realize that all currently existing organisms will eventually become extinct.

Tracing the evolutionary history of an organism back to its origins is a very difficult task because many of its ancestors no longer exist. Keep in mind that the fossil record is incomplete and provides only limited information

about the biology of the organism represented in that record. We may know a lot about the structure of the bones and teeth of an extinct ancestor, but know almost nothing about its behavior, physiology, and natural history. Biologists must use a great deal of indirect evidence to piece together the series of evolutionary steps that led to a current species. Figure 10.10 is typical of evolutionary trees

that help us understand how time and structural changes are related in the evolution of birds, mammals, and reptiles.

Although divergence is the basic pattern in evolution, several other patterns of evolution have been identified. One special pattern, characterized by a rapid increase in the number of kinds of closely related species, is known as **adaptive radiation.** Adaptive radiation

Adaptive Radiation of Darwin's Finches

An example of adaptive radiation is found among the finches of the Galápagos Islands, 1,000 kilometers (600 miles) west of Ecuador in the Pacific Ocean. These birds were first studied by Charles Darwin. Since these islands are volcanic, the assumption is that they have always been isolated from South America and originally lacked finches and other land-based birds. It is thought that one kind of finch arrived from South America to colonize the islands and that adaptive radiation from the common ancestor resulted in the many different kinds of finches found on the islands today. Some of these finches took roles normally filled by other kinds of birds such as woodpeckers. Some even became warblerlike, and one uses a cactus spine as a tool to probe for insects, just as woodpeckers use their beaks.

results in an evolutionary explosion of new species from a common ancestor. There are basically two situations that are thought to favor adaptive radiation. One is a condition in which an organism invades a previously unexploited environment. For example, at one time there were no animals on the land masses of the earth. The amphibians were the first vertebrate animals able to spend part of their lives on land. Food was plentiful and predators were scarce, so a variety of different kinds of amphibians evolved rapidly and exploited several kinds of lifestyles.

A second set of conditions that can favor adaptive radiation is when a type of organism evolves a new set of characteristics that enable it to displace organisms that previously filled niches in the environment. For example, although amphibians were the first vertebrates to occupy land, they were replaced by reptiles because reptiles had characteristics, such as dry skin and an egg that could develop on land.

Adaptive Radiation When Darwin discovered the finches of the Galápagos Islands, he thought they might all have derived from one ancestor that arrived on these relatively isolated islands. If they were the only birds to inhabit the islands, they could have evolved very rapidly into the many different types shown here. The drawings show the specialized beaks for eating different kinds of food.

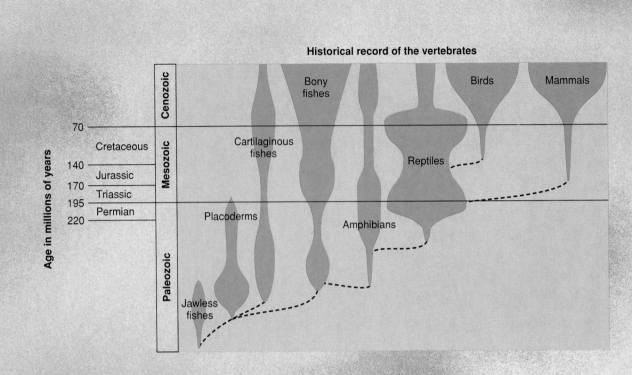

Figure 10.11 **Adaptive Radiation in Terrestrial Vertebrates** The amphibians were the first vertebrates to live on land. They were replaced by the reptiles, which were better adapted. The reptiles, in turn, were replaced by the adaptive radiation of birds and mammals. (*Note:* The width of the colored bars indicates the relative number of species present.)

These characteristics allowed them to replace most of the amphibians, which could only live in relatively moist surroundings where they would not dry out and where their aquatic eggs could develop. The adaptive radiation of reptiles was extensive. They invaded most terrestrial settings and even evolved forms that flew or lived in the sea. Subsequently, many reptiles

CONCEPT CONNECTIONS

- **Adaptive radiation** is a specific evolutionary pattern in which there is a rapid increase in the number of kinds of closely related species.

- **Convergent evolution** is an evolutionary pattern in which widely different organisms facing similar environmental pressures show similar characteristics.

- **Divergent evolution** is a basic evolutionary pattern in which individual speciation events cause many branches in the evolution of a group of organisms.

were replaced by the mammals, who went through a similar radiation. Figure 10.11 shows the sequence of radiations that occurred within the vertebrate group. The number of species of amphibians and reptiles has declined, while the number of species of birds and mammals has increased.

Convergent Evolution through Natural Selection

When organisms of widely different backgrounds develop similar characteristics, we see an evolutionary pattern known as **convergent evolution**. This particular pattern often leads people to misinterpret the evolutionary history of organisms. For example, many kinds of plants that live in desert situations have thorns and lack leaves during much of the year. Superficially they appear similar, but they are often

quite different from one another. They have not become one species, although they may resemble one another to a remarkable degree. The presence of thorns and the absence of leaves are adaptations to a desert environment—the thorns discourage herbivores, and the absence of leaves reduces water loss. Another example involves animals that survive by catching insects while flying. Bats, swallows, and dragonflies all obtain food in this manner. They all have wings, but they are derived from the modification of different structures (figure 10.12). At first glance, they may appear to be very similar and perhaps closely related, but detailed study of their wings and other structures shows that they are quite different kinds of animals. They have simply converged in structure, type of food eaten, and method of obtaining food. Likewise, whales, sharks, and tuna appear to be similar but are different kinds of animals that all happen to live in the open ocean.

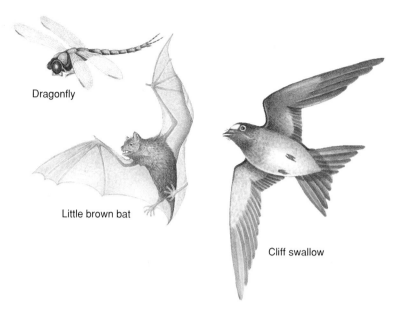

Figure 10.12 **Convergent Evolution** All of these animals have evolved wings as a method of movement, and capture insects for food as they fly. However, they have completely different evolutionary origins.

Rates of Evolution

Although it is commonly thought that evolutionary change occurs over long periods of time, you should understand that rates of evolution can vary greatly. One factor that influences the rate of evolution is reproductive potential. For example, a population of fruit flies with high reproductive capabilities will respond to environmental changes more rapidly than an elephant population possessing low reproductive potential.

A second factor that influences the rate of evolution is the rate of environmental change. Remember that natural selection is driven by the environment. If the environment is changing rapidly, one would expect rapid changes in the organisms present. Periods of rapid change also result in extensive episodes of extinction. During some periods in the history of the earth when little environmental change was taking place, the rate of evolutionary change was probably slow. Nevertheless, when we talk about evolutionary time, we are generally thinking in thousands or millions of years.

When we examine the fossil record, we can see gradual changes in physical features of organisms over time. The accumulation of these changes could result in such extensive change from the original species that we would consider the current organism to be a different species from its ancestor. This is such a common feature of the evolutionary record that biologists refer to this kind of evolutionary change as **gradualism** (figure 10.13a).

Charles Darwin's view of evolution was based on the gradual change in species he perceived from his studies of geology and natural history. However, as early as the 1940s, some biologists challenged this idea. They pointed out that the fossils of some species were virtually unchanged over millions of years. If gradualism were the only explanation of species evolution, then gradual changes in the fossil record of a species would be found. However, some organisms seem to have appeared from nowhere in the fossil record. They appeared suddenly and showed rapid change from the time they first appeared. In 1972, two biologists,

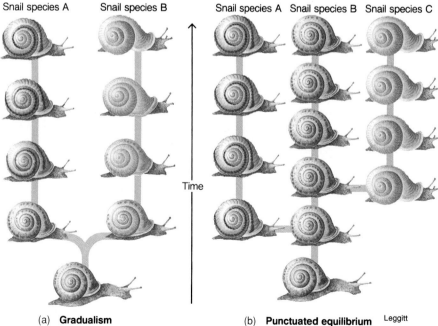

Figure 10.13 **Gradualism vs. Punctuated Equilibrium** Gradualism (a) proposes that the evolution of new species results from the accumulation of small changes over a long period of time. Punctuated equilibrium (b) proposes that the evolution of new species results from a large number of changes in a short period of time.

Niles Eldredge of the American Museum of Natural History and Stephen Jay Gould of Harvard University, proposed the idea of **punctuated equilibrium.** This hypothesis suggests that evolution occurs in spurts of rapid change followed by long periods with little evolutionary change (figure 10.13b).

SUMMARY

At one time people thought that all organisms had remained unchanged from the time of their creation. Jean Baptiste de Lamarck suggested that change did occur and thought that acquired characteristics could be passed from generation to generation. Charles Darwin and Alfred Wallace proposed the theory of natural selection as the mechanism for evolution.

Populations are usually genetically diverse. Mutations and sexual reproduction tend to introduce genetic variety into a population. Natural selection is driven by genetic diversity and the differential survival and reproductive rates of individuals with differing genes. This leads to changes in gene frequencies and evolution.

Evolution is basically a divergent process upon which other patterns can be superimposed.

Adaptive radiation is a very rapid divergent evolution, while convergent evolution involves the development of superficial similarities among widely different organisms. The rate at which evolution has occurred probably varies. The fossil record shows periods of rapid change interspersed with periods of little change. This has caused some scientists to look for mechanisms that could cause the sudden appearance of large numbers of new species in the fossil record and to challenge the traditional idea of slow, steady change accumulating enough differences to cause a new species to be formed.

CHAPTER GLOSSARY

acquired characteristics (ă-kwīrd′ kar″ak-ter-iss′tiks) A characteristic of an organism gained during its lifetime, not determined genetically, and therefore not transmitted to the offspring.

adaptive radiation (uh-dap′tiv ra-de-a′shun) A specific evolutionary pattern in which there is a rapid increase in the number of kinds of closely related species.

convergent evolution (kon-vur′jent ev-o-lu′shun) An evolutionary pattern in which widely different organisms show similar characteristics.

divergent evolution (di-vur′jent ěv-o-lu′shun) A basic evolutionary pattern in which individual speciation events cause many branches in the evolution of a group of organisms.

evolution (ěv-o-lu′shun) The genetic adaptation of a population of organisms to its environment.

gene (jēn) Inheritable units that determine the physical, chemical, and behavioral characteristics of an organism.

genetic recombination (jě-net′ik re-kom-bĭ-na′shun) The gene mixing that occurs during sexual reproduction.

gradualism (grad′u-al-izm) The theory stating that evolution occurred gradually with an accumulated series of changes over a long period of time.

mutation (miu-ta′shun) A change in the genetic material.

natural selection (nat′chu-ral se-lek′shun) A broad term used in reference to the various mechanisms that encourage the passage of beneficial genes and discourage the passage of harmful or less valuable genes to future generations.

punctuated equilibrium (pung′chu-a-ted e-kwĭ-lib′re-um) The theory stating that evolution occurs in spurts, between which there are long periods with little evolutionary change.

selecting agent (se-lek′ting a′jent) Any factor that affects the probability that a gene will be passed to the next generation.

theory of natural selection (the′o-re uv nat′chu-ral se-lek′shun) In a species of genetically differing organisms, the organisms with the genes that enable them to survive better in the environment and thus reproduce more offspring than others will transmit more of their genes to the next generation.

CONCEPT MAP TERMINOLOGY

Construct a diagram to represent the relationships that exist among the following concepts:

adaptive radiation
convergent evolution
divergent evolution
evolution

gradualism
natural selection
punctuated equilibrium

LABEL•DIAGRAM•EXPLAIN

Discuss how human intervention has unintentionally resulted in the selection of traits that we would view as unfavorable in pests or parasites (favorable to the pest or parasite). Your discussion should

1) cite a specific example.
2) describe how the process of natural selection has led to the evolution of a more resilient population.
3) identify the selecting agent.

Multiple Choice Questions

1. Evolution involves changes that occur:
   a. within individuals during their lifetimes
   b. within populations over generations of time
   c. as a result of the inheritance of acquired characteristics
   d. all of the above
2. The inheritance of acquired characteristics was proposed by:
   a. Jean Baptiste de Lamarck
   b. Georges-Louis Buffon
   c. Charles Darwin
   d. Alfred Wallace
3. The fittest organism in a population is the:
   a. organism that successfully produces the most offspring
   b. strongest and fastest organism
   c. organism that lives longest
   d. most intelligent organism
4. Mutation(s):
   a. involve changes in genes
   b. introduce new genes into populations
   c. is the mechanism by which all existing genes were formed
   d. all of the above
5. Which of the following contribute to genetic variety?
   a. mutations and sexual reproduction
   b. acquired characteristics
   c. excess reproduction
   d. all of the above
6. The several species of finches that developed on the Galápagos Islands are an example of:
   a. convergent evolution
   b. divergent evolution
   c. adaptive radiation
   d. inheritance of acquired characteristics

7. Under natural conditions without modern medicine, which of the following human conditions has the *least* impact on fitness?
   a. Tay-Sachs, a congential (from time of birth) disorder in which neurons degenerate
   b. Amenorrhea, the inability to menstruate
   c. Alzheimer's disease, a neurological disorder that generally afflicts individuals over the age of 65
   d. Valvular stenosis, a birth defect resulting in a narrow valve that regulates blood flow to the heart
8. The idea that species remain unchanged for millions of years and then rapidly evolve within a short period of time is:
   a. convergent evolution
   b. divergent evolution
   c. gradualism
   d. punctuated equilibrium
9. Unrelated birds who nest in holes produce eggs with white shells as opposed to eggs with speckled or colored shells. This is an example of:
   a. divergent evolution
   b. punctuated equilibrium
   c. adaptive radiation
   d. convergent evolution
10. Selecting agents:
    a. influence an organism's fitness
    b. are environmental factors that favor certain characteristics
    c. influence the likelihood that certain characteristics will be passed on to future generations
    d. all of the above

Questions with Short Answers

1. Why are acquired characteristics of little interest to evolutionary biologists?
2. Why is over-reproduction necessary for evolution?
3. Give two examples of selecting agents and explain how they function.
4. Describe two differences between convergent evolution and adaptive radiation.
5. What is the difference between gradualism and punctuated equilibrium?

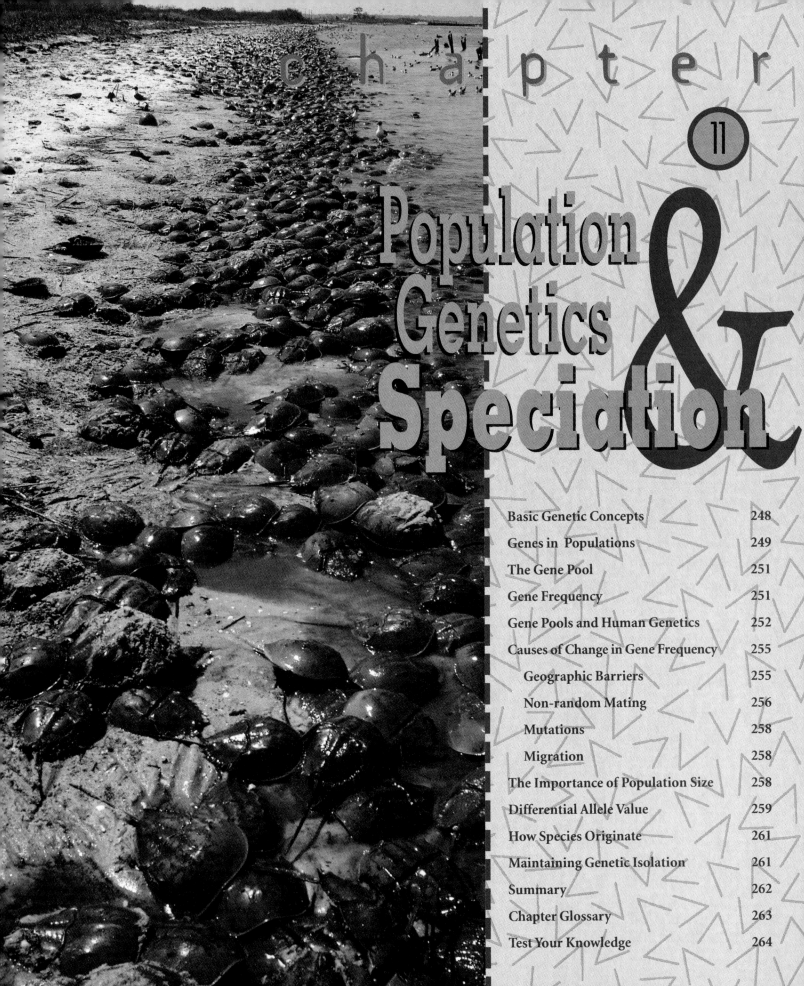

chapter

11

Population Genetics & Speciation

Horseshoe crabs, Delaware Bay, New Jersey.

Population Genetics & Speciation

- Understand differences among the terms gene, allele, dominant allele, and recessive allele.

- Distinguish among population, species, and subspecies.

- Recognize that a gene pool contains more alleles than does an individual.

- Describe the occurrence of an allele in a population in terms of gene frequency.

- Recognize the importance of mutation, sexual reproduction, population size, and migration in gene frequency.

- Understand the importance of reproductive isolation in speciation.

- Recognize the steps necessary for speciation to occur.

Basic Genetic Concepts

In previous chapters we have surveyed the vast diversity of living things and learned that organisms are classified (kingdom, phylum, class, order, family, genus, and species) according to similarities among individuals within given groups. Yet, even within a species a great deal of diversity exists between individuals. Consider the students in your biology class. Chances are that your class contains tall students and short students, students with light hair and students with dark hair, and students with less pronounced physical features and students with prominent physical features. You are probably aware that these human traits can be passed on to the next generation. The study of factors that determine such characteristics and how they are inherited is **genetics.** Biologists study the passage of traits from one generation to another at several levels. In future chapters, you will learn about the genetics of individual organisms as well as study genetics from cellular and molecular perspectives. This chapter addresses the study of genetics as it applies to populations and the formation of new species through natural selection.

Fundamental to the field of genetics is the concept of the gene. **Genes** are the carriers of information that determine an organism's traits. Your height, skin color, and blood type are all the result of the chemical activities directed by your genes. Within each of your cells are two copies of each type of gene. One copy you received from your mother's egg cell and one copy you received from your father's sperm cell. These genes need not be identical.

For example, a person with AB blood type received one A blood type gene from one parent and one B blood type gene from the other parent. Different forms of a gene for a single trait are called **alleles.**

Often when an organism possesses two different alleles for a given trait, one allele is expressed and the other is not. For example, there are two alleles controlling the shape of your hairline at the forehead (figure 11.1). One form of the gene causes the hair to grow down into a point at the middle of the forehead (called a widow's peak) and the other form causes the hair to grow straight across the forehead. The straight hairline allele will not be expressed if the allele for widow's peak is also present. If a person has two alleles for widow's peak, the person will exhibit the pointed hairline trait. If the person has one allele for widow's peak and one allele for straight hairline, the person will have a widow's peak. Only if a person has two alleles for straight hairline will the individual express that trait. The allele for widow's peak hairline is called *dominant* and the allele for straight hairline is called *recessive.*

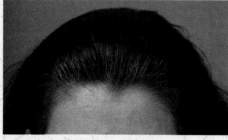

(a)

(b)

Figure 11.1   Genes Control Structural Features   Whether you have widow's peak (a) or a straight hairline (b) depends on the genes you have inherited. As genes express themselves, their actions affect the development of various tissues and organs. How genes control this complex growth pattern and why certain genes function differently from others is yet to be clarified.

Consider a conversation among friends in which one friend *dominates* the conversation. By monopolizing the conversation, this person prevents the others in the group from expressing their ideas. Alleles work much the same way. A **dominant allele** expresses itself and hides the activity of other alleles for the trait. A **recessive allele**, when present with a dominant allele, does not express itself; it is concealed by the effect of the other allele. Recessive traits only express themselves when the organism has two copies of the recessive allele.

Capital letters (*A*) are used to designate dominant alleles and lowercase letters (*a*) symbolize recessive alleles (*AA* = two dominant alleles; *Aa* = one dominant allele and one recessive allele; *aa* = two recessive alleles). The term recessive has nothing to do with the value of the gene—it simply describes how it is expressed when in combination with another allele.

CONCEPT CONNECTIONS

gene The inheritable units that in part determine an organism's characteristics.

allele An alternative form of a gene.

dominant allele A gene form that can hide the expression of other gene forms.

recessive allele A gene form that can be hidden by other (dominant) gene forms. The expression of recessive traits is seen only in organisms with two copies of the recessive allele.

Genes in Populations

Biologists in the field of **population genetics** study similarities or differences of genes between populations. This information is used as the basis for classifying organisms into species and studying evolutionary change. A **species** is a group of organisms of the same kind that are genetically similar. The members of a species usually look alike, although there are some exceptions. For example, all the dogs in the world are of the same species, but a Great Dane, a bulldog, a terrier, and a Pekingese do not look much alike (figure 11.2). However, mating can occur between two quite different-appearing types of dogs. In some species, males may appear much different than females. Many male birds have bright showy feathers whereas their female counterparts have feathers that allow the birds to blend in with the environment (figure 11.3). It would be unwise to classify organisms solely on the basis of physical appearance. Frequently the genes that determine external differences represent only a small percentage of an organism's total

Figure 11.2 Genetic Variety in Dogs   Although these four breeds of dogs look different, they all have the same number of chromosomes and are capable of interbreeding. Therefore they are members of the same species. The considerable difference in appearance is evidence of the genetic variety among breeds.

genetic makeup. In some flowers, for example, a change in a single gene can result in the doubling of petal number.

Reproductive capability is frequently used as a simple measure to determine genetic similarity and to classify organisms into species, because only organisms with great genetic similarity can successfully reproduce with one another. Based on this rationale, a species is described as a group of organisms that have the potential to interbreed *naturally* to produce *fertile* offspring. This working definition applies for most sexually reproducing organisms but must be interpreted to embrace some exceptions. The definition contains three key ideas. First, a species is a worldwide population of organisms. An individual—you, for example—is not a species. You can only be a member of the group that is called a species. The human species consists of over 5 billion people, whereas the endangered California condor species consists of only about fifty birds.

A second part of the species concept addresses the ability of individuals within the group to produce fertile offspring. Another way to look at this is to think about the movement of genes from one generation to the next or from one area to another. Two or more groups of organisms that demonstrate gene exchange between them are a single species. Likewise, two or more populations that are not able to exchange genes are considered different species. An example will clarify this working definition: The mating of a male donkey and a female horse produces young that grow to be adult mules, incapable of reproduction (figure 11.4). Since mules are sterile, there can be no gene exchange between horses and donkeys and they are considered separate species.

The third part of the species definition has to do with mating. For organisms to belong to the same species, matings must have the potential to occur *naturally*. Lions and tigers can be mated in zoos to produce offspring. However, this does not happen in nature and the offspring are not likely to be fertile. Wolves have been known to mate with dogs and produce fertile pups, but this mating does not happen often in nature. The domestic dog and wolf are therefore considered separate species.

The concepts of population and species are interwoven: Recall that a population is considered to be all the organisms of the same species found within a specific geographic region. Population is primarily concerned with

Figure 11.3 Sexual Differences The males and females of many birds look different from one another yet are the same species. Their physical differences might suggest that they are of different species.

Horse

Donkey

Mule

Figure 11.4 Hybrid Sterility Even though they don't do so in nature, donkeys and horses can be mated. The offspring is called a mule and is sterile. Because the mule is sterile, the donkey and the horse are considered to be different species.

numbers of organisms in a particular place at a particular time. The material in this chapter incorporates both concepts. It deals with populations and how they differ from each other genetically. For example, why are there more blue-eyed people in Scandinavia than in Spain, or why do some populations have a high frequency of blood type A?

The Gene Pool

Previously, we related the species concept to genetic similarity. You know, however, that not all individuals of a species are genetically identical. Each individual has a unique genetic makeup consisting of that organism's genes. Typically, each individual has a maximum of two forms of each gene. In a population, however, there may be many more than two alleles for a specific characteristic. For example, in humans the gene for blood type can be in three forms: blood type A, blood type B, and blood type O, but no one person can have more than two forms of this gene. All of the alleles of all of the organisms in a population make up a **gene pool**. Since, theoretically, all individuals of a species are able to exchange genes, we can think of all the genes of all the individuals of the same species as a giant gene pool. A gene pool is like a gum ball machine containing red, blue, yellow, and green gum balls (alleles). For a quarter and a turn of the knob, two gum balls are dispensed from the machine. Two red gum balls, a red and a blue, a yellow and a green, or any of the other possible color combinations may result from any one gum ball purchase. A child buying gum balls will receive no more than two of the four possible gum ball colors and only one of ten possible color combinations. Similarly, individuals can have no more than two of the many alleles for a given gene contained within the gene pool and only one of several possible combinations of alleles.

Gene Frequency

Individuals are usually not found evenly distributed but occur in clusters as a result of several factors, such as geographic barriers that restrict movement or the local availability of

Wolf-Dog Hybrids

In the past few years, a business has developed in the breeding and selling of wolf-dog hybrids. A **hybrid** is the offspring of two different genetic lines. Among owners of these animals, the wolf-dog hybrid is prized for its physical resemblance to a wild wolf. Unfortunately, these animals also inherit some of the wolf's innate behaviors and have been responsible for the injury and death of several small children. When a wolf-dog hybrid in captivity sees a small child running and screaming it may attack as a wolf would attack a vulnerable prey animal in distress. In the wild, such an attack of a child by a wolf would be extremely unusual.

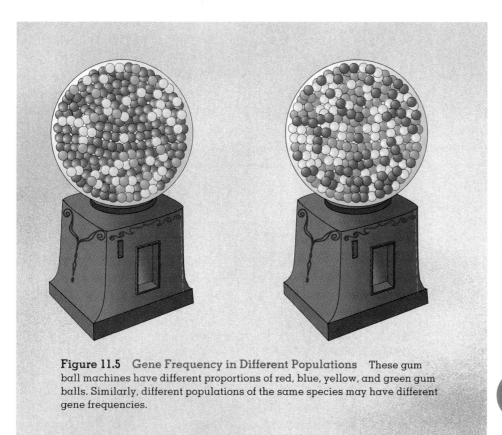

Figure 11.5 Gene Frequency in Different Populations These gum ball machines have different proportions of red, blue, yellow, and green gum balls. Similarly, different populations of the same species may have different gene frequencies.

resources. These clusters of individuals are local populations. The frequency of occurrence of an allele in the gene pool of one local population may differ from the frequency of occurrence of the same allele in the gene pool of other local populations. This is like having one gum ball machine with 400 red, 300 blue, 200 yellow, and 100 green gum balls and a second gum ball machine with 100 red, 200 blue, 300 yellow, and 400 green gum balls (figure 11.5). Each machine contains 1,000 gum balls, but the proportion of each color differs. How often an allele is found in a population is known as its **gene frequency**. Gene frequency is typically stated in terms of a percentage or decimal fraction and is a mathematical statement of how frequently a particular allele shows up in a population. For example, 10 percent or 0.1 means that 10 out of every 100 genes for a given trait in a gene pool are of a particular allele. It is possible for two local populations to have all the same genes, but with very different gene frequencies.

Let's consider an example of gene frequency. All humans are of the same species and, therefore, constitute one large gene pool. There are, however, many distinct local populations scattered across the surface of the earth. These localized populations show many distinguishing characteristics that have been passed on from generation to generation. In Africa, genes for dark skin, tightly curled hair, and a flat nose have very high frequencies. In Europe, the frequencies of genes for light skin, straight hair, and a narrow nose are the highest. In Asia, genes for moderately colored skin, straight hair, and broad noses are most common (figure 11.6). Many other genes show differences in frequency from one race to another, but these three characteristics are easy to see.

Gene Pools and Human Genetics

The human gene pool consists of a number of subgroups called *races*. The particular characteristics that set one race apart from another originated many thousands of years ago before travel was as common as it is today, and we still associate certain racial types with certain geographic areas. Although there is much more movement of people and a mixing of racial types today, people still tend to have children with others who are of the same racial background who live in the same locality. This nonrandom mate selection can sometimes bring together two individuals, each of whom has genes that are relatively rare. Information about human gene frequencies within specific subpopulations can be very important to anyone who wishes to know the probability of having children who could have certain genetic diseases. This is particularly common if both individuals are descended from a common ancestral tribe.

Tay-Sachs disease, for example, causes degeneration of the nervous system and early death of children. Since it is caused by a recessive allele, both parents must pass the allele to their child in order for the child to have the disease. By knowing the frequency of the allele in the background of both parents, we can determine the probability of their having a child with this disease. Jews with ancestors from northern and central Europe have a higher fre-

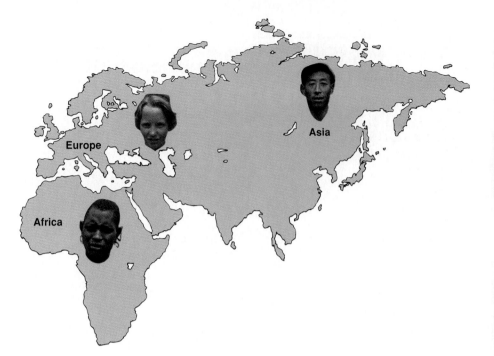

Figure 11.6 **Gene Frequency Differences among Humans** Different physical characteristics displayed by people from different parts of the world are an indication that gene frequencies differ as well.

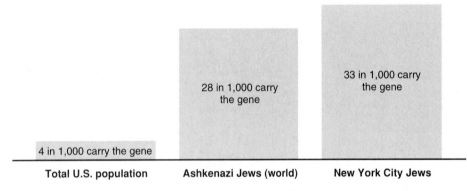

28 in 1,000 carry the gene

33 in 1,000 carry the gene

4 in 1,000 carry the gene

Total U.S. population   **Ashkenazi Jews (world)**   **New York City Jews**

Frequency of Tay-Sachs gene in three populations

Figure 11.7 **The Frequency of Tay-Sachs Gene** The frequency of a gene can vary from one population to another. Genetic counselors use this information to advise people of their chances of having specific genes and of passing them on to their children.

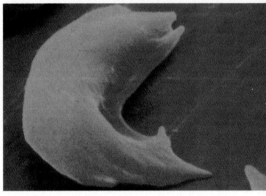

(a)

(b)

Figure 11.8 **Normal and Sickle-Shaped Cells** Sickle-cell anemia is caused by a recessive allele that changes one amino acid in the structure of the oxygen-carrying hemoglobin molecule in red blood cells. (a) Normal cells are disk-shaped. (b) The abnormal hemoglobin molecules tend to stick to one another and distort the shape of the cell when the cells are deprived of oxygen.

quency of this recessive allele than do people of any other group (figure 11.7). Therefore, people of this particular background should be aware of the probability that they can have children who will develop Tay-Sachs disease. Likewise, sickle-cell anemia is more common in people of specific African ancestry (figure 11.8). Anemia is a condition in which the blood has low oxygen-carrying capacity. One in twelve African Americans carries the recessive allele for sickle-cell anemia. Americans of

Mediterranean descent are at greater risk of carrying the allele for thalassemia, another type of anemia. These and other cases make it very important that trained genetic counselors have information about the frequencies of alleles in specific human populations so they can help couples with genetic questions.

People tend to think that gene frequency has something to do with the dominance or recessiveness of genes. This is not true. Often in a population, recessive genes are more frequent

than their dominant counterparts. Straight hair, blue eyes, and light skin are all recessive characteristics, yet they are quite common in the populations of certain European countries.

The frequency of an allele in a population is determined by the value that the gene has to the organisms possessing it. Dark-skin genes are very valuable to people living under the bright sun in tropical Africa. These genes are less valuable to people living in the less intense sunlight of the cooler European countries.

Differences in individuals and local populations are easy to recognize in humans but may go unnoticed in plant and animal populations. All house mice, for instance, may appear identical to the untrained eye, but genetic differences between individuals and separate populations do exist. Likewise, genetic differences occur among individuals in virtually all sexually reproducing species, from maple trees to fruit flies. Figure 11.9 indicates the relationship of genes to individuals, individuals to local populations, and local populations to the entire

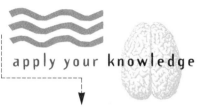

apply your knowledge

Eliminating Genes, Eliminating People

An African-American couple has just learned that their child suffers from sickle-cell anemia, yet both parents have normal hemoglobin. How is this possible? If this child and all other individuals who suffer from this disease refrain from having children of their own, will the sickle-cell allele be eliminated from the human gene pool? Support your answer.

Albinism is a condition caused by a recessive allele that prevents the development of pigment in the skin and other parts of the body. Albinos need to protect their skin and eyes from sunlight. The allele has a frequency of about 0.00005. What is the likelihood that the members of a couple would both carry the gene? Why might two cousins or two members of a small tribe be more likely to both have the gene than two nonrelatives from a larger population? If an island population has its first albino baby in history, why might it have suddenly appeared? Would it be possible to eliminate this gene from the human population? Would it be desirable to do so?

Figure 11.9 Genes, Local Populations, and Gene Pools Each individual shown here has a specific genotype. Local breeding populations differ from one another in the frequency of each gene, but each population has all of the different genes. The gene pool includes all of the individuals present. Assume that A = long tail, a = short tail, B = brown color, b = white color, C = large size, and c = small size. Notice how the different frequencies of genes affect the appearance of the organisms in the different local populations.

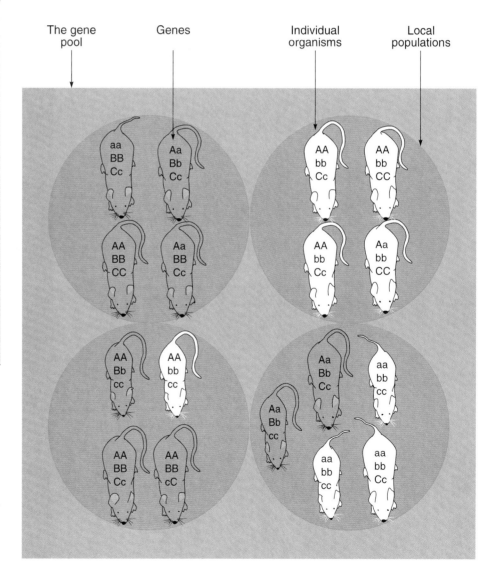

The gene pool   Genes   Individual organisms   Local populations

species gene pool. Note, for example, that while all the populations contain the same kinds of genes, the relative number of alleles *A* and *a* differ from one local population to another.

Causes of Change in Gene Frequency

Within a population, genes are repackaged into new individuals from one generation to the next. Often in local populations very few alleles will be subtracted or new alleles added. As a result, a widely distributed species will consist of a number of separate groups with different allele frequencies. These separate groups are known as **races, breeds, strains,** or **varieties.** All these terms are used to describe different forms of organisms that are all members of the same species. Some of the terms are used more frequently than others, depending on one's field of interest. For example, dog breeders use the term *breed,* plant breeders use the term *variety,* microbiologists use the term *strain,* and anthropologists use the term *race.* When populations within a species have clearly recognizable genetic differences and are geographically isolated they are frequently classified as **subspecies.**

Geographic Barriers

When a portion of a species becomes totally isolated from the rest of the gene pool by some geographic change, such as the formation of a mountain range, river valley, desert, or ocean, we say that the local population is in geographic isolation from the rest of the species. The geographic features that keep the different portions of the species from exchanging genes are called **geographic barriers.** The uplifting of mountains, the rerouting of rivers, and the formation of deserts all may separate one portion of a gene pool from another. For example, two kinds of squirrels are found on opposite sides of the Grand Canyon. Some people consider them to be separate species, while others consider them to be different isolated subspecies of the same species (figure 11.10). Even small changes may cause geographic isolation in species that have little ability to move. A fallen tree, a plowed field, or even a new freeway may effectively isolate populations within species.

Sometimes a single population will be separated into two populations by the formation of a new barrier, but other times a new

Genetic Variety in Domestic Plants and Animals

Humans often work with small, select populations of plants and animals in order to artificially construct specific gene combinations that are useful or desirable. Cabbage, kale, kohlrabi, brussels sprout, cauliflower, and broccoli all originated from a single species of plant. Each has been selectively bred to enhance the development of a particular plant structure. Likewise, we can produce domesticated animals and plants with genes for rapid growth, high reproductive capacity, resistance to disease, or other desirable characteristics. Plants are particularly easy to work with in this manner since individuals with desirable traits can often be reproduced asexually. Potatoes, apple trees, strawberries, and many other plants can be reproduced by simply cutting the original plant into a number of parts and allowing these parts to sprout roots, stems, and leaves. Plants reproduced asexually have exactly the same genes and are usually referred to as **clones**.

Cloning Plants

Kaibab Squirrel   Aberts Squirrel

Figure 11.10 Geographic Isolation   These two squirrels are closely related but are found on opposite sides of the Grand Canyon. Many biologists feel they were at one time members of a single population that was split in two by the canyon. Subsequently they have developed genetic and structural differences.

FYI

Extinction by the Removal of Geographic Barriers   Human activities such as digging canals, and the introduction of non-native organisms, have removed many barriers that previously separated populations. The Tecopa pupfish illustrates this point. The Tecopa pupfish once lived only in the warm waters of the outflows of two hot springs near Tecopa, California, where the population was isolated from all other subspecies of pupfish. In the course of developing bath houses at the hot springs, a water diversion channel was built that unintentionally allowed a neighboring subspecies of pupfish, the Amargosa pupfish, access to the Tecopa hot springs. Eventually the pure Tecopa pupfish genetic line was lost by hybridization with the Amargosa pupfish. The Tecopa pupfish became the first organism to become officially extinct under the U.S. Endangered Species Act of 1973. Many other species have followed. Presently, the gray duck of New Zealand is in danger of extinction because of hybridization with the introduced American mallard.

population will be established by the migration of individuals across a preexisting barrier. Let us consider the second scenario. Suppose that a small group of individuals crosses a preexisting barrier to establish a new population in a region where the species did not previously exist. For example, a few seeds from a mainland plant may be carried by wind, water current, or migrating birds to a remote island. The collection of alleles from a small founding population is likely to be different from the alleles present in the larger parent population from which they came. After all, a few individuals leaving a population would be unlikely to carry copies of all the alleles found within the original population. They may even carry an unrepresentative mixture of alleles. Once a small founding population establishes itself, it tends to maintain its collection of alleles different from other local populations because the organisms mate only among themselves. Conversely, if a barrier is removed, gene exchange will be reestablished and differences between local populations will be reduced and may even be eliminated.

Once a local population has developed, a particular mixture of genes will be present in the population, and that mixture will tend to maintain itself unless something operates to change the frequencies. In other words, gene frequencies do not change without reason. If (1) mating between individuals is random, if (2) no mutations or (3) migration occurs, if the (4) population is large and (5) all individuals with all allele combinations survive and reproduce equally, gene frequencies will not change. If any of these five conditions is not met, however, gene frequency will change and populations with differing gene frequencies may develop.

Non-random Mating

Gene frequencies within a population are maintained by random mating, whereas gene frequencies change as a result of non-random mating. Many trees employ wind as a means of pollen dispersal. Such a reproductive strategy would appear to involve random mating since the tree has little control over the pollination process. Yet the chances of pollination occurring between trees that are in the same geographic area are much greater than the chances of pollination occurring between distant trees. Therefore mating is in fact not random at all, and local gene frequencies, different from those of other local populations, are maintained. In

some populations, individuals may be highly selective in choosing mates. In animal populations, some individuals may be chosen as mates more frequently than others. Obviously, those that are frequently chosen have an opportunity to pass on more copies of their genes than those that are rarely chosen. Attributes of the more frequently chosen may involve general characteristics, such as body size or aggressiveness, or specific conspicuous characteristics attractive to the opposite sex.

Male moose, for example, establish and defend territories where female moose will breed. A male moose will chase out all other males but not females (figure 11.11*a*). Some males have large territories, some have small territories, and some are unable to establish territories. Although it is possible for any male to mate, it has been demonstrated that those who have no territory are least likely to mate. Those who defend large territories may breed with two or more females in their territories. It is unclear exactly why females will choose one male's territory over another, but the fact is that some males are chosen as mates and others are not. In other animals, it appears that the females select males that display specific conspicuous characteristics. Male peacocks have very conspicuous tail feathers. Those with spectacular tails are more likely to mate and have offspring (figure 11.11*b*).

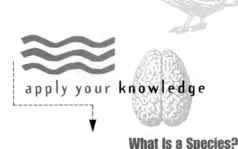

What Is a Species?

The gray duck, native to New Zealand, and the mallard, native to North America, are able to mate and produce fertile offspring. They have done so since humans introduced the mallard to New Zealand. Prior to the artificial introduction, the mallard and gray duck were extremely unlikely to interbreed due to geographic separation of their home ranges. Considering this information and the species definition presented in this text, do you think the mallard and gray duck should be considered the same species? Support your answer. Do you perceive any problems with the species definition?

Extinction by Hybridization The gray duck of New Zealand is in danger of becoming extinct because it interbreeds with the introduced mallard. Before the introduction of the mallard, the gray duck was considered a separate species.

(a)

(b)

Figure 11.11 Sexual Selection (*a*) Big, strong bull moose like this one are more likely to attract mates than are weak or immature bull moose. The females (cows) select which bulls they will mate with. (*b*) Peacocks with large colorful tails are more likely to attract mates than those without such adornment.

Maintaining Genetic Variety in Captive Populations

Many captive populations of animals in zoos are in danger of dying out because of severe inbreeding (breeding with near relatives), resulting in reduced genetic variety. Most zoo managers have recognized the importance of increasing variety in their animals and have instituted programs of loaning breeding animals to distant zoos in an effort to increase genetic variety. In effect, they are simulating natural migration so that new genes can be introduced into distant populations.

Many domesticated plants and animals also have significantly reduced genetic variety. Corn, wheat, rice, and other crops are in danger of losing their genetic variety. The establishment of gene banks in which wild or primitive relatives of domesticated plants are grown is one way that a source of genes can be kept for later introduction if domesticated varieties are threatened by new diseases or environmental changes.

Gene Banks These plants are being grown as part of a program to preserve the genes of primitive varieties of wheat. Although most of these plants are not commercially valuable, they may contain genes that would be extremely useful for improving our domesticated wheat plants in the future.

Mutations

Mutations introduce new genes into a population. If a mutation produces an unfavorable allele, that allele will remain uncommon in the population. Very rarely a mutation will occur that is valuable to the organism. For example, at some time in the past, mutations occurred in certain insect species that made some individuals tolerant to the insecticide dichloro-diphenyl-trichloroethane (DDT), even though the chemical had not yet been invented. These alleles remained very rare in these insect populations until DDT was used. Then these genes became valuable to the insects that carried them. Insects that lacked the gene for tolerance died when they came in contact with DDT and the DDT-tolerant individuals were left to reproduce the species. The DDT-tolerant gene, therefore, became much more common within these populations.

In order for mutations to be important in the evolution of organisms, they must occur in the cells that produce eggs and sperm. Mutations to the cells of the skin or liver will only affect those specific cells and will not be passed to the next generation.

Migration

The *migration* of individuals is also an important way for alleles to be added to or subtracted from a local population. Whenever an organism leaves one population and enters another, it subtracts its genes from the population it left and adds them to the population it joins. If it possesses a rare allele it may significantly affect the allelic frequency of both populations. The extent of migration need not be great. As long as alleles are entering or leaving a population, the gene pool will change.

The Importance of Population Size

The *size of the population* has a lot to do with how effective these mechanisms are at generating variety in a gene pool. The smaller the population, the less genetic variety it can contain. Therefore, migrations, mutations, and accidental death can have great effects on the genetic makeup of a small population.

Sometimes a large local population is reduced to a small population by some type of natural event—a severe winter, drought, or hurricane, for example. Like a founding population, the individuals that remain after the population decline (sometimes referred to as a

California Condor

apply your knowledge

The California Condor: Enough Variety To Survive?

The California condor species contains about twenty individuals. Suppose that through successful breeding practices and habitat reclamation, wildlife biologists are able to reintroduce the condor into the wild and increase their population to 10,000 individuals. How would the genetic diversity of this condor population compare to the genetic diversity of the California condor population that lived 3,000 years ago?

population bottleneck) may not contain all of the alleles or the same proportion of alleles that were present in the original population. Although the population may again grow with improved environmental conditions, a change in gene frequency will have resulted. Extreme cases of this can be seen at the species level for organisms whose populations have severely declined as a result of human activities. The northern elephant seal, for example, was hunted almost to extinction. In the 1890s only approximately twenty individuals remained. Today the northern elephant seal population has grown to over 30,000, yet these seals exhibit almost no genetic diversity.

Differential Allele Value

Since organisms within a population are not genetically identical, some individuals may possess combinations of genes that are of greater value for survival in the local environment. As a result, some individuals find the environment less hostile than others do. The individuals with unfavorable combinations of genes leave the population more often, either by death or migration. Therefore, two local populations that occupy two sites that differ greatly would be expected to consist of individuals having different alleles suited to local conditions. In other words, whenever there is a difference in the fitness of two alleles, selective agents will cause gene fre-

Figure 11.12 Blind Cave Fish This fish lives in caves where there is no light. Its eyes do not function and it has very little color in its skin. Because of its unusual habitat, genes for eyes and skin color are not important. If at some time in the past these genes were lost or mutated, the loss did not negatively affect the organism; hence, the present population has high frequencies of genes for the absence of color and eye function.

quencies to change, and we can say that evolution has occurred. For example, vision is a valuable trait to most fish. A blind fish living in a lake is at a severe disadvantage and in most cases will not survive to pass its genes on to future generations. A blind fish living in a cave where there is no light, however, is not at the same disadvantage and the genes causing blindness are more likely to persist in the allele pool (figure 11.12).

CONCEPT CONNECTIONS

Two or more isolated populations change gene frequencies as a result of

- non-random mating
- mutations
- migrations
- initial small population size
- differential allele value

BIO *feature*

Loss of Variety Through Selective Breeding

Our primary food plants are derived from wild ancestors with combinations of genes that allowed them to compete successfully with other organisms in their environment. However, when humans use selective breeding within small populations to increase the frequency of certain desirable genes in our food plants, other valuable genes are lost from the gene pool. When we select specific good characteristics, we often get bad ones along with them. Therefore, these "special" plants and animals require constant attention. Insecticides, herbicides, cultivation, and irrigation practices are all used to aid the plants and animals that we need to maintain our dominant food-producing position in the world. In effect, these plants are able to live only under conditions that people carefully maintain. Furthermore, we plant vast expanses of the same plant, creating tremendous potential for extensive crop loss from diseases.

Whether we are talking about clones or hybrids, there is the danger of the environment destroying the population. Since these organisms are so similar, most of them will be affected in the same way. If the environmental change is a new variety of disease to which the organism is susceptible, the whole population may be killed or severely damaged.

Monoculture   This wheat field is an example of monoculture, a kind of agriculture in which large areas are exclusively planted with a single crop. Monoculture makes it possible to use large farm machinery, but it also creates conditions that can encourage the spread of disease.

How Species Originate

A question that intrigues us is where did all the different species come from? Species development is an ongoing process. It begins when two or more isolated populations change gene frequencies as a result of non-random mating, mutations, migrations, initial small population size, or differential allele value. These factors will ultimately produce genetically different populations and in time may generate new species.

Even after many generations of geographic isolation, these separate groups may still be able to exchange genes (mate and produce fertile offspring). They have not yet accumulated enough genetic differences to prevent reproductive success. This may result in regional subspecies that are significantly modified structurally, physiologically, or behaviorally. The differences among some subspecies eventually may be so great that there is reduced reproductive success when the members of different subspecies mate.

Speciation is the process of generating new species. This process occurs only if gene flow between isolated populations does not occur even after barriers are removed. In other words, the process of speciation can begin with the geographic isolation of a portion of the species, but new species are generated only when isolated populations become separate from one another *genetically*. Speciation is really a three-step process. It begins with geographic isolation, is followed by the action of selective agents that choose specific genetic combinations as being valuable for the environment of each subspecies, and ends with the genetic differences becoming so great that reproduction between the two groups is impossible.

Maintaining Genetic Isolation

Organisms that allow mating across species lines will not be very successful because most cross-species matings result in no offspring or offspring that are sterile. Part of the speciation process typically involves the development of **reproductive isolating mechanisms** or **genetic isolating mechanisms.** These mechanisms prevent cross-species matings. A great many types of genetic isolating mechanisms are recognized.

In central Mexico, two species of robin-sized birds called *towhees* live in different envi-

(a)

(b)

Figure 11.13 Spacial Isolation The spotted towhee (*a*) and the collared towhee (*b*) exemplify ecological isolation. The collared towhee lives in mountainside pine forests; the spotted towhee makes its home in the oak forests of lower elevations. (*b* © R. N. Bowres/VIREO.)

ronmental settings. The collared towhee lives on the mountainsides in the pine forests, while the spotted towhee is found at lower elevations in oak forests. Geography presents no barriers to these birds (figure 11.13). They are perfectly capable of flying to each other's habitats, but they do not. Similarly, plants that grow in wet soil are different species than plants that grow nearby in drier soils. Because of their **habitat preference** or **ecological isolation,** mating between these two similar species does not occur.

Seasonal isolation refers to differences in the time of the year at which reproduction takes place. It is an effective genetic isolating

mechanism. Some plants flower only in the spring of the year, while other species that are closely related flower in midsummer or in the fall; therefore, the two groups are not likely to pollinate one another. Among many species of insects there is a similar spacing of the reproductive periods of closely related species; therefore they do not overlap.

Inborn behavior patterns that prevent breeding between species result in **behavioral isolation.** The mating calls of frogs and crickets are highly specific. The sound pattern produced by the males is species-specific and invites only females of the same species to mate. The females have a built-in response to the particular species-specific call (figure 11.14).

The courtship behavior of birds involves both sound and visual signals that are species-specific. For example, groups of male prairie chickens gather on meadows shortly before dawn in the early summer and begin their dances. The air sacs on either side of the neck are inflated to show bright red skin. They move their feet up and down very rapidly, while their wings are spread out and quiver. This combination of sight and sound attracts females. When they arrive, the males compete for the opportunity to mate. Other species of birds conduct their own distinctive courtship displays. The differences among the displays are great enough that a female can recognize a male of her own species.

Behavioral isolating mechanisms such as these occur among other types of animals as well. The strutting of a peacock, the fin display of Siamese fighting fish, and the croaking of male frogs are all examples of behaviors that help to prevent different species from interbreeding.

(a)

(b)

Figure 11.14   Animal Communication by Displays   Most animals use behaviors to communicate with others of the same species. The croaking of a male frog is specific to its species and is different from the behaviors of males of other species (a). The visual displays of Siamese fighting fish are also used to communicate with others of the same species (b).

SUMMARY

Genetics is the study of the transmission of traits from one generation to the next, and the gene is the basic unit of inheritance. Most organisms contain two copies of each type of gene, although the copies need not be identical. Differing forms of a gene are called alleles. A dominant allele expresses itself and masks the expression of recessive alleles, whereas a recessive allele only expresses itself when two copies are present in an individual.

All organisms with similar genetic information and the potential to interbreed are members of the same species. A species usually consists of several local groups of individuals known as populations. Groups of interbreeding organisms are members of a gene pool. Although individuals are limited in the number of alleles they can contain, within the population there may be many different kinds of alleles for a trait. Local populations may have different gene frequencies from one another and are called races, varieties, breeds, strains, or subspecies.

Populations are usually genetically diverse. Organisms with wide geographic distribution often show different gene frequencies in different parts of their range. Selecting agents act to change the gene frequencies of the population if mating is non-random, mutations occur, migration occurs, population size is small, and some genes are more favorable than others. Typically, after generations of time, the genes of the more favored individuals will make up a greater proportion of the gene pool.

The process of speciation usually involves the geographic separation of a species into two or more isolated populations. While they are separated, each population adapts to its environment. If this generates enough change, the two populations may become so different that they cannot interbreed. Similar organisms that have recently evolved into separate species normally have mechanisms to prevent interbreeding. Some of these are habitat preference, seasonal isolation, and behavioral isolation.

CHAPTER GLOSSARY

alleles (a-lēlz) Alternative forms of a gene for a particular characteristic.

behavioral isolation (be-hav'yu-ral i-so-la'shun) A genetic isolating mechanism that prevents interbreeding between species because of differences in behavior.

clones (klōnz) All of the individuals reproduced asexually that have exactly the same genes.

dominant allele (dom'in-ant a-lēl) A gene form that expresses itself and hides the effect of other gene forms for that trait.

ecological isolation (e-kŏ-loj'ĭ-kal i-so-la'shun) A genetic isolating mechanism that prevents interbreeding between species because they live in different areas; also called **habitat preference.**

gene frequency (jēn fre'kwen-se) A measure of the number of times that a gene occurs in a population. The percentage of sex cells that contain a particular gene.

gene pool (jēn pool) All the genes of all the individuals of a species.

genes (jēnz) The inheritable units that in part determine an organism's characteristics.

genetics (je-net'iks) The study of genes, how genes produce characteristics, and how characteristics are inherited.

geographic barriers (je-o-graf'ik băr'yurz) Geographic features that keep different portions of a species from exchanging genes.

hybrid (hy'brid) The offspring of two different genetic lines produced by sexual reproduction.

mutation (miu ta'shun) A change in genetic information.

population genetics (pop"u-la'shun je-net'iks) The study of population gene frequencies.

recessive alleles (re-se'siv a-lēlz) A gene that is hidden by other gene forms.

reproductive isolating mechanism (rē-prō-duk'tiv i-so-la' ting me'kan-izm) A mechanism that prevents interbreeding between species; also called **genetic isolating mechanism.**

seasonal isolation (se'zun-al i-so-la'shun) A genetic isolating mechanism that prevents interbreeding between species because their reproductive periods differ.

speciation (spe-she-a'shun) The process of generating new species.

species (spe'shēz) A group of organisms of the same kind that have the ability to interbreed naturally and produce offspring that are also capable of reproducing.

subspecies (sub'spe-shēz) A number of more or less separate groups within the same gene pool that differ from one another in gene frequency; also called **races, breeds, strains,** or **varieties.**

CONCEPT MAP TERMINOLOGY

Construct a diagram to show the relationships among the following concepts.

behavioral isolation
differential allele value
ecological isolation
geographic barriers
migrations

mutations
non-random mating
seasonal isolation
small population size
speciation

LABEL • DIAGRAM • EXPLAIN

Describe how subspecies may develop.

Briefly describe how each of the following factors influences gene frequency.

non-random mating
mutations
migrations
small population size
differential allele value

Multiple Choice Questions

1. Which one of the following matings involves organisms belonging to the same species?
   a. *donkey* × *horse* → produce mule; offspring sterile
   b. *domestic dog* × *wolf* → offspring fertile; mating not likely to occur in nature
   c. *lion* × *tiger* → offspring sterile, mating will not occur in nature
   d. *blue goose* × *snow goose* → offspring fertile; interbreed where ranges overlap

2. Which of the following terms is least like the others?
   a. species
   b. breed
   c. strain
   d. variety

3. Organisms with the same genes are:
   a. clones
   b. alleles
   c. hybrids
   d. strains

4. Gene frequency describes:
   a. how valuable an allele is to a population
   b. how often an allele is found in a population
   c. all of the genes found within an organism
   d. the number of species possessing a given gene

5. Dominant alleles are:
   a. always the most common alleles in the population
   b. the most valuable alleles to the organism
   c. concealed by the presence of other alleles
   d. alleles that express themselves and inhibit the expression of other alleles

6. Which one of the following will *not* cause gene frequencies to change?
   a. migrations
   b. mutations
   c. random mating
   d. differential gene value

7. All of the genes shared by a population are its:
   a. gene frequency
   b. dominant alleles
   c. recessive alleles
   d. gene pool

8. Various species of lightning bugs have their own characteristic patterns of light flashes. Females flash an answering code to the males' flash. Mating follows the appropriate flash recognition. This is an example of a genetic isolating mechanism called:
   a. habitat preference
   b. behavioral isolation
   c. seasonal isolation
   d. geographic isolation

9. Some related species of plants cannot interbreed by pollinating each other because sexual structures of the one species mature at a different time than those of the other. This genetic isolating mechanism is called:
   a. habitat preference
   b. behavioral isolation
   c. seasonal isolation
   d. geographic isolation

10. A canal is built that divides a population of land snails. The snails of the east side of the canal are no longer able to interbreed with the snails on the west side of the canal. The canal represents:
    a. habitat preference
    b. behavioral isolation
    c. seasonal isolation
    d. geographic isolation

Questions with Short Answers

1. What is the difference between a gene and an allele?

2. What is the difference between a dominant allele and a recessive allele?

3. Give an example of a species containing several separate strains, breeds, varieties, or races.

4. How do the terms species, subspecies, and population differ?

5. What is gene frequency?

chapter

12

The Origin and Evolution of Life

Trilobite fossils, Lucas County, Ohio.

The Origin and Evolution of Life

learning objectives

- Describe the formation of our solar system, including Earth.

- Beginning with the gases in the early atmosphere, trace the events that led to the first living cells.

- Describe the events leading to the evolution of the first autotrophs.

- Explain the evolution of the first eukaryotic cells.

- Recognize the significance of adaptive radiations and mass extinctions in the history of life.

- Know the characteristics of primates.

- Describe the evolutionary history of humans.

Early Earth

One explanation of the origin of the solar system is the **solar nebula theory.** This theory proposes that the solar system was formed from a large cloud of gases that developed 10 billion to 20 billion years ago (figure 12.1). A gravitational force was created by the collection of particles within this cloud that caused other particles to be pulled from the outer edges to the center. As gravity caused the particles to collect, they formed a very large disk. Tremendous amounts of heat and pressure were created as materials condensed to form our sun in the center of this disk.

As the particles within the solar nebula were pulled into the center to form the sun, other gravitational centers developed. Clouds of dust and gases collected in these areas and became the planets of our solar system, including Earth. Many scientists think Earth was formed at least 4.5 billion years ago (figure 12.2). A large amount of heat was generated as the particles became concentrated to form the planet. While not as hot as the sun, the materials of Earth formed a hot liquid core that became encased by a solid *mantle* and thin outer *crust* as it cooled. These layers of the earth were created as

(a)

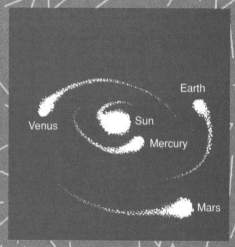

(b)

Figure 12.1 Formation of Our Solar System When particles in a gas cloud came together they generated gravity, which pulled additional gas particles into the center. The condensation of matter in the center gave rise to our sun (*a*). In other regions, smaller gravitational forces caused the formation of the sun's planets (*b*).

FYI

The Earth's Early Atmosphere	Today's Atmosphere	
Ammonia (NH_3)	Nitrogen (N_2)	78%
Methane (CH_4)	Oxygen (O_2)	21%
Carbon dioxide (CO_2)	Carbon dioxide (CO_2)	0.03%
Water vapor (H_2O)	Water vapor (H_2O) and	
	other compounds	0.07%

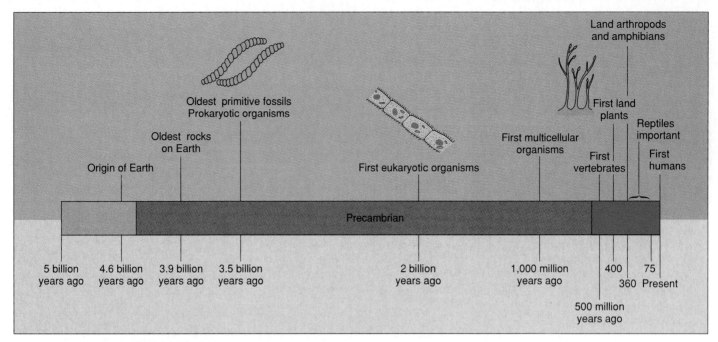

Figure 12.2   Evolutionary Time Line

Figure 12.3   Formation of the Oceans   As the Earth cooled, the water vapor in the atmosphere condensed and collected in the lower areas of the Earth, resulting in the oceans.

denser materials concentrated in the center of the earth and lighter materials moved toward the surface.

Over hundreds of millions of years, Earth slowly changed. Volcanic activity caused the release of water vapor, carbon dioxide, methane, ammonia, and hydrogen; Earth's early atmosphere was formed. This atmosphere was very different from the atmosphere we have today. Most notably, oxygen (O_2), the molecule often associated with life, was absent. The water vapor of the warm planet created a dense cloud cover. As the Earth cooled, the water vapor in the atmosphere condensed into droplets of rain that ran over the land and collected to form the oceans (figure 12.3).

The First Organic Molecules

In the 1920s a Russian biochemist, Alexander I. Oparin, and a British biochemist, J. B. S. Haldane, working independently, proposed the idea that the first organic molecules were formed in this early atmosphere that lacked molecular oxygen. Organic molecules are the

Organic Molecules on Earth Today

Chemists divide molecules into two categories: inorganic and organic. Inorganic molecules are chemical combinations of small numbers of atoms. Examples of common inorganic molecules include water (H_2O), carbon dioxide (CO_2), ammonia (NH_3), table salt (NaCl), battery acid (H_2SO_4), and alkaline drain cleaner (NaOH). Organic molecules are large chemical combinations of atoms containing carbon in rings or chains. There are four major categories of organic molecules: carbohydrates, lipids, proteins, and nucleic acids.

Appendix A

Water | Carbon dioxide | Sulfuric acid | Sodium hydroxide

Inorganic Molecules

Carbohydrates

Common name: sugar.

Atomic components: carbon (C), hydrogen (H), and oxygen (O).

Function in cells: energy and structural material.

Examples: glucose, sucrose, lactose, corn starch, glycogen, cellulose.

Glucose
$C_6H_{12}O_6$

Organic Molecules

Lipids

Common name: fats, oils, and steroids.

Atomic components: carbon (C), hydrogen (H), and oxygen (O); sometimes phosphorus (P) and nitrogen (N).

Function in cells: energy storage, some hormones, shock absorbers, cell membrane components, and cell communication.

Examples: body fat, cholesterol, testosterone, estrogen, corn oil, and lard.

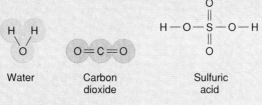

Fat

Proteins

Common name: protein.

Atomic components: carbon (C), hydrogen (H), nitrogen (N), and oxygen (O); sometimes sulfur (S).

Function in cells: some hormones, enzymes, structural material, cell membrane components, and cell communication.

Examples: muscle fiber, lactase, hair, and oxytocin.

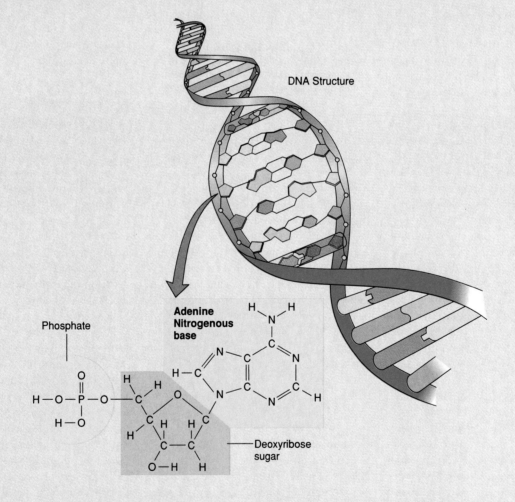

Tripeptide

Nucleic acids

Common name: DNA and RNA.

Atomic components: carbon (C), hydrogen (H), oxygen (O), phosphorus (P), and nitrogen (N).

Function in cells: genetic material and protein synthesis.

Examples: component of chromosomes (deoxyribonucleic acid) and component of ribosomes (ribonucleic acid).

DNA Structure

building blocks of life. The formation of organic molecules was a necessary step toward the development of life on Earth. According to Oparin and Haldane's hypothesis, lightning, heat from volcanoes, and ultraviolet radiation furnished energy necessary for the molecules in the atmosphere to combine and produce larger organic molecules that were washed by rain into the oceans. This accumulation of organic molecules is thought to have occurred over half a billion years, resulting in oceans that were a dilute organic soup.

Evidence supporting the ideas of Oparin and Haldane was presented by Stanley Miller in 1953. Miller constructed a model of the early Earth (figure 12.4). In a glass apparatus he placed distilled water to represent the early oceans. Earth's early atmosphere was duplicated by adding hydrogen, methane, and ammonia. Electrical sparks provided energy needed to produce organic compounds. By heating parts of the apparatus and cooling others, he simulated the rains that are thought to have fallen into the early oceans. After a week of operation, he removed a water sample from the apparatus. When this water was analyzed, it was found to contain many simple organic compounds.

Miller demonstrated that simple organic molecules like sugars and amino acids can be produced in the absence of living organisms, but he did not prove Oparin and Haldane's hypothesis to be true. Miller merely demonstrated that if Oparin and Haldane's assumptions were true, then their explanation concerning the origin of the first organic molecules was feasible.

A key element of the scientific process is to challenge ideas rather than to blindly accept explanations. More than forty years have passed since Stanley Miller's experiment, and the origin of the first organic molecules has been greatly debated. Some scientists believe that methane and ammonia were never plentiful in the atmosphere, and therefore Miller's experimental conditions

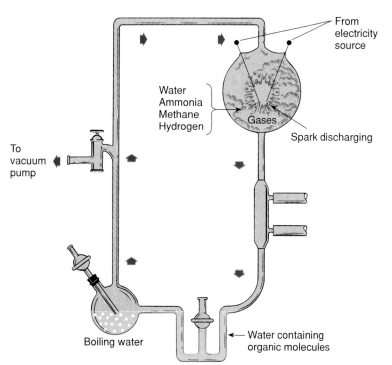

Figure 12.4 Miller's Apparatus
Stanley Miller developed this apparatus to demonstrate that the spontaneous formation of complex organic molecules could take place in a reducing atmosphere.

never existed. One alternate model proposes that the comets, meteorites, and asteroids that bombarded our young planet brought with them the organic molecules needed for life to originate. Therefore, synthesis from smaller compounds was not necessary.

When the first organic molecules originated they must have been in great enough concentration for reactions to occur among them, yielding larger, more complex organic molecules. Imagine that you have 10,000 bricks scattered across the land area of one county. You would have enough bricks to build a wall, but it could not be built until the bricks were gathered in one location. Likewise, large organic molecules could not be assembled in the ocean's dilute organic soup until their building blocks (simple organic molecules) were concentrated.

Several ideas have been proposed for the concentration of simple organic molecules. A portion of the early ocean could have been separated from the main ocean by geological changes. The evaporation of water from this pool could have concentrated the

molecules, which might have led to the manufacture of larger organic molecules. It has also been proposed that freezing may have been the means of concentration. When a mixture of alcohol and water is placed in a freezer, the water freezes and the alcohol becomes concentrated into a small portion of liquid. A similar process could have occurred on Earth's early surface. A third hypothesis proposes that clay particles may have been a factor; small particles of clay have electrical charges that can attract and concentrate organic molecules. Once simple organic molecules are brought together, iron and zinc atoms on a clay particle's surface can facilitate reactions that bind simple organic molecules together.

The First Cells

Geologists and biologists typically measure the history of life by looking back from the present. Therefore, time scales are given in "years ago." It has been estimated that the formation of simple organic molecules in the atmosphere began about 4 billion years ago and lasted approximately 1.5 billion years. The oldest known fossils of living cells are thought to have formed 3.5 billion to 3.8 billion years ago (figure 12.5). The question is, how do you get from the formation of complex organic molecules to the first primitive cells in half a billion years?

Several hypotheses are proposed for the formation of nonliving structures that led to the development of the first cells. Oparin speculated that a structure consisting of a collection of organic molecules surrounded by a film of water molecules could have been a predecessor to the first cell. This arrangement of water molecules, while not a membrane, could have functioned as a physical barrier between organic molecules and their surroundings. These structures have been synthesized in the laboratory. They can selectively absorb chemicals from the surrounding water and incorporate them

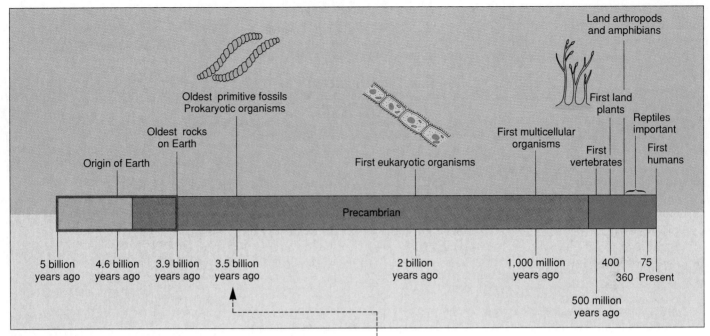

Figure 12.5 Evolutionary Time Line First prokaryotic organisms.

into their structure. Also, the chemicals within the water molecule container have a specific arrangement—they are not random collections of molecules. Some contain molecules called *enzymes* that direct a specific type of chemical reaction. No one claims that these structures are alive, but they do exhibit some lifelike traits: they are able to grow and divide if the environment is favorable.

A second structure that has been produced in the laboratory and may be similar to the predecessor of the first cell consists of a collection of organic molecules surrounded by a double-layered outer boundary. These structures can be formed when proteinlike molecules are placed in boiling water and slowly allowed to cool. Some of the proteinlike molecules produce a double-boundary structure that encloses to form a sphere. Although these membranes do not contain the same materials as a cell membrane, they do exhibit some membranelike characteristics. The spheres are able to absorb and release water and display a type of internal movement similar to that exhibited by cells. In addition, these cell-like structures contain molecules that function like enzymes, and use molecular energy to fuel

chemical reactions. They can absorb material from the surrounding medium and form buds, which result in a second generation. Given these characteristics, some investigators think that the protein spheres are very similar to the first living cells.

Both of these hypotheses assume that proteins, which are essential to cell structure and function, were created prior to the development of the molecular code (DNA), which is synthesized with the help of proteins. Yet the function of the molecular code is to direct the production of proteins in living cells. Similar to the question "What came first, the chicken or the egg?" we can ask "What came first, proteins or the genetic code?" Studies have demonstrated that certain **ribo**nucleic **a**cid (RNA) molecules are able to function as both a genetic code (like DNA) and facilitate reactions (like enzymatic proteins). RNA is a molecule similar to DNA which, among other things, functions as genetic material in some viruses today. One hypothesis suggests that the first cells had RNA genes that were capable of self-copying, and DNA molecules that require protein to function evolved later.

Although there are several possible explanations as to how the first organic molecules

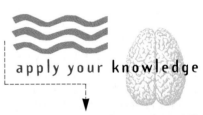

apply your knowledge

Spontaneous Generation at Home!?

Take a handful of straw, hay, grass clippings, or leaves and boil them to make an organic soup broth. After it has cooled, pour it through a strainer into a clear glass bottle. Add a teaspoon of soil and plug it with a cotton ball or piece of cloth. Place in a sunny spot. Observe this bottle over the next two weeks and note the changes that occur. If you have a microscope or magnifying lens, sample the materials with an eyedropper and try to identify the organisms.

This exercise does not show the spontaneous generation of living organisms from nonliving material, nor does it result in primitive living cells. Develop a hypothesis on the source and nature of the organisms growing in this culture.

originated and what types of molecules developed first, all explanations have one thing in common—they all assume an atmosphere lacking in oxygen (O_2). It is widely believed that *spontaneous generation*, the origin of living things from nonliving material, did at one time happen, but that it could not occur in today's oxygen-rich atmosphere. The laboratory synthesis of cell-like structures helps us to understand how the first primitive living cells *might* have developed. However, it leaves a large gap in our understanding because it does not explain how these first cells might have become the highly complex living cells we have today.

Heterotrophs and Autotrophs

The first primitive cells were most likely heterotrophs, which require an organic energy source from the environment. At first, this would not have been a problem. The organic molecules that had been accumulating in the oceans for millions of years would have served as an ample source of food for the early heterotrophs. These simple single-celled organisms were capable of undergoing simple chemical reactions, but made little modification to the food molecules from their environment. They most readily utilized environmental molecules that could be used directly. As the population of heterotrophs increased through reproduction, the supply of organic material would have been consumed faster than it was being produced in the atmosphere and could have eventually become exhausted. However, some of the heterotrophs may have contained a mutated form of the genetic code that would have allowed them to convert material that was previously not directly usable into a compound that could be used. Heterotrophs with this mutation could have survived, while those without it would have become extinct as the compounds they used for food became scarce. Some scientists suggest that through a series of mutations in the early heterotrophs, a more complex series of biochemical reactions originated within some cells. Such cells could use chemical reactions to convert inorganic molecules into usable organic compounds. These kinds of mutations may have led the way to the development of autotrophs, organisms like present-day blue-green bacteria, that are capable of converting inorganic materials and sun energy into organic molecules.

An Atmosphere Containing Molecular Oxygen (O_2)

Ever since its formation, Earth has undergone constant change. At first, it was too hot to support an atmosphere. Later, as it cooled and gases issued forth from volcanoes, an atmosphere void of molecular oxygen was formed. This atmosphere created an environment suitable for the formation of organic molecules and the spontaneous generation of life. The emergence of heterotrophic cells led to the development of autotrophic cells, which converted inorganic molecules from the environment into organic compounds. Next, autotrophs developed simple pathways for trapping light energy and storing it in the chemical bonds of organic molecules. More complex metabolic activities evolved that produced autotrophs capable of photosynthesis. This resulted in the release of molecular oxygen into Earth's atmosphere.

The formation of molecular oxygen was a significant change to our planet's atmosphere and the evolution of life for two reasons. First, spontaneous generation ended with the accumulation of molecular oxygen in the atmosphere. This is because the presence of molecular oxygen tends to make organic molecules break down. Also, oxygen molecules reacted with each other to form ozone (O_3). Ozone collected in the upper atmosphere and acted as a screen to prevent most of the ultraviolet radiation from reaching Earth's surface. The lack of ultraviolet light stopped the spontaneous formation of organic molecules. It also reduced the number of mutations in primitive cells that had produced a great variety in cellular forms.

Second, the development of an atmosphere containing molecular oxygen (O_2) opened the door for the evolution of *aerobic* organisms, organisms that utilize oxygen to convert food molecules into usable energy. All life forms prior to the development of an atmosphere containing molecular oxygen were of necessity *anaerobic*, and thus able to function in an environment without molecular oxygen. However, these organisms were less efficiently able to extract energy from food molecules than were aerobic organisms.

The Origin of Eukaryotic Cells

Today, heterotrophic, autotrophic, aerobic, and anaerobic lifestyles are all represented among present-day bacteria. Bacteria may be very similar to the early life forms that inhabited the Earth. Recall that bacteria are prokaryotic cells that lack nuclear membranes and many complex organelles composed of membranes. All other forms of life are composed of eukaryotic cells possessing a nuclear membrane and other membranous organelles, including the energy-converting organelles mitochondria (present in both autotrophs and heterotrophs) and chloroplasts (present in autotrophs only).

Biologists generally think that eukaryotes evolved from prokaryotes. The **endosymbiotic theory** attempts to explain this evolution. This theory suggests that present-day eukaryotic cells evolved from the combining of several different types of primitive cells. It is thought that some organelles found in eukaryotic cells may have originally been freeliving prokaryotes. The evidence for this is found in the fact that mitochondria and chloroplasts contain bacteria-like DNA and ribosomes, control their own reproduction, and synthesize their own enzymes. These bacterial cells could have established a mutually beneficial relationship with another primitive cell type that contained a nuclear membrane (figure 12.6).

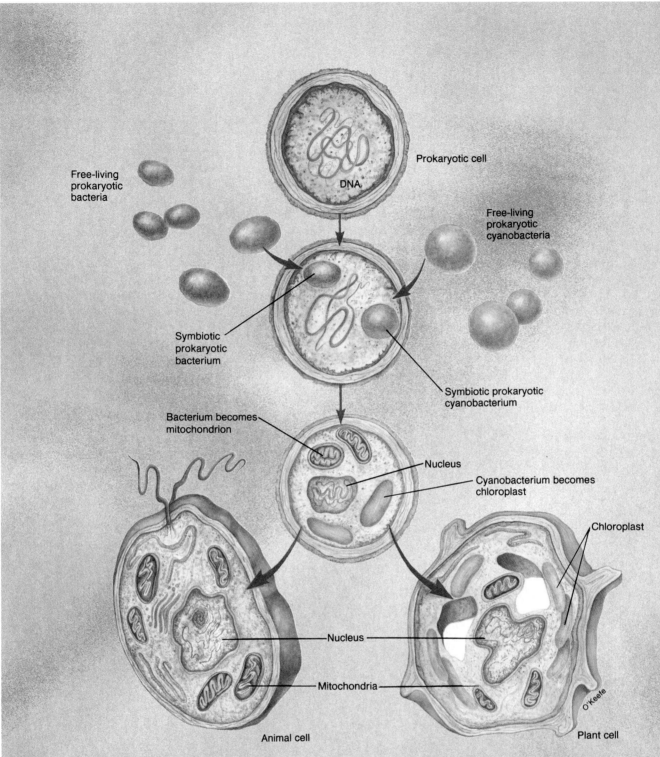

Figure 12.6 The Endosymbiotic Theory This theory proposes that eukaryotic cells developed when some free-living prokaryotic bacteria and cyanobacteria (blue-green algae) developed the ability to live inside other bacteria in a symbiotic relationship with a host cell. The bacteria subsequently developed into mitochondria and the cyanobacteria developed into chloroplasts. From this beginning developed all of the many kinds of eukaryotic organisms (plants, animals, protozoa, fungi, algae).

If these cells adapted to one another and were able to survive and reproduce better as a team, it is possible that this relationship may have evolved into present-day eukaryotic cells. If this relationship had included only a nuclear membrane-containing cell and aerobic bacteria, the newly evolved cell would have been similar to present-day heterotrophic protozoa, fungi, and animal cells, which contain mitochondria. If this relationship had included both aerobic bacteria and photosynthetic bacteria, the newly formed cell would have been similar to present-day autotrophic algae and plant cells, which contain both mitochondria and chloroplasts.

Regardless of the type of cell (prokaryotic or eukaryotic), or whether the organisms are heterotrophic or autotrophic, all organisms have a common basis: DNA is the universal genetic material, protein serves as structural material, chemical reactions are regulated in a similar fashion, and the same energy molecules are utilized. Although there is a wide variety of organisms, they all are composed of cells built from the same basic molecular building blocks. Therefore, it is probable that all life had a single origin and that the variety of living things seen today evolved from the first primitive cells.

Evolutionary Time Line

For two billion years, life on earth consisted of simple prokaryotic organisms (bacteria). The first eukaryotic organisms appeared 1.8 billion years ago (figure 12.7). These organisms, like the bacteria, were unicellular. The part of the Earth's history dominated by unicellular organisms (the Age of Bacteria) is generally referred to as the Precambrian Era. There is little fossil record of Precambrian unicellular life, although it comprises a span of time much greater than the entire history of multicellular plants and animals. The first multicellular organisms appeared 700 million years ago at the end of the Precambrian Era. During the Cambrian Period of the Paleozoic Era, an explosion of multicellular organisms occurred. Marine invertebrates were abundant, with individuals from most present-day phyla existing at that time.

Several other "explosions," or *adaptive radiations*, followed. Note on the geological time chart shown in the Bio Feature that each era was dominated by a different major form of vegetation. The Paleozoic Era was dominated by nonvascular and primitive vascular plants; the Mesozoic Era by cone-bearing evergreens; and the Cenozoic by the flowering plants with which we are most familiar. Likewise, many periods are associated with specific animal groups. Among vertebrates, the Devonian Period is considered the Age of Fishes and the Carboniferous Period the Age of Amphibians. The Mesozoic Era is considered the Age of Reptiles and the Cenozoic Era is considered the Age of Mammals. In each instance, dominance of a particular animal group resulted from adaptive radiation events.

Amphibians, for example, most likely evolved from a lobe-finned fish of the Devonian Period (figure 12.8). This organism possessed two important adaptations: lungs and paired lobed fins that allowed the organism to pull itself onto land and travel to new water holes during times of drought (figure 12.9). Selective pressures caused these fins to evolve into legs and the first

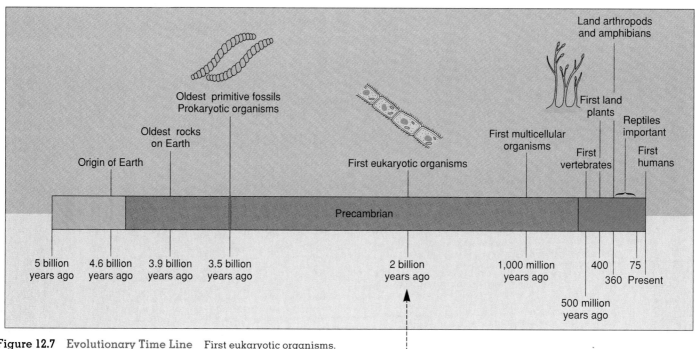

Figure 12.7 Evolutionary Time Line First eukaryotic organisms.

BIO feature

Geological Time Chart

A **geological time chart** shows a chronological history of living organisms based on the fossil record. The largest geological time units are called eras. From earliest to most recent, the geological eras are Precambrian, Paleozoic, Mesozoic, and Cenozoic. Each of these eras is subdivided into smaller time units called periods. For example, Jurassic is a period of the Mesozoic Era that began 180 million years ago.

Precambrian   *pre* = before, *cambria* = another name for the country of Wales where early fossils were first found

Paleozoic   *paleo* = prehistoric, *zoic* = animal

Mesozoic   *meso* = middle, *zoic* = animal

Cenozoic   *ceno* = new, *zoic* = animal

Era	Period	Epoch	Millions of Years Ago	Important Events
Cenozoic (Age of Mammals)	Quaternary	Recent	0.01	Modern humans
		Pleistocene	2	Early humans
	Tertiary	Pliocene	6	Ape radiation
		Miocene	23	Abundant grazing mammals
		Oligocene	38	Angiosperms dominant
		Eocene	54	Mammalian radiation
		Paleocene	65	First placental mammals
Mesozoic (Age of Reptiles)	Cretaceous		135	Climax of reptiles; first angiosperms; extinction of ammonoids
	Jurassic		180	Reptiles dominant; first birds; first mammals
	Triassic		225	First dinosaurs; cycads and conifers dominant
Paleozoic	Permian		275	Widespread extinction of marine invertebrates; expansion of primitive reptiles
	Carboniferous		345	Great swamp trees (coal forests); amphibians prominent
	Devonian		395	Age of fishes; first amphibians
	Silurian		435	First land plants; eurypterids prominent
	Ordovician		500	Earliest known fishes
	Cambrian		600	Abundant marine invertebrates; trilobites and brachiopods dominant; algae prominent
Precambrian time			>3,000	Soft-bodied primitive life

Geologic Time Chart

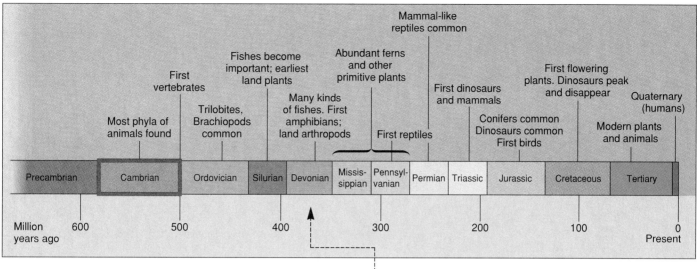

Figure 12.8 **Evolutionary Time Line** The first terrestrial animals.

Figure 12.9 **Fish Ancestors of Amphibians** Probably the first vertebrate to spend time on land was a kind of fish that had stubby, leglike fins that were useful in moving short distances on land. The coelacanth, shown here, is still found in deep portions of the ocean and is believed to be a descendent of the group of fish that gave rise to amphibians.

amphibian came into being. During this lengthy time, land masses were colonized by vegetation but very few animals. The first vertebrates to spend part of their lives on land found a variety of unexploited niches resulting in the rapid evolution of new amphibian species and their dominance during the Carboniferous Period.

For 40 million years amphibians were the only vertebrate animals on land. During this time, mutations continued to occur, and valuable modifications were passed on to fu-ture generations that eventually led to the development of reptiles. One change allowed the male to deposit sperm directly within the female. Because the sperm could directly enter the female and remain in a moist interior, it was no longer necessary for the animals to return to the water to mate, as amphibians still must do. However, developing young still required a moist environment for early growth. A second modification, the amniotic egg, solved this problem. An amniotic egg, like a chicken egg, protects the developing young from injury and dehydration while allowing for the exchange of gases with the external environment. A third adaption, the development of protective scales and relatively impermeable skin, protected reptiles from dehydration. With these adaptations reptiles were able to out-compete amphibians in most terrestrial environments. Amphibians that did survive were the ancestors of present-day frogs, toads, and salamanders. With extensive adaptive radiation, reptiles took to the land, sea, and air. A particularly successful group of reptiles was the dinosaurs (figure 12.10). The length of time that dinosaurs dominated the earth, more than 100 million years, was greater than the length of time from their extinction to the present (figure 12.11).

As reptiles diversified, some developed characteristics common to other classes of vertebrates found today, such as warm-bloodedness, feathers, and hair. Warm-blooded reptiles with scales modified as feathers for insulation eventually evolved into organisms capable of flight. Through natural selection, reptilian characteristics were slowly eliminated and characteristics typical of today's modern birds (multiple adaptations to flight, keen senses, and complex behavioral instincts) developed. Archaeopteryx, the first bird, had characteristics typical of both birds and reptiles.

Also evolving from reptiles were the mammals. The first reptiles with mammalian

Figure 12.10 The Age of the Dinosaurs Dinosaurs were the dominant vertebrates for millions of years.

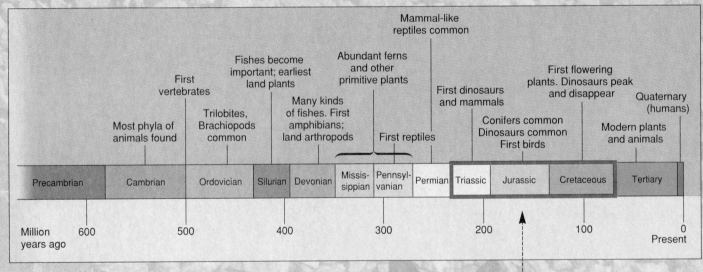

Figure 12.11 Evolutionary Time Line The rise and extinction of dinosaurs.

BIO feature

Mass Extinctions

A **mass extinction** involves the disappearance of thousands of species of organisms during a short period of time. They are thought to be brought about by dramatic changes in either the physical (nonliving) or living components of the environment. During the transition between the Permian and Triassic Periods, for example, a mass extinction of marine life occurred. Some scientists speculate that **continental drift,** the movement of the Earth's crustal plates, was responsible for this mass extinction. Two large continents came together during the Permian Period to form one giant continent, Pangaea. This union of land masses would have reduced offshore habitat and created competition by bringing together

organisms that had evolved independently, resulting in many extinctions.

The disappearance of dinosaurs and many other organisms during the Cretaceous Period also represents a mass extinction. One hypothesis explaining the extinction of dinosaurs proposes that one or more asteroids struck the earth, causing tidal waves, dust clouds, or both. The tidal waves would have destroyed coastal life and the dust clouds would have reduced photosynthesis, altered temperatures, and produced acid precipitation. Not all scientists agree with this explanation of the dinosaurs' demise. Some propose that extensive volcanic activity could have produced the same consequences.

characteristics appeared in the Permian Period, although the first true mammals did not appear until the Triassic Period (figure 12.12). These organisms remained relatively small in number and size until after the mass extinction of the reptiles. These extinctions opened many niches and allowed for the subsequent adaptive radiation of mammals. As with the other adaptive radiations, mammals possessed unique characteristics that made them better adapted to the changing environment. The characteristics include insulating hair, constant body temperature, internal development of young, milk gland development, and increased parental care of young.

Reproductive Strategies in Mammals Reflect Their Evolutionary Past

One small group of mammals, including the duck-billed platypus and the spiny anteater, still possess the reptilian trait of laying amniotic eggs, though they also possess the mammalian traits of fur and milk production (figure 12.13). It appears that this group of organisms is not an ancestor to other mammalian species but is rather an offshoot of the mammalian lineage. Marsupial mammals like

Continental Drift Although we think of the land as being firm and well anchored, it is clear that the continents are drifting. These drawings re-create the sequence of events thought to have led to the current position of the continents.

(a)

(b)

(c)

(d)

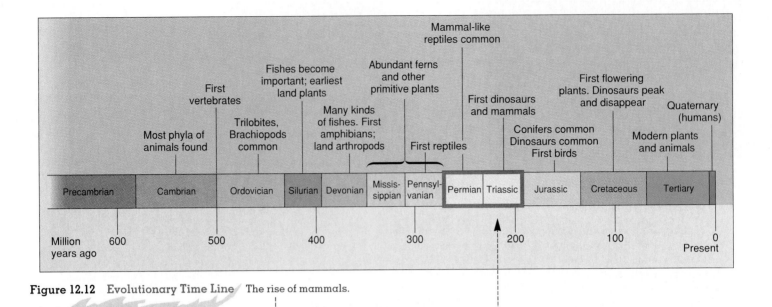

Figure 12.12 Evolutionary Time Line The rise of mammals.

opossums and kangaroos possess another primitive mammalian reproductive strategy. Their offspring begin development inside the mother and are then born prematurely. The young crawl into the mother's pouch where they attach to a nipple and complete development. The most common mammalian reproductive strategy is represented by the placental mammals. In these organisms, a placenta develops through which nutrients, oxygen, and waste products are exchanged between the mother and fetus. Placental mammals are born in a later stage of development than are marsupials (figure 12.14).

It is important to keep in mind that other organisms have experienced similar evolutionary patterns. Insects, for example, have experienced two major periods of adaptive radiation. Marine invertebrate organisms and plants have also had periods of rapid speciation and dominance.

(a)   (b)

Figure 12.13 Egg-Laying Mammals The duckbilled platypus (a) and the echidna (spiny anteater) (b) lay eggs and incubate them. However, they have hair and provide milk for their offspring.

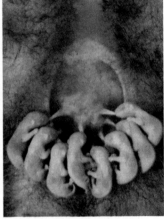

Figure 12.14 Marsupials The young of marsupials are born in a very immature state and crawl to the mother's pouch where they attach to nipples and feed for an extended period of time.

BIO *feature*

Warm-blooded Flying Reptiles

Unlike today's reptiles, some reptiles of the Mesozoic Era may have been warm-blooded organisms. This seems especially likely for the flying reptiles, which would have had very high metabolic needs. Two types of reptiles developed the ability to fly. More than 120 species of one group, the pterosaurs, have been identified by fossil record. These organisms varied greatly and probably occupied the diverse niches held by today's birds. Pterosaur species had wingspans ranging from 46 centimeters (18 inches) to 12 meters (39 feet). The wings themselves were probably composed of membrane with reinforcing fibers. Eventually the pterosaurs became extinct. The second type of flying reptile had scales modified as long feathers. This kind of organism is thought to have evolved into the first known bird of the Jurassic Period, *Archaeopteryx*. *Archaeopteryx* had many reptilian characteristics absent in modern birds, such as teeth and a jointed tail.

Flying Reptiles

Cockroach

Evolution of Primates

Discussions of the evolution of life inevitably lead to questions concerning the origin of our own species and our place in the scheme of life. Humans belong to the Primate order. The first primates evolved during the early Cenozoic Era from small tree-dwelling shrew-like animals (figure 12.15). Today, most primates continue to be tree dwellers and have evolved characteristics well suited for this way of life. Primates possess highly mobile limbs with rotational movement at the wrist, elbow, and shoulder joints. Digits frequently have individual mobility, and the thumbs and large toes are in most cases *opposable,* that is, able to move toward the other digits. In addition, primates have flattened nails that replace claws and their digits are highly sensitive to tactile stimuli (figure 12.16). Each of these traits enhances the primates' ability to maneuver through trees and grasp limbs. The muzzle of primates is reduced compared to other mammals, eyes are located at the front of the head, and the brain is large and highly developed. Having two eyes with overlapping vision and an enlarged visual region of the brain gives primates depth perception, which also enhances maneuverability through the forest canopy (figure 12.17). Generally, primates produce only one or two offspring per pregnancy, and the length of time to maturation is great. This increases the opportunity for mastering complex learned behaviors.

The primate order is divided into two suborders: Prosimii and Anthropoidea. The prosimians include tarsiers and lemurs, which have a foxlike appearance with little resemblance to the more familiar primates (figure 12.18). They do, though, have the primate traits of flattened nails and thumbs and big toes that move independently of the other digits. Anthropoids are thought to have evolved from prosimians. The anthropoids include Old World monkeys, New World monkeys, and hominoids. The hominoid group includes humans and apes (gibbons, orangutans, gorillas, and chimpanzees) (figure 12.19). For years the scientific community has debated which apes are more closely related to humans. DNA evidence currently points to chimpanzees as humans' closest living relatives. It is thought that the chimpanzees and humans shared a common ancestor 6 million years ago.

Evolution of Modern Humans

An obvious distinction between humans and other primates is that humans are *bipedal,* and thus walk upright on two legs. Although our species does not inhabit tree canopies, many of our most valued human traits originated in our tree-dwelling ancestors and have evolved according to the demands of our more recent ancestors' environments. The fossil record strongly points to Africa as the place of origin of our first humanlike ancestors. These organisms originated at a time of climatic change in Africa. Forest habitat was declining and land became comparatively more favorable for survival. Hence, the ability to survive on the ground and bipedal locomotion evolved.

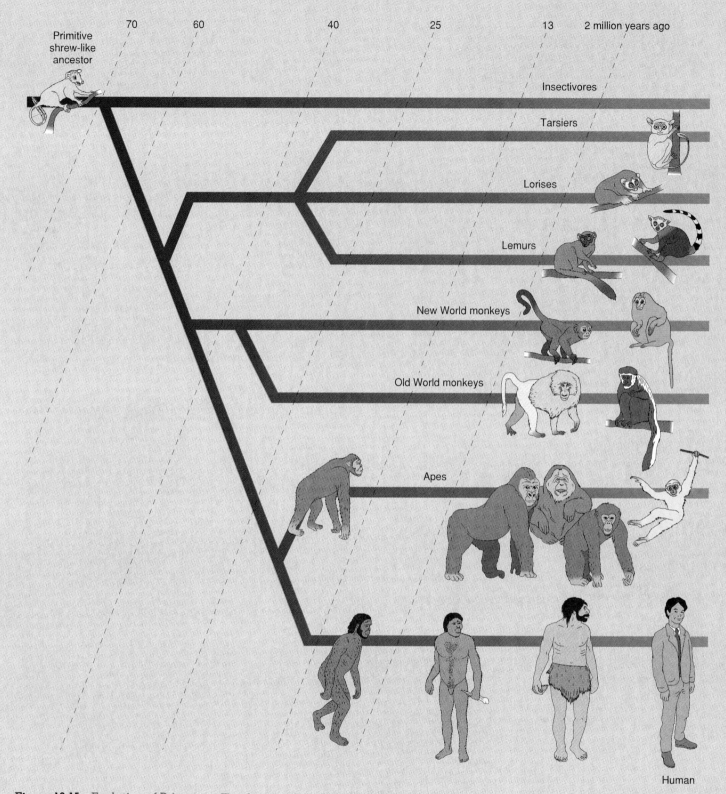

70   60   40   25   13   2 million years ago

Primitive
shrew-like
ancestor

Insectivores

Tarsiers

Lorises

Lemurs

New World monkeys

Old World monkeys

Apes

Human

Figure 12.15 Evolution of Primates   This diagram shows how the major
branches of primate evolution are thought to have developed.

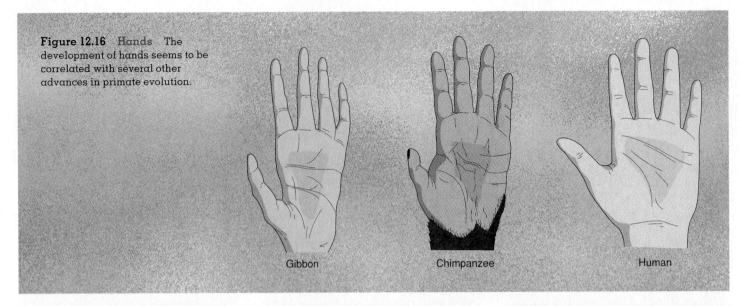

Figure 12.16 **Hands** The development of hands seems to be correlated with several other advances in primate evolution.

Gibbon

Chimpanzee

Human

Figure 12.18 **Prosimians** Prosimians are considered to be most similar to the ancestors of primates.

Figure 12.17 **Binocular Vision** Most primates have both eyes on the front of the face. This allows for good depth perception.

Figure 12.19 Hominoids: (a) Gorilla,
(b) Gibbon, (c) Orangutan, (d) Chimpanzee,
(e) Human. These are representatives of the
hominoid group of primates.

The oldest hominid fossils date back to 4.4 million years ago and belong to the species *Australopithecus ramidus.* The fossil remains of these individuals who lived in what is today Ethiopia are scant. However, evidence suggests that these fossils lie closer than any others yet found to the split between the chimpanzee and human lineages. Much more fossil evidence remains of the species *Australopithecus afarensis,* which lived 4 million years ago and most likely descended from *A. ramidus* (figure 12.20). These organisms had brains somewhat larger than a chimpanzee. Their faces were apelike with large jaws and less pronounced foreheads and chins. The appendages of these organisms would have been suitable for tree climbing, although they stood fully erect. *Australopithecus afarensis* was the ancestor of other australopithecus species as well as the ancestor of the first known species of our genus, *Homo habilis.*

Homo habilis lived 2 million years ago and was more similar to modern humans than australopithecus species. These organisms had a larger body size and brain capacity, their pelvis and leg bones were more humanlike, and they made simple tools. *Homo habilis* lived alongside *Australopithecus.* The *Australopithecus* lineage died out, but *H. habilis* evolved into an even more humanlike species, *Homo erectus,* approximately 1.5 million years ago (figure 12.21). *Homo erectus* surpassed *H. habilis* in body size and brain capacity. This species used fire as well as tools. Their tribes had social structure and they were probably group hunters of large game. *Homo erectus* was highly successful and migrated from Africa into Europe and Asia one million years ago.

Modern humans, *Homo sapiens,* are believed to have evolved from *H. erectus,* although the specifics of this origin are disputed. The **out of Africa hypothesis** proposes that modern humans evolved in Africa from *H. erectus* and then, like their predecessors, migrated throughout Europe and Asia, where they out-competed the more primitive species. A second hypothesis, known as the **multi-regional continuity hypothesis,** suggests that *H. sapiens* evolved throughout

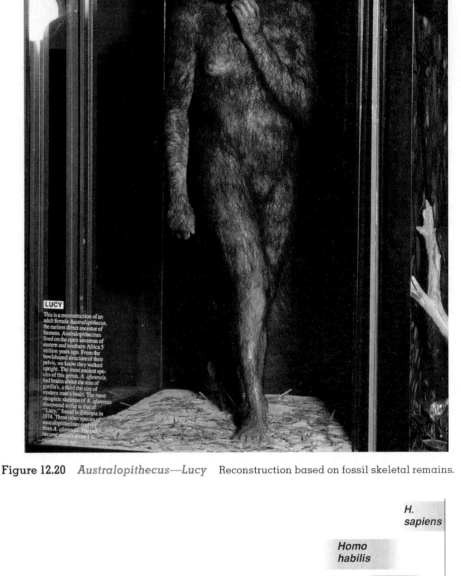

Figure 12.20 *Australopithecus—Lucy* Reconstruction based on fossil skeletal remains.

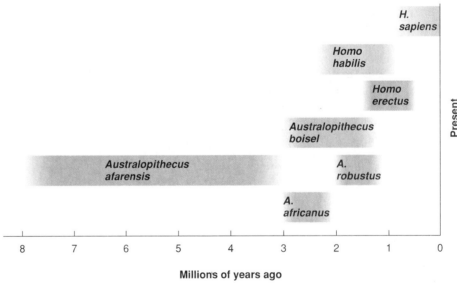

Millions of years ago

Figure 12.21 **Human Evolution** Currently, of several different humanlike organisms from the past, only *Homo sapiens* remains. Notice that between 1 million and 2 million years ago several humanlike organisms coexisted.

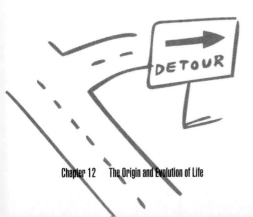

the Old World from the existing populations. Breeding between existing populations allowed the human species to evolve in concert throughout the Old World, while geographic barriers worked to block gene exchange. This resulted in the distinct traits of the human races we see today.

SUMMARY

The current theory of the origin of life proposes that primitive Earth's environment led to the spontaneous formation of organic molecules that became organized into primitive anaerobic heterotrophic cells. These basic units of life led to the development of autotrophic cells and eventually aerobic heterotrophic cells. Eukaryotic cells may have arisen from associations between prokaryotic cells. Since the emergence of life on Earth, life-forms have been constantly changing and different life-forms have dominated at different times. Life as we know it has only existed on earth a very short time, as has our species itself. It is believed that the predecessors of modern humans evolved in Africa. The out of Africa hypothesis proposes that *H. sapiens* evolved from other hominids in Africa and migrated to other parts of the world. The multi-regional continuity hypothesis proposes that *H. sapiens* evolved in several locations throughout the Old World.

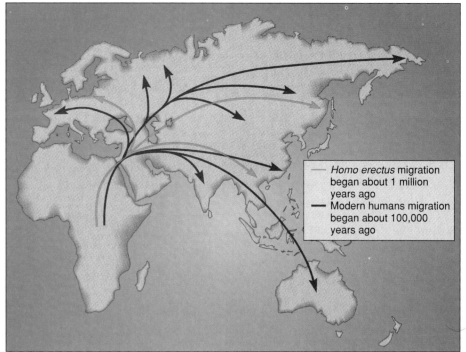

(a) Out of Africa hypothesis

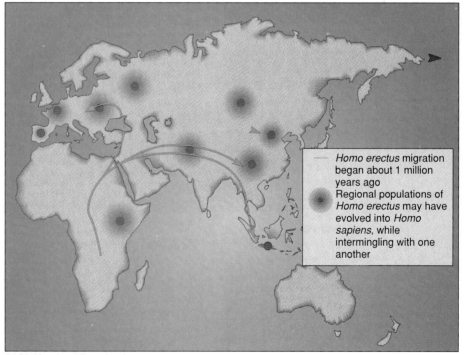

(b) Multiregional hypothesis

How Did Early Humans Migrate?

CHAPTER GLOSSARY

carbohydrate (kar-bo-hi′drāt) One class of organic molecules composed of carbon, hydrogen, and oxygen. The basic building block of a carbohydrate is a simple sugar.

continental drift (kan″ten-en′tal drift) The movement of the Earth's crustal plates.

endosymbiotic theory (en″do-sim-be-ot′ik the′o-re) A theory suggesting that present-day eukaryotic cells evolved from the combining of several different types of primitive prokaryotic cells.

geological time chart (je-a-la′jik-el tīm chart) A chart of the chronological history of living organisms based on the fossil record.

lipids (lī′pids) Large organic molecules that do not easily dissolve in water; classes include fats, phospholipids, and steroids.

mass extinction (mas ik-stink′shen) The disappearance of thousands of species of organisms during a short period of time. Mass extinctions are thought to be brought about by dramatic changes in either the physical or living components of the environment.

multi-regional continuity hypothesis (mul′ti-re′jen-el kan″ten-oo′e-te hi-poth′e-sis) The idea that *Homo sapiens* evolved throughout the Old World from *Homo erectus* populations that had previously migrated from Africa to Europe and Asia.

nucleic acids (nu-kle′ik ă′sids) Complex molecules that store and transfer information within a cell. They are constructed of fundamental units known as nucleotides.

out of Africa hypothesis (out uv a′fri-ka hi-poth′e-sis) The idea that *Homo sapiens* evolved in Africa from *Homo erectus* and then migrated throughout Europe and Asia, where they out-competed *Homo erectus* populations that had previously migrated from Africa to Europe and Asia.

protein (pro′tēn) Macromolecules made up of amino acid subunits.

solar nebula theory (so′ler ne′byu-la the′o-re) A theory proposing that the solar system was formed from a large cloud of gases that developed 10 billion to 20 billion years ago.

CONCEPT MAP TERMINOLOGY

Using the information in the text, construct a family tree that includes the following:

Anthropoidea

chimpanzees

gibbons

gorillas

Hominoids

humans

lemurs

New World monkeys

Old World monkeys

orangutans

primate order

Prosimii

LABEL • DIAGRAM • EXPLAIN

Choose four physical traits that are characteristic of primates. Compare these four traits to the corresponding traits of a dog. Explain how the differences make a primate adapted to mobility in trees while a dog is adapted to mobility on land.

How are the out of Africa and multi-regional continuity hypotheses similar? How do they differ?

Multiple Choice Questions

1. Which one of the following was most likely absent from the Earth's early atmosphere?
   a. ammonia, NH_3
   b. water vapor, H_2O
   c. carbon dioxide, CO_2
   d. oxygen, O_2

2. Which one of the following is an organic molecule?
   a. H_2O
   b. H_2SO_4
   c. NaCl
   d. $C_3H_6O_3$

3. The major categories of organic molecules are:
   a. acids, bases, and salts
   b. fats, oils, and steroids
   c. nutrients, vitamins, and water
   d. carbohydrates, lipids, proteins, and nucleic acids

4. Stanley Miller's experiment demonstrated how:
   a. simple organic molecules can be produced in the absence of living organisms
   b. complex organic molecules can be produced in the absence of living organisms
   c. living organisms produce organic molecules
   d. the first life forms developed

5. The probable sequence of the evolution of cells was:
   a. autotroph → anaerobic heterotroph → aerobic heterotroph
   b. anaerobic heterotroph → aerobic heterotroph → autotroph
   c. anaerobic heterotroph → autotroph → aerobic heterotroph
   d. aerobic heterotroph → autotroph → anaerobic heterotroph

6. An explanation of the origin of the first eukaryotic cells is presented by:
   a. Oparin and Haldane's hypothesis
   b. Stanley Miller's experiment
   c. the endosymbiotic theory
   d. the out of Africa hypothesis

7. The Age of Bacteria was the:
   a. Precambrian Era
   b. Paleozoic Era
   c. Mesozoic Era
   d. Cenozoic Era

8. Birds are most closely related to:
   a. flying fish
   b. insects
   c. bats
   d. reptiles

9. Humans' closest living relatives according to DNA evidence are:
   a. orangutans
   b. monkeys
   c. gorillas
   d. chimpanzees

10. The idea that *Homo erectus* migrated from Africa to Europe and Asia and then evolved throughout the Old World into modern humans is known as the:
    a. out of Africa hypothesis
    b. multi-regional continuity hypothesis
    c. endosymbiotic theory
    d. solar nebula theory

Questions with Short Answers

1. What was Oparin and Haldane's hypothesis?

2. What did each of the following represent in Stanley Miller's experiment?
   sparks:

   distilled water:

hydrogen, methane, and ammonia:

heating and cooling the experimental apparatus:

3. How might the organic molecules of the early oceans have become concentrated?

4. What did the first cells use as a source of energy?

5. How did the molecular oxygen in the atmosphere develop?

6. What is the endosymbiotic theory?

Cell Division and Heredity

All living things reproduce. They either reproduce by dividing one organism into two or they are involved in the more complicated process of sexual reproduction. In fact, most organisms can reproduce in both ways. Both kinds of reproduction involve the division of cells. Chapter 13 deals with one of these processes—mitosis—and chapter 14 deals with the other—meiosis. The genes contained within the cells are in the form of DNA molecules. These DNA molecules are passed from generation to generation during reproduction. Chapter 15 describes many human characteristics and how they are inherited, while chapter 16 introduces the structure of DNA and how it stores and transmits information.

Genetic diversity in tulip field.

part

4

Cell Division and Heredity

▶ **How does the body heal wounds?**

▶ **What are genes and how do they control organisms?**

▶ **How is the production of eggs and sperm different from the division of skin cells?**

▶ **Where are genes located?**

Cell Division: Making More of the Same

Cell division (mitosis) in root cells of broad bean.

Cell Division: Making More of the Same

learning objectives

- List the purposes of cell division.
- Define differentiation.
- Explain how cancer is caused and treated.
- Explain the cell cycle.
- State the processes that occur during interphase.
- Name the stages of mitosis and explain what is happening during each stage.

The Importance of Cell Division

In an earlier chapter we learned that cells are the basic unit of living things. Eukaryotic cells have a nucleus, a nuclear membrane, and organelles. You began life as a single cell formed from the union of a sperm and an egg. That single cell divided into two cells, they divided into four, which divided into eight, and so on and so on. Your cells continued to divide until now, as an adult, your body is made of several trillion cells.

The process by which one cell divides to make two is called cell division. The cells created from the first cell are called the **daughter cells.** Cell division enables organisms to grow, replace dead cells with new ones, and repair damaged tissue.

During cell division, two events occur. First, the genetic information of a cell, contained in the DNA in its nucleus, is equally distributed to two daughter nuclei in a process called **mitosis.** We will learn more about mitosis later in the chapter. During the second event, the cytoplasm of the original cell separates into two new cells. This process is called

cytokinesis, or cell splitting. Each new cell gets one of the two daughter nuclei, which means that each new cell has a complete set of the original cell's genetic information. Most of the time mitosis and cytokinesis occur without a hitch, but sometimes there are problems, as we will see later in the chapter.

In growing organisms, such as infants, seedlings, or embryos, most cells are capable of many divisions and divide at a rapid rate.

In older organisms, many cells, as a result of differentiation, lose their ability to divide, and the frequency of cell division throughout the organism decreases. As the organism ages, the lower frequency of cell division and the natural loss of DNA from chromosomes may affect many bodily processes, including healing. Older people may have so few cells capable of dividing that a broken bone may never heal.

CONCEPT CONNECTIONS

mutation Any change in the genetic information of a cell; it may result in the cell showing characteristics other than those of the parent cell.

FYI

When people exercise to build their muscles, they do not increase the number of muscle cells by stimulating cell division. The increase in muscle mass is due to an increase in the number of contractile protein fibers in each cell.

Muscle Development

Regulating Cell Division

Understanding mitosis can help you understand certain biological problems and how to solve them. Cells divide at different rates, and each kind of cell has its own regulated division rhythm. Regulation of the cycle can come from inside or outside the cell. When human white blood cells are grown outside the body under special conditions, they develop a regular cell-division cycle. The cycle is determined by the DNA of the cells. However, white blood cells in the human body may increase their rate of mitosis as a result of outside influences. Disease organisms entering the body (i.e., bacteria and viruses), tissue damage, and changes in cell DNA all may alter the rate at which white blood cells divide. An increase in white blood cells in response to the invasion by disease-causing organisms is valuable because these white blood cells are capable of destroying the organisms.

Many kinds of cells divide regularly to replace cells that are lost. Simple acts like showering, dressing, and undressing result in thousands of cells being scraped off the body's surface. Likewise, swallowed food rubs away the cells lining the surface of the mouth, esophagus, stomach, and intestines. As they are lost, they must be replaced with new cells. Altogether, you lose about one million cells per second; this means that the replacement process of cell division is beginning a million times in your body at any given moment.

When damage occurs to the skin or other tissues, the rate of cell division increases, resulting in repair of the damaged region. If skin cells are destroyed by a cut or abrasion, an increased rate of cell division produces new cells to repair the damage and heal the wound. When a bone is broken, the break heals because cells in the region of the break divide, increasing the number of cells available to knit the broken pieces together.

White Blood Cells Engulf and Destroy Bacteria

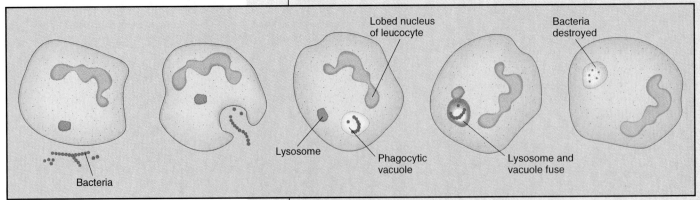

BIO *feature*

Cells Become Specialized

It may surprise you to know that every cell in your body has the same genetic composition. At conception, you received a set of genes from your father in his sperm, and a set of genes from your mother in her egg. This single cell that resulted contained your genetic makeup. As your cells divided again and again, all of them received the same genetic information contained in that very first cell.

This is not to say that all the cells in your body are the same. You have nerve cells, muscle cells, bone cells, skin cells, and many other types, all of which look very different. Each type is specialized for a certain function. How can cells with the same genes be so different? Think of the genes in your cells as recipes in a cookbook. One hundred people can cook from the same book, but they can each prepare a different dish. If you use the recipe for chocolate cake, you ignore the directions for making tuna salad or fried chicken.

Your cells do something similar. Certain cells use, or trigger, only certain genes, and ignore the rest. Muscle cells, for example, use the genes that instruct them to produce proteins that contract. Pancreas cells use the genes that instruct them to form digestive enzymes. This process of forming specialized cells in a multicellular organism is called **differentiation.**

In an embryo, all cells continually divide. Later in life, some cells, such as muscle and nerve cells, lose the ability to divide. They remain permanently in a nondividing condition. Other cells, such as skin cells and cells that line the digestive tract, continue dividing throughout the life of the organism.

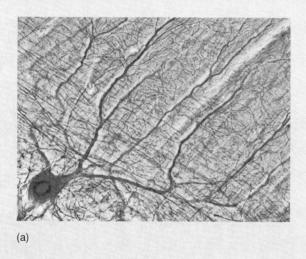

(a)

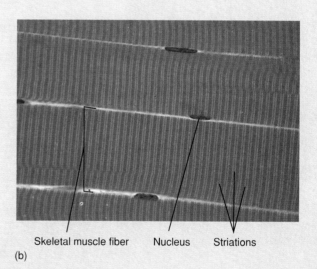

Skeletal muscle fiber   Nucleus   Striations

(b)

Cells and Tissues   Groups of similar cells that function together are known as tissues. The cells within tissues have special structures and abilities. (a) Nerve cells convey messages, (b) muscle cells contract, (c) bone cells secrete mineral material, and (d) skin cells form a protective layer.

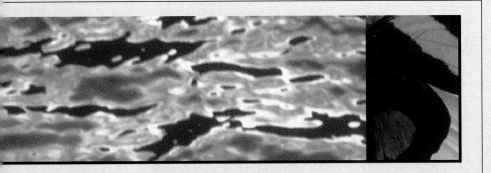

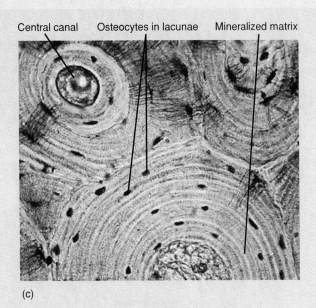

Central canal   Osteocytes in lacunae   Mineralized matrix

(c)

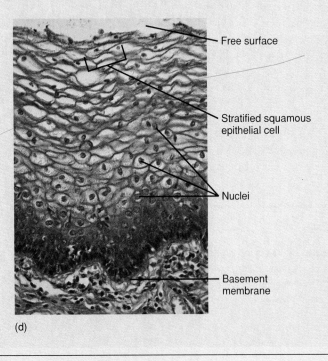

Free surface

Stratified squamous
epithelial cell

Nuclei

Basement
membrane

(d)

Cell Division and Cancer

As we have seen, cells become specialized for a particular function. Each cell type has its cell-division process regulated so that it does not interfere with the activities of other cells or the whole organism. Some cells, however, may go back to an embryonic state and begin to divide as if they were "newborn" or undifferentiated cells. Sometimes this division occurs in an uncontrolled fashion.

When such uncontrolled division occurs, a group of cells forms what is known as a tumor (figure 13.1). A *benign* tumor is a cell mass that does not fragment and spread beyond its original area of growth. A benign tumor can become harmful by growing large enough to interfere with normal body functions. Some tumors are *malignant.* Cells of these tumors move from the original site (metastasize) and establish new colonies in other regions of the body. **Cancer** is an abnormal growth of cells that has a malignant potential.

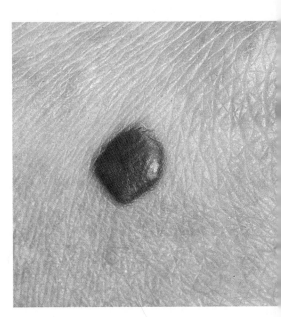

Figure 13.1 Skin Cancer Malignant melanoma is a type of skin cancer. It forms because some pigmented skin cells divide repeatedly, giving rise to an abnormal mass of pigmented skin cells. This kind of cancer is particularly dangerous because the cells break off and spread to other parts of the body (metastasize).

Some Known Causes of Cancer:

Radiation:
   X- rays and Gamma Rays
   Ultraviolet light (UV-B, the
      cause of sunburn)

Carcinogenic chemicals:
   tobacco
   nickel
   arsenic
   benzene
   dioxin
   asbestos
   uranium
   tar
   cadmium
   chromium
   polyvinyl chloride (PVC)

Diet:
   alcohol
   smoked meats and fish
   food containing nitrates
      (e.g., bacon)

Viruses:
   Hepatitis B virus (HBV) and
      liver cancer
   Herpes simplex (HSV) type II
      and uterine cancer
   Epstein-Barr virus and
      Burkitt's lymphoma
   Human T-cell lymphotropic
      virus (HTLV-1) and
      lymphomas and leukemias

Hormonal imbalances:
   diethylstilbestrol (DES)
   oral contraceptives

**Types of genetic and familial
cancers:**
   chronic myelogenous leukemia
   acute leukemias
   retinoblastomas
   certain skin cancers
   breast
   endometrial
   colorectal
   stomach
   prostate
   lung

CONCEPT CONNECTIONS

cancer The unchecked reproduction of cells by cell division; the cells resemble immature or undifferentiated cells.

tumor A mass of undifferentiated cells not normally found in a certain portion of the body.

malignant tumor A nonencapsulated growth of tumor cells that is harmful; they may spread or invade other parts of the body.

benign tumor An encapsulated growth of tumor cells that tend to reproduce slowly and seldom cause harm if they are removed before they compress vital organs, interfering with their function.

metastasis The spread of malignant cells throughout the body that causes the formation of other tumors.

Once cancer has been detected, the tumor must be eliminated. There are three common forms of treatment. If the cancer is confined to a few specific locations, it may be possible to surgically remove it. Many cancers of the skin or breast are dealt with in this manner. However, in some cases surgery is impractical. If the tumor is located where it can't be removed without destroying healthy tissue, surgery may not be used. For example,

removing certain brain cancers can severely damage the brain. In such cases, two other methods may be used to treat cancer: chemotherapy and radiation.

Chemotherapy uses various types of chemicals to destroy mitotically dividing cancer cells. This treatment may be used even when physicians do not know exactly where the cancer cells are located. However, it has negative effects on normal cells. It lowers the body's immune reaction because it decreases the body's ability to reproduce new white blood cells by mitosis. Chemotherapy interferes with the body's normal defense mechanisms. Therefore, cancer patients undergoing chemotherapy must be given antibiotics. The antibiotics help them defend against dangerous bacteria that might invade their body. Other side effects include intestinal disorders and loss of hair, which are caused by damage to healthy cells in the intestinal tract and the skin that divide by mitosis.

The Cell Cycle

All cells go through a basic life cycle that involves a period of division alternating with a period of growth, but they vary in the amount

CONCEPT CONNECTIONS

chemotherapy The use of chemicals to help treat a disease; chemicals used include anticancer drugs, antibiotics, and hormones.

FYI

How Cancer Cells Spread **Cancer cells break off from the original tumor and enter the bloodstream. When they get stuck to the inside of the smallest blood vessels, called capillaries, they move through the walls of the blood vessels and invade the tissue. There they begin to reproduce by mitosis. This tumor causes new blood vessels to grow into the cancerous site; these new vessels will carry nutrients to the growing mass. They can also bring other spreading cancer cells to the new tumor site.**

BIO*feature*

Cancer's Seven Warning Signals

1. A change in bowel or bladder habits
2. A sore that does not heal
3. Unusual bleeding or discharge
4. A thickening or lump in the breast or elsewhere
5. Indigestion or difficulty in swallowing
6. An obvious change in a wart or mole
7. A nagging cough or hoarseness

Thirty percent of Americans will eventually have cancer. Cancer will strike three out of four families. If you notice one of the preceding symptoms, see your doctor.

To become better informed about cancer, contact your local Cancer Society office or write the American Cancer Society, Inc., 90 Park Ave., New York, NY 10016. Request a copy of the latest publication, *Cancer Facts and Figures.*

BIO*feature*

Radiation Therapy

Radiation therapy uses large amounts of X rays or gamma rays. Since this treatment damages surrounding healthy cells, it is used only very cautiously when surgery is impractical. Radiation therapy can be effective because cancer cells are undifferentiated and undergoing mitosis on a regular basis in comparison to normal, differentiated cells.

Nondividing cells pause with their genetic material uncoiled and randomly arranged in their nuclei making a thin,

difficult target to hit with radiation. During division, the genetic material of cancer cells and other dividing cells becomes tightly coiled, presenting a thicker target than the DNA of nondividing, differentiated cells. Radiation therapy must be used very discreetly since the same rays that can kill cancer cells can create mutations in normal cells, causing them to become cancer cells.

Chemotherapy and radiation treatment are often used to control leukemia, a type of cancer in which white blood cells divide

in an uncontrolled manner. These treatments take advantage of the fact that leukemia cells undergo more mitotic divisions than normal, surrounding cells. Since leukemia cells spend more time dividing than normal bone marrow cells do, they have a greater chance of being killed by radiation. Physicians, therefore, often prescribe cobalt therapy for certain cancer patients with widely dispersed cancers such as leukemia. Cobalt is a radioactive element that releases gamma radiation.

Using Total Body Radiation to Control Leukemia

An uncontrolled increase in the rate of division in white blood cells causes a kind of cancer known as leukemia. This condition causes a general weakening of the body, because the excess number of white blood cells diverts necessary nutrients from other cells of the body and interferes with their normal activities.

Patients with leukemia have cancer of blood-forming cells located in their bone marrow, but not all of these blood-forming cells are cancerous. It is possible to separate the cancerous from the noncancerous bone marrow cells. A radiation therapy method prescribed for some patients involves the removal of some of their bone marrow and isolation of the noncancerous cells for laboratory growth. After these healthy cells have been cultured and increased in number, the patient's whole body is exposed to high doses of radiation sufficient to kill all the cancerous cells remaining in the bone marrow. Since this treatment is potentially deadly, the patient is isolated from all harmful substances and infectious microbes. They are fed sterile food, drink sterile water, and breathe sterile air while being closely monitored and treated with antibiotics. The lab-cultured, noncancerous cells are injected back into the patient. As if they had a memory, they migrate back to their origin in the bone marrow, establish residence, and begin cell division all over again.

Radiation Sickness at Chernobyl

As a treatment for cancer, radiation is dangerous for the same reasons that it is beneficial. In cases of extreme exposure to radiation, people develop what is called *radiation sickness*. This is what occurred to many people living near the Chernobyl nuclear power plant in Ukraine (formerly part of the Soviet Union) when on April 26, 1986, it malfunctioned and released harmful radiation into the environment.

The symptoms of radiation sickness include loss of hair, bloody vomiting and diarrhea, and a reduced white blood cell count. These symptoms occur because rapidly dividing cells in parts of the body where mitosis is common are killed. The lining of the intestine, for example, is constantly being lost as food travels through and it must be replaced by the process of mitosis. Hair growth is the result of the continuous division of cells at the roots. White blood cells are also continuously reproduced in the bone marrow and lymph nodes. When radiation strikes these rapidly dividing cells and kills them, the lining of the intestine wears away and bleeds, hair falls out, and few new white blood cells are produced to defend the body against infection.

of time they spend in each stage. A diagram of a cell's life cycle may help you to understand it (figure 13.2). Once begun, cell division is a continuous process without a beginning or an end. It is a cycle in which cells continue to grow and divide. There are two stages to the life cycle of a eukaryotic cell: (1) the nondividing interphase during which DNA is copied and, (2) the dividing stage in which mitosis and cytokinesis occur.

Interphase is the stage between cell divisions. During interphase, the cell grows in volume as it produces ribosomes, enzymes, and other cell components. The cell is engaged in the metabolic activities typical of that cell. For example, muscle cells contract, nerve cells conduct impulses, and glandular cells manufacture and secrete their products. In addition, DNA replication occurs in preparation for the distribution of genes to daughter cells. Final preparations are also made for mitosis with the synthesis of special proteins that will help guide the chromosomes during cell division.

During interphase, the nuclear membrane is intact and the individual chromosomes are not visible (figure 13.3). The individual strands are too thin and tangled to be seen. **Chromosomes** are composed of proteins and DNA. The DNA is the molecule that contains the cell's genetic information. These parent chromosomes replicate during interphase, producing identical copies called **chromatids.** There are two chromatids for each chromosome. It is these chromatids that will be distributed to the next generation of daughter cells during mitosis.

The Stages of Mitosis

All stages in the life cycle of a cell are continuous; there is no exact point where one stage begins and another ends. However, to more easily understand the process of cell division, scientists have subdivided the process into four typical, easily identifiable stages. These stages are like looking at a series of only four game highlight photographs used to explain the entire game of baseball. You know that these photos do not show the entire game, but only illustrate the most typical events that go on during the game. A lot of other things happen in between these pictures. The four pictures of mitosis are named prophase, metaphase, anaphase, and telophase.

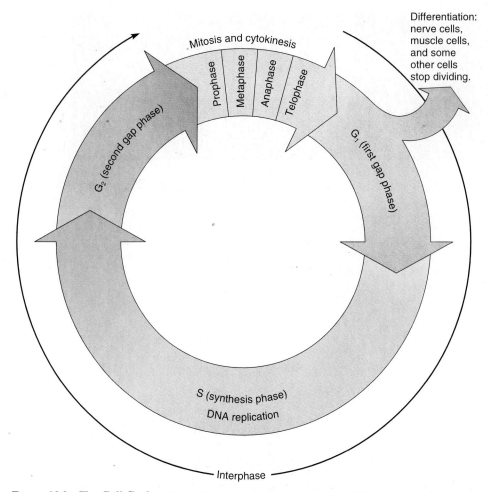

Figure 13.2 **The Cell Cycle** During the cell cycle, there is a period of time known as interphase during which many kinds of cells double their DNA and make other preparations for division. During mitosis, two nuclei are formed from the original, and the cell contents (cytoplasm) are divided in two (cytokinesis), with each of the new cells having a nucleus. Once muscles and some organs, such as the brain, have reached adult size, their cells do not divide. They are permanently in interphase. The time periods shown vary depending on the type of cell and the age of the organism.

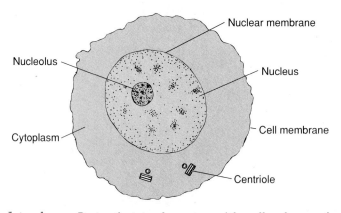

Figure 13.3 **Interphase** During the interphase stage of the cell cycle, growth and the production of necessary organic compounds occurs. If the cell is going to divide, DNA replication also occurs during interphase. The individual chromosomes that carry the DNA are not visible, but a distinct nucleus surrounded by nuclear membrane is present. One or more bodies known as nucleoli (singular: nucleolus) are usually present within the nucleus.

BIO feature

Some of What We Know So Far

For years geneticists tried to deduce the location of genes on chromosomes. This was difficult to do with humans because of the large number of chromosomes in human DNA and the ethics of doing planned matings in humans. Modern techniques have enabled researchers to locate several genes on specific chromosomes. Several examples are given here.

Chromosome Number	Genetic Trait
1	Factor V blood clotting factor for venous thrombosis
2	Hereditary nonpolyposis colorectal cancer
4	Huntington's disease (HD)
5	Cytokines, immune system chemicals
7	Cystic fibrosis
10	Phenylketonuria (PKU)
10	Hirschsprung's disease, a digestive disorder
15	Angelman and Prader-Willi syndromes
17	Susceptibility to psoriasis
19	Gene associated with Alzheimer's disease
21	Gene associated with Alzheimer's disease
21	Lou Gehrig's disease/amyotrophic lateral sclerosis (ALS)
X	Duchenne muscular dystrophy
Y	Gonadal dysgenesis

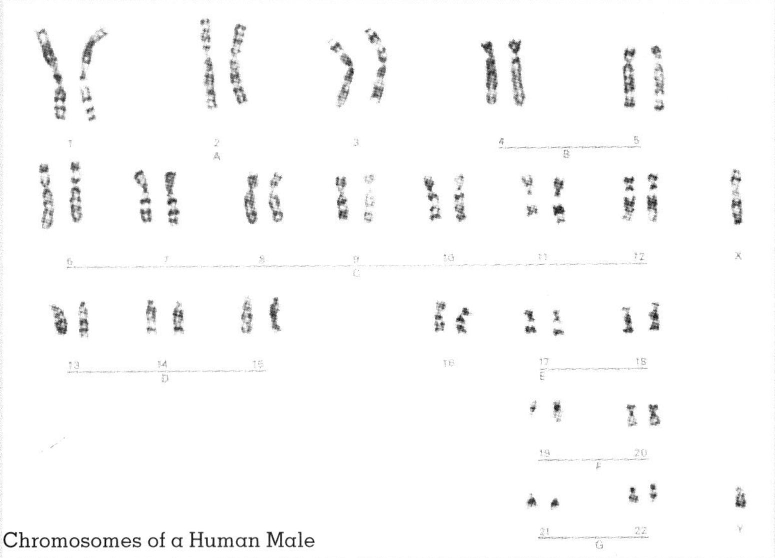

Chromosomes of a Human Male

Following the Traits

In the diagrams in this text, a few genes are shown as they might occur on human chromosomes. The diagrams show fewer chromosomes and fewer genes on each chromosome than are actually present in human cells. Normal human cells have 10 billion nucleotides arranged into 46 chromosomes, each chromosome with thousands of genes. In this book, smaller numbers of genes and chromosomes are used to make it easier to follow the events that happen in cell division.

Prophase

Prophase is the first stage of mitosis. One of the first noticeable changes is that the individual chromosomes become visible (figure 13.4). The thin, tangled DNA and proteins present during interphase gradually coil and thicken, becoming visible as separate chromosomes. The DNA portion of the chromosomes carries genes that are arranged in a specific order. Each chromosome carries its own set of genes that is different from the sets of genes on other chromosomes.

As prophase proceeds, and as the chromosomes become more visible, we recognize that each chromosome is made of two thickened, parallel parts lying side by side. Each parallel part is called a chromatid (figure 13.5). These chromatids were formed during interphase, when DNA was copied. The two identical chromatids are attached at a place on the DNA called the **centromere.**

Several other events occur as the cell proceeds to the late prophase stage (figure 13.6). One of these events is the duplication of the centrioles. Remember that some cells contain centrioles, which are hollow protein tubes located just outside the nucleus. As they duplicate, they move to the poles of the cells. As the centrioles move to the poles, a *spindle* is formed. The **spindle** is an array of pipelike, protein molecules extending from pole to pole that is used in the movement of chromosomes.

As prophase is occurring, the nuclear membrane gradually falls apart. It is present at the beginning of prophase but it disappears by the time this stage is completed. The nucleoli

Figure 13.4 Early Prophase At the beginning of the prophase stage, chromosomes begin to appear as thin tangled threads, and the nucleolus and nuclear membrane are clearly visible. Structures known as centrioles separate and begin to move to opposite poles of the cell. As they separate, a series of fibers known as spindle fibers are formed between the two centrioles. The complete structure formed by the spindle fibers is called a spindle.

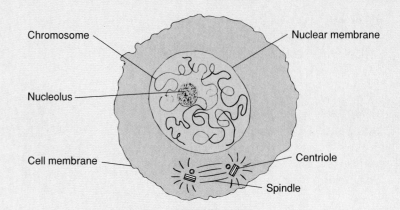

Figure 13.5 Chromosome Structure Chromosomes are the structures in cells that are constructed of protein and DNA. DNA is the genetic material of the cell. During interphase, the amount of protein and DNA is doubled and the chromosome then consists of two separate structures called chromatids joined at a point called the centromere. Since the DNA contains the genetic data, different genes are located along the length of the chromosomes of a cell. (We are just beginning to determine on which chromosomes specific human genes are located. The examples presented here are for illustrative purposes only.)

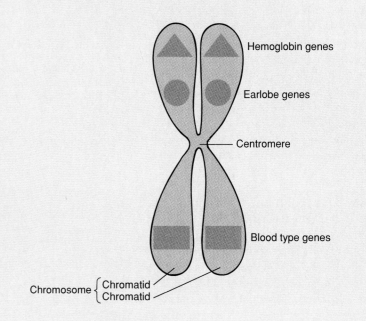

Figure 13.6 Late Prophase By late prophase, the chromosomes are visible as two chromatids connected at a centromere. The nuclear membrane has disintegrated, and the nucleolus is no longer visible. In addition, the centrioles have moved farther apart, producing the arrangement of fibers known as the spindle.

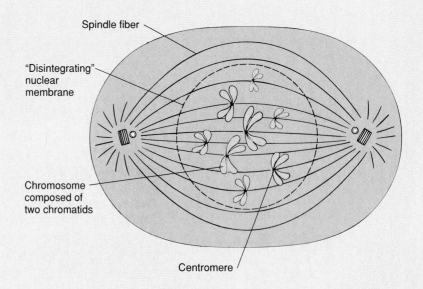

within the nucleus also disappear. As the nuclear membrane falls apart, the chromosomes are free to move anywhere within the cytoplasm of the cell. As this movement occurs, the cell enters the next stage of mitosis.

Metaphase

During **metaphase,** there is no nuclear membrane to separate the chromosomes from the cytoplasm, and the spindle, which started to form during prophase, is completed. The centrioles are at the poles, and the spindle extends between them. At the beginning of metaphase, the chromosomes become attached to the spindle fibers at their centromeres.

As the cell enters metaphase, the chromosomes are distributed randomly throughout the cytoplasm. As metaphase proceeds, the chromosomes move until all their centromeres align themselves along the middle of the cell (figure 13.7). At this stage in mitosis, each chromosome still consists of two chromatids. In a human cell, there are 46 chromosomes, or 92 chromatids, during metaphase.

Anaphase

Anaphase is the third stage of mitosis. The nuclear membrane is still absent and the spindle extends from pole to pole. In anaphase, each chromosome splits at the centromere, and the two chromatids that make up the chromosome separate as they move along the spindle fibers toward the poles (figure 13.8). After this separation of chromatids occurs,

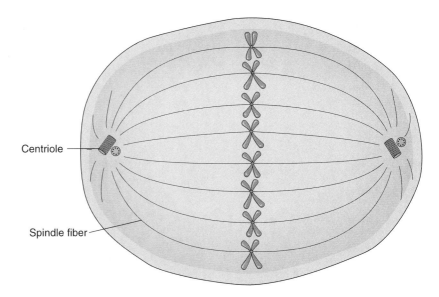

Figure 13.7 Metaphase During metaphase, the chromosomes, which are attached to spindle fibers by their centromeres, travel along the spindle and align at the equatorial plane. Notice that each chromosome still consists of two chromatids.

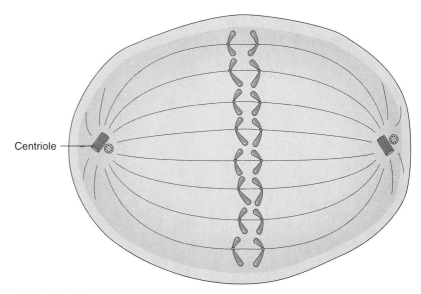

Figure 13.8 Anaphase During anaphase, the two chromatids of each chromosome separate as the centromeres divide. The chromatids move along the spindle fibers toward opposite poles. At this time the chromatids may be called daughter chromosomes.

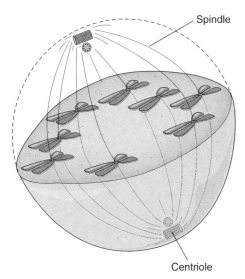

A Polar View of Metaphase A polar view shows that the chromosomes are actually spread out on the equatorial plane, not arranged in a line down the middle of the cell.

FYI

From a Different Angle If we view a cell in the metaphase stage from the side (figure 13.7), it is an equatorial view. In this view, the chromosomes appear as if they were in a line. If we view the cell from the pole, it is a polar view. The chromosomes are seen on the equatorial plane. Chromosomes viewed from this direction look like hot dogs scattered on a plate.

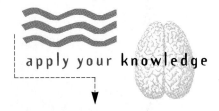

A chemical known as *colchicine* is extracted from the seeds of a small, crocuslike plant. This chemical is used in biological laboratories because it can prevent the formation of the spindle. Which parts of the cell cycle would proceed normally and which parts would be altered if this chemical were used on cells? If you know that the cells are not killed by colchicine and begin mitosis normally, what changes might occur in the number of chromosomes of the next cell generation, and how might this change the metabolism of the cell? Do you think colchicine could be used to control cancer? What could it do? What problems could result from its use as a chemotherapeutic agent?

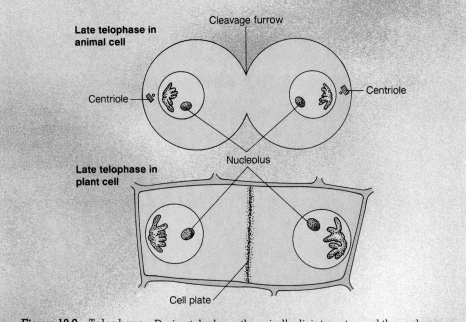

Figure 13.9 Telophase During telophase, the spindle disintegrates and the nuclear membrane re-forms around the chromosomes. Nucleoli reappear inside the nucleus. Two separate daughter cells are formed as a result of the division of the cytoplasm. This division, called cytokinesis, differs in plants and animals.

the chromatids are called *daughter chromosomes*. Daughter chromosomes contain identical genetic information.

Examine figure 13.8 closely and notice that the four daughter chromosomes moving to one pole have exactly the same genetic information as the four moving to the opposite pole. It is the alignment of the chromosomes in metaphase and their separation in anaphase that causes this distribution. At the end of anaphase, there are two identical groups of chromosomes, one group at each pole.

Telophase

Telophase is the last stage in mitosis. During telophase, daughter nuclei are formed. Each set of chromosomes becomes enclosed by a nuclear membrane and the nucleoli reappear. Now the cell has two identical daughter nuclei (figure 13.9). In addition, the spindle breaks up and disappears from view. With the formation of the daughter nuclei, mitosis, the first process in cell division, is completed, and the second process, cytokinesis, can occur. Cytokinesis splits the cytoplasm of the original cell and forms two smaller daughter cells that

have identical genetic information. Both daughter cells can grow, replicate their DNA, and enter another round of cell division to continue their cell cycle.

Chromosome Models

Models can help us understand complex biological events such as mitosis. You can create model chromosomes very easily by using various lengths of colored string, thread, or yarn to simulate four pairs of chromosomes in a cell. Each pair should be different from the other pairs in color, length, or both. Begin your modeling with each chromosome in its replicated form (i.e., two chromatids per chromosome). Attach the two chromatids with a loose twist. Manipulate your model chromosomes through the stages of mitosis. If you have performed the actions properly, you should end up with two daughter cells, each with one set of chromosomes.

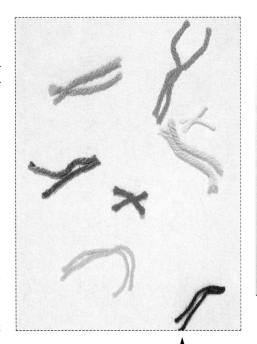

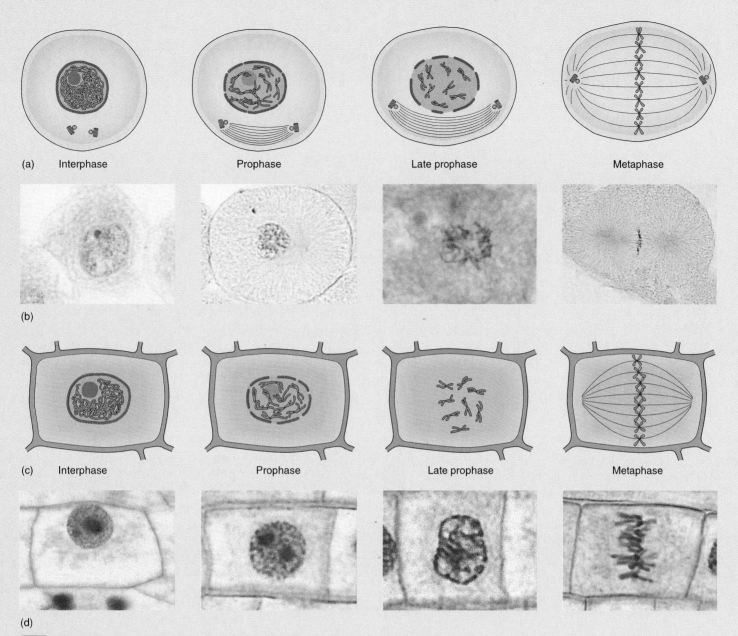

Figure 13.10 **A Comparison of Plant and Animal Mitosis** (a) Drawings of mitosis in an animal cell. (b) Photographs of mitosis in a whitefish blastula. (c) Drawings of mitosis in a plant cell. (d) Photographs of mitosis in an onion root tip.

Plant-Cell and Animal-Cell Differences

Cell division is slightly different in plant cells than in animal cells. One difference concerns the centrioles (figure 13.10). Centrioles are essential in animal cells, but they are not found in plant cells. However, by some process, plant cells do produce a spindle. There is also a difference in the way cytokinesis is carried out (figure 13.11). In animal cells, cytokinesis results from a cleavage furrow. This is an indentation or squeezing of the cell membrane of an animal cell that pinches the cytoplasm into two parts as if a string were tightened about its middle. Cytokinesis begins at the cell membrane and proceeds to the center. In plant cells, a cell plate is produced in between the newly forming daughter nuclei and grows to the surrounding cell membrane, resulting in a new piece of cell wall that separates the two daughter cells.

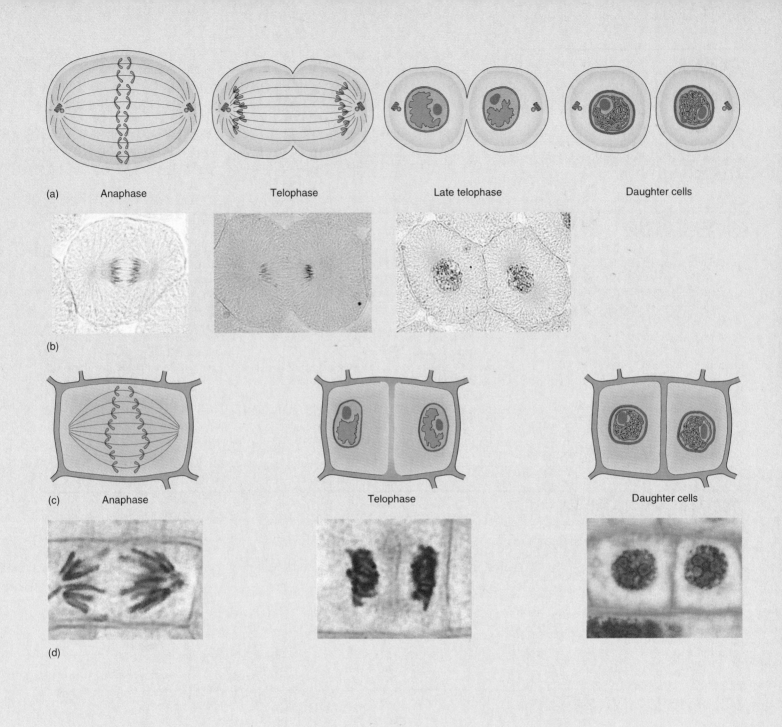

(a)   Anaphase   Telophase   Late telophase   Daughter cells

(b)

(c)   Anaphase   Telophase   Daughter cells

(d)

Figure 13.11   A Comparison of Cytokinesis in Plants and Animals   In animal cells there is a pinching in of the cytoplasm that eventually forms two daughter cells, each with a nucleus. Daughter cells in plants are formed when a cell plate forms between the two nuclei and separates the cytoplasm into two cells.

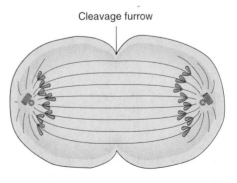

Cleavage furrow

Early telophase, animal cell

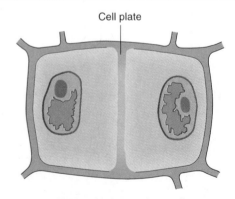

Cell plate

Early telophase, plant cell

SUMMARY

Cell division is necessary for growth, repair, and reproduction. The regulation of cell division is important if organisms are to remain healthy. Uncontrolled cell division may result in cancer and disruption of the total organism's well-being.

Cells go through a cell cycle that includes cell division (mitosis and cytokinesis) and interphase. Interphase is the period of growth and preparation for division. Mitosis is divided into four stages: prophase, metaphase, anaphase, and telophase. During mitosis, two daughter nuclei are formed from one parent nucleus. These nuclei have identical sets of chromosomes and genes that are exact copies of those of the parent. Although the process of mitosis has been presented as a series of phases, it is actually a continuous, flowing process from prophase through telophase. Following mitosis, cytokinesis divides the cytoplasm, and the cells return to interphase.

CHAPTER GLOSSARY

anaphase (an'a-faze) The third stage of mitosis, characterized by splitting of the centromeres and movement of the chromosomes to the poles.

cancer (kan'sur) A tumor that is malignant.

centromere (sen'tro-mere) The region where two chromatids are joined.

chromatid (kro'mah-tid) One of two identical parts of a chromosome attached at the centromere.

chromosomes (kro'mo"-somz) Complex structures within the nucleus composed of various kinds of proteins and DNA that contain the cell's genetic information.

cytokinesis (si-to-ki-ne'sis) Division of the cytoplasm of one cell into two new cells.

daughter cells (daw' tur sels) Two cells formed by cell division.

differentiation (dif"fur-ent-she-a'shun) The process of forming specialized cells within a multicellular organism.

interphase (in' tur-faze) The stage between cell divisions in which the cell is engaged in metabolic activities.

metaphase (me'tah-faze) The second stage in mitosis, characterized by alignment of the chromosomes at the equatorial plane.

mitosis (mi-to'sis) A process that results in equal and identical separation and distribution of chromosomes into two newly formed nuclei.

prophase (pro'faze) The first phase of mitosis during which individual chromosomes become visible.

spindle (spin'dul) An array of microtubules extending from pole to pole; used in the movement of chromosomes.

telophase (tel'uh-faze) The last phase in mitosis, characterized by the formation of daughter nuclei.

CONCEPT MAP TERMINOLOGY

Construct a concept map to represent the relationships among the following concepts.

cancer
centromere
chromatid
chromosome
cytokinesis
metaphase

mitosis
radiation
radiation sickness
spindle fiber
whole body radiation

LABEL•DIAGRAM•EXPLAIN

Draw and label a cell as it proceeds through mitosis.

Multiple Choice Questions

1. In mitosis, the centromeres split during:
   a. anaphase
   b. prophase
   c. metaphase
   d. interphase
2. The unchecked reproduction of cells by mitosis, which resembles immature or undifferentiated cells, is called:
   a. cancer
   b. metastasis
   c. differentiation
   d. cytokinesis
3. The normal state of chromosomes in prophase is as:
   a. daughter chromosomes
   b. chromosomes composed of two chromatids
   c. chromatids composed of two chromosomes
   d. chromosomes consisting of single chromatids
4. An uncontrolled increase in the rate of mitosis in white blood cells causes a kind of cancer known as:
   a. leukemia
   b. lymphoma
   c. retinoblastoma
   d. asthma
5. Cell cytoplasmic division occurs during the _____ stage of mitosis.
   a. anaphase
   b. telophase
   c. metaphase
   d. prophase
6. The symptoms of radiation sickness typically include loss of hair, bloody vomiting and diarrhea, and:
   a. an increased white blood cell count
   b. diabetes
   c. a reduced white blood cell count
   d. sickle-cell anemia
7. During which phase of the cell cycle does DNA replication occur?
   a. interphase
   b. anaphase
   c. metaphase
   d. fertilization

8. Nerve cells do not normally undergo mitosis. This means that:
   a. the brain is unimportant
   b. your brain cannot grow
   c. cytokinesis will be common in nerve cells
   d. brain cells do not synthesize proteins from DNA
9. Which of the following is *not* a treatment for cancer?
   a. radiation
   b. chemicals that interfere with mitosis
   c. antibiotics
   d. surgery
10. The process of forming specialized cells within a multicellular organism is known as:
   a. differentiation
   b. cytokinesis
   c. mutation
   d. metastasis

Questions with Short Answers

1. Name the four stages of mitosis and describe what occurs in each stage.

2. What is meant by cell cycle?

3. List the factors that cause a cell to become cancerous.

4. At what phase of mitosis does the DNA become most visible?

5. What are the differences between plant and animal mitosis?

6. Why can X-ray treatment be used to control cancer?

7. What is the purpose of mitosis?

8. Explain what is happening in a person with leukemia.

9. What types of activities occur during interphase?

10. What happens when a cell differentiates?

14

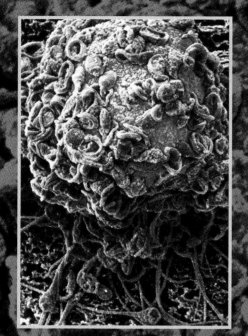

Meiosis: Sex-Cell Formation

Human sperm on egg (3,400 magnification).

Boys and Girls—Males and Females

Professional biologists and others use symbols to represent the males and females of a species. The symbol for the male (♂) represents the shield and spear of the Greek god Zeus, and the female symbol (♀) represents the vanity mirror of the goddess Aphrodite. While their origins were certainly sexist, the symbols have persisted in Western culture and are universally recognized. Biologists also use a variety of other terms to refer to males and females (e.g., *boys* and *girls,* + and − strains), all of which can become pretty confusing after a while. To keep things clear in your mind, just remember that the males are the individuals that donate or contribute genetic material to the females, which accept or receive the genetic material.

Sexual Reproduction

A quick glance at a family picture, and most people recognize that while both parents contribute genetic information to their children, all the children are not the same. The differences seen among the children are the result of events that occurred during the formation of egg and sperm cells. The organisms that show the greatest variety are those that have developed a method of reproduction that enables them to shuffle and exchange genetic information.

Adult organisms have two sets of genetic data, one inherited from each parent. They received these two sets of genetic data through a process called *sexual reproduction.* **Sexual reproduction** is the formation of a new individual as a result of the union of **sperm** and **egg,** which are called the sex cells or **gametes.**

Before sexual reproduction can occur, each parent must reduce his or her genetic information in half. During sex cell formation, these halves, or sets, of chromosomes are mixed up into new combinations. This is somewhat similar to shuffling cards and dealing out hands: the shuffling and dealing assure

CONCEPT CONNECTIONS

sperm The sex cells produced by male organisms.

eggs The sex cells produced by female organisms.

gametes A general term sometimes used to refer to either eggs or sperm.

gametogenesis The cellular process that is responsible for generating gametes.

fertilization The uniting of an egg and sperm (gametes).

zygote The cell that results from the union of an egg and a sperm; also known as the fertilized egg.

that each hand will be different. An organism with two sets of genetic data, or chromosomes, can produce many combinations of chromosomes when it produces sex cells, just as many different hands can be dealt from one pack of cards. When one of these sex cells unites with another, a new organism containing two sets

FYI

homos = the same; *logos* = portion
chromo = colored; *some* = body

of genetic information is formed. This new organism's information will be different than the parents' and occasionally will produce combinations that are better than that found in either parent. This is the real value of sexual reproduction.

Haploid and Diploid Sex Cells

After fertilization, the **zygote,** which results from the union of an egg and a sperm, divides repeatedly by mitosis to form the complete organism. Notice in the following figure that the zygote, and the cells that form by mitosis from the zygote, have two sets of chromosomes. In contrast, the male gamete (sperm) and the

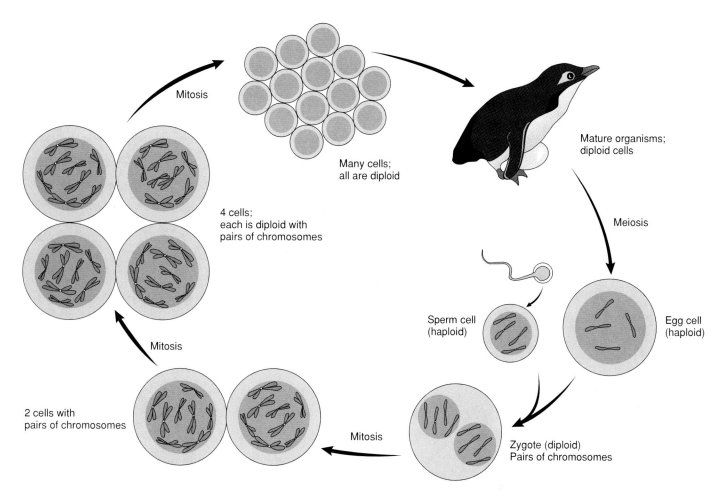

Life Cycle Since the zygote is formed from the union of two sex cells, it contains two sets of chromosomes and is said to be diploid. All the cells produced by mitosis likewise have the diploid number of chromosomes. In preparation for sexual reproduction, the number of chromosomes must be reduced by half, or become haploid, so that fertilization will result in the original number of eight chromosomes in the new individual. (The actual number of chromosomes typical of penguins has not been shown.)

female gamete (egg) each contain only one set of chromosomes. These sex cells are said to be **haploid.** The haploid number of chromosomes is noted as n. A zygote, as we have noted, contains two sets of chromosomes and is said to be **diploid.** The diploid number of chromosomes is noted as $2n$ ($n + n = 2n$). Diploid cells have one set of chromosomes from each parent. Remember, a chromosome is composed of two chromatids, each containing DNA. The two chromatids are attached to each other at a point called the centromere (see chapter 13). In the nucleus of a diploid cell, the chromosomes occur as pairs called **homologous chromosomes;** that is, they have similar genes throughout their length. One of the chromosomes of each homologous pair was donated by the father, the other by the mother.

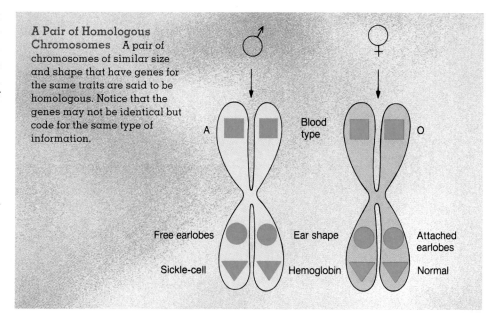

A Pair of Homologous Chromosomes A pair of chromosomes of similar size and shape that have genes for the same traits are said to be homologous. Notice that the genes may not be identical but code for the same type of information.

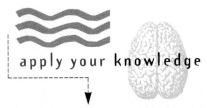
Chromosome Numbers

See if you can answer the following questions about haploid and diploid chromosome numbers.

An organism with a haploid number of 8 will have a diploid number of _____.

If an organism has a diploid number of 18, how many chromosomes will there be in its sex cells?

An organism with a diploid number of 8 will have a haploid number of _____.

Different species vary in the number of chromosomes they contain. Table 14.1 lists several species and their haploid and diploid chromosome numbers. Some of the numbers have been left blank. Be sure to fill these in with the correct answers.

table 14.1

Chromosome Numbers

Organism	Haploid Number	Diploid Number
Mosquito	3	6
Fruit fly	4	8
Housefly	6	12
Toad	18	—
Cat	19	—
Human	—	46
Hedgehog	23	—
Chimpanzee	24	48
Horse	32	64
Dog	—	78
Onion	—	16
Kidney bean	11	—
Rice	12	24
Tomato	12	24
Potato	24	48
Tobacco	24	48
Cotton	—	52

Keeping the Numbers Straight

It is necessary for organisms that reproduce sexually to form gametes having only one set of chromosomes. If gametes contained two sets of chromosomes, the zygote resulting from fertilization would have four sets of chromosomes. If this were to continue in each new generation, the number of sets of chromosomes would increase, and the result could create such genetic confusion that the cell would die. However, this does not happen; the number of chromosomes remains constant generation after generation.

Since cell division by mitosis and cytokinesis results in cells that have the same number of chromosomes as the parent cell, as we saw in chapter 13, two questions arise: how are sperm and egg cells formed, and how do they get only one-half the chromosomes of the diploid cell? The answers lie in the process of **meiosis,** a specialized type of cell division that reduces the chromosome number from diploid ($2n$) to haploid (n). The major function of meiosis is to produce cells that have one set of genetic information, not two sets. That way, when

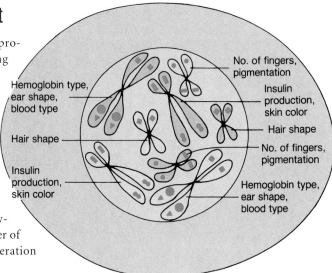

- ■ Blood type O
- ■ Blood type A
- ● Attached earlobes
- ● Free earlobes
- ▼ Normal hemoglobin
- ▼ Sickle-cell hemoglobin
- ▬ Normal pigment
- ▬ Albino (no pigment)
- ● 5 fingers
- ● 6 fingers
- ◆ Curly hair
- ◆ Straight hair
- ⦙ Light skin color
- ⦙ Dark skin color
- ▮ Normal insulin
- ▮ Diabetes

Figure 14.1 Homologous Chromosomes in a Cell In this diagram of a cell, the eight chromosomes are scattered in the nucleus. Even though they are not arranged in pairs, note that there are four pairs of homologous chromosomes. Check to be sure you can pair them up using the list of characteristics.

fertilization occurs between haploid gametes, the zygote will have two sets of chromosomes, as did each parent.

In order to simplify the process of meiosis, we have chosen to show only 8 chromosomes (figure 14.1). In reality, humans

have 46 chromosomes (23 pairs). The haploid number of chromosomes for the cell illustrated is four, and these haploid cells contain only one complete set of four chromosomes. As you can see, there are eight chromosomes in this sample cell—four from the mother and four from the

BIO *feature*

Where the Action Is

Only a few cells go through the process of meiosis. These cells occur in specialized organs capable of producing haploid cells. In animals, the organs in which meiosis occurs are called **gonads.** The female gonads that produce eggs are called **ovaries.** The male gonads that produce sperm are called **testes.** Flowering plants also contain organs that produce haploid cells. In plants, the **pistil** produces eggs or ova, and the **stamen** produces pollen, which contains sperm.

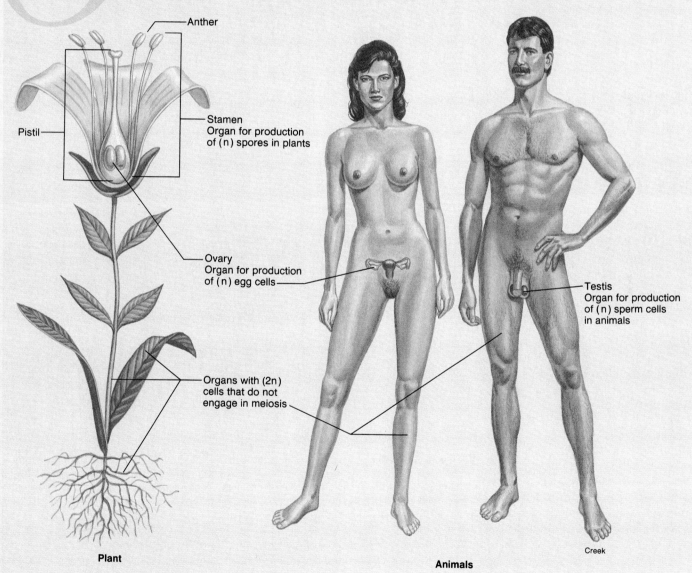

Anther

Pistil

Stamen
Organ for production
of (n) spores in plants

Ovary
Organ for production
of (n) egg cells

Organs with (2n)
cells that do not
engage in meiosis

Testis
Organ for production
of (n) sperm cells
in animals

Creek

Plant

Animals

Haploid and Diploid Cells in Organisms   The bodies of many organisms are composed of cells with the diploid number of chromosomes. However, both plants and animals have structures that produce cells with a haploid number of chromosomes. The male anther in plants and the testes in animals produce haploid male cells called sperm. In both plants and animals, the ovaries produce haploid female cells called eggs.

father (figure 14.1). Though they are not arranged in pairs, you can see that there are four pairs of homologous chromosomes:

- two long chromosomes consisting of chromatids attached near the center
- two long chromosomes consisting of chromatids attached near one end
- two short chromosomes consisting of chromatids attached near one end
- two short chromosomes consisting of chromatids attached near the center.

Thus we can talk about the number of chromosomes in two ways. We can say that our hypothetical diploid cell has eight chromosomes, or we can say that it has four pairs of homologous chromosomes.

Haploid cells, on the other hand, do not have homologous chromosomes. They have only one of each type of chromosome. The whole point of meiosis is to distribute the chromosomes and the genes they carry so that each daughter cell gets one member of each homologous pair. In this way, each daughter cell gets one complete set of genetic information.

The Mechanics of Meiosis: Meiosis I

Meiosis is preceded by an interphase stage when DNA replication occurs. In a sequence of events called *meiosis I,* members of homologous pairs of chromosomes divide into two complete sets. This is sometimes called a **reduction division** because in this stage the chromosome number is reduced from diploid to haploid, that is, one of each pair ends up in a daughter cell. The division begins with chromosomes composed of two chromatids. Like the stages in mitosis, the sequence of events in meiosis I are artificially divided into phases. These phases are named prophase I, metaphase I, anaphase I, and telophase I.

Prophase I

During prophase I, the cell is preparing itself for division (figure 14.2). The chromatin material coils and thickens into chromosomes, the nucleoli disappear, the nuclear membrane disintegrates, and the spindle begins to form. In animals the spindle is formed when the centrioles move to the poles. In plant cells there are no centrioles, but the spindle does form.

There is an important difference between the prophase stage of mitosis and prophase I of meiosis. During prophase I, homologous chromosomes come to lie next to each other in a process called **synapsis.** While the chromosomes are synapsed, a unique event called *crossing-over* can occur. **Crossing-over** is the exchange of equivalent sections of DNA on homologous chromosomes. We will fit crossing-over into the whole picture of meiosis later.

Metaphase I

The synapsed pair of homologous chromosomes now move into position on the equatorial plane of the cell. In this stage, the centromere of each chromosome attaches to the spindle. The synapsed homologous chromosomes move to the equator of the cell as single units. How they are arranged on the equator (which one is on the left and which one is on the right) is determined by chance. In the cell in figure 14.3, three blue chromosomes from the father and one pink chromosome from the mother are lined up on the left. Similarly, one blue chromosome from the father and three pink chromosomes from the mother are on the right. They could have aligned themselves in several other ways. For instance, they could have lined up as shown in the rectangular box at the right in figure 14.3.

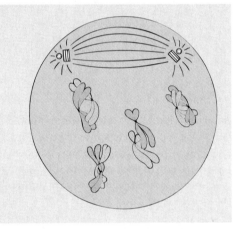

Figure 14.2 Prophase I During prophase I, the cell is preparing for division. A unique event that occurs in prophase I is the pairing up (synapsis) of homologous chromosomes. Notice that the nuclear membrane is no longer apparent and that the pairs of homologous chromosomes are free to move about the cell.

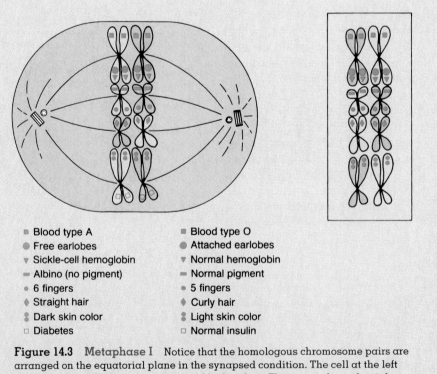

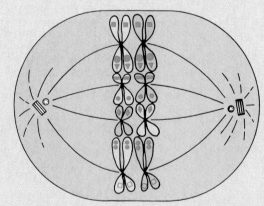

■ Blood type A	■ Blood type O
● Free earlobes	● Attached earlobes
▼ Sickle-cell hemoglobin	▼ Normal hemoglobin
▬ Albino (no pigment)	▬ Normal pigment
● 6 fingers	● 5 fingers
◆ Straight hair	◆ Curly hair
‖ Dark skin color	‖ Light skin color
□ Diabetes	□ Normal insulin

Figure 14.3 Metaphase I Notice that the homologous chromosome pairs are arranged on the equatorial plane in the synapsed condition. The cell at the left shows one way the chromosomes could be lined up. The rectangle on the right shows a second arrangement. How many other ways can you diagram?

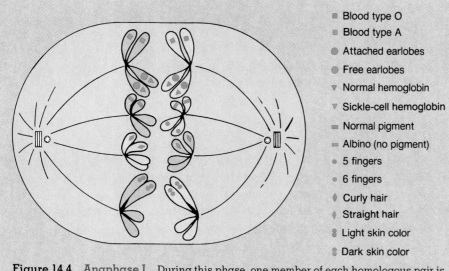

- ▨ Blood type O
- ▨ Blood type A
- ● Attached earlobes
- ◉ Free earlobes
- ▽ Normal hemoglobin
- ▽ Sickle-cell hemoglobin
- ▬ Normal pigment
- ▭ Albino (no pigment)
- ◦ 5 fingers
- ◦ 6 fingers
- ◊ Curly hair
- ◊ Straight hair
- ▯ Light skin color
- ▯ Dark skin color

Figure 14.4 Anaphase I During this phase, one member of each homologous pair is separated (segregated) from the other member of the pair. Notice that the centromeres connecting the two chromatids of the chromosomes do not split.

Anaphase I

Anaphase I is the stage during which homologous chromosomes separate (figure 14.4). During this stage, the chromosome number is reduced from diploid to haploid. The two members of each pair of homologous chromosomes move away from each other toward opposite poles. The direction each takes is determined by how each pair was originally arranged on the spindle. Each chromosome is independently attached to a spindle fiber at its centromere. Unlike the anaphase stage of mitosis, the centromeres that hold the chromatids together *do not divide* during anaphase I of meiosis. Each chromosome still consists of two chromatids. Because the homologous chromosomes, and the genes they carry, are being separated from one another, this process is called **segregation**. The way in which a single pair of homologous chromosomes segregates (separates) does not influence how other pairs of homologous chromosomes segregate. That is, each pair segregates independently of other pairs. This is known as **independent assortment** of chromosomes.

Telophase I

Telophase I consists of changes that return the cell to an interphase condition (figure 14.5). The chromosomes uncoil and become long, thin threads, the nuclear membrane re-forms around them, and nucleoli reappear. During this activity, cytokinesis divides the cytoplasm into two separate cells.

Because of meiosis I, the total number of chromosomes is divided equally, and each daughter cell has one member of each homologous chromosome pair. This means that the genetic data each cell receives is one half of the total, but each cell still has a complete set of the genetic information. Each individual chromosome is still composed of two chromatids joined at the centromere, and the chromosome number is reduced from diploid (2n) to haploid (n). In the cell we have been using as our example, the number of chromosomes is reduced from eight to four. The four pairs of chromosomes have been distributed to the two daughter cells.

Depending on the type of cell, there may be a time following telophase I when the cell engages in normal metabolic activity that corresponds to an interphase stage. However, the chromosomes do not replicate before the cell enters meiosis II. Figure 14.6 shows the events in meiosis I.

The Mechanics of Meiosis: Meiosis II

Meiosis II includes four phases: prophase II, metaphase II, anaphase II, and telophase II. Each of the two daughter cells formed during meiosis I continue through meiosis II, so that usually four cells result from the two divisions.

Prophase II

Prophase II is similar to prophase in mitosis: the nuclear membrane disintegrates, nucleoli disappear, and the spindle apparatus begins to form. However, it differs from prophase I because these cells are haploid, not diploid (figure 14.7).

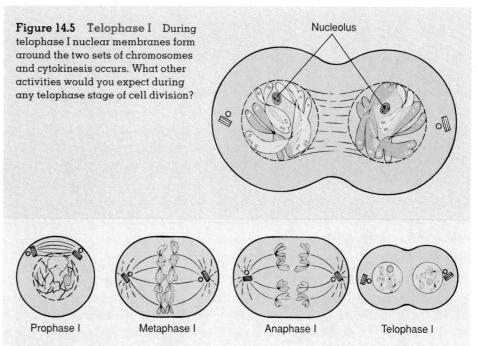

Figure 14.5 Telophase I During telophase I nuclear membranes form around the two sets of chromosomes and cytokinesis occurs. What other activities would you expect during any telophase stage of cell division?

Nucleolus

Prophase I   Metaphase I   Anaphase I   Telophase I

Figure 14.6 Meiosis I The stages of meiosis I result in a reduction of the number of chromosomes by one half. Meiosis I is therefore often called a reduction division. This reduces the number of chromosomes from the diploid number in the parent cell to the haploid number in each of the two daughter cells.

Metaphase II

The metaphase II stage is typical of any metaphase stage because the chromosomes attach by their centromeres to the spindle at the equatorial plane of the cell. Since pairs of chromosomes are no longer together in the same cell, each chromosome moves as a separate unit (figure 14.8).

Anaphase II

Anaphase II differs from anaphase I because the centromere of each chromosome splits in two, and the chromatids, now called *daughter chromosomes,* move to the poles (figure 14.9). Remember, there are no paired homologous chromosomes in this stage; therefore, segregation and independent assortment cannot occur.

Telophase II

During telophase II, the haploid cell returns to a nondividing condition. As cytokinesis occurs, new nuclear membranes form, chromosomes uncoil, nucleoli re-form, and the spindles disappear (figure 14.10). This stage is followed by differentiation; the four cells mature into gametes—either sperm or eggs. The events of meiosis II are compared to the events of meiosis I in figure 14.11.

In many organisms, egg cells are produced in such a manner that three of the four cells resulting from meiosis in a female disintegrate. However, since the one that survives is randomly chosen, the likelihood of any one particular combination of genes being formed is not affected. The whole point of learning the mechanism of meiosis is to see how variation happens. Now we can look at variation and how it comes about.

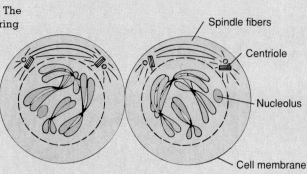

Figure 14.7 Prophase II The two daughter cells are preparing for the second division of meiosis. Since there are no homologous chromosomes in each of these cells it is not possible for synapsis and crossing-over to occur. However, the chromosomes become shorter and thicker and both cells prepare for division.

Spindle fibers
Centriole
Nucleolus
Cell membrane

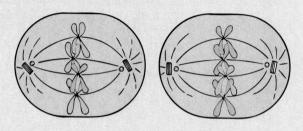

Figure 14.8 Metaphase II During metaphase II, each chromosome lines up on the equatorial plane. Each chromosome is composed of two chromatids joined at a centromere. How does metaphase II of meiosis compare to metaphase of mitosis?

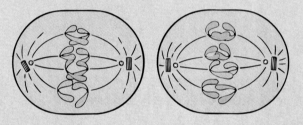

Figure 14.9 Anaphase II The anaphase II stage is very similar to the anaphase of mitosis. The centromere of each chromosome divides and the chromatids separate from each other. The separate chromatids are often referred to as daughter chromosomes.

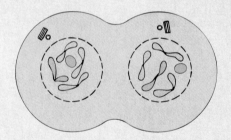
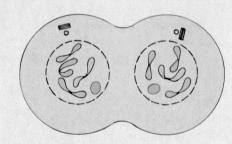

Figure 14.10 Telophase II Telophase II results in four cells each having the haploid number of chromosomes. What other events would you expect to occur during telophase?

Figure 14.11 Meiosis II During meiosis II, the centromeres split and each chromosome divides into separate chromatids. Four haploid cells result from the original, diploid parent cell that began meiosis.

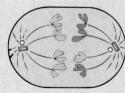
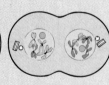

Prophase I   Metaphase I   Anaphase I   Telophase I

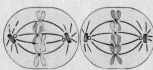
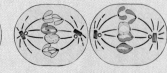

Prophase II   Metaphase II   Anaphase II   Telophase II

315

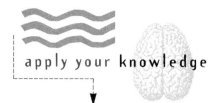
Chromosome Models

Models can help us understand complex biological events such as meiosis. You can create model chromosomes very easily by using various lengths of colored string, thread, or yarn to simulate the twenty-three pairs of homologous chromosomes in a human cell. Each homologous pair should be different from the other pairs in color, length, or both. Begin your modeling with each chromosome in its replicated form (i.e., two chromatids per chromosome). Attach the two chromatids with a loose twist. Manipulate your twenty-three pairs of model chromosomes through the stages of meiosis I and II. If you have performed the actions properly, you should end up with four cells, each haploid (*n* = 23).

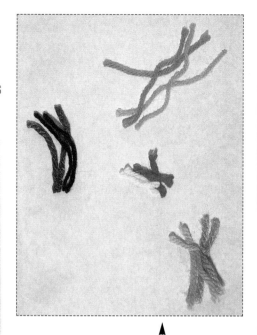

Why Are We All So Different?

The formation of a haploid cell by meiosis and the combination of two haploid cells to form a diploid cell by sexual reproduction results in variety in the offspring. Five factors influence genetic variation in offspring: crossing-over, segregation, independent assortment, mutations, and fertilization.

Crossing-Over

Crossing-over is the exchange of a section of one homologous chromosome with an equivalent section from the other homologous chromosome. This exchange results in a new gene combination on both of the chromatids involved in the cross. Crossing-over occurs during meiosis I while homologous chromosomes are synapsed. To separate a section of a chromosome, bonds are broken at the same spot on both chromatids, and the two pieces switch places. After switching places, the two pieces of DNA are bonded back together.

Examine figure 14.12 carefully to note precisely what occurs during crossing-over. This figure shows a pair of homologous chromosomes close to each other. Notice that each gene occupies a specific place on the chromosome. This is the *locus,* the place on a chromosome where a gene is located. For the

sake of simplicity, only a few loci are labeled on the chromosomes used as examples. Actually, each chromosome is composed of thousands of genes.

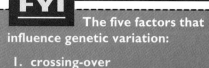
The five factors that influence genetic variation:

1. **crossing-over**
2. **segregation**
3. **independent assortment**
4. **mutations**
5. **fertilization**

What does crossing-over have to do with the possible kinds of cells that result from meiosis? Consider figure 14.13. Notice that without crossing-over, only two kinds of genetically different gametes result. Two of the four gametes have one type of chromosome, while the other two have the other type of chromosome. With crossing-over, four genetically different gametes are formed.

With just one crossover, we double the number of kinds of gametes possible from meiosis. Since crossing-over can occur at almost any point along the length of the chromosome, great variation is possible. In fact, crossing-over can occur at a number of different points on the same chromosome; that is, there can be several crossovers per chromosome pair (figure 14.14).

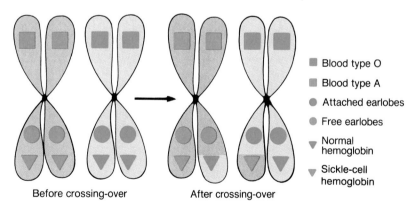

Before crossing-over   After crossing-over

■ Blood type O
■ Blood type A
● Attached earlobes
● Free earlobes
▼ Normal hemoglobin
▼ Sickle-cell hemoglobin

Figure 14.12 Synapsis and Crossing-Over While pairs of homologous chromosomes are in synapsis, one part of one chromatid can break off and be exchanged for an equivalent part of its homologous chromatid. List the new combination of genes on each chromatid that has resulted from the crossing-over.

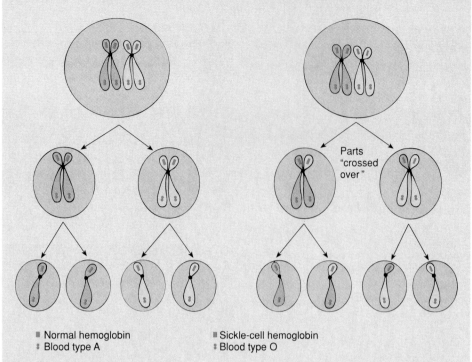

■ Normal hemoglobin
⁑ Blood type A

■ Sickle-cell hemoglobin
⁑ Blood type O

Figure 14.13 **Variations Resulting from Crossing-Over** The cells on the left resulted from meiosis without crossing-over; those on the right had one crossover. Compare the combinations of genes found in the cells resulting from meiosis in both cases.

Crossing-over helps explain why a child can show a mixture of family characteristics. In figure 14.15, if the violet chromosome was the chromosome that your mother received from your grandmother, you could receive some genetic information not only from your grandmother but also from your maternal grandfather. Now think about what's happening in your body. The chromosomes in your gametes contain a mixture of genes from your mother and father as a result of crossing-over. When crossing-over occurs during the meiotic process, pieces of genetic material are exchanged between the chromosomes. This means that genes that were originally on the same chromosome become separated. They are moved to their synapsed homologue, and therefore into different gametes. The closer two genes are to each other on a chromosome (i.e., the more closely they are *linked*), the more likely it is that they will stay together and not be separated during crossing-over. If they are not separated, then they will be inherited together. The farther

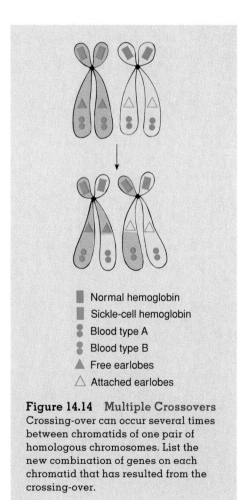

■ Normal hemoglobin
■ Sickle-cell hemoglobin
⁑ Blood type A
⁑ Blood type B
▲ Free earlobes
△ Attached earlobes

Figure 14.14 **Multiple Crossovers** Crossing-over can occur several times between chromatids of one pair of homologous chromosomes. List the new combination of genes on each chromatid that has resulted from the crossing-over.

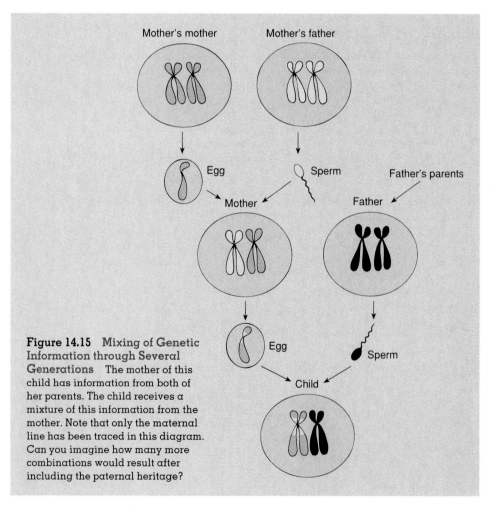

Figure 14.15 **Mixing of Genetic Information through Several Generations** The mother of this child has information from both of her parents. The child receives a mixture of this information from the mother. Note that only the maternal line has been traced in this diagram. Can you imagine how many more combinations would result after including the paternal heritage?

The Human Genome Project

The human genome project was first proposed in 1986 and is one of the most ambitious projects ever undertaken in the biological sciences. The goal is nothing less than the complete characterization of the genetic makeup of humans. If the effort is successful, scientists will have produced a map of each of the twenty-three kinds of human chromosomes that will show the names and places of all of our genes. This international project involving about one hundred laboratories is expected to take fifteen years. Work began in many of these labs in 1990. Powerful computers are used to store and share the enormous amount of information derived from the analyses of human DNA. To get an idea of the size of this project, consider this: A human Y chromosome (one of the smallest of the human chromosomes) is estimated to be composed of 28 million chemical subunits called *nitrogenous bases.* The larger X chromosome may be composed of 160 million bases. It is the sequence of these bases that determines the genes.

Two kinds of work are progressing simultaneously. Physical maps are being constructed by determining the location of specific "markers" (known sequences of bases) and their closeness to genes. A kind of chromosome map already exists that pictures patterns of colored bands on chromosomes, a result of chromosome-staining procedures. Using these banded chromosomes, the markers can then be related to these colored bands on a specific region of a chromosome.

The second goal, to determine the exact order of nitrogenous bases of the DNA for each chromosome, will take longer. Techniques exist for determining base sequences, but it is a time-consuming job to sort out the several billion bases that may be found in any one chromosome. It is estimated, for example, that the base sequences and exact positions of over 100,000 genes are yet to be determined.

When the maps are completed for all of the human chromosomes, it will be possible to examine a person's DNA and identify genetic abnormalities. This could be extremely useful in diagnosing diseases and providing genetic counseling to carriers of genetic defects who are considering having children. This kind of information would also create possibilities for new gene therapies. Once it is known where an abnormal gene is located and how it differs in base-sequence form from normal DNA sequence, steps could be taken to correct the abnormality. There is also fear, however, that as knowledge of our genetic makeup becomes easier to determine, some people may attempt to use this information for profit or political power. This is a real concern, since some insurance companies refuse to insure people at "genetic risk." Refusing to provide coverage would save these companies the expense of future medical bills incurred by "less-than-perfect" people. Another fear is that attempts may be made to "breed out" certain genes and people from the human population in order to create a "perfect race."

James D. Watson and Francis H. C. Crick

apart two genes are, the more likely it is that they will be separated during crossing-over. This fact enables biologists to construct chromosome maps (figure 14.16).

Segregation

After crossing-over has taken place, segregation occurs. This involves the separation of homologous chromosomes and their movement to the poles. Let's say a person has a gene for insulin production on one chromosome and a gene for diabetes on the homologous chromosome. (One type of diabetes is caused by a mutation in the gene for insulin production.) Such a person would produce enough insulin to be healthy. When this pair of chromosomes segregates during anaphase I, one daughter cell receives a chromosome with an allele for insulin production and the second daughter cell receives a chromosome with an

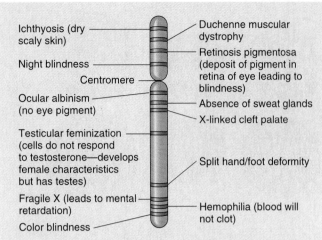

Ichthyosis (dry scaly skin)
Night blindness
Centromere
Ocular albinism (no eye pigment)
Testicular feminization (cells do not respond to testosterone—develops female characteristics but has testes)
Fragile X (leads to mental retardation)
Color blindness
Duchenne muscular dystrophy
Retinosis pigmentosa (deposit of pigment in retina of eye leading to blindness)
Absence of sweat glands
X-linked cleft palate
Split hand/foot deformity
Hemophilia (blood will not clot)

Figure 14.16 Genes Known to Be on the X Chromosome This gene map shows the approximate position of several genes known to be on the X chromosome.

Diabetes
Normal insulin
5 fingers
6 fingers

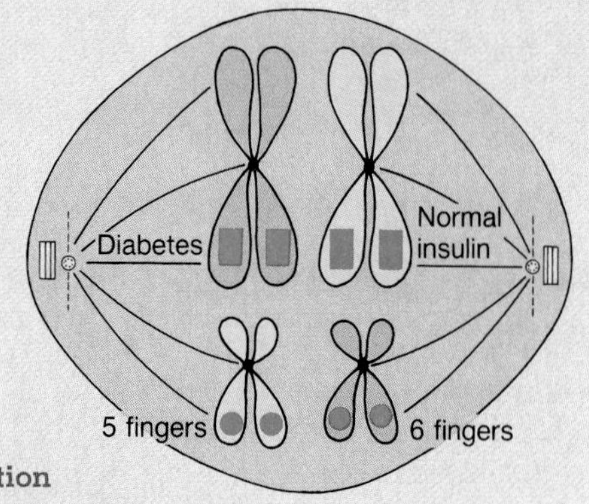

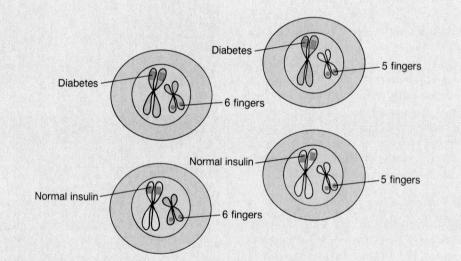

Figure 14.17 The Independent Orientation of Homologous Chromosome Pairs The orientation of one pair of chromosomes on the equatorial plane does not affect the orientation of a second pair of chromosomes. This means there are several ways the homologous pairs can separate during anaphase I. This results in great variety among the haploid cells produced by meiosis.

allele for diabetes. The process of segregation causes the genes to be separated from one another so that they have an equal chance of being transmitted to the next generation. If the mate also has one allele for insulin production and one allele for diabetes, the mate also produces two kinds of gametes.

Both of the parents have normal insulin production. If one or both of them contributed an allele for normal insulin production during fertilization, the offspring would produce enough insulin to be healthy. However if by chance both parents contributed the gamete with the allele for diabetes, the child would be a diabetic. Thus, parents may produce offspring with traits different from their own. In this variation, no new genes are created; they are simply redistributed in a fashion that allows for allele combinations in the offspring to be different from the parents' combinations. This will be explored in greater detail in chapter 15.

Diabetes
5 fingers
Diabetes
6 fingers
Normal insulin
5 fingers
Normal insulin
6 fingers

Figure 14.18 Variation Resulting from Independent Assortment When a cell has two pairs of homologous chromosomes, four kinds of haploid cells can result from independent assortment. How many kinds of haploid cells could result if the parental cell had three pairs? Four pairs?

Independent Assortment

So far in discussing variety we have only dealt with one pair of chromosomes, which allows two varieties of gametes. Now let's consider how variation increases when we add a second pair of chromosomes (figure 14.17).

In figure 14.17, chromosomes carrying insulin-production information always separate from each other. The second pair of chromosomes containing information for the number of fingers also separates. Since the pole to which a chromosome moves is a chance event, half of the time the chromosomes divide so that insulin production and six-fingeredness move in one direction, while diabetes and five-fingeredness move in the opposite direction. The other half of the time, insulin production and five-fingeredness go together, while diabetes and six-fingeredness go to the other pole. With four chromosomes (two pair), four kinds of gametes are possible (figure 14.18). With three pairs of homologous chromosomes, there are eight possible kinds of cells with respect to chromosome combinations resulting from meiosis.

The number of possible chromosomal combinations of gametes is found by the expression 2^n, where n equals the number of pairs of homologous chromosomes. With three pairs of homologous chromosomes,

n equals 3, and so $2^n = 2^3 = 2 \times 2 \times 2 = 8$. With twenty-three pairs of chromosomes, as in the human cell, $2^n = 2^{23} = 8,388,608$. More than 8 million kinds of chromosome combinations are possible from a single human parent. This huge variation is possible because each pair of homologous chromosomes assorts independently of the other pairs of homologous chromosomes (independent assortment). In addition to this variation, crossing-over creates new gene combinations, and mutation can cause the formation of new genes, thereby increasing variation even more.

Mutation

When children use the word "mutant," what usually comes to mind is some strange, grotesque monster. In the movies this "thing" may have been exposed to some sort of radiation or chemical that caused a hideous, overnight transformation (figure 14.19). This portrayal is fiction! A mutation is a change in the genetic material of an organism. Most mutations do not show themselves as the "hulking" changes depicted in movies. In many cases, they don't even show themselves until the next generation. Remember, only if the mutation occurs in a gamete will it be passed on to the next generation. In other words, the genes change in

Figure 14.19 Mutations A common understanding of the word *mutation* is that of malformed monsters. However, the proper use of the word refers to changes in the DNA of an organism.

the parent, they are passed to the children during sexual reproduction, and the change is seen in the children. Therefore the children, because of mutations, may have features that have never been seen in either of the parents or their families. There are two types of mutations: point mutations and chromosomal mutations. In point mutations, there is a change in a specific portion of the DNA that results in the organism producing a different kind of protein. In chromosomal mutations, whole genes are moved within a chromosome or between chromosomes. By causing the production of different proteins, both types of mutations increase variation.

Fertilization

Because of the large number of possible gametes resulting from independent assortment, segregation, mutation, and crossing-over, an incredibly large number of types of offspring can result. Since human males can produce millions of genetically different sperm and females can produce millions of genetically different eggs, the number of kinds of offspring possible is, for all practical purposes, infinite. With the possible exception of identical twins, every human that has ever been born is genetically unique.

Nondisjunction

In the normal process of meiosis, diploid cells have their number of chromosomes reduced to haploid. This involves segregating homologous chromosomes into separate cells during the first meiotic division. Occasionally, a pair of homologous chromosomes does not segregate properly during gamete formation, also called *gametogenesis,* and both chromosomes of a pair end up in the same gamete. This abnormal kind of division is known as **nondisjunction.** As you can see in figure A, two cells are missing a chromosome and the genes that were carried on it. The other cells have a double dose of chromosomes. Apparently, the genes of an organism are balanced against one another. A double dose of some genes and a single dose of others results in abnormalities that may lead to the death of the cell. Some of these abnormal cells, however, do live and develop into sperm or eggs. If one of these abnormal sperm or eggs with an extra chromosome unites with a normal gamete, the offspring will have an abnormal number of chromosomes. There will be three of one of the kinds of chromosomes instead of the normal two, a condition called **trisomy** (*tri* = three; *some* = body). All the cells that develop by mitosis from that zygote will also be trisomic.

Figure A   Nondisjunction during Gametogenesis   When a pair of homologous chromosomes fails to separate properly during meiosis I, gametogenesis results in gametes that have an abnormal number of chromosomes. Notice that two of the highlighted cells have an additional chromosome while the other two are deficient by that same chromosome.

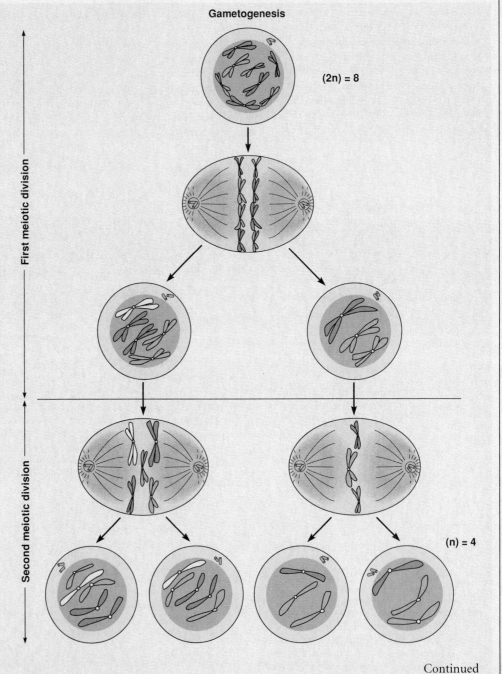

Gametogenesis

First meiotic division

Second meiotic division

(2n) = 8

(n) = 4

Continued

BIO *feature*

Nondisjunction (continued)

It is possible to examine cells and count chromosomes. Among the easiest cells to view are white blood cells. They are dropped onto a microscope slide so that the cells are broken open and the chromosomes separated. The chromosomes are stained with a dye that makes them more visible. Photographs are taken of chromosomes from cells in the metaphase stage of mitosis.

The chromosomes in the pictures can then be cut out, arranged, and compared to known samples (figure B).

One example of the effects of nondisjunction is the condition known as **Down syndrome** (at one time called mongolism). A *syndrome* is a combination of features or symptoms of a disease. If an egg cell with two number 21 chromosomes has

been fertilized by a sperm containing the typical one copy of chromosome number 21, the resulting zygote would have three copies of chromosome 21 and would have a total of forty-seven chromosomes (twenty-four from the female parent plus twenty-three from the male parent) (figure C). The child who developed from this fertilized egg would have forty-seven chromosomes in

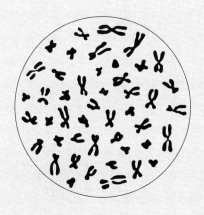 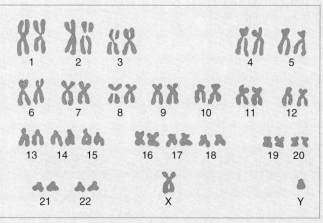

Figure B Human Male Chromosomes The randomly arranged human male chromosomes shown on the left were photographed from metaphase cells spattered onto a microscope slide. Those on the right were cut out and arranged into pairs of homologous chromosomes. Each chromosome has been given an identification number.

Figure C Chromosomes from an Individual Displaying Down Syndrome Notice that each pair of chromosomes has been numbered and that the person from whom these chromosomes were taken has an extra chromosome number 21. A person with this condition displays a variety of physical characteristics, including slightly slanted eyes, flattened facial features, a large tongue, and a tendency toward short stature and fingers. Some individuals also display mental retardation.

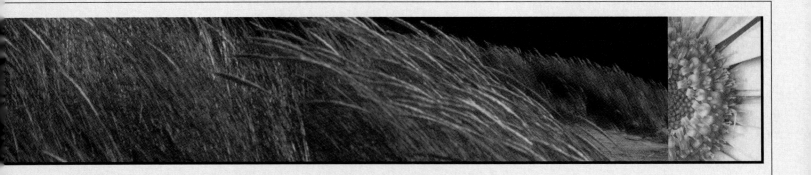

every cell of his or her body as a result of mitosis, and the symptoms characteristic of Down syndrome. These could include thickened eyelids, some mental impairment, and faulty speech (figure D). Premature aging is probably the most significant impact of this genetic disease.

The parents' age is an important consideration in the occurrence of trisomies such as Down syndrome. Gametogenesis begins in women before they are born, but cells destined to become eggs are put on hold during meiosis I. Beginning at puberty and ending at menopause, one of these cells completes the meiotic process monthly. This means that eggs released for fertilization later in life are older than those released earlier in life. Therefore, the chances of abnormalities such as nondisjunction increase as the age of the mother increases. The frequency of occurrence of nondisjunction at different ages in women is illustrated in figure E. Notice that the frequency of nondisjunction increases very rapidly after age thirty-seven. For this reason, many physicians encourage couples to have their children in their early to mid-twenties and not in their late thirties or early forties. Physicians normally encourage older couples to have the cells of their fetus checked to see if they have the normal chromosome number.

Figure D Down Syndrome Every cell in a downic child's body has one extra chromosome. This frequently results in some degree of mental retardation. However, with special care, planning, and training, people with this syndrome can lead happy, productive lives.

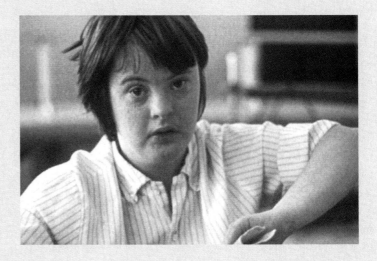

Figure E Nondisjunction as a Function of a Mother's Age Notice that as the age of the female increases, the rate of nondisjunction increases only slightly, until the age of approximately 37. From that point on, the rate increases drastically.

Number of births with Down syndrome per 100,000

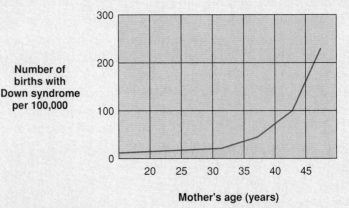

Mother's age (years)

The Birds and the Bees . . . and the Alligators

The determination of the sex of an individual depends on the kind of organism you are! For example, in humans, the physical features that result in maleness are triggered by a gene on the Y chromosome. Lack of a Y chromosome results in an individual that is female. In other organisms, sex may be determined by other combinations of chromosomes or environmental factors.

Organism	Sex Determination
Birds	Chromosomally determined; XY individuals are female.
Bees	Males (the drones) are haploid and females (workers or queens) are diploid.
Certain species of alligators, turtles, and lizards	Egg incubation temperatures cause hormonal changes in the developing embryo; higher temperatures cause the developing brain to favor the individual becoming a female. (Placing a drop of the hormone estrogen on the developing egg also causes the embryo to become a female!)
Boat shell snails	Males can become females but will remain male if they mate and stay in one spot.
Shrimp, orchids, and some tropical fish	Males convert to females; on occasion females convert to males. Reason? Probably to maximize breeding!
African reed frog	Females convert to males. Reason? Probably to maximize breeding!

Chromosomes and Sex Determination

Sexual characteristics are determined by genes in the same manner as other types of characteristics. In many organisms, sex-controlling genes are located on specific chromosomes known as **sex chromosomes.** All other chromosomes not involved in controlling the sex of an individual are known as **autosomes.** In humans and all other mammals, and in some other organisms, the sex of an individual is controlled by the presence of a certain

chromosome combination. The genes that control maleness in humans are located on a small chromosome known as the *Y chromosome*. This Y chromosome behaves as if it and another larger chromosome, known as the *X chromosome*, were homologs. Males have one X and one Y chromosome. Females have two X chromosomes. Some animals have their sex determined in a completely different way. In bees, for example, the females are diploid and the males are haploid. Other plants and animals have still other chromosomal mechanisms for determining their sex.

SUMMARY

Sexual reproduction produces offspring with combinations of genetic information different than that of the parents. This is made possible by the events that occur during meiosis and fertilization. Meiosis is a specialized process of cell division resulting in the production of four cells, each of which has the haploid number of chromosomes. The total process involves two sequential divisions during which one diploid cell reduces to four haploid cells. Since the chromosomes act as carriers for genetic information, genes separate into different sets during meiosis. Crossing-over, segregation, and independent assortment allow hidden characteristics to be displayed and characteristics donated by the mother and the father to be mixed in new combinations.

Together, crossing-over, segregation, and independent assortment ensure that all sex cells are unique. Variety is also generated by mutations. When fertilization unites unique sex cells to form a zygote, the zygote will also be one of a kind. The sex of many kinds of organisms is determined by specific chromosome combinations. In humans, females have two X chromosomes, while males have an X and a Y chromosome.

FYI

A Comparison of Mitosis and Meiosis

Mitosis	Meiosis
1. One division completes the process.	1. Two divisions are required to complete the process.
2. Chromosomes do not synapse.	2. Homologous chromosomes synapse in prophase I.
3. Homologous chromosomes do not cross over.	3. Homologous chromosomes do cross over.
4. Centromeres divide in anaphase.	4. Centromeres divide in anaphase II, but not in anaphase I.
5. Daughter cells have the same number of chromosomes as the parent cell ($2n \rightarrow 2n$ or $n \rightarrow n$).	5. Daughter cells have half the number of chromosomes as the parent cell ($2n \rightarrow n$).
6. Daughter cells have the same genetic information as the parent cell.	6. Daughter cells are genetically different from the parent cell.
7. Results in growth, replacement of worn-out cells, and repair of damage.	7. Results in sex cells.

CHAPTER GLOSSARY

autosomes (aw′to-somz) Chromosomes that are composed of genes that determine general body features.

crossing-over (kro′sing o′ver) The exchange of a part of a chromatid from one chromosome with an equivalent part of a chromatid from a homologous chromosome.

diploid (dip′loid) A cell that has two sets of chromosomes: one set from the maternal parent and one set from the paternal parent.

Down syndrome (down sin′drom) A genetic disorder resulting from the presence of an extra chromosome number 21. Symptoms include slightly slanted eyes, flattened facial features, a large tongue, and a tendency toward short stature and fingers. Some individuals also display mental retardation.

eggs (egs) The haploid sex cells produced by sexually mature females.

fertilization (fer″ti-li-za′shun) The joining of haploid nuclei, usually from an egg and a sperm cell, resulting in a diploid cell called the zygote.

gamete (gam′ēt) A haploid sex cell.

gametogenesis (ga-me″to-jen′e-sis) The generating of gametes; the meiotic cell-division process that produces sex cells.

gonad (go′nad) In animals, the organs in which meiosis occurs.

haploid (hap′loid) Having a single set of chromosomes resulting from the reduction division of meiosis.

homologous chromosomes (ho-mol′o-gus kro′mo-somz) A pair of chromosomes in a diploid cell that contain similar genes at corresponding loci throughout their length.

independent assortment (in″de-pen′dent a-sort′ment) The segregation, or assortment, of one pair of homologous chromosomes, independently of the segregation, or assortment, of any other pair of chromosomes.

meiosis (mi-o′sis) The specialized pair of cell divisions that reduce the chromosome number from diploid ($2n$) to haploid (n).

nondisjunction (non″dis-junk′shun) An abnormal meiotic division that results in sex cells with too many or too few chromosomes.

ovaries (o′var-ez) The female sex organs that produce haploid sex cells (the eggs or ova).

pistil (pis′til) The sex organ in plants that produces eggs or ova.

reduction division (re-duk′shun di-vi′zhun) A type of cell division in which daughter cells get only half the chromosomes from the parent cell.

segregation (seg″re-ga′shun) The separation and movement of homologous chromosomes to the poles of the cell.

sex chromosomes (seks kro′mo-somz) Chromosomes that contain the genes that control development of sex characteristics.

sexual reproduction (sek′shu-al re″pro-duk′shun) The propagation of organisms involving the union of gametes from two parents.

sperm (spurm) The haploid sex cells produced by sexually mature males.

stamen (sta′men) The male reproductive structure of a flower.

synapsis (sin-ap′sis) The condition in which the two members of a pair of homologous chromosomes come to lie close to one another.

testes (tes′tēz) The male sex organs that produce haploid cells (the sperm).

trisomy (tris′oh-me) An abnormal number of chromosomes (3) resulting from the nondisjunction of homologous chromosomes during meiosis; for example, as in Down syndrome.

zygote (zi′gōt) A diploid cell that results from the union of an egg and a sperm.

CONCEPT MAP TERMINOLOGY

Construct a concept map to represent the relationships among the following concepts.

age
anaphase I
anaphase II
Down syndrome
female gamete
fertilization
male gamete
meiosis
metaphase I
metaphase II
nondisjunction
synapse

LABEL•DIAGRAM•EXPLAIN

In mitosis and meiosis there are events that are common to both processes. List events that occur in both cell division processes. Include each stage in your answer.

There are also differences between the two processes. Describe these differences.

Multiple Choice Questions

1. The exchange of genetic material (genes) between segments of homologous chromosomes results in:
   a. new gene combinations
   b. zygotes
   c. diploid cells
   d. segregation of genes

2. A process that occurs during prophase I is:
   a. segregation
   b. synapsis
   c. reduction division
   d. independent assortment

3. The diploid number of chromosomes is found in cells during:
   a. prophase II
   b. telophase I
   c. anaphase II
   d. prophase I

4. The fact that each homologous pair of chromosomes in humans separates and moves to the poles without being influenced by the other pairs is
   a. segregation
   b. disintegration
   c. independent assortment
   d. fertilization

5. A new nuclear membrane is formed in:
   a. anaphase I
   b. prophase II
   c. telophase I
   d. anaphase II

6. In mitosis, the centromeres split during anaphase; in meiosis they split during:
   a. anaphase I
   b. telophase I
   c. prophase II
   d. anaphase II

7. Diploid cells are formed by:
   a. synapsis
   b. reduction division
   c. fertilization
   d. independent assortment

8. An organism having a diploid number of 12 forms gametes having:
   a. 6 chromosomes
   b. 12 chromosomes
   c. 18 chromosomes
   d. 24 chromosomes

Questions with Short Answers

1. List three differences between mitosis and meiosis.

2. How do haploid cells differ from diploid cells?

3. What are the major sources of variation in the process of meiosis?

4. Can a haploid cell undergo meiosis?

5. What is unique about prophase I?

6. Why is meiosis necessary in organisms that reproduce sexually?

7. Define the terms *zygote, fertilization,* and *homologous chromosomes.*

8. How much variation as a result of independent assortment can occur in cells with the following diploid numbers: 2, 4, 6, 8, and 22?

9. Diagram the metaphase I stage of a cell with the diploid number of 8.

10. Diagram fertilization as it would occur between a sperm and an egg with the haploid number of 3.

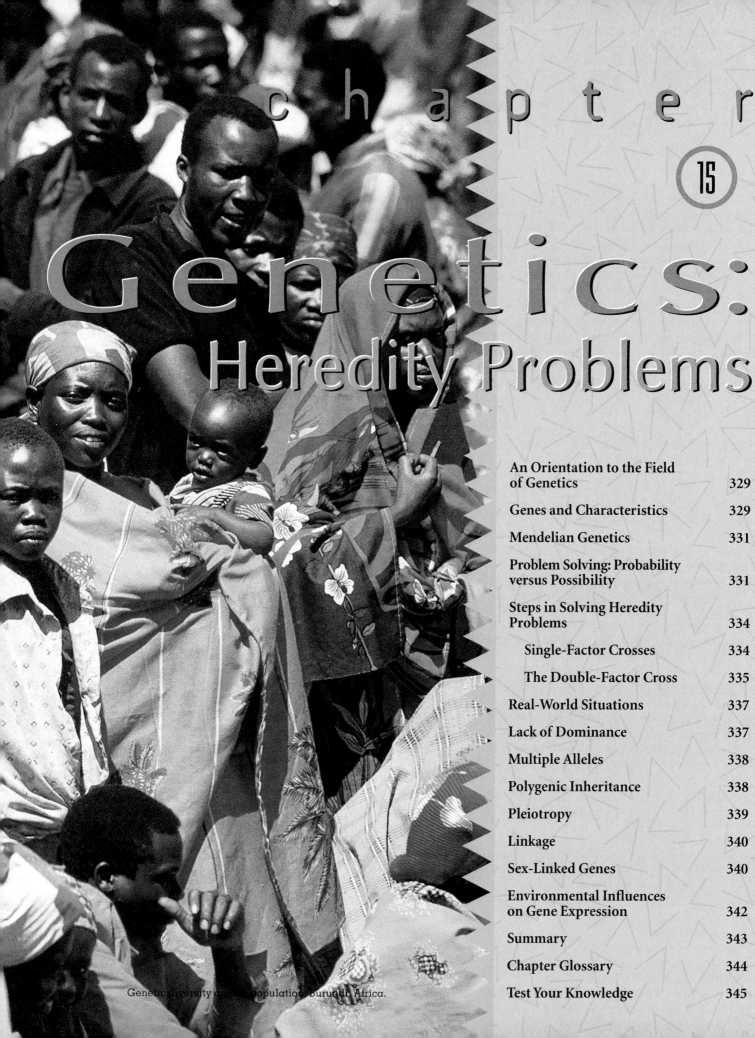

chapter

15

Genetics:
Heredity Problems

Genetic diversity in a human population, Burundi, Africa.

- Be able to work single-factor and double-factor genetic problems dealing with traits that show dominance, recessiveness, and lack of dominance.

- Be able to work genetic problems dealing with multiple alleles, polygenic inheritance, and X-linked characteristics.

- Explain how environmental conditions influence an organism's phenotype.

An Orientation to the Field of Genetics

Why do you have a particular blood type or hair color? Why do some people have the same skin color as their parents, while others have a skin color different from that of their parents? These questions can be better answered if you understand how genes work. A *gene* is a portion of DNA that determines an organism's characteristics. Through meiosis and reproduction, these genes can be transmitted from one generation to another.

The study of genes, how genes produce characteristics, and how the characteristics are inherited is the field of biology called *genetics*. The first person to systematically study inheritance and formulate laws about how characteristics are passed from one generation to the next was an Augustinian monk named Gregor Mendel (1822–1884). Mendel's work was not generally accepted until 1900, when three men, working independently, rediscovered some of the ideas that Mendel had formulated more than thirty years earlier. Because of his early work, the study of the pattern of inheritance that follows the laws formulated by Gregor Mendel is often called **Mendelian genetics.**

To be able to understand how a genetically determined trait is transmitted from one generation to the next, and to solve genetics problems, you need to know some basic terminology. One term that you have already encountered is *gene.* Mendel thought of a gene as a particle, like a single bead on a string that could be passed from parents to their offspring (also called children, descendants, or progeny). Today we know that a gene is actually one set of instructions for the manufacture of a protein or RNA molecule and is composed of specific sequences of chemical subunits called DNA nucleotides. The particle concept is not entirely inaccurate, however, because genes are located one after the other on specific chromosomes.

Genes and Characteristics

The **genome** of an individual is the set of all the genes necessary to specify that organism's complete list of characteristics. A diploid ($2n$) cell has two genomes and a haploid cell (n) has one genome. The **genotype** of an organism is a listing of the genes present in that organism. It consists of the cell's DNA code; therefore, you cannot see the genotype of an organism. It is impossible to know the complete genotype of most organisms, but it is often possible to figure out the genes present that determine a particular characteristic. Recall that different forms of a gene are alleles and that alleles may be dominant (A) or recessive (a). An organism containing two alleles of the same type is called **homozygous** (AA or aa), while an organism with two different alleles is called **heterozygous** (Aa). The way each combination of genes expresses itself is known as the **phenotype** of the organism. For example, in humans the ability to curl your tongue is a genetic trait. A person's genotype could be TT, Tt, or tt, where T represents the dominant trait, the ability to roll the sides of the tongue up into a tube shape (figure 15.1). The t represents the recessive allele, the inability to curl the tongue. The phenotypes would be TT = ability to curl the tongue, Tt = ability to curl the tongue, and tt = inability to curl the tongue.

CONCEPT CONNECTIONS

homozygous A diploid organism that has two identical alleles for a particular characteristic.

heterozygous A diploid organism that has two different allelic forms of a particular gene.

homologous Chromosomes that are the same length, one donated by the male parent and another donated by the female parent; their centromeres are located in the same position and they have comparable alleles.

Figure 15.1 **Tongue Rolling** The ability to roll the tongue is controlled by a dominant allele.

Expressing Yourself

When an allele "expresses" itself, it makes enough product to be seen as a physical trait. An allele that does not express itself may not produce enough product to be identified or may produce a defective product that fails to show itself as the trait typical for that allele. Sometimes an allele is present in an individual but just doesn't do anything. Sometimes alleles express themselves at different times during a person's life; that is, they may not "turn on" while you are young but may begin to function later in life. A good example of this delayed type of genetic expression is seen in people who have never been allergic to ragweed but become allergic to it later in life: they get hay fever. An individual who has an allele for a trait that does not show itself is said to be a **carrier** of that allele.

For various reasons, certain alleles may not express themselves. Sometimes the physical environment determines whether or not an allele functions. For example, some cats have coat-color alleles that do not reveal themselves unless the temperature of the skin falls below a certain point. Often, the only parts of a cat that become cool enough to allow the alleles to express themselves are the tips of the ears and the feet. Consequently, these areas differ in color from the rest of the cat's body. Another example, in humans, is the presence of genes for freckles that do not show themselves fully unless the person's skin is exposed to sunlight.

(a)

No sunlight   Exposed to sun

(b)

The Environment and Gene Expression The expression of many genes is influenced by the environment. (*a*) The gene for dark hair in the cat is sensitive to temperature and expresses itself only in the parts of the body that stay cool. (*b*) The gene for freckles expresses itself more fully when a person is exposed to sunlight.

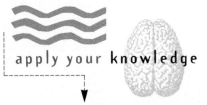
Genotype or Phenotype?

Check the category that applies to each of the following items.

		Genotype	Phenotype
1.	Brown hair	[ ]	[ ]
2.	*aa*	[ ]	[ ]
3.	*Bb*	[ ]	[ ]
4.	Hairy knuckles (mid-digital hair)	[ ]	[ ]
5.	*Ff Gg HH pp*	[ ]	[ ]
6.	Intelligent/stupid	[ ]	[ ]
7.	Pretty/handsome	[ ]	[ ]
8.	*Ss*	[ ]	[ ]

Mendelian Genetics

Mendelian genetics involves the study of the transfer of genes from one generation to another and the ways in which the genes received from the parents influence the traits of the offspring. Before you go on, be certain that you understand how meiosis works and how the gametes formed from this process are combined by fertilization (see chapter 14). It is in meiosis that the two alleles in a pair of genes segregate. Although we will talk about the segregation of genes in this chapter, remember that it is the chromosomes and not the individual genes that actually segregate.

Notice that the chromosomes in figure 15.2 have two different alleles for hemoglobin: one for sickle-cell hemoglobin and one for normal hemoglobin. By meiosis, the gamete the parent contributes may contain an allele for sickle-cell hemoglobin or an allele for normal hemoglobin. By fertilization, the offspring will receive two alleles for hemoglobin production, but only one from each parent. For most of the remainder of this chapter, we will deal with how to determine which genes are passed on by the parents and how to determine the genetic makeup of the offspring resulting from fertilization.

Problem Solving: Probability versus Possibility

In order to solve heredity problems, you must have an understanding of probability. **Probability** is the chance that an event will happen, and is often expressed as a percent or a fraction. *Probability* is not the same as *possibility*. It is possible to toss a coin and have it come up heads. But the probability of getting a head is more precise than just saying it is possible to get a head. The probability of getting a head is one out of two (1/2 or 0.5 or 50%), because there are two sides to the coin, only one of which is a head. Probability can be expressed as a fraction:

$$\text{Probability} = \frac{\text{the number of events that can produce a given outcome}}{\text{the total number of possible outcomes}}$$

What is the probability of cutting a deck of cards and getting the ace of hearts? The number of times that the ace of hearts can occur is one. The total number of possible outcomes (number of cards in the deck) is fifty-two. Therefore, the probability of cutting an ace of hearts is 1/52.

What is the probability of cutting an ace? The total number of aces in the deck is four, and the total number of cards is fifty-two. Therefore, the probability of cutting an ace is 4/52 or 1/13.

It is also possible to determine the probability of two independent events occurring together. *The probability of two or more events occurring simultaneously is the product of their individual probabilities.* If you throw a pair of dice,

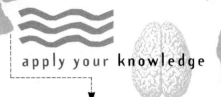
Play the Game and Take Your Chances!

What is the probability of drawing a face card (king, queen, jack) from a deck of bridge cards?

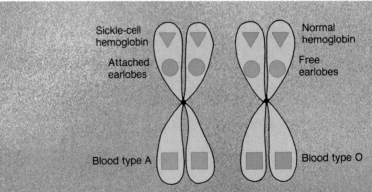

Figure 15.2 **A Pair of Homologous Chromosomes** Homologous chromosomes contain genes for the same characteristics at the same place. Note that the attached-earlobe allele is located at the ear-shape locus on one chromosome, and the free-earlobe allele is located at the ear-shape locus on the other member of the homologous pair of chromosomes. The other two genes are for hemoglobin structure (alleles normal and sickled) and blood type (alleles type A and type O). We are only just learning on which chromosomes most human genes are located. It is hoped that the Human Genome Project described in chapter 9 will resolve this problem. The examples presented here are for illustrative purposes only.

Mendel's Laws of Heredity: A Historical Perspective

Heredity problems are concerned with determining which alleles are passed from the parents to the offspring and how likely it is that various types of offspring will be produced. The first person to develop a method of predicting the outcome of inheritance patterns was Gregor Mendel, who performed experiments on the inheritance of certain characteristics in sweet pea plants. He developed a line of reasoning that enables us to determine how various traits are inherited by studying their occurrence in parents and their offspring. From his work, Mendel concluded which traits were dominant and which were recessive. Some of his results follow.

Characteristic	Alleles
Plant height	Tall and dwarf
Pod shape	Full and constricted
Pod color	Green and yellow
Seed surface	Round and wrinkled
Seed color	Yellow and green
Flower color	Purple and white

Dominant	Recessive
Tall	Dwarf
Full	Constricted
Green	Yellow
Round	Wrinkled
Yellow	Green
Purple	White

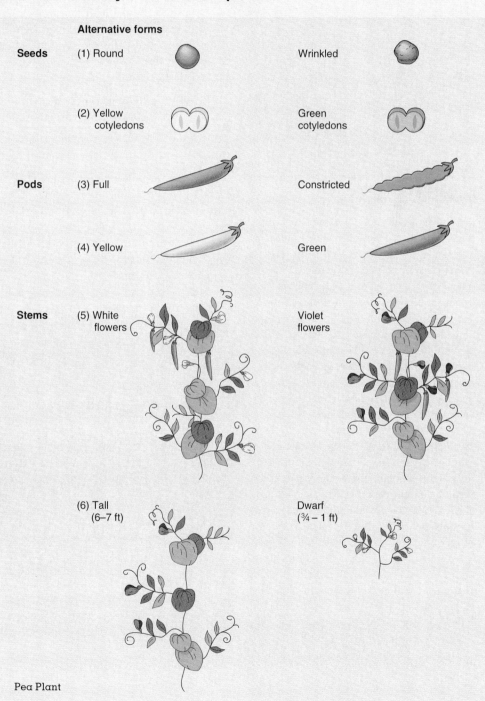

Alternative forms

Seeds

(1) Round   Wrinkled

(2) Yellow cotyledons   Green cotyledons

Pods

(3) Full   Constricted

(4) Yellow   Green

Stems

(5) White flowers   Violet flowers

(6) Tall (6–7 ft)   Dwarf (¾ – 1 ft)

Pea Plant

What made Mendel's work unique was that he studied only one trait at a time. Previous investigators had tried to follow numerous traits at a time. This made the total set of characteristics so cumbersome to work with that no clear idea could be formed of how the offspring inherited traits. Mendel used traits with clear-cut alternatives, such as purple or white flower color, yellow- or green-colored seed pods, and tall or dwarf pea plants. He was very lucky to have chosen pea plants in his study because they naturally self-pollinate. When self-pollination occurs in pea plants over many generations, it is possible to develop a population of plants that is homozygous for a number of characteristics. Such a population is known as a *pure line*.

Mendel took a pure line of pea plants having purple flower color, removed the male parts (anthers), and discarded them so that the plants could not self-pollinate. He then took anthers from a pure-breeding white-flowered plant and pollinated the antherless purple flower. When the pollinated flowers produced seeds, Mendel collected, labeled, and planted them. When these seeds germinated and grew, they eventually produced flowers.

You might be surprised to learn that all of the plants resulting from this cross had purple flowers. One of the prevailing hypotheses of Mendel's day would have predicted that the purple and white colors would have blended,

resulting in flowers that were lighter than the parental purple flowers. Another hypothesis would have predicted that the offspring would have had a mixture of white and purple flowers. The unexpected result—all of the offspring produced flowers like those of one parent and no flowers like those of the other—caused Mendel to examine other traits as well and form the basis for much of the rest of his work. He repeated his experiments using pure strains for other traits. Pure-breeding tall plants were crossed with pure-breeding dwarf plants. Pure-breeding plants with yellow pods were crossed with pure-breeding plants with green pods. The results were all the same: the offspring showed the characteristic of one parent and not the other.

Next, Mendel crossed the offspring of the white–purple cross (all of which had purple flowers) with each other to see what the third generation would be like. Had the characteristic of the original white-flowered parent been lost completely? This second-generation cross was made by pollinating among themselves the purple flowers that had one white parent. The seeds produced from this cross were collected and grown. When these plants flowered, three-fourths of them produced purple flowers and one-fourth produced white flowers.

After analyzing his data, Mendel formulated several genetic laws to describe how characteristics are passed from one generation to the next and how they are expressed in an individual.

Mendel's law of dominance When an organism has two different alleles for a trait, the allele that is expressed, overshadowing the expression of the other allele, is said to be *dominant*. The allele whose expression is overshadowed is said to be *recessive*.

Mendel's law of segregation When gametes are formed by a diploid organism, the alleles that control a trait separate from one another into different gametes, retaining their individuality.

Mendel's law of independent assortment Members of one gene pair separate from each other independently of the members of other gene pairs on other chromosomes.

At the time of Mendel's research, biologists knew nothing of chromosomes or DNA or of the processes of mitosis and meiosis. Mendel assumed that each gene was separate from other genes. It was fortunate for him that each characteristic he picked to study was found on a separate chromosome. If two or more of these genes had been located on the same chromosome (*linked genes*), he probably would not have been able to formulate his laws. The discovery of chromosomes and DNA have led to modifications in Mendel's laws, but it was Mendel's work that formed the foundation for the science of genetics.

apply your **knowledge**

A Throw of the Dice?

What is the probability of throwing three dice and having them all come up as fours?

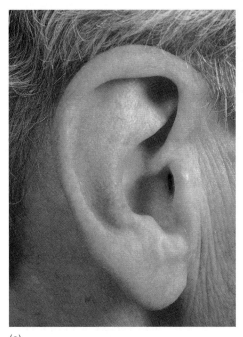

(a)

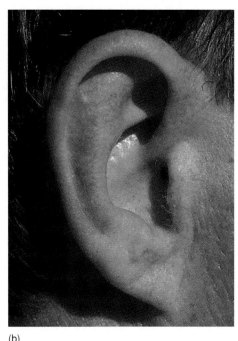

(b)

Figure 15.3 Structure of the Earlobe   Whether your earlobe is "free" (a) or "attached" (b) depends on the genes you have inherited. Some peoples' earlobes do not separate from the sides of their heads while others' do.

it is possible that both will be fours. What is the probability that both will be fours? The probability of one die being a four is 1/6. The probability of the other die being a four is also 1/6. Therefore, the probability of throwing two fours is

$$1/6 \times 1/6 = 1/36$$

Steps in Solving Heredity Problems

Single-Factor Crosses

The first type of problem we will work is the easiest type, a single-factor cross. A **single-factor cross** is a genetic cross or mating in which a single characteristic is followed from one generation to the next.

In humans, the allele for free earlobes is dominant and the allele for attached earlobes is recessive (figure 15.3). If both parents are heterozygous (have one allele for free earlobes and one allele for attached earlobes), what is the probability that they can have a child with free earlobes? with attached earlobes?

Solving a heredity problem involves five basic steps:

Step 1: Assign a symbol for each allele.
Usually a capital letter is used for a dominant allele and a lowercase letter for a recessive allele.

Use the symbol E for free earlobes and e for attached earlobes.

E = free earlobes e = attached earlobes

Genotype		Phenotype
EE	=	free earlobes
Ee	=	free earlobes
ee	=	attached earlobes

Step 2: Determine the genotype of each parent and indicate a mating.
Since both parents are heterozygous, the male genotype is Ee and the female genotype is also Ee. The $\times$ between them is used to indicate a mating.

$$Ee \times Ee$$

Step 3: Determine all the possible kinds of gametes each parent can produce.
Remember that gametes are haploid; therefore, they can only have one allele for a trait instead of the two present in the diploid cell. Since the male has both the free-earlobe allele and the attached-earlobe allele, half of his gametes will contain the free-earlobe allele and the other half will contain the attached-earlobe allele. Since the female has the same genotype, the genotype of her gametes will be the same as his.

To determine probabilities in genetic crosses, biologists use a *Punnett square.* A

Punnett square is a box that allows you to calculate the probability of genotypes and phenotypes of the offspring of a particular cross. Remember, because of the process of meiosis, each gamete receives only one allele for each characteristic listed. Therefore, the male will give either an E or an e; the female will also give either an E or an e. The possible gametes produced by the male parent are listed on the left side of the square, while the female gametes are listed on the top. In our example, the Punnett square would show a single dominant allele and a single recessive allele from the male on the left side. The alleles from the female would appear on the top (figure 15.4).

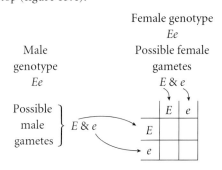

Step 4: Determine all the gene combinations that can result when these gametes unite.
To determine the possible combinations of alleles that could occur as a result of this mating,

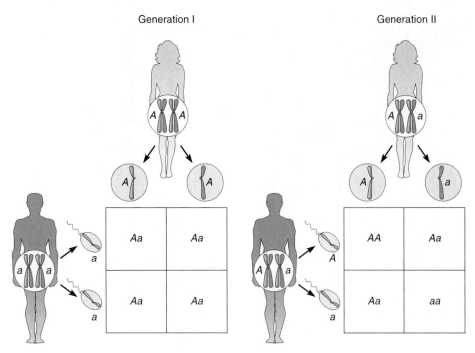

Generation I   Generation II

Figure 15.4 Punnet Square Set-Up The Punnet square is a way of illustrating the possible gene combinations that can occur during fertilization.

simply fill in each of the empty squares with the alleles that can be donated from each parent. Determine all the gene combinations that can result when these gametes unite.

Step 5: Determine the phenotype of each possible gene combination.

In this problem, three of the offspring, *EE, Ee,* and *Ee,* have free earlobes. One offspring, *ee,* has attached earlobes. Therefore, the answer to the problem is that the probability of having offspring with free earlobes is 3/4; for attached earlobes, it is 1/4.

Now let's work a problem in which one parent is heterozygous and the other is homozygous for a trait. Some people are unable to convert the amino acid phenylalanine into the amino acid tyrosine. The buildup of phenylalanine in the body prevents the normal development of the nervous system. Such individuals suffer from phenylketonuria (PKU) and may become mentally retarded. The normal condition is to convert phenylalanine to tyrosine. It is dominant over the condition for

PKU. If one parent is heterozygous and the other parent is homozygous for PKU, what is the probability that they will have a child who is normal? A child with PKU?

Step 1:

Use the symbol *N* for normal and *n* for PKU.

N = normal n = PKU

Genotype		Phenotype
NN	=	normal metabolism of phenylalanine
Nn	=	normal metabolism of phenylalanine
nn	=	PKU disorder

Step 2:

$$Nn \times nn$$

Step 3:

	n
N	
n	

Step 4:

	n
N	*Nn*
n	*nn*

Step 5:

In this problem, one-half of the offspring will be normal, and one-half will have PKU.

The Double-Factor Cross

A **double-factor cross** is a genetic study in which two pairs of alleles are followed from the parental generation to the offspring. This problem is worked in basically the same way as a single-factor cross. The main difference is that in a double-factor cross you are working with two different characteristics from each parent.

It is necessary to use Mendel's law of independent assortment when working double-factor problems. Recall that according to this law, members of one allelic pair separate from each other independently of the members of other pairs of alleles. This happens during meiosis when the chromosomes segregate. (Mendel's law of independent assortment applies only if the two pairs of alleles are located on separate chromosomes. We will assume this is so in double-factor crosses.)

In humans, the allele for free earlobes dominates the allele for attached earlobes. The allele for dark hair dominates the allele for light hair. If both parents are heterozygous for

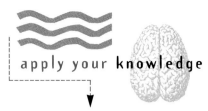

A Matter of Taste

Some humans inherit the ability to taste the chemical phenylthiocarbamide (PTC), and others are unable to taste PTC. Ask your instructor to furnish you with a supply of paper impregnated with PTC. Place a piece of this paper on your tongue. If you experience a bitter taste, you are a taster. If you do not, you are a nontaster. Take enough strips to test those members of your family you can readily contact. These may include your grandparents, parents, siblings, children, aunts, and uncles. After you have tested the members of your family, construct a pedigree. Is it possible for you to determine the genotypes of your family members? Is the ability to taste PTC a dominant or a recessive trait? Could it be a case of lack of dominance? Might it be linked to another trait?

In a pedigree, circles represent females and squares represent males. Symbols of parents are connected by a horizontal mating line, and the offspring are shown on a horizontal line below the parents. Individuals who are tasters are represented by a solid symbol, nontasters are represented by an open symbol. A typical pedigree might look something like this one, which shows three generations. Diagram your own pedigree. You will need to modify the pedigree shown here to fit your particular family situation.

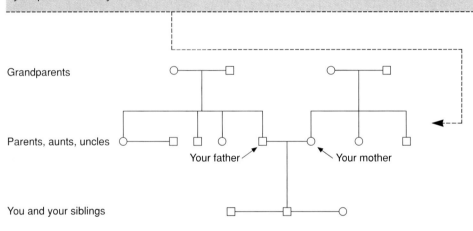

earlobe shape and hair color, what types of offspring can they produce, and what is the probability for each type?

Step 1:

Use the symbol *E* for free earlobes and *e* for attached earlobes. Use the symbol *D* for dark hair and *d* for light hair.

E = free earlobes   *D* = dark hair
e = attached earlobes   *d* = light hair

Genotype		Phenotype
EE	=	free earlobes
Ee	=	free earlobes
ee	=	attached earlobes
DD	=	dark hair
Dd	=	dark hair
dd	=	light hair

Step 2:

Determine the genotype for each parent and show a mating. The male genotype is *EeDd*, the female genotype is *EeDd*, and the × between them indicates a mating.

$$EeDd \times EeDd$$

Step 3:

Determine all the possible gametes each parent can produce and write the symbols for the alleles in a Punnett square. Since there are two pairs of alleles in a double-factor cross, each gamete must contain one allele from each pair—one from the earlobe pair (either *E* or *e*) and one from the hair color pair (either *D* or *d*). In this example, each parent can produce four different kinds of gametes. The four squares on the left indicate the gametes produced by the male; the four on the top indicate the gametes produced by the female.

To determine the possible gene combinations in the gametes, select one allele from one of the pairs of alleles and match it with one allele from the other pair of alleles. Then match the second allele from the first pair of alleles with each of the alleles from the second pair. This may be done as follows:

EeDd

	ED	*Ed*	*eD*	*ed*
ED				
Ed				
eD				
ed				

Step 4:

Determine all the gene combinations that can result when these gametes unite. Fill in the Punnett square.

	ED	*Ed*	*eD*	*ed*
ED	*EEDD*	*EEDd*	*EeDD*	*EeDd*
Ed	*EEDd*	*EEdd*	*EeDd*	*Eedd*
eD	*EeDD*	*EeDd*	*eeDD*	*eeDd*
ed	*EeDd*	*Eedd*	*eeDd*	*eedd*

Step 5:

Determine the phenotype of each possible gene combination. In this double-factor problem there are sixteen possible ways in which gametes could combine to produce offspring. There are four possible phenotypes in this cross. They are represented in the chart on page 337.

The probability of having a given phenotype is

9/16 free earlobes, dark hair

3/16 free earlobes, light hair

3/16 attached earlobes, dark hair

1/16 attached earlobes, light hair

For our next problem, let's say a man with attached earlobes is heterozygous for hair color and his wife is homozygous for free earlobes and light hair. What can they expect their offspring to be like?

This problem has the same characteristics as the previous problem. Following the same steps, the symbols would be the same, but the parental genotypes would be as follows:

$$eeDd \times EEdd$$

The next step is to determine the possible gametes that each parent could produce and place them in a Punnett square. The male parent can produce two different kinds of gametes, *eD* and *ed*. The female parent can only produce one kind of gamete, *Ed*.

	Ed
eD	
ed	

If you combine the gametes, only two kinds of offspring can be produced:

	Ed
eD	EeDd
ed	Eedd

They should expect either a child with free earlobes and dark hair or a child with free earlobes and light hair.

Figure 15.5 Snapdragons These snapdragons show a lack of dominance. The white flowers are homozygous recessive, the red are homozygous dominant, and the pink are heterozygous.

Real-World Situations

So far we have considered a few straightforward cases in which a characteristic is determined by simple dominance and recessiveness between two alleles. Other situations, however, may not fit these patterns. Some genetic characteristics are determined by more than two alleles; moreover, some traits are influenced by gene interactions and some traits are inherited differently, depending on the sex of the offspring.

Lack of Dominance

In the cases that we have considered so far, one allele of the pair was clearly dominant over the other. Although this is common, it is not always the case. In some combinations of alleles, there is a **lack of dominance.** This is a situation in which two unlike alleles both express themselves, neither being dominant. A classic example involves the color of the petals of snapdragons. There are two alleles for the color of these flowers (figure 15.5). Because neither allele is recessive, we cannot use the traditional capital and small letters as symbols for these alleles. Instead, the allele for white petals is given the symbol F^W, and the one for red petals is given the symbol F^R. There are three possible combinations of these two alleles:

Genotype		Phenotype
$F^W F^W$	=	white flower
$F^R F^R$	=	red flower
$F^R F^W$	=	pink flower

Notice that there are only two different alleles, red and white, but there are three phenotypes, red, white, and pink. Both the red-flower allele and the white-flower allele partially express themselves when both are present, and this results in pink.

Heredity problems dealing with lack of dominance are worked according to the five steps used in other problems. In the following lack-of-dominance problem, only one trait is involved; therefore, it is worked as a single-factor problem. If a pink snapdragon is crossed with a white snapdragon, what phenotypes can result, and what is the probability of each phenotype?

Genotype		Phenotype	Symbol
EEDD or *EEDd* or *EeDD* or *EeDd*	=	free earlobes and dark hair	= *
EEdd or *Eedd*	=	free earlobes and light hair	= ∧
eeDD or *eeDd*	=	attached earlobes and dark hair	= "
eedd	=	attached earlobes and light hair	= +

	ED	Ed	eD	ed
ED	EEDD *	EEDd *	EeDD *	EeDd *
Ed	EEDd *	EEdd ∧	EeDd *	Eedd ∧
eD	EeDD *	EeDd *	eeDD "	eeDd "
ed	EeDd *	Eedd ∧	eeDd "	eedd +

Step 1:

F^W = white flowers F^R = red flowers

Genotype		Phenotype
F^WF^W	=	white flower
F^WF^R	=	pink flower
F^RF^R	=	red flower

Step 2:

$$F^RF^W \times F^WF^W$$

Step 3:

	F^W
F^R	
F^W	

Step 4:

	F^W
F^R	F^WF^R pink
F^W	F^WF^W white

Step 5:

This cross results in two different phenotypes—pink and white. No red flowers can result because this would require that both parents be able to contribute at least one red allele. The white flowers are homozygous for white, and the pink flowers are heterozygous.

Multiple Alleles

So far we have discussed only traits that are determined by two alleles. However, there can be more than two alleles for a trait. The fact that some characteristics are determined by three or more alleles is called **multiple alleles.** However, any one individual can have at most two of the alleles for the characteristic. A good example of a characteristic that is determined by multiple alleles is the ABO blood type. There are three alleles for blood type:

I^A	=	blood type A
I^B	=	blood type B
i	=	blood type O

Blood types A and B show *codominance* when they are together in the same individual, but both are dominant over the O allele. These three alleles can be combined as pairs in six different ways, resulting in four different phenotypes:

Genotype		Phenotype
I^AI^A	=	blood type A
I^Ai	=	blood type A
I^BI^B	=	blood type B
I^Bi	=	blood type B
I^AI^B	=	blood type AB
ii	=	blood type O

Multiple-allele problems are worked as single-factor problems. Some examples are in the practice problems at the end of this chapter.

Polygenic Inheritance

Thus far we have considered phenotypic characteristics that are determined by alleles at a specific, single place on homologous chromosomes. However, some characteristics are determined by the interaction of genes at several different loci (on different chromosomes or at different places on a single chromosome). This is called **polygenic inheritance.** A number of different pairs of alleles may combine their efforts to determine a characteristic. Skin color in humans is a good example of this inheritance pattern. According to some experts, genes for skin color are located at a minimum of three loci. At each of these loci, the allele for dark skin is dominant over the allele for light skin. Therefore, a wide variety of skin colors is possible depending on how many dark-skin alleles are present (figure 15.6). Polygenic inheritance is very common in determining characteristics that are quantitative in nature. In the skin-color example, and in many others as well, the characteristics cannot be categorized in terms of *either/or,* but the

Locus 1	d^1d^1	d^1D^1	d^1D^1	D^1D^1	D^1d^1	D^1d^1	D^1D^1
Locus 2	d^2d^2	d^2d^2	d^2D^2	D^2d^2	D^2d^2	D^2D^2	D^2D^2
Locus 3	d^3d^3	d^3d^3	d^3d^3	d^3d^3	D^3D^3	D^3D^3	D^3D^3
Total number of dark-skin genes	0	1	2	3	4	5	6
	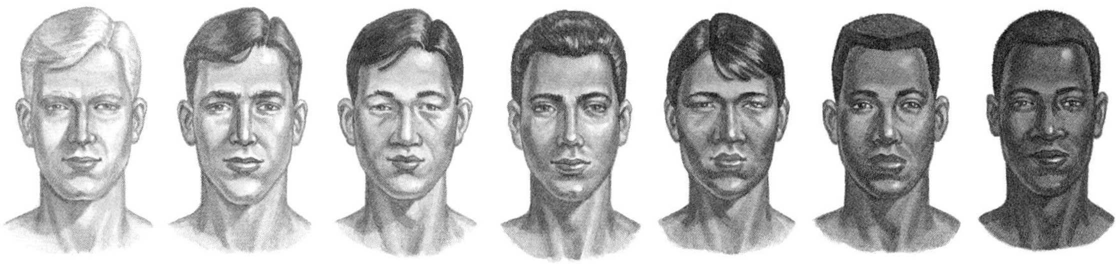 Very light			Medium			Very dark

Figure 15.6 Polygenic Inheritance Skin color in humans is an example of polygenic inheritance. The darkness of the skin is determined by the number of dark-skin genes a person inherits from his or her parents.

variation in phenotypes can be classified as *how much* or *what amount*. For instance, people show great variations in height. There are not just tall and short people—there is a wide range. Some people are as short as one meter, and others are taller than two meters. This quantitative trait is probably determined by a number of different genes. Intelligence also varies significantly, from those who are severely retarded to those who are geniuses. Many of these traits may be influenced by outside environmental factors such as diet, disease, accidents, and social factors. These are just a few examples of polygenic inheritance patterns.

Pleiotropy

A gene often has a variety of effects on the phenotype of an organism. In fact, every gene probably affects or modifies the expression of many different characteristics exhibited by an organism. This is called *pleiotropy*. **Pleiotropy** is a term used to describe the multiple effects that a gene may have on the phenotype of an organism. For example, the gene for sickle-cell hemoglobin has two major effects. One is good and one is bad. Having the allele for sickle-cell hemoglobin can result in abnormally shaped red blood cells. This occurs because the hemoglobin molecules are synthesized with the wrong amino acid sequence. These abnormal hemoglobin molecules tend to attach to one another in long, rodlike chains when oxygen is in short supply. These rodlike chains distort the shape of the red blood cells into a sickle shape. When these abnormal red blood cells change shape, they clog small blood vessels. The sickled red cells are also destroyed more rapidly than normal cells. This results in a shortage of red blood cells, causing anemia and an oxygen deficiency in the tissues that have become clogged. People with sickle-cell anemia may experience pain, swelling, and damage to organs such as the heart, lungs, brain, and kidneys.

Although sickle-cell anemia is usually lethal in the homozygous condition, it can be beneficial in the heterozygous state. A person with a single sickle-cell allele is more resistant to malaria than a person without this gene because the malarial protozoan cannot survive in

The Inheritance of Eye Color

It is commonly thought that eye color is inherited in a simple dominant/recessive manner. Brown eyes are considered to be dominant over blue eyes. The real pattern of inheritance, however, is considerably more complicated than this. Eye color is determined by the amount of a brown pigment, known as *melanin*, present in the iris of the eye. If there is a large quantity of melanin present on the anterior surface of the iris, the eyes are dark. Black eyes have a greater quantity of melanin than brown eyes.

If there is not a large amount of melanin present on the anterior surface of the iris, the eyes will appear blue, not because of a blue pigment but because blue light is returned from the iris. The iris appears blue for the same reason that deep bodies of water tend to appear blue. There is no blue pigment in the water, but blue wavelengths of light are returned to the eye from the water. People

appear to have blue eyes because the blue wavelengths of light are reflected from the iris.

Just as black and brown eyes are determined by the amount of pigment present, colors such as green, gray, and hazel are produced by the various amounts of melanin in the iris. If a very small amount of brown melanin is present in the iris, the eye tends to appear green, whereas relatively large amounts of melanin produce hazel eyes.

Several different genes are probably involved in determining the quantity and placement of the melanin and, therefore, in determining eye color. These genes interact in such a way that a wide range of eye color is possible. Eye color is probably determined by polygenic inheritance, just as skin color and height are. Some newborn babies have blue eyes that later become brown. This is because they have not yet begun to produce melanin in their irises at the time of birth.

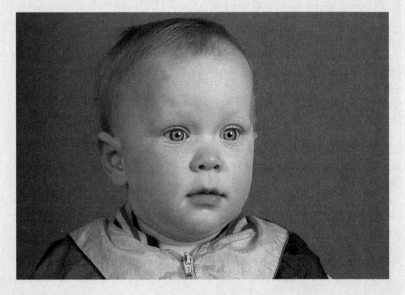

these cells. A heterozygous person may not demonstrate any ill effects, but under laboratory conditions with low oxygen, there is a change in the red blood cells. Three genotypes can exist (Hb^A = normal hemoglobin, Hb^S = sickle-cell hemoglobin):

Genotype		Phenotype
$Hb^A Hb^A$	=	normal hemoglobin and nonresistance to malaria
$Hb^A Hb^S$	=	normal hemoglobin and resistance to malaria
$Hb^S Hb^S$	=	resistance to malaria but death from sickle-cell anemia

Sickle-cell anemia was originally found throughout the world in places where malaria was common. Today, however, this genetic disease can be found anywhere in the world. In the United States, it is most common among black populations whose ancestors came from equatorial Africa.

Let's look at another example of pleiotropy. In this example, a single gene affects many different chemical reactions that depend on the way a cell metabolizes the amino acid phenylalanine (figure 15.7). People normally have a gene for the production of an enzyme that converts the amino acid phenylalanine to tyrosine. If this gene is functioning properly, phenylalanine will be converted to tyrosine, which will be available to be converted into thyroxine and melanin by other enzymes. If the enzyme that normally converts phenylalanine to tyrosine is absent, toxic materials can accumulate and result in a loss of nerve cells, causing mental retardation. Because less tyrosine is produced, there is also less of the growth hormone thyroxine, resulting in abnormal body growth. Because tyrosine is necessary to form the pigment melanin, people who have this condition have lighter skin color. The one abnormal allele produces three different phenotypic effects: mental retardation, abnormal growth, and light skin.

Linkage

Pairs of alleles located on nonhomologous chromosomes separate independently of one another during meiosis when the chromosomes separate into sex cells. Since each chromosome has many genes on it, these genes tend to be inherited as a group. Genes located on the same chromosome that tend to be inherited together are called a **linkage group.** The closer two genes are to each other on a chromosome, the more probable it is that they will be inherited together. The process of crossing over, which occurs during prophase I of meiosis, is a physical exchange of pieces of chromosomes and may split up these linkage groups. Crossing over happens between homologous chromosomes donated by the mother and the father, and results in a mixing of genes.

Sex-Linked Genes

In humans and some other animals there are two types of sex chromosomes: the X chromosome and the Y chromosome. The Y chromosome is much shorter than the X chromosome and probably has no genes for traits found on the X chromosome. One portion of the Y chromosome contains a male determining gene. Females are normally produced when two X chromosomes are present. Males are usually produced when one X chromosome and one Y chromosome are present.

Genes found on the X chromosome are said to be **X-linked.** Because the Y chromosome is shorter than the X chromosome, it does not have many of the alleles that are found on the comparable portion of the X chromosome. Therefore, in a man, the presence of a single allele on his only X chromosome will be expressed, regardless of whether it is dominant or recessive. The SRY gene has been confirmed to be Y-linked in humans. This gene controls the differentiation of the embryonic gonad to a male testis. By contrast, more than one hundred genes are linked to the

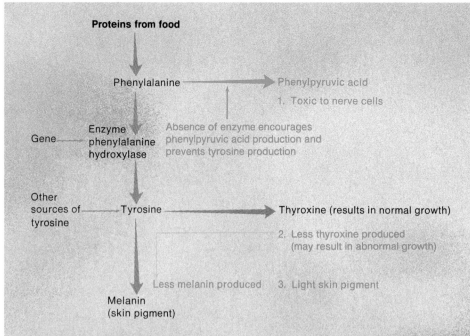

Figure 15.7 Pleiotropy Pleiotropy is a condition in which a single gene has more than one effect on the phenotype. This diagram shows how normal pathways work (these are shown in black). If the enzyme phenylalanine hydroxylase is not produced because of an abnormal gene, there are three major results: (1) mental retardation, because phenylpyruvic acid kills nerve cells; (2) abnormal body growth, because less of the growth hormone thyroxine is produced; and (3) pale skin pigmentation, because less melanin is produced (abnormalities are shown in color). It should also be noted that if a woman who has PKU becomes pregnant, her baby is likely to be born retarded. While the embryo may not have the genetic disorder, the phenylpyruvic acid produced by the pregnant mother will damage the developing brain cells. This is called maternal PKU.

BIO feature

We Don't Know "Y"

Why is the Y chromosome so small in so many species (including mammals, birds, and certain species of fish, reptiles, and insects)? This makes individuals with an XY combination genetically deficient in comparison to XX individuals. In other words, male humans are genetically deficient in comparison to females!

Is there an advantage to a species in having one of their sex chromosomes deficient in genes? One hypothesis answers "Yes!" Consider the idea that, with genes for supposedly "female" characteristics eliminated from the Y chromosome, crossing-over and recombining with "female" genes on the X chromosome during meiosis could help to keep sex traits separated. Males would be males and females would stay females. The chances of "male" and "female" genes getting mixed onto the same chromosome would be next to impossible since they would not even exist on the Y chromosome. For example, in common tropic fish (the guppy), males have genes for bright colors and fancy fins. The females use these genetic traits to recognize the males as suitable mates. Female guppies lack these genes. If both the X and Y chromosomes had the genes for color and fancy fin shapes and were to cross over during meiosis, both males and females could acquire these genes and it would be impossible to tell a male guppy from a female guppy.

X chromosome. Some of these X-linked genes are color blindness, hemophilia, brown teeth, and a form of muscular dystrophy. To better understand an X-linked gene, let's now use the five steps of solving genetics problems to work an X-linked problem.

In humans, the gene for normal color vision is dominant and the gene for color blindness is recessive. Both genes are X-linked. A male who has normal vision mates with a female who is heterozygous for normal color vision. What type of children can they have in terms of these traits, and what is the probability for each type?

Step 1:

Because this condition is linked to the X chromosome, it has become traditional to symbolize the allele as a superscript on the letter X. Since the Y chromosome does not contain a homologous allele, only the letter Y is used.

X^N	=	normal color vision
X^n	=	color-blind
Y	=	male (no gene present)

Genotype		Phenotype
$X^N Y$	=	male, normal color vision
$X^n Y$	=	male, color-blind
$X^N X^N$	=	female, normal color vision
$X^N X^n$	=	female, normal color vision
$X^n X^n$	=	female, color-blind

Step 2:

Male's genotype = $X^N Y$ (normal color vision)

Female's genotype = $X^N X^n$ (normal color vision)

$$X^N Y \times X^N X^n$$

Step 3:

The genotype of the gametes are listed in the Punnett square:

	X^N	X^n
X^N		
Y		

Step 4:

The genotypes of the probable offspring are listed in the body of the Punnett square:

	X^N	X^n
X^N	$X^N X^N$	$X^N X^n$
Y	$X^N Y$	$X^n Y$

Step 5:

The phenotypes of the offspring are determined:

normal female	carrier female
normal male	color-blind male

1/4 normal female   1/4 carrier female

1/4 normal male   1/4 color-blind male

As we noted earlier, a carrier is any individual who is heterozygous for a trait. In this situation, the recessive allele is hidden. In X-linked situations, only the female can be heterozygous because the males lack one of the X chromosomes. Heterozygous females will exhibit the dominant trait (normal vision in this problem), but have the recessive allele hidden. If a male has a recessive allele on the X chromosome, it will be expressed because there is no other allele on the Y chromosome to dominate it. If a heterozygous carrier (must be female) has sons, she should expect half of them to be color-blind and half to have normal color vision. For these reasons, there are many more color-blind males than there are color-blind females.

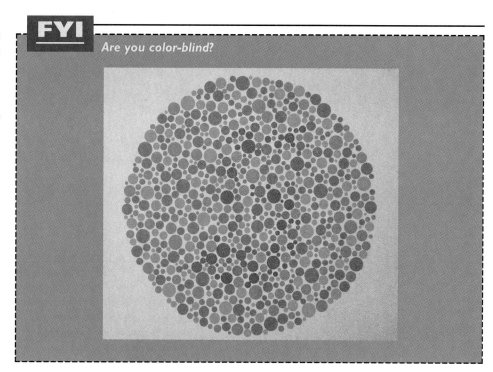

Environmental Influences on Gene Expression

The specific phenotype an organism exhibits is determined by the interplay of the genotype of the individual and the conditions the organism encounters as it develops. Therefore, it is possible for two organisms with identical genotypes (identical twins) to differ in their phenotypes. All genes must express themselves through the manufacture of proteins. These proteins may be structural or enzymatic, and the enzymes may be more or less effective depending on the specific biochemical conditions that exist when the enzyme is in operation. The expression of the genes will vary depending on the environmental conditions that exist while the gene is operating.

Maybe you assumed that the dominant allele would always be expressed in a heterozygous individual. It is not so simple! Here, as in other areas of biology, there are exceptions. For example, the allele for six fingers is dominant over the allele for five fingers in humans. Some people who have received the allele for six fingers have a fairly complete sixth finger; in others, it may appear as a little stub. An-other example is a dominant allele that may cause the formation of a little finger that cannot be bent as a normal little finger can be. However, not all people who are believed to have inherited this dominant allele have a stiff little finger. In some cases, the dominant characteristic is not expressed or perhaps only shows on one hand. Thus, there may be variation in the degree to which a dominant allele expresses itself, and in some cases it may not be expressed at all. Other genes may be interacting with these dominant alleles, causing the variation in expression. It is important to recognize that the environment affects the expression of our genes in many ways.

Both internal and external environmental factors can influence the expression of genes. For example, at conception, a male receives genes that will eventually determine the pitch of his voice. However, these genes are expressed differently after puberty. At puberty, male sex hormones are released. This internal environmental change results in the deeper male voice. A male who does not produce these hormones retains a higher-pitched voice in later life. A comparable situation in females occurs when an abnormally functioning adrenal gland causes the release

of large amounts of male hormones. This results in a female with a deeper voice.

Another influence on gene expression is called *gene imprinting*. This occurs when a gene donated by one of the parents has its expression altered by a gene donated by the other parent. Certain genes donated by the father affect those donated by the mother, and certain genes donated by the mother affect those donated by the father. It is also true that certain genes function properly only when donated by the father, and other genes function properly only when donated by the mother. Although the mechanics of the process are not yet well understood, it is known that paternal and maternal genes contribute in different ways to the developing embryo. Two forms of mental retardation illustrate gene imprinting in humans—Angelman syndrome and Prader-Willi syndrome. Prader-Willi is transmitted from the mother and Angelman is transmitted from the father. Both are disorders of chromosome 15. Patients with Angelman syndrome display excessive laughter, jerky movements, and other symptoms of physical and mental retardation. Those with Prader-Willi syndrome show mental retardation, extreme obesity, short stature, and unusually small hands and feet.

Many external environmental factors can influence the phenotype of an individual. One such factor is diet. Diabetes mellitus, a metabolic disorder in which glucose is not properly metabolized and is passed out of the body in the urine, has a genetic basis. Some people who have a family history of diabetes are thought to have inherited the trait for this disease. Evidence indicates that they can delay the onset of the disease by reducing the amount of sugar in their diet. This change in the external environment influences gene expression in much the same way that temperature influences the expression of color production in cats or sunlight affects the expression of freckles in humans, as we saw at the beginning of the chapter.

SUMMARY

Genes are units of heredity composed of specific lengths of DNA that determine the characteristics an organism displays. Specific genes are at specific loci on specific chromosomes.

BIO *feature*

Baldness and the Expression of Genes

It is a common misconception that males have genes for baldness and females do not. Male pattern baldness is a sex-influenced trait in which both males and females may possess alleles coding for baldness. These genes are turned on by high levels of the hormone testosterone. This is another example of an internal gene regulating factor.

The phenotype displayed by an organism is the result of the effect of the environment on the ability of the genes to express themselves. Diploid organisms have two genes for each characteristic. The alternative genes for a characteristic are called alleles. There may be many different alleles for a particular characteristic. Those organisms with two identical alleles are homozygous for a characteristic; those with different alleles are heterozygous for a characteristic. Some alleles are dominant over other alleles that are said to be recessive.

Sometimes two alleles will express themselves, and often a gene will have more than one recognizable effect on the phenotype of the organism. Some characteristics may be determined by several different pairs of alleles. In humans and some other animals, males have an X chromosome with a normal number of genes and a Y chromosome with fewer genes. Although they are not identical, they behave as a pair of homologous chromosomes. Since the Y chromosome is shorter than the X chromosome and has fewer genes, many of the recessive characteristics present on the X chromosome appear more frequently in males than in females, who have two X chromosomes.

CHAPTER GLOSSARY

carrier (ka´re-er) Any individual having a hidden, recessive allele.

double-factor cross (dub´l fak´tur kros) A genetic study in which two pairs of alleles are followed from the parental generation to the offspring.

genome (je´nōm) A set of all the genes necessary to specify an organism's complete list of characteristics.

genotype (je´no-tīp) The catalog of genes of an organism, whether or not these genes are expressed.

heterozygous (he˝ter-ō-zi´gus) A diploid organism that has two different allelic forms of a particular gene.

homozygous (ho˝mō-zi´gus) A diploid organism that has two identical alleles for a particular characteristic.

lack of dominance (lak uv dom´in-ans) The condition of two unlike alleles both expressing themselves, neither being dominant.

law of dominance (law uv dom´in-ans) When an organism has two different alleles for a trait, the allele that is expressed and overshadows the expression of the other allele is said to be dominant. The allele whose expression is overshadowed is said to be recessive.

law of independent assortment (law uv in˝de-pen´dent a-sort´ment) Members of one allelic pair will separate from each other independently of the members of other allele pairs.

law of segregation (law uv seg˝re-ga´shun) When gametes are formed by a diploid organism, the alleles that control a trait separate from one another into different gametes, retaining their individuality.

linkage group (lingk´ij grup) Genes located on the same chromosome that tend to be inherited together.

locus (lo´kus) The spot on a chromosome where an allele is located (pl. *loci*).

Mendelian genetics (men-de´le-an je-net´iks) The pattern of inheriting characteristics that follows the laws formulated by Gregor Mendel.

multiple alleles (mul´ti-pul a-le lz´) A term used to refer to conditions in which there are several different alleles for a particular characteristic, not just two.

phenotype (fe n´o-tīp) The physical, chemical, and behavioral expression of the genes possessed by an organism.

pleiotropy (pli-ot´ro-pe) The multiple effects that a gene may have on the phenotype of an organism.

polygenic inheritance (pol˝e-jen´ik in-her´i-tans) The concept that a number of different pairs of alleles may combine their efforts to determine a characteristic.

probability (prob˝a-bil´i-te) The chance that an event will happen, expressed as a percent or fraction.

Punnett square (pun´net sqwar) A method used to determine the probabilities of allele combinations in a zygote.

single-factor cross (sing´ul fak´tur kros) A genetic study in which a single characteristic is followed from the parental generation to the offspring.

X-linked gene (eks-lingkt jen) A gene located on one of the sex-determining X chromosomes.

 # EXPLORATIONS   Interactive Software

Heredity in Families

This interactive exercise allows students to explore how the character and location of a genetic trait influences how it is inherited in families. The variables are dominance/recessiveness of the allele, and X-linkage vs. autosomal location of the allele. From a bank of 30 actual pedigrees, one is selected at random, and the two variables analyzed. The program scores the answer selected, and then presents another pedigree from the bank. As the 30 pedigrees are examined in random order, the program keeps score of the analyses. These pedigrees are often all a human geneticist has to work with in attempting to assess the dominance and chromosomal location of a trait.

1. In analyzing family pedigrees, why are sex-linked alleles expressed so much more frequently in male offspring than in females?
2. Why do you imagine some genetic disorders like sickle-cell anemia and hemophilia are common, while others are rare?
3. In a pedigree, what affect does sex linkage have on your ability to determine if a trait is being influenced by a single gene?
4. What is the minimal evidence you would accept that a trait appearing in a family is indeed hereditary (that is, caused by an allele), rather than environmentally induced (that is, not due to an allele)?

Cystic Fibrosis

This interactive exercise explores the relationship between membrane transport and concentrations of materials inside and outside a cell. The student studies the effects of ion concentrations on osmosis by seeing what happens when the transport of chloride ions is blocked, due to a faulty channel protein, as occurs in the condition known as cystic fibrosis.

1. What is the effect of increased extracellular chloride ion concentration on water movement into the cell?
2. What is the effect upon water movement of disabling the chloride channel?
3. Can artificially decreasing the chloride ion concentration in extracellular fluids counteract cystic fibrosis?

CONCEPT MAP TERMINOLOGY

Construct a concept map to represent the relationships among the following concepts.

allele
chromosomes
DNA
dominant
locus
meiosis
recessive
single-factor inheritance

LABEL • DIAGRAM • EXPLAIN

At a very large family reunion people were mentioning the fact that everyone present had curly black hair and dark brown eyes. Several hours later a long-forgotten cousin, Fred, arrived with his wife Sarah and their ten children. Great-grandmother could not understand why Fred had straight red hair and green eyes; no one in the family ever had these traits. She also wanted to know why Sarah and (1) five of the children had curly black hair and brown eyes; (2) three children had curly red hair and green eyes; and (3) two of the children had Fred's straight red hair and brown eyes.

You are a college-educated member of the family. Explain to grandmother how this could occur.

Multiple Choice Questions

1. *BbCCDd* is the genotype of an organism. How many different types of gametes can this organism produce?
   a. 1
   b. 2
   c. 3
   d. 4

2. An example of a phenotype is:
   a. AB type blood
   b. allele for type A and B blood
   c. a sperm with an allele for A blood
   d. lack of iron in the diet causing anemia

3. Body size is determined by the interaction of numerous alleles. This is an example of:
   a. autosomes
   b. pleiotropy
   c. single-factor crosses
   d. polygenic inheritance

4. The probability of guessing the correct answer to this question is:
   a. four
   b. one-fourth
   c. sixteen
   d. 1/4 times 1/4

5. If both parents are heterozygous, the probability that the recessive trait will appear in the offspring is:
   a. 1/2
   b. 1/4
   c. 3/4
   d. zero

6. In order for a recessive X-linked trait to appear in a female, she must inherit a recessive allele from:
   a. neither parent
   b. both parents
   c. her father only
   d. her mother only

7. In certain flowers it was noticed that if a flower had red petals they were usually small. If the petals were white, pink, or orange they were usually large. This could be the result of:
   a. lack of dominance
   b. pleiotropy
   c. linkage groups
   d. polygenic inheritance

Questions with Short Answers

1. How many kinds of gametes are possible with each of the following genotypes?
   a. *Aa*
   b. *AaBB*
   c. *AaBb*
   d. *AaBbCc*

2. What is the probability of getting the gamete *ab* from each of the following genotypes?
   a. *aabb*
   b. *Aabb*
   c. *AaBb*
   d. *AABb*

3. What is the probability of each of the following sets of parents producing the given genotypes in their offspring?

	Parents	Offspring Genotype
a.	*AA × aa*	*Aa*
b.	*Aa × Aa*	*Aa*
c.	*Aa × Aa*	*aa*
d.	*AaBb × AaBB*	*AABB*
e.	*AaBb × AaBB*	*AaBb*
f.	*AaBb × AaBb*	*AABB*

4. If an offspring has the genotype *Aa,* what possible combinations of parental genotypes can exist?

5. In humans, the allele for albinism is recessive to the allele for normal skin pigmentation.
   a. What is the probability that a child of a heterozygous mother and father will be an albino?
   b. If a child is normal, what is the probability that it is a carrier of the recessive albino allele?

6. In certain pea plants, the allele *T* for tallness is dominant over *t* for shortness.
   a. If a homozygous tall and homozygous short plant are crossed, what will be the phenotype and genotype of the offspring?
   b. If both individuals are heterozygous, what will be the phenotypic and genotypic ratios of the offspring?

7. Smoos are strange animals with one of three shapes: round, cuboidal, or pyramidal. If two cuboidal smoos mate, they always have cuboidal offspring. If two pyramidal smoos mate, they always produce pyramidal offspring. If two round smoos mate, they produce all three kinds of offspring. Assuming only one locus is involved, answer the following questions.
   a. How is smoo shape determined?
   b. What would be the phenotypic ratio if a round and cuboidal smoo were to mate?

8. What is the probability of a child having type AB blood if one of the parents is heterozygous for A blood and the other is heterozygous for B? What other genotypes are possible in these children?

9. A color-blind woman marries a man with normal vision. They have ten children, six boys and four girls.
   a. How many are likely to be normal?
   b. How many are likely to be color-blind?

10. A light-haired man has blood type O. His wife has dark hair and blood type AB, but her father had light hair.
   a. What is the probability that this couple will have a child with dark hair and blood type A?
   b. What is the probability that they will have a light-haired child with blood type B?
   c. How many different phenotypes could their children show?

11. Certain kinds of cattle have two alleles for coat color: *R* = red, and *r* = white. When an individual cow is heterozygous, it is spotted with red and white (roan). When two red alleles are present, it is red. When two white alleles are present, it is white. The allele *L,* for lack of horns, is dominant over *l,* for the presence of horns.
   a. If a bull and a cow both have the genotype *RrLl,* how many possible phenotypes of offspring can they have?
   b. How probable is each phenotype?

12. Hemophilia is a disease that prevents the blood from clotting normally. It is caused by a recessive allele located on the X chromosome. A boy has the disease; neither his parents nor his grandparents have the disease. What are the genotypes of his parents and grandparents?

The Molecular Basis of Heredity: DNA&RNA

DNA finger printing for genetic and medical research.

Getting the Message from Genetic Material

The phenotype of an organism is the physical, chemical, and behavioral expression of its genetic material. But just what is the chemical makeup of a chromosome and what does it mean to "express" a gene? In order for a person to "show" or "express" an allele for attached earlobes, what must happen in an embryo's cells that cause them to grow so that the earlobes attach to the side of the face instead of hanging free?

learning objectives

- Recognize the structure of DNA and RNA.

- Distinguish between DNA, nucleoprotein, chromatin, and chromosome.

- Diagram the DNA replication process.

- Diagram the DNA transcription process.

- Explain the need for enzymes in the maintenance of living things.

- Describe what happens when an enzyme and a substrate combine.

- Relate the three-dimensional structure of an enzyme to its ability to catalyze a reaction.

- Diagram the process of translation.

- Describe the processes involved in recombinant DNA procedures and genetic medicine.

In the nucleus of every eukaryotic cell is a very important library of instructions in the form of the chromosomes. This library contains all the directions necessary to create the structures and regulate the chemical reactions of an entire individual. It is like a reference library of books on how to make an organism. The books (genes) cannot be "checked out." It is a reference library only. However, you can make copies of the books, that is, they can be photocopied for use outside the nucleus. The originals always stay in the library. It may seem impossible that the directions for growth and development of an entire organism can be stored in structures as small as chromosomes. This can be done because the directions are coded in the chemical structure of molecules called **DNA (deoxyribonucleic acid)** and **RNA (ribonucleic acid)**. These two kinds of molecules are often referred to as **nucleic acids** since they are common in the nucleus of cells.

The Central Dogma

Once scientists began to understand the chemical makeup of the nucleic acids, they attempted to understand how DNA and RNA relate to cell structure and activities. The concept that resulted is known as the *central dogma*, main belief, or "source of all information." It is most easily written in this form:

Appendix A

What this says is that DNA is the genetic material of the cell and (reading to the left) it is capable of reproducing itself, a process called *DNA replication*. Reading to the right, DNA is capable of supervising the manufacture of RNA (a process known as *transcription*), which in turn is involved in the production of protein molecules, a process known as *translation*.

DNA ⟵ (replication) ⟵ DNA ⟶ (transcription) ⟶ RNA ⟶ (translation) ⟶ Proteins

structural

regulator

The Molecular Basis of Heredity

DNA & RNA

Molecular Paleontologists—The New Archaeologists

Molecular biologists are revolutionizing archaeology and paleontology with new ways of identifying the genetic material, DNA, found in specimens of plants, animals, and microorganisms from previous geological periods. These methods enable researchers to better understand the evolutionary relationships among long-extinct species and their rates of evolution. They also provide clues to the racial makeup and the kinship, distribution, and migration patterns of our ancient ancestors. The old picture of archaeologists digging for dinosaur bones is giving way to a new image. We now see them performing DNA analysis on the genetic "remains" of preserved fossils found in the field and in museums. One of the oldest samples of mammal DNA to be analyzed was from a woolly mammoth, frozen in the permafrost of Siberia. Molecular paleontologists have also successfully copied and analyzed DNA from cells taken from the fossil Quagga, an animal that resembles a cross between a horse and a zebra, extinct for more than 100 years; numerous 2,400-year-old Egyptian mummies, including the famous Pharaoh Tutankhamen; and 18-million-year-old magnolia leaves.

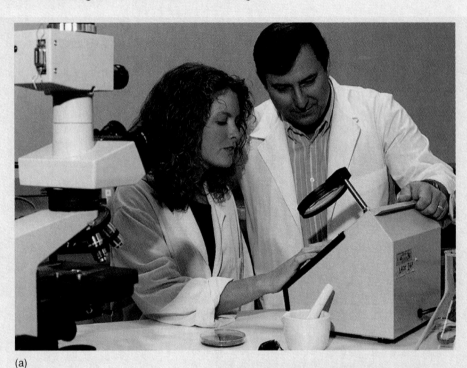

(a)

Molecular Paleontologists: The New Archaeologists

(a) California Polytechnic State University microbiologist Raul Cano and graduate student Jody Johnsbaugh analyze DNA of bacteria revived from the abdomen of an extinct bee trapped in amber for 25 to 40 million years. (b) This photo shows a chunk of amber containing ancient spider fragments.

(b)

The Molecular Structure of DNA

Nucleic acids (DNA and RNA) are large and complex molecules that store and transfer information within a cell. The nucleic acids are made up of subunits called **nucleotides.** Each nucleotide is composed of a *sugar* molecule (S) containing five carbon atoms, a *phosphate* group (P), and a molecule containing nitrogen that will be referred to as a *base* (B) (figure A).

There are eight common types of nucleotides available in a cell for building nucleic acids. These eight nucleotides differ from one another in the kind of sugar- and nitrogen-containing parts they possess (figure B). Because of these differences, it is possible to classify nucleic acids into two main groups: *ribonucleic acid* (RNA) and *deoxyribonucleic acid* (DNA). The nucleotides can contain two sizes of nitrogenous bases. The larger molecules are **adenine** (A) and **guanine** (G), which differ in the kinds of atoms attached to their double-ring structure. The smaller nitrogen-containing molecules are **cytosine** (C), **thymine** (T), and **uracil** (U). Each of these differs from the others in the kinds of atoms attached to its single-ring structure. These differences in size are important, as you will see later.

In cells, DNA is the nucleic acid that functions as the original blueprint for the synthesis of proteins. It contains the sugar deoxyribose; phosphates; and adenine, guanine, cytosine, and thymine. RNA is a type of nucleic acid that is directly involved in the synthesis of protein. It contains the sugar ribose; phosphates; and adenine, guanine, cytosine, and uracil. There is no thymine in RNA and no uracil in DNA.

The nucleic acids are made as a chain of nucleotides that can be compared to a comb. The protruding "teeth" (A, T, G, C, or U) are connected to a common "backbone" (sugar and phosphate molecules). This is the basic structure of both RNA and DNA (figure C).

DNA and RNA differ in one other respect. DNA is actually a double molecule. It consists of *two* flexible comblike strands held together between their protruding teeth. The two strands are twisted about each other in a coil or double helix (plural *helices*)

(a)　　　　　(b)　　　　　(c)　　　　　(d)

Figure A   Nucleotide Structure   All nucleotides are constructed in the basic way shown in (a). The nucleotide is the basic structural unit of all nucleic acid molecules. Notice in (c) that the phosphate group is written in "shorthand" form as a P inside a circle. Part (d) is a stylized version of a nucleotide that will be used throughout the chapter. Remember that this style is only representative of the complex organic molecule shown in (a).

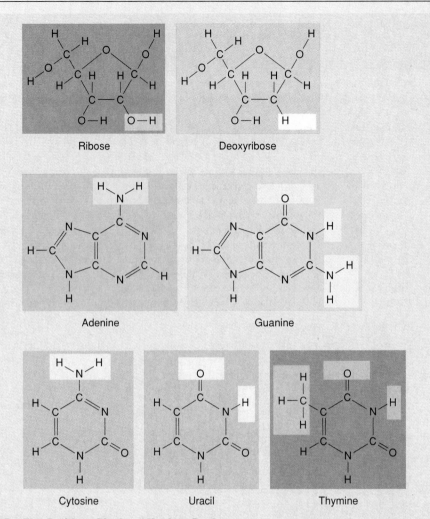

Ribose

Deoxyribose

Adenine

Guanine

Cytosine

Uracil

Thymine

Figure B   The Building Blocks of Nucleic Acids   All nucleic acids are composed of two organic components: a five-carbon sugar molecule and a nitrogenous base. Notice the difference (highlighted in color) between the two sugar molecules in part (a). The nitrogenous bases are divided into two groups according to their size. The large *purines*—adenine and guanine molecules—differ from each other in their attached groups (in color), as do the three smaller *pyrimidine* nitrogenous bases—cytosine, thymine, and uracil—in (c). The two types of nucleic acids—DNA and RNA—are composed of these eight building blocks. Note that each building block has a sugar, base, and phosphate component. These nucleotides are color-coded throughout the chapter so that you can recognize the difference between DNA and RNA.

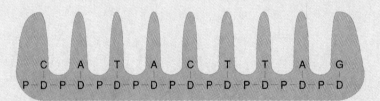

Figure C   A Single Strand of DNA   A single strand of DNA resembles a comb. The molecule is much longer than pictured here and is composed of a sequence of linked nucleotides.

Continued

The Molecular Structure of DNA (continued)

(figure D). The two combs of the molecule are held together by weak chemical forces. The four kinds of teeth always pair in a definite way: adenine (A) with thymine (T), and guanine (G) with cytosine (C). Notice that the large molecules (A and G) pair with the small ones (T and C), thus keeping the two complementary (matched) strands parallel. Three weak chemical forces are formed between guanine and cytosine:

$$G \vdots \vdots \vdots C$$

and two between adenine and thymine:

$$A \vdots \vdots \vdots T$$

You can "write" a message in the form of a stable DNA molecule by combining the four different DNA nucleotides (A, T, G, C) in particular sequences. Notice in figure C, that it is possible to make sense out of the sequence of nitrogen-containing portions of the molecule. If you "read" them from left to right in groups of three, you can read three words: CAT, ACT, and TAG. In this case, the four DNA nucleotides are being used as an alphabet to construct three-letter words. In order to make sense out of such a code, it is necessary to read in one direction. Reading the sequence in reverse does not always make sense, just as reading this paragraph in reverse would not make sense.

The genetic material of humans and other eukaryotic organisms is coiled DNA, which has proteins attached along its length.

These coiled DNA strands with attached proteins, which become visible during mitosis and meiosis, are called **nucleoproteins,** or **chromatin fibers.** The protein and DNA are not arranged randomly, but come together in a highly organized pattern. When eukaryotic chromatin fibers coil into condensed, highly knotted bodies, they are seen easily through a microscope after staining with dye. Condensed like this, a chromatin fiber is referred to as a chromosome (figure E*b*). The genetic material in bacteria is also double-stranded (duplex) DNA, but the ends of the molecule are connected to form a loop and they do not form condensed chromosomes (figure E*c*).

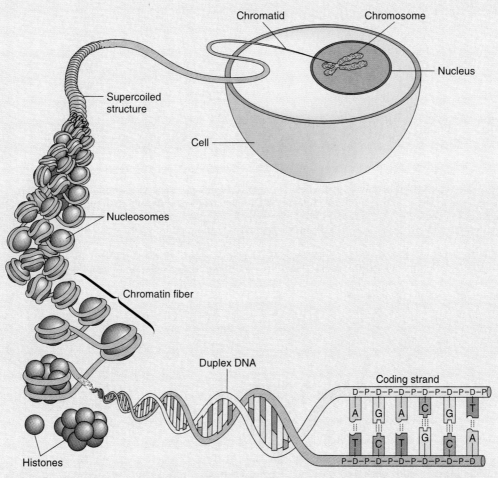

Figure D   Double-stranded DNA   The nuclei of eukaryotic cells contain double-stranded DNA, which takes the form of a three-dimensional helix. One strand is a chemical code (the coding strand) that contains the information necessary to control and coordinate the activities of the cell. The two strands are bound together by weak bonds formed between the protruding nitrogenous bases according to the base-pairing rule: A pairs with T and C pairs with G. The length of the DNA molecule is measured in numbers of "base pairs"—the number of rungs on the ladder.

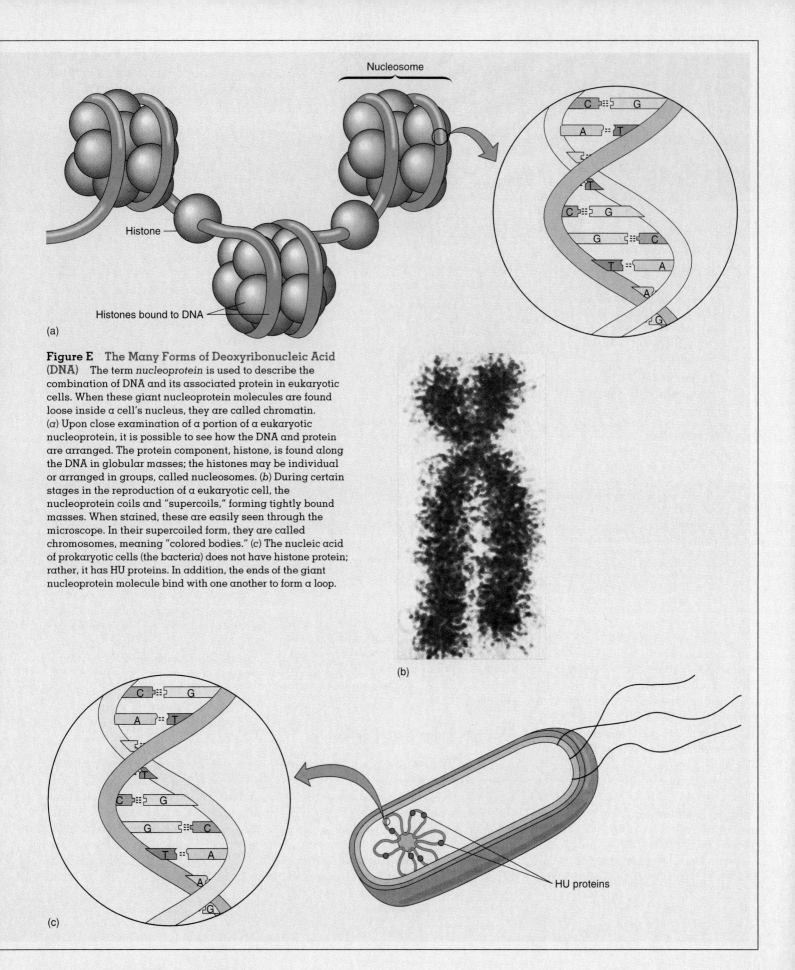

**Figure E   The Many Forms of Deoxyribonucleic Acid
(DNA)**   The term *nucleoprotein* is used to describe the
combination of DNA and its associated protein in eukaryotic
cells. When these giant nucleoprotein molecules are found
loose inside a cell's nucleus, they are called chromatin.
(*a*) Upon close examination of a portion of a eukaryotic
nucleoprotein, it is possible to see how the DNA and protein
are arranged. The protein component, histone, is found along
the DNA in globular masses; the histones may be individual
or arranged in groups, called nucleosomes. (*b*) During certain
stages in the reproduction of a eukaryotic cell, the
nucleoprotein coils and "supercoils," forming tightly bound
masses. When stained, these are easily seen through the
microscope. In their supercoiled form, they are called
chromosomes, meaning "colored bodies." (*c*) The nucleic acid
of prokaryotic cells (the bacteria) does not have histone protein;
rather, it has HU proteins. In addition, the ends of the giant
nucleoprotein molecule bind with one another to form a loop.

What Kind of Information Is Coded by DNA?

Each piece of DNA is different because each strand has a different chemical code. Coded DNA serves as a central cell library. Tens of thousands of messages are in this storehouse of information. This information tells the cell such things as:

1. how to produce enzymes required for the digestion of nutrients
2. how to manufacture enzymes that will metabolize the nutrients and eliminate harmful wastes
3. how to repair and assemble cell parts
4. how to reproduce healthy offspring
5. when and how to react to favorable and unfavorable changes in the environment
6. how to regulate all of life's essential functions

If any of these functions are not performed, the cell will die. The importance of maintaining DNA in a cell becomes clear when we consider cells that have lost their DNA. For example, human red blood cells lose their nuclei as they become specialized for carrying oxygen and carbon dioxide throughout the body. Without DNA they are unable to manufacture the essential cell components needed to sustain themselves. The cells continue to exist for about 120 days, using enzyme molecules manufactured earlier in their lives. As the enzymes are lost, the cells die.

DNA replication occurs in cells in preparation for mitosis and meiosis. Without this process, daughter cells would not receive the library of information required to sustain life. The transcription process results in the formation of a strand of RNA that is a copy of a piece of the DNA. Some of the RNA molecules become involved directly in various biochemical processes while others are involved in directing the manufacture of proteins, a process called translation. Structural proteins are used by the cell as building materials (feathers, hair) while regulatory proteins are used to direct and control chemical reactions. The most common regulatory proteins are **enzymes,** which control the kinds of chemical reactions within organisms and the rates at which they take place. We will discuss enzymes in greater detail later in the chapter. Some hormones, such as insulin, are also proteins.

Making Copies: DNA Replication

Since all cells must maintain a complete set of genetic material, there must be a doubling of DNA before cells can divide. **DNA replication** is the process of duplicating the genetic material prior to its distribution to daughter cells. Accuracy of duplication is essential in order to guarantee the continued existence of that type of cell. Should the daughters not receive exact copies, they may be unable to manufacture the structural and regulatory proteins essential for their survival. The DNA replication process requires many enzymes.

1. The DNA replication process begins as an enzyme breaks the attachments between the two strands of DNA. In eukaryotic cells, this occurs in hundreds of different spots along the length of the DNA (figure 16.1a).
2. Moving along the DNA, the enzyme "unzips" the halves of the DNA (figure 16.1b and c).
3. Proceeding in opposite directions on each side, the enzyme *DNA polymerase* moves down the length of the DNA, attaching new DNA nucleotides into position (figure 16.1d).
4. The enzyme that speeds the addition of new nucleotides to the growing chain works along with another enzyme to make sure that no mistakes are made. If the wrong nucleotide appears to be headed for a match, the enzyme will reject it in favor of the correct nucleotide. If a mistake is made and a wrong nucleotide is paired into position, specific enzymes have the ability to replace it with the correct one (figure 16.1e).
5. Replication proceeds in both directions, appearing as "bubbles" (figure 16.1e).
6. The complementary molecules (A : : : T, G : : : C) pair with the exposed nitrogenous bases of both DNA strands by forming new weak bonds (figure 16.1f).
7. Once properly aligned, a bond is formed between the sugars and phosphates of the newly positioned nucleotides. A strong sugar and phosphate backbone is formed in the process (figure 16.1g).
8. This process continues until all the replication "bubbles" join (figure 16.1h). Figure 16.2 summarizes this process.

A new complementary strand of DNA forms on each of the old DNA strands, resulting in the formation of two double-stranded DNA molecules. In this process, the exposed halves of the original DNA serve as a pattern for the formation of the new DNA. As the two new DNA molecules are completed, they twist into their double-helix shape.

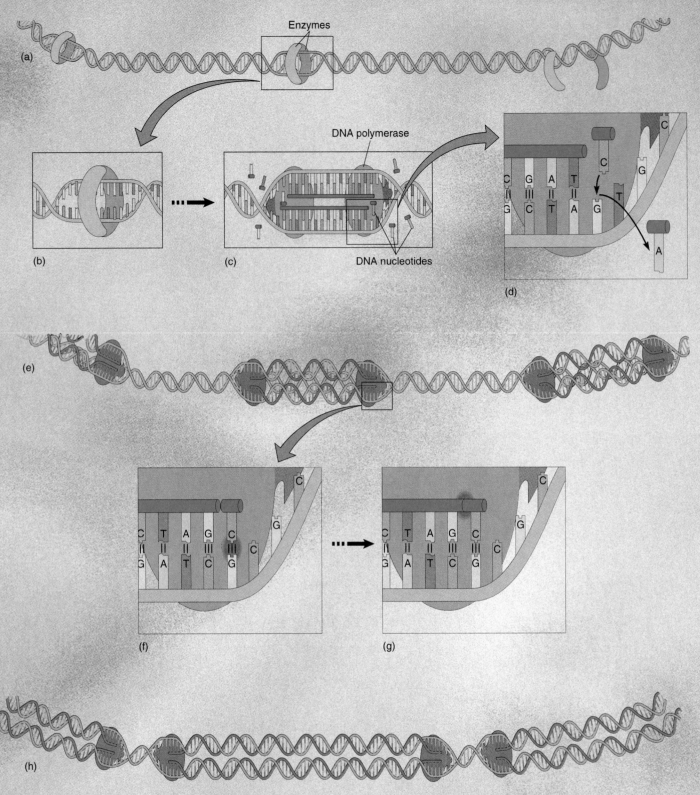

(a)

Enzymes

(b) → (c)

DNA polymerase

DNA nucleotides

(d)

(e)

(f) ┅→ (g)

(h)

Figure 16.1 DNA Replication

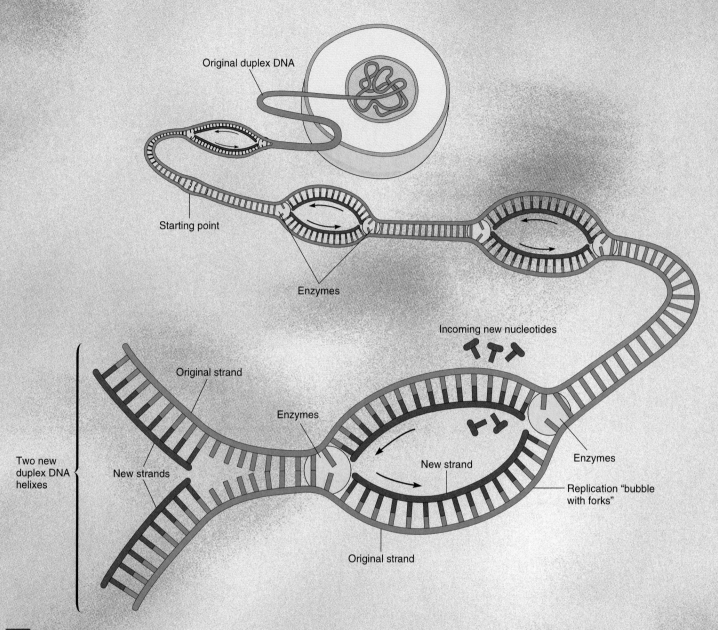

Original duplex DNA

Starting point

Enzymes

Incoming new nucleotides

Two new duplex DNA helixes

Original strand

New strands

Enzymes

New strand

Enzymes

Replication "bubble with forks"

Original strand

Figure 16.2 DNA Replication Summary In eukaryotic cells, the "unzipping" enzymes attach to the DNA at numerous points, breaking the bonds that bind the complementary strands. As the DNA replicates, numerous replication "bubbles" or "forks" appear along the length of the DNA. Eventually all the forks come together, completing the replication process.

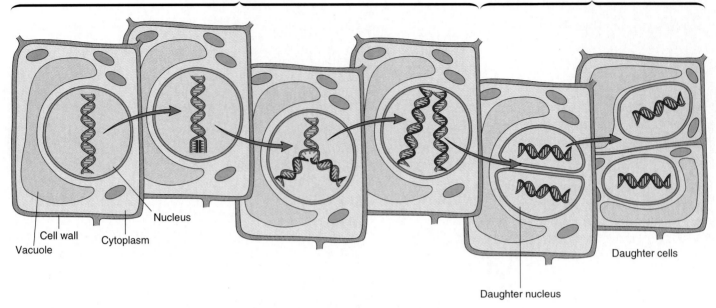

Nucleus

Cell wall
Cytoplasm
Vacuole

Daughter cells

Daughter nucleus

Figure 16.3 **Replication and Cell Division** These are the generalized events in the nucleus of a eukaryotic cell during the process of DNA replication, which precedes the division of the cell into two daughter cells. Notice that after replication the cell has two double helices; they are identical to each other and identical to the original strands. Each daughter cell gets one of these copies.

The completion of the process yields two double helices that are identical in their nucleotide sequences. The DNA replication process is highly accurate. It has been estimated that only one error is made for every 2 billion nucleotides. A human cell contains 46 chromosomes consisting of about 5,000,000,000 (5 billion) base pairs. This averages to 2.5 errors per cell! Don't forget that this figure is an estimate, and while some cells may have 2.5 errors per replication, others may have more, and some may have no errors at all. It is also important to note that some errors may be major and deadly, while others are insignificant. Since this error rate is so small, DNA replication is considered by most to be essentially error-free.

Following DNA replication, the cell now contains twice the amount of genetic information and is ready to begin the process of distributing one set of genetic information to each of its two daughter cells.

The distribution of DNA involves splitting the cell and distributing a set of genetic information to the two new daughter cells during mitosis. In this way, each new cell has the necessary information to control its activities. The mother cell ceases to exist when it divides its contents between the two smaller daughter cells (figure 16.3).

Kinds of Proteins: Structural and Regulatory

Proteins are made by direction of cellular information in the DNA library. They are chains made up of links known as **amino acids.** There  **Appendix A** are about twenty amino acids that are important to cells. Any amino acid can bond with any other amino acid. You can imagine that by using twenty different amino acids as building blocks, you can construct millions of different combinations. Each of these combinations is called a protein. The linking of a specific sequence of amino acids to form a protein is controlled by the genetic information of an organism.

The thousands of kinds of proteins can be placed into two categories. Some proteins are important for maintaining the shape of cells and organisms—they are usually referred to as **structural proteins.** The proteins that make up the cell membrane, muscle cells, tendons, and blood cells are examples of structural proteins. The other kinds of proteins, **regulator proteins,** are used as tools. The tools can make certain chemical reactions occur very rapidly, and each one is specifically designed for a particular job. These regulator proteins include enzymes and certain hormones. These molecules help control the chemical activities of cells and organisms. Some examples of enzymes are the digestive enzymes in the stomach and the mouth and cellular enzymes that control the release of energy in cells. Two hormones that are regulator proteins are insulin and oxytocin. The shape of both tools and building materials is very important. If they do not have the correct shape, they cannot be used to perform the proper chemical reactions or construct the cell.

A string of amino acids may be referred to as a polypeptide, though larger molecules are called proteins. A protein is likely to twist into a particular shape. For example, some

CONCEPT CONNECTIONS

replication The process of manufacturing copies of DNA.

transcription The process of manufacturing RNA from the surface of DNA. Three forms of RNA that may be produced are mRNA, rRNA, and tRNA.

translation The assembly of individual amino acids into a polypeptide or protein.

FYI

Amino acids: The basic building block of all protein molecules. Twenty common forms of amino acid are found in all living things. While all amino acids have a similar basic structure, each of the twenty forms can be identified by a unique group of atoms attached to this basic structure. Amino acids are released during digestion of protein nutrients. Cells may also construct amino acids from carbon, hydrogen, oxygen, nitrogen, and sulfur.

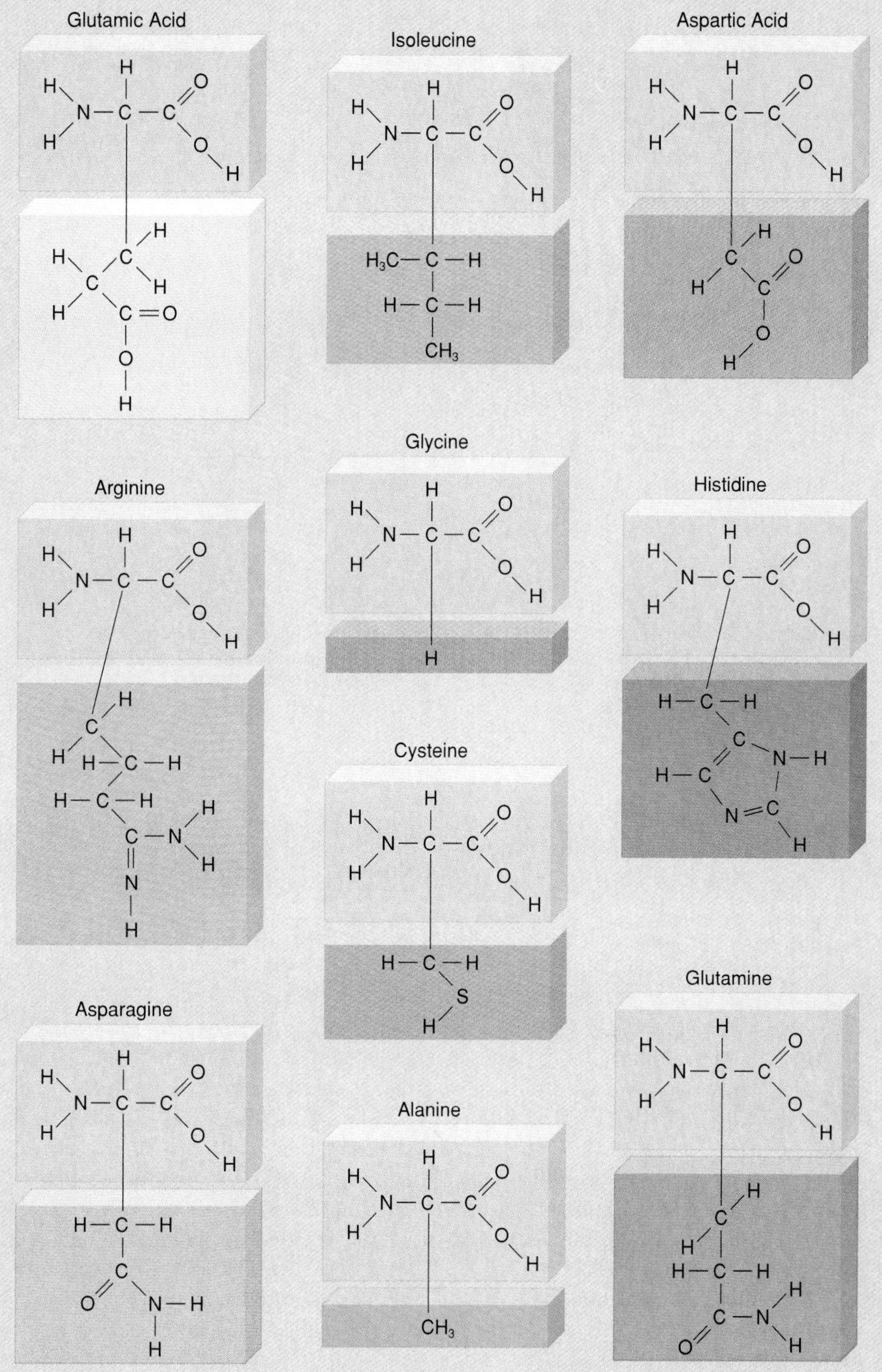
Glutamic Acid
Isoleucine
Aspartic Acid
Glycine
Arginine
Histidine
Cysteine
Glutamine
Asparagine
Alanine

Different Amino Acids

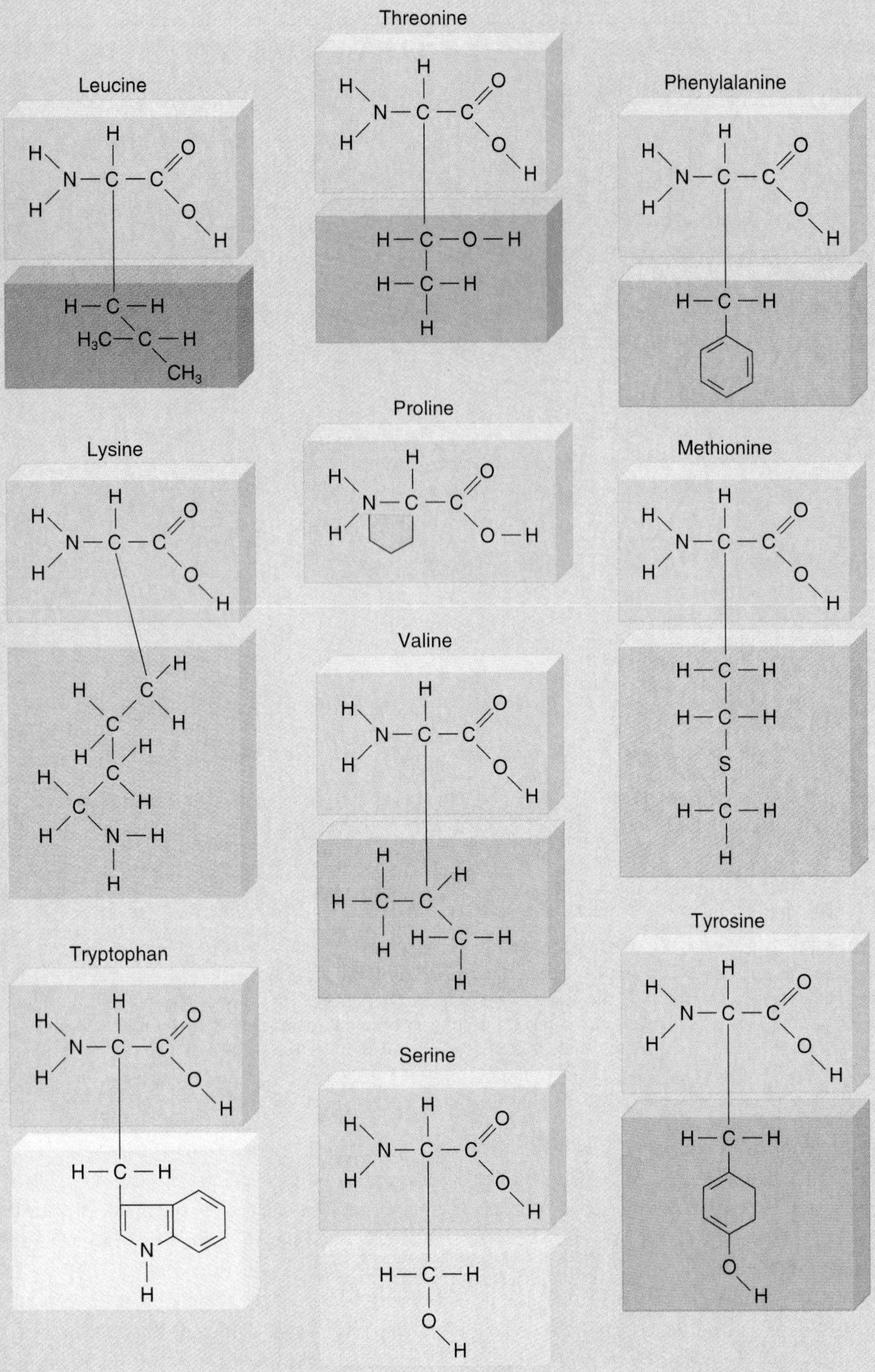

Leucine

Threonine

Phenylalanine

Lysine

Proline

Methionine

Valine

Tryptophan

Serine

Tyrosine

Tyrosine

proteins (e.g., hair) take the form of a coiled telephone cord. Other proteins form bonds that cause them to make several flat folds that resemble a pleated skirt. Silk, the protein made by moths for their cocoons, is such a protein.

It is also possible for a single protein to contain one or more coils and pleated sheets along its length. As a result, these different portions of the molecule can interact to form an even more complex three-dimensional structure. This occurs when the coils and pleated sheets twist and combine with each other. A good example of this kind of structure can be seen when a coiled phone cord becomes so twisted that it folds around and back on itself in several places. Myoglobin, the oxygen-holding protein found in muscle cells, displays this kind of structure.

FYI

The Classes of Immunoglobulins

IgG (*Gamma*) is found in body fluids and is particularly effective in combating microorganisms and their toxins.

IgM (*Mu*) is found in the blood serum and is very effective as a first-line defense against bacterial blood infections.

IgA (*Alpha*) is found in the blood and secretions and serves as a defense against initial microbial invasion.

IgE (*Epsilon*) is found in the blood and body fluids and is associated with certain types of allergies (hypersensitivities), for example, hay fever and asthma.

IgD (*Delta*) is also in the blood; however, its function is not clear.

How Enzymes Work

If organisms are to survive, they must obtain large amounts of energy and building materials in a very short time. Experience tells us that the sugar sucrose in candy bars contains the energy needed to keep us active, as well as building materials to help us grow (sometimes to excess!). Yet, left uneaten, a candy bar could last millions of years before it is broken down by random chemical processes, releasing its energy and building materials. Living things cannot wait that long. To sustain life, biochemical reactions must occur at extremely rapid rates.

One way to increase the rate of any chemical reaction is to increase the temperature. In general, the higher the temperature, the faster the reactions will occur. However, this method has a major drawback when it comes to living things. If the temperature is too high, the organism will die because cellular proteins are altered. This is of practical concern to people who are experiencing a fever. Should the fever stay too high for too long, major disruptions of cellular biochemical processes can be fatal.

There is a way of increasing the rate of chemical reactions without increasing the temperature. This involves using substances called *catalysts.* A **catalyst** is a chemical that speeds the reaction but is not used up in the reaction. It can be recovered unchanged when the reaction is complete. Catalysts function by lower-

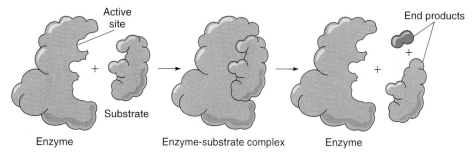

Active site

End products

Substrate

Enzyme   Enzyme-substrate complex   Enzyme

Figure 16.4   Enzyme-Substrate Complex Formation   During an enzyme-controlled reaction, the enzyme and substrate come together to form a new molecule—the enzyme-substrate complex molecule. This molecule exists for only a very short time. During that time, bonds are changed. The result is the formation of a new molecule or molecules called the end products of the reaction. Notice that the enzyme comes out of the reaction intact and ready to be used again.

ing the amount of energy needed to start the reaction. Cells make specific proteins that act as catalysts. A protein molecule that acts as a catalyst to change the rate of a reaction is called an **enzyme.** Enzymes can be used over and over again until they are worn out. The instructions for the manufacture of all enzymes are found in the genes (DNA) of every cell. Organisms make their own enzymes.

As the instructions for the production of an enzyme are read from the genetic material, a specific sequence of amino acids is linked together. This process takes place with the assistance of cell structures called ribosomes. Once linked together, the chain of amino acids folds and twists to form a globular molecule. The nature of its three-dimensional shape allows this enzyme to lower the amount of energy

necessary to get a specific reaction going. Each enzyme has a very specific three-dimensional shape that is, in turn, very specific to the kind of molecule with which it can combine. The enzyme physically fits the molecule. This molecule to which the enzyme attaches itself (the *reactant*) is known as the **substrate.** When the enzyme attaches itself to the substrate molecule, a new, temporary molecule is formed (figure 16.4). When a substrate is combined with its enzyme, its chemical bonds are less stable and more likely to be altered.

You might think of an enzyme as a tool that makes a job easier and faster. For example, the use of an open-end crescent wrench can make the job of removing or attaching a nut and bolt go much faster than that same job done by hand. In order to accomplish this job,

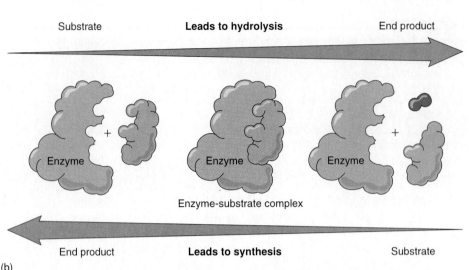

Substrate   **Leads to hydrolysis**   End product

Enzyme + Enzyme Enzyme +

Enzyme-substrate complex

End product   **Leads to synthesis**   Substrate

(b)

(a)

Figure 16.5 It Fits, It's Fast, and It Works (*a*) While it could be done by hand, the wheel from this bicycle can be removed more efficiently with an open-end crescent wrench. The wrench is adjusted and attached, temporarily forming a nut-bolt-wrench complex. Turning the wrench loosens the bonds holding the nut to the bolt and the two are separated. The use of the wrench makes the task much easier. Keep in mind that the same wrench used to disassemble the bicycle can be used to reassemble it. Enzymes function in the same way. (*b*) An enzyme will "adjust itself" as it attaches to its substrate, forming a temporary enzyme-substrate complex. The presence and position of the enzyme in relation to the substrate lowers the activation energy required to alter the bonds. Depending on the circumstances (what job needs to be done), the enzyme might be involved in synthesis (constructive) or hydrolysis (destructive) reactions.

the proper wrench must be used. Just any old tool (screwdriver or hammer) won't work! Similarly, the enzyme must physically attach itself to the substrate; thus there is a specific binding site or attachment site on the enzyme surface. Figure 16.5 illustrates the specificity of both wrench and enzyme. Note that the wrench and enzyme are recovered unchanged after they have been used. This means that the enzyme and wrench can be used again. Eventually, like wrenches, enzymes wear out and need to be replaced. The cell makes new enzymes using the instructions provided by the cell's genes. Generally, only very small quantities of enzymes are necessary because they work so fast and can be reused.

    Both enzymes and wrenches are specific in that they have a particular shape that matches the shape of their substrates. Note that both the enzyme and wrench are flexible. The enzyme can bend or fold to fit the substrate just as the wrench, to a limited extent, can be "adjusted" to fit the nut. While the enzyme does mold itself to a substrate, enzymes do not have the ability to fit all substrates. Enzymes are specific to a certain substrate or group of very similar substrate molecules. One enzyme cannot speed the

FYI

Enzymes and "Stonewashed" Jeans **The popularity of stonewashed jeans grew dramatically in the late '60s. To get the stonewashed effect, the denim was actually washed in machines along with stones. The stones rubbed against the denim, wearing the blue dye off the surface of the material. But the stones also damaged the cotton fibers. The damage shortened the life of the fabric, a feature that many consumers found unacceptable. Now, to create the stonewashed look and still maintain strong cotton fibers, enzymes are used that "digest" or hydrolyze the blue dye of the surface of the fabric. Since the enzyme is substrate or dye specific, the cotton fibers are not harmed by the enzymes.**

rate of all types of biochemical reactions. Rather, a special enzyme is required to control the rate of each type of chemical reaction occurring in an organism.

Making Proteins: The Genetic Code

As noted earlier, DNA functions in the manner of a reference library that does not allow its books to circulate. Information from the originals must be copied. Thus, a major function of DNA is to make a single-stranded, complementary RNA copy of DNA. This operation is called **transcription,** which means

to convert information from one form to another within the same language. When you read aloud, you are transcribing from the written to the spoken form, but the message is in the same language. In this case, the data is copied from DNA structure into RNA structure. The genetic information stored as a DNA chemical code is carried in the form of an RNA copy to other parts of the cell. It is the RNA copy that is used to guide the assembly of amino acids into structural and regulatory proteins. The building of proteins from amino acid building blocks is often referred to as **protein synthesis.** Since the protein language is

different from the DNA and RNA language, this process is often referred to as **translation.** Without the process of transcription, genetic information would be useless in directing cell functions. Although many types of RNA are synthesized from the genes, the three most important are messenger RNA (mRNA), transfer RNA (tRNA), and ribosomal RNA (rRNA).

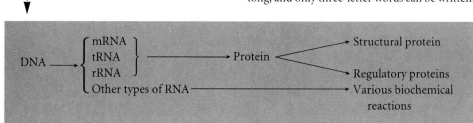

Transcription begins when the two halves of the DNA are separated by an enzyme, exposing the A, T, G, C sequences of the two strands. Transcription occurs on only one of the two DNA strands, which serves as a pattern for the synthesis of RNA. But which strand is copied? Where does transcription start and when does it stop? Where along the sequence of thousands of bases does the chemical code for the manufacture of a particular enzyme begin and where does it end? If transcription begins randomly, the resulting RNA may not be a copy of a useful code, and the enzyme product may be useless or deadly to the cell. To answer these questions, it is necessary to explore the nature of the genetic code itself.

We know that genetic information is in chemical-code form in the DNA molecule. When the coded information is used or *expressed*, it guides the assembly of particular amino acids into structural and regulatory proteins. If DNA is a molecular language, then each nucleotide in this language can be thought of as a letter within a four-letter alphabet. Each word, or code, is always three letters (nucleotides) long, and only three-letter words can be written.

A **codon** is a triple-letter sequence that codes for one of the twenty common amino acids. The number of codons in this language is limited because there are only four nucleotides, and these nucleotides are used only in groups of three, a triplet sequence. The order of these three letters is just as important in DNA language as the order of letters is in our language. We recognize that CAT is not the same as TAC. If all the possible three-letter codes were written using only the four DNA nucleotides for letters, there would be a total of sixty-four combinations.

$$4^3 = 4 \times 4 \times 4 = 64$$

When codes are found at a particular place along a coding strand of DNA, and the sequence has meaning, the sequence is called a **gene.**

"Meaning" in this case refers to the fact that the gene can be transcribed into an RNA molecule, which in turn may control the assembly of individual amino acids into a protein.

Several things occur when a gene is transcribed into RNA (figure 16.6).

1. The process begins as one portion of the enzyme RNA polymerase breaks the attachments between the two strands of DNA; the enzyme "unzips" the two strands of the DNA.
2. A second portion of the enzyme RNA polymerase attaches at a particular spot on the DNA called the *start codon.* It proceeds in one direction along one of the two DNA strands, attaching new RNA nucleotides into position until it reaches a *stop codon.* The enzymes then assemble RNA nucleotides into a complete, single-stranded RNA copy of the gene. There is no thymine in RNA molecules; it is replaced by uracil. Therefore, the start codon in DNA (TAC) would be paired by RNA polymerase to form the RNA codon AUG.
3. The enzyme that speeds the addition of new nucleotides to the growing chain works along with another enzyme to make sure that no mistakes are made.
4. When transcription is complete, the newly assembled RNA is separated from its DNA template and made available for use in the cell; the DNA recoils into its original double-helix form.

As previously mentioned, three general types of RNA are produced by transcription: messenger RNA, transfer RNA, and ribosomal RNA. Each kind of RNA is made from a specific gene and performs a specific function. **Messenger RNA (mRNA)** is a straight-chain copy of a gene that describes the exact sequence in which amino acids should be attached together to form a protein.

Transfer RNA (tRNA) molecules are responsible for picking up particular amino acids and transferring them to the ribosome for assembly into the protein. All tRNA molecules are shaped like a cloverleaf. One end of the tRNA is able to attach to a specific amino acid. Toward the midsection of the molecule, a triplet sequence of bases can be paired with a codon on mRNA.

Ribosomal RNA (rRNA) is a highly coiled molecule and is used along with protein

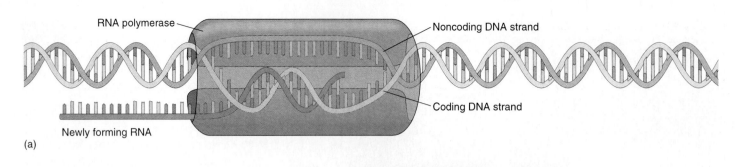

RNA polymerase

Noncoding DNA strand

Coding DNA strand

Newly forming RNA

(a)

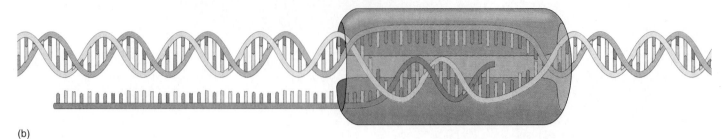

(b)

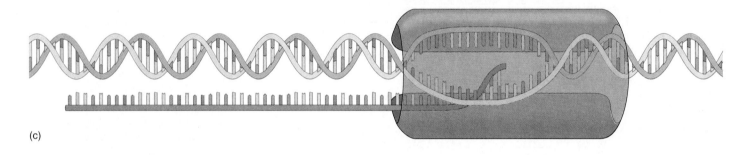

(c)

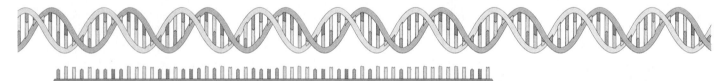

RNA

(d)

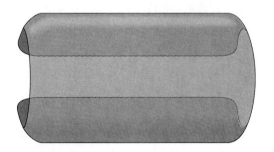

Figure 16.6 Transcription of an RNA Molecule   This summary illustrates the basic events that occur during the transcription of one side (the coding strand) of double-stranded DNA. The enzyme attaches to the DNA at a point that allows it to break the hydrogen bonds that bind the complementary strands. As this enzyme, RNA polymerase, moves down the DNA, new complementary RNA nucleotides are base-paired on one of the exposed strands and linked together, forming a new strand that is complementary to the nucleotide sequence of the DNA. The newly formed (transcribed) RNA is then separated from its DNA complement. Depending on the DNA segment that has been transcribed, this RNA molecule may be a messenger RNA (mRNA), a transfer RNA (tRNA), a ribosomal RNA (rRNA), or an RNA molecule used for other purposes within the cell.

Antibodies: One of the Body's Defense Mechanisms

Several types of vaccinations are regularly given to stimulate the formation of molecules called *antibodies* that protect against infectious diseases. The DPT vaccination immunizes against diphtheria, pertussis (whooping cough), and tetanus; the MMR against measles, mumps, and rubella; and either the OPV (oral polio vaccine) or the IPV (inactivated polio vaccine) against polio. The chemicals produced by the body that are responsible for this type of defense are complex antibody proteins bonded to each other to form globular molecules (see figure below). These antibodies are technically called *immunoglobulins,* (*immuno* = resistant; *globulin* = globular protein).

Immunoglobulins are commonly referred to as antibodies since they were once thought to be little bodies that defended against (*anti-*) infectious agents such as bacteria and viruses. There are five classes of these "Y-shaped" molecules, each identified by a particular letter of the Greek alphabet. IgG, (*Gamma*), is the most abundant immunoglobulin in body fluids, and is particularly effective in combating microorganisms and their toxins.

Each class of immunoglobulin can contain an infinite variety of unique protein molecules that are able to combine with specific agents recognized to be foreign (nonself) to the body. Immunoglobulins are always globular proteins manufactured by

particular white blood cells known as B cells and plasma cells. The presence of nonself agents stimulates the production of this form of resistance (immunity). When an immunoglobulin reacts with the foreign substance, a sequence of events occurs that results in the destruction of the molecules. In this way, nonself agents such as bacteria, viruses, toxic molecules, and cancer cells can be eliminated from the body. The production of protective immunoglobulins can be initiated by vaccination. The vaccination stimulates production of immunoglobulins without causing the disease. Thus, the person who received the vaccine is protected against a future encounter with a disease-causing organism.

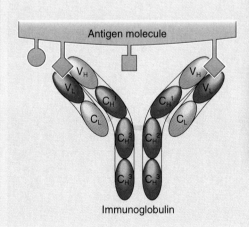

Antigen molecule

Immunoglobulin

Shape Is Important: Sickle-Cell Anemia and Protein Structure

The structure of a protein is closely related to its function. Any changes in the arrangement of amino acids within a protein can have far-reaching effects on its function. For example, normal hemoglobin found in red blood cells consists of two kinds of protein chains called the alpha and beta chains. The beta chain is 146 amino acids long. If just *one* of these amino acids is replaced by a different one, the hemoglobin molecule may not function properly. A classic example of this results in a condition known as *sickle-cell anemia*. In this case, the sixth amino acid in the beta chain is replaced by a different amino acid. This minor change causes the hemoglobin to fold differently, and the red blood cells that contain this altered hemoglobin assume a sickle shape when the body is deprived of an adequate supply of oxygen.

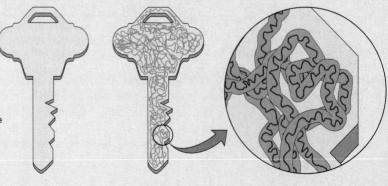

Environmental Influences on Protein Shape

Changing environmental conditions influence the shapes of proteins. The accompanying illustration shows the importance of the three-dimensional shape of protein. In addition, energy in the form of heat or light may break up the protein molecules. When this occurs, the chemical and physical properties of the protein are changed and the protein is said to be **denatured.** A common example of this occurs when the gelatinous, clear portion of an egg is cooked and the protein changes to a white solid.

Some medications are proteins and must be protected from denaturation so as not to lose their effectiveness. Insulin is an example of a regulatory protein that must maintain its shape in order to function. Insulin is produced by the pancreas and controls the amount of glucose in the blood. If insulin production is too low, or if the molecule is improperly constructed, glucose molecules are not removed from the bloodstream at a fast enough rate. The excess sugar is then eliminated in the urine. Other symptoms of excess sugar in the blood include excessive thirst and even loss of consciousness. The disease caused by improperly functioning insulin is known as *diabetes.* For protection, medications such as insulin should not be exposed to temperatures that cause the protein to be denatured.

The Three-Dimensional Shape of Proteins The arrangement of amino acids results in side chains that are available to bond with other side chains. The results are three-dimensional proteins that have a specific surface geometry. We frequently compare this three-dimensional shape to the three-dimensional shape of a key.

table 16.1

Amino Acids

Amino Acid	Abbreviation
Alanine	Ala
Arginine	Arg
Asparagine	AspN
Aspartic acid	Asp
Cysteine	Cys
Glutamic acid	Glu
Glutamine	GluN
Glycine	Gly
Histidine	His
Isoleucine	Ileu
Leucine	Leu
Lysine	Lys
Methionine	Met
Phenylalanine	Phe
Proline	Pro
Serine	Ser
Threonine	Thr
Tryptophan	Try
Tyrosine	Tyr
Valine	Val

There are twenty common amino acids used in the protein synthesis operation of a cell. Each has a known chemical structure.

FYI The triplet sequence on tRNA that is complementary to a codon of mRNA is called an **anticodon.**

molecules in the structure of all ribosomes. Ribosomes are tiny cellular structures that, along with tRNA and mRNA, help in the synthesis of proteins.

The mRNA molecule is a coded message written in the biological world's universal nucleic acid language. The code is read in one direction beginning at the "start" signal. The information is used to assemble amino acids into proteins by a process called translation. The word **translation** refers to the fact that nucleic acid language is being changed to protein language. To translate mRNA language

table 16.2

The Amino Acid-Nucleic Acid Dictionary

Amino Acid	mRNA Codons	Amino Acid	mRNA Codons
Alanine	GCU	Lysine	AAA
	GCC		AAG
	GCA	Methionine	AUG
	GCG		
Arginine	CGU	Phenylalanine	UUU
	CGC		UUC
	CGA	Proline	CCU
	CGG		CCC
	AGA		CCA
	AGG		CCG
Asparagine	AAU	Serine	UCU
	AAC		UCC
Aspartic acid	GAU		UCA
	GAC		UCG
Cysteine	UGU		AGU
	UGC		AGC
Glutamic acid	GAA	Threonine	ACU
	GAG		ACC
Glutamine	CAA		ACA
	CAG		ACG
Glycine	GGU	Tryptophan	UGG
	GGC	Tyrosine	UAU
	GGA		UAC
	GGG	Valine	GUU
Histidine	CAU		GUC
	CAC		GUA
Isoleucine	AUU		GUG
	AUC	Terminator	UAA
	AUA		UAG
Leucine	UUA		UGA
	UUG	Initiator	AUG
	CUU		
	CUC		
	CUA		
	CUG		

A dictionary can come in handy for learning any new language. This one is used to translate nucleic acid language into protein language.

into protein language, a dictionary is necessary. Remember, the four letters in the nucleic acid alphabet yield sixty-four possible three-letter words. The protein language has twenty words in the form of twenty common amino acids, shown in table 16.1. Thus, there are more than enough nucleic acid code words for the twenty amino acid molecules.

Table 16.2 is an amino acid–nucleic acid dictionary. Notice that more than one

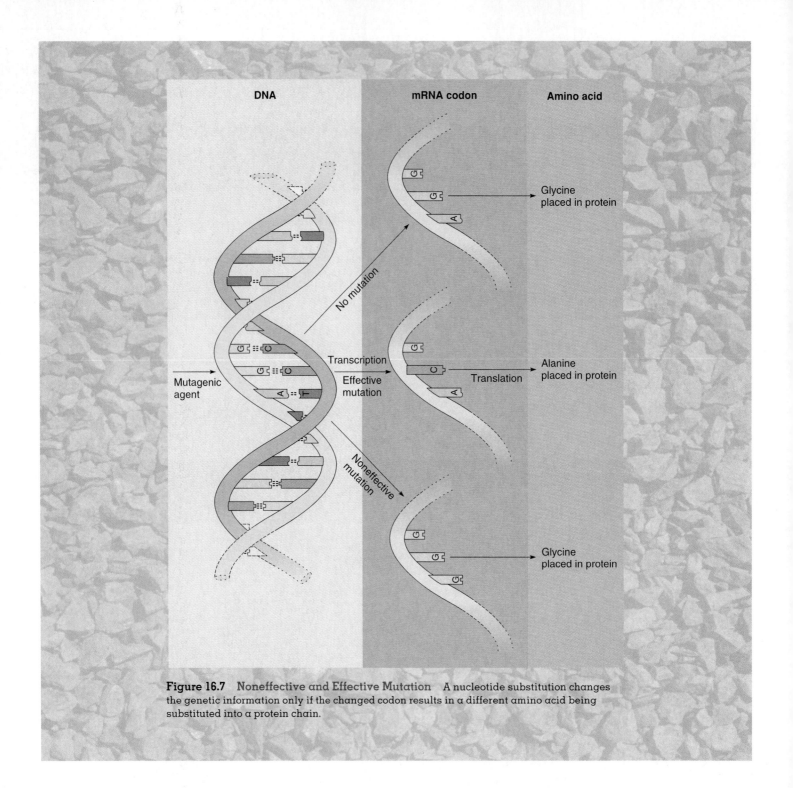

Figure 16.7 Noneffective and Effective Mutation   A nucleotide substitution changes the genetic information only if the changed codon results in a different amino acid being substituted into a protein chain.

codon may code for the same amino acid. You might think that this is needless repetition, but such "synonyms" can have survival value. If, for example, the gene or the mRNA becomes damaged in a way that causes a particular nucleotide to change to another type, the chances are still good that the proper amino acid will be read into its proper position. But not all such changes can be compensated for by the codon system, and an altered protein may be produced (figure 16.7). Changes can occur that cause great harm. Some damage is so extensive that the entire strand of DNA is broken, resulting in improper protein synthesis, or a total lack of synthesis. Any change in DNA is called a **mutation.**

The construction site of the protein molecules (i.e., the translation site) is in the ribosome. A ribosome is composed of two subunits, one large and one small. All of the parts (ribosomes, mRNA, tRNA, and amino acids) come together long enough for certain reactions to occur. The basic steps of translation are summarized on pages 370 to 372.

BIO *feature*

HIV Infection and Reverse Transcriptase

AIDS is an acronym for *a*cquired *i*mmun*od*eficiency *s*yndrome. It is caused by *h*uman *i*mmunodeficiency *v*iruses (HIV). These viruses are members of the *retrovirus* family. HIV is a spherical virus with an outer membrane surrounding a protein coat with an RNA core. Its genetic material is RNA, not DNA. HIV genes are carried

from one generation to the next as RNA molecules. This is not the case in humans and most other organisms where DNA is the genetic material.

Human sequence:

$$DNA \xrightarrow{\hspace{1cm}} RNA \xrightarrow{\hspace{1cm}} protein$$
$$\textit{transcriptase}$$

Having entered a suitable, susceptible host cell, however, HIV must convert its RNA to a DNA genome in order to attach to the host cell's chromosome. Only then can it become an active, disease-causing parasite. This conversion of RNA to DNA is contrary, reverse, or *retro* to the RNA-forming process controlled by the enzyme

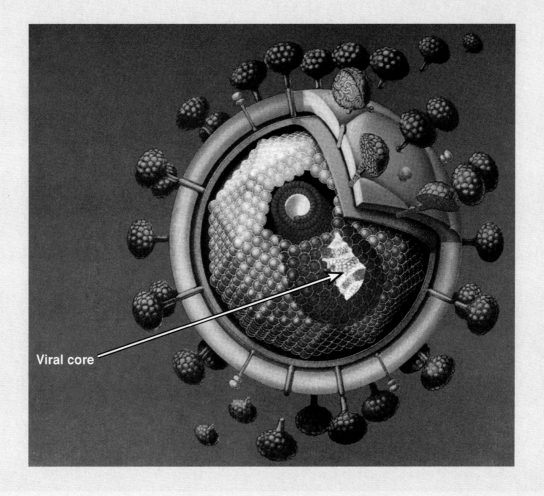

Viral core

transcriptase. Humans do not have the genetic capability to manufacture *reverse transcriptase,* the enzyme necessary to convert RNA to DNA. Only some kinds of viruses have this ability.

Retrovirus sequence:

$$RNA \xrightarrow{\text{reverse transcriptase}} DNA$$

Once the virus RNA has been reverse transcribed into DNA, the host cell's enzymes carry out gene replication and protein synthesis for the virus. We can diagram the process this way:

This process has two important implications. First, the presence of reverse transcriptase in a human can be looked upon as an indication of retroviral infection, because this enzyme is not manufactured by human cells. However, since HIV is only one type of several retroviruses, the presence of the enzyme in an individual indicates some type of retroviral infection, but not necessarily an

HIV infection. Second, interference with reverse transcriptase will frustrate the virus's attempt to attach to the host chromosome. This may prevent disease caused by viral replication within the host cell. The drugs AZT (azidothymidine, or Zodovudine) and DDC (dideoxycytosine) disrupt the operation of reverse transcriptase and are used to prolong the lives of AIDS patients.

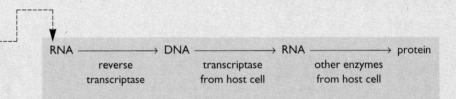

$$RNA \xrightarrow[\substack{\text{reverse} \\ \text{transcriptase}}]{} DNA \xrightarrow[\substack{\text{transcriptase} \\ \text{from host cell}}]{} RNA \xrightarrow[\substack{\text{other enzymes} \\ \text{from host cell}}]{} protein$$

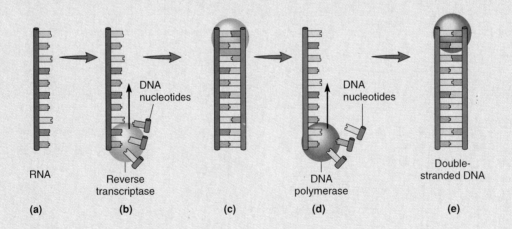

RNA

Reverse transcriptase

DNA nucleotides

(a)   (b)   (c)   (d)   (e)

DNA nucleotides

DNA polymerase

Double-stranded DNA

1. An mRNA molecule is placed in the small portion of a ribosome so that six nucleotides (two codons) are locked into position (figure 16.8a).

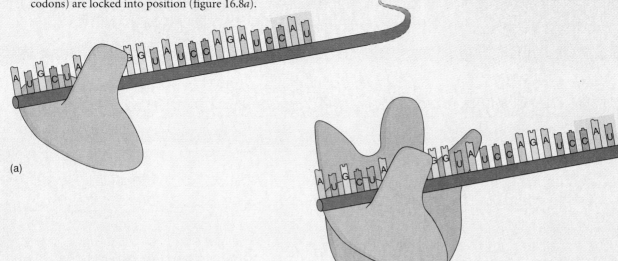

(a)

2. The larger ribosomal unit is added to the ribosome/mRNA combination (figure 16.8b).

(b)

3. A tRNA with bases that match the second mRNA codon attaches to the mRNA. The tRNA is carrying a specific amino acid. Once attached, a second tRNA carrying another specific amino acid moves in and attaches to its complementary mRNA codon right next to the first tRNA/amino acid complex (figure 16.8c).

(c)

4. The two tRNAs properly align their two amino acids so that they may be chemically attached to one another (figure 16.8d).

(d)

🎞 **Figure 16.8 a–j** Basic Steps of Translation.

5. Once the two amino acids are connected to one another, the first tRNA detaches from its amino acid and mRNA codon and leaves (figure 16.8e).

(e)

6. The ribosome moves along the mRNA to the next codon (the first tRNA is set free to move through the cytoplasm to attach to and transfer another amino acid) (figure 16.8f).

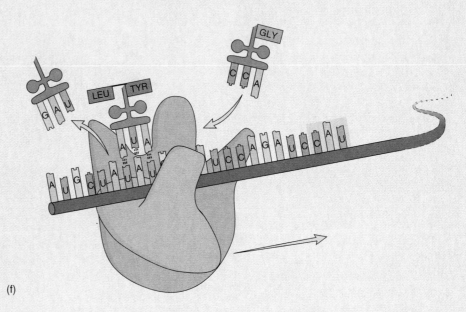

(f)

7. The next tRNA/amino acid unit enters the ribosome and attaches to its codon next to the first set of amino acids (figure 16.8g).

(g)

8. The tRNAs properly align their amino acids so that they may be chemically attached to one another, forming a chain of three amino acids (figure 16.8*h*).

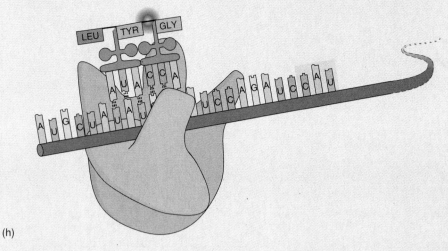

(h)

9. Once three amino acids are connected to one another, the second amino acid is released from its tRNA and mRNA (this tRNA is set free to move through the cytoplasm to attach to and transfer another amino acid) (figure 16.8*i*).

(i)

10. The ribosome moves along the mRNA to the next codon and the tRNA is released, free to move through the cytoplasm to attach to and transfer another amino acid (figure 16.8*j*).
11. This process repeats until all of the amino acids needed to form the protein have attached to one another in the proper sequence. This amino acid sequence was encoded by the DNA gene.
12. Once the final amino acid is attached to the growing chain of amino acids, all the molecules (mRNA, tRNA, and newly formed protein) are released from the ribosome. The "all done" codons signal this action.
13. The ribosome is again free to become involved in another protein-synthesis operation.
14. The newly synthesized chain of amino acids (the new protein) leaves the ribosome to begin its work. However, the protein may need to be altered by the cell before it will be ready for use.

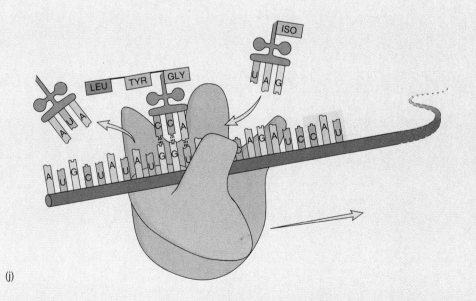

(j)

Figure 16.9 shows an example of protein synthesis.

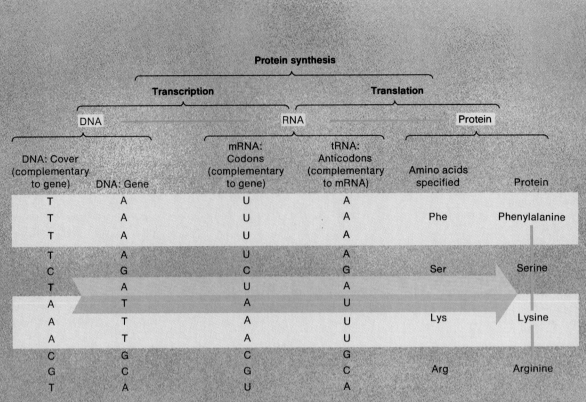

Figure 16.9 **Protein Synthesis** Several steps are involved in protein synthesis: (1) mRNA is manufactured from a DNA molecule in the transcription operation; (2) the mRNA enters the cytoplasm and attaches to ribosomes; (3) the tRNA carries amino acids to the ribosome and positions them in the order specified by the mRNA in the translation operation; (4) the amino acids are combined chemically to form a protein.

Manipulating DNA to Our Advantage

One aspect of **genetic engineering,** the science of gene manipulation, is splicing genes from one organism into another, resulting in a new form of DNA called *recombinant DNA.* This process uses enzymes that are naturally involved in DNA replication, along with special bacterial enzymes that cut and splice DNA. When genes are spliced from one organism into host cells of another organism, the host cells replicate these new, "foreign" genes and synthesize their proteins. Biologists begin gene splicing by isolating DNA from an organism that contains the desired gene; for example, from human cells they may isolate the gene for making insulin. If the gene is short enough and its base sequence is known, it actually may be made in the laboratory from separate nucleotides. However, many genes are too long and complex. Therefore, the entire DNA sequence is cut from the chromosome with enzymes. DNA segments have successfully been cut from rats, frogs, bacteria, humans, and many other organisms.

This isolated gene is usually spliced into microbial (yeast, viral, or bacterial) DNA, although animal genes have been spliced into plants and genes have been transferred between plants and between animals. The host DNA is opened up, and enzymes are used to insert the new DNA into the host DNA. Once inside the host cell, the genes may be replicated along with the rest of the DNA to clone the "foreign" gene, or they may begin to synthesize the encoded protein.

As this highly sophisticated procedure has been refined, it has become possible to quickly and accurately splice genes from a variety of species into host bacteria or other cells, making possible the synthesis of large quantities of medically important products.

In most cells, the mRNA travels through more than one ribosome at a time. When viewed with the electron microscope, this appears as a long thread (the mRNA) with several dark knots (the ribosomes) along its length.

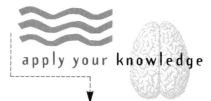

Where Do You Stand?

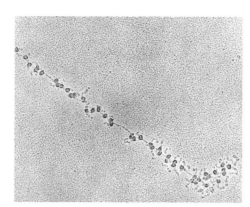

A Polyribosome   Any single mRNA molecule can be translated on a series of ribosomes. The mRNA and its several attached translating ribosomes are called a polyribosome. The mRNA moves along and through the ribosomes in a way similar to an audiocassette tape moving between the heads of a cassette player. Within the polysome complex, the first ribosome to bind to the mRNA completes the translation of a polypeptide first, and the last ribosome to bind to the mRNA synthesizes the polypeptide last. The mRNA can be "played" again.

The field of bioengineering is advancing quickly. The first bioengineering efforts focused on the development of genetically altered crops that displayed improvements over past varieties, such as increased resistance to infectious disease. The second wave of research involved manipulating DNA, which resulted in improved food handling and processing, such as slower ripening of tomatoes.

Currently, crops are being genetically manipulated to manufacture large quantities of specialty chemicals such as antibiotics, steroids, and other biologically useful organic chemicals. While small amounts of these products have been produced from genetically engineered microorganisms, turnips, potatoes, and tobacco can generate tens or hundreds of kilograms of specialty products per year. Researchers have shown, for example, that turnips can produce interferon (an antiviral agent), tobacco can create antibodies to fight human disease, oilseed rape plants can serve as a source of human brain hormones, and potatoes can synthesize human serum albumin that is indistinguishable from the genuine human blood protein. The work of genetic engineers may sound exciting and positive, but many ethical questions must be addressed in this field. Divide into small groups and identify and discuss five ethical issues associated with bioengineering.

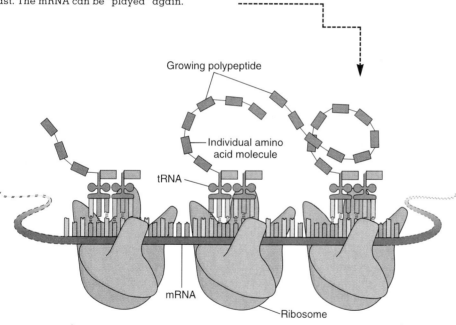

Growing polypeptide

Individual amino acid molecule

tRNA

mRNA

Ribosome

Polysome complex

BIO *feature*

Genetic Medicine

Genetic engineering is making its way into the medical profession. **Genetic medicine** is the art and science of manipulating genes for the diagnosis and treatment of disease and the maintenance of health. Some human diseases are the result of genes that function improperly. Others are the result of genes that do not function at all. As such diseases are identified, biomedical engineers work to find cures or therapies. These involve amplifying the operation of existing genes or adding functional genes to defective cells.

One experimental form of gene therapy is a treatment for a deadly form of skin cancer called *malignant melanoma*. Some people experience a severe form of melanoma because they do not produce enough of a chemical called *tumor necrosis factor*. TNF is a powerful cancer-shrinking chemical. At normal levels, TNF helps to destroy melanoma cells. In an attempt to control this cancer, a healthy gene for TNF is attached to lipid-containing particles called *liposomes* and injected directly into the patient's body. The genes are taken in by cells and integrated into the DNA. The new, functional genes produce enough TNF to control the melanoma. Another bioengineered chemical, *interleukin-2* (IL-2), has been used to stimulate the genes that make TNF, causing them to produce increased amounts. These enhancement therapies are used cautiously since TNF can cause potentially toxic side effects.

Another genetic disease is the "boy-in-the-bubble" disease or *SCID, severe combined immunodeficiency*. A person with SCID lacks the ability to synthesize the enzyme adenosine deaminase. ADA is crucial to the formation of white blood cells known as T-cells. T-cells play a key role in defending the body against infection. (HIV/AIDS patients also suffer from a decreased number of T-cells.) Genetic medicine is being used to increase the production of ADA in SCID patients, which increases the number of infection-fighting T-cells.

Several steps are required in the process. Defective T-cells are removed from a SCID patient and cultured in the lab to increase their numbers. ADA genes are extracted from healthy cells and attached to the genes of special viruses. These viruses have the ability to act as *vectors:* they deliver genes into target cells. When a virus enters a deficient T-cell, viral enzymes splice the ADA gene into the T-cell chromosomes. The virus is destroyed when the ADA gene is transferred and causes no harm. The cultured T-cells with functioning ADA genes are then reintroduced into the patient. Patients that have undergone this gene therapy have shown increased numbers of T-cells. This may seem to be a cure-all, but T-cells do not live long. Therefore, when the genetically altered T-cells die, immunodeficiency symptoms reappear. A genetic cure will occur only if ADA genes can be spliced into the DNA of the white blood cells that produce T-cells. This would ensure a continuous supply of correctly functioning T-cells.

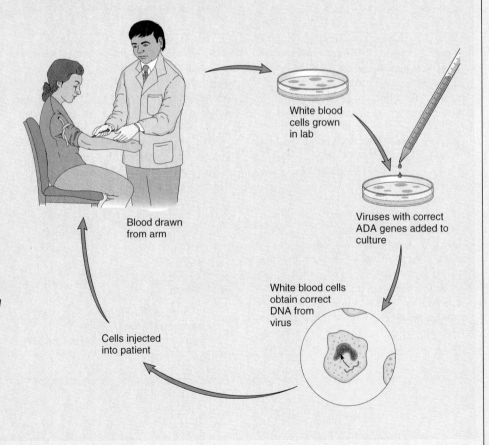

Blood drawn from arm

White blood cells grown in lab

Viruses with correct ADA genes added to culture

White blood cells obtain correct DNA from virus

Cells injected into patient

PCR and Genetic Fingerprinting

In 1989 the American Association for the Advancement of Science named DNA polymerase "Molecule of the Year." The value of this enzyme in the polymerase chain reaction is so great that it could not be ignored. But just what is the polymerase chain reaction (PCR), how does it work, and what can you do with it?

The PCR is a laboratory procedure that enables scientists to copy selected segments of DNA. Using PCR, scientists can get enough DNA for analysis and identification from a single cell! Having many copies of a target sequence of nucleotides enables biochemists to more easily work with DNA. This is like increasing the one "needle in a haystack" to such large numbers (*100 billion in a matter of hours*) that they're not hard to find, recognize, and work with. The types of specimens that can be used include semen, hair, blood, bacteria, protozoa, viruses, mummified tissue, and frozen cells. The process requires the DNA specimen, free DNA nucleotides, synthetic "primer" DNA, DNA polymerase, and simple lab equipment, such as a test tube and a source of heat.

Having decided which target sequence of nucleotides (which "needle") is to be replicated, scientists heat the specimen of DNA to separate the two halves. Molecules of synthetic "primer" DNA are added to the specimen. These primer molecules are specifically designed to attach to the ends of the target sequence. Next, a mixture of nucleotides is added so that they can become the newly replicated DNA. The primer, attached to the DNA

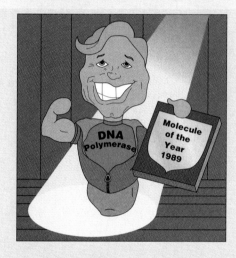

and added nucleotides, serves as the substrate for the DNA polymerase. Once added, the polymerase begins making its way down the length of the DNA from one attached primer end to the other. The enzyme attaches the new DNA nucleotides to the strand, replicating the molecule as it goes. It stops when it reaches the other end, having produced a new copy of the target sequence. Since the enzyme will continue to operate as long as substrates are available, the process continues, and in a short time there are billions of small pieces of DNA, all replicas of the target sequence.

So what, you say? Well, consider the following. Using the PCR, scientists have been able to

1. more accurately diagnose such diseases as sickle-cell anemia, cancer, Lyme disease, AIDS, and Legionnaires' disease;

2. perform highly accurate tissue typing for matching organ-transplant donors and recipients;

3. help resolve criminal cases of rape, murder, assault, and robbery by matching suspect DNA to that found at the crime scene. Some states have used the PCR to compile lists of sex offenders;

4. detect specific bacteria in environmental samples;

5. monitor the spread of genetically engineered microorganisms in the environment;

6. check water quality by detecting bacterial contamination from feces;

7. identify viruses in water samples;

8. identify disease-causing protozoa in water;

9. determine specific metabolic pathways and activities occurring in microorganisms;

10. determine races, distribution patterns, kinships, migration patterns, evolutionary relationships, and rates of evolution of long-extinct species;

11. accurately settle paternity suits;

12. confirm identity in amnesia cases;

13. identify a person as a relative for immigration purposes;

14. provide the basis for making human antibodies in specific bacteria;

15. possibly provide the basis for replicating genes that could be transplanted into individuals suffering from genetic diseases; and

16. identify nucleotide sequences peculiar to the human genome (an application currently under way as part of the Human Genome Project).

Figure 16.10 Genetic Engineering Crops with unique characteristics have been developed by manipulatig their DNA.

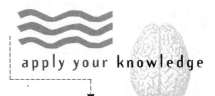

A Case for the Jury

For example, recombinant DNA procedures are responsible for the bacterial production of human insulin, used to control diabetes; interferon, an antiviral agent; human growth hormone, used to stimulate growth in children lacking this hormone; and somatostatin, a brain hormone also implicated in growth. The possibilities that open up with the manipulation of DNA are revolutionary. These methods enable cells to produce molecules that they would not normally make. Some research laboratories have even spliced genes into laboratory-cultured human cells. Should such a venture prove to be practical, genetic diseases such as sickle-cell anemia could be controlled. The process of recombinant DNA gene splicing also enables cells to be more efficient at producing molecules that they normally synthesize. Some of the likely rewards are (1) production of additional, medically useful proteins; (2) mapping of the locations of genes on human chromosomes; (3) more complete understanding of how genes are regulated; (4) production of crop plants with increased yields; and (5) development of new varieties of garden plants (figure 16.10).

The discovery of the structure of DNA more than forty years ago seemed very far removed from the practical world. We are just now realizing the importance of the "pure" research. Many companies are involved in recombinant DNA research with the aim of alleviating or curing disease.

An 18-year-old college student reported that she had been raped by someone she identified as a "large, tanned white man." A fellow student in her biology class fitting that description was said by eyewitnesses to have been in the area at approximately the time of the crime. The suspect was apprehended and upon investigation was found to look very much like someone who lived in the area and who had a previous record of criminal sexual assaults. Samples of semen from the woman's vagina were taken during a physical exam following the rape. Cells were also taken from the suspect. He was brought to trial but found to be innocent of the crime based on evidence from the criminal investigations laboratory. His alibi that he had been working alone on a research project in the biology lab held up. Without PCR-genetic fingerprinting, the suspect would surely have been wrongly convicted, based solely on circumstantial evidence provided by the victim and the "eyewitnesses."

Place yourself in the position of the expert witness from the criminal laboratory who performed the PCR-genetic fingerprinting tests on the two specimens. The prosecuting attorney has just asked you to explain to the jury what led you to the conclusion that the suspect could not have been responsible for this crime. Remember, you must explain this to a jury of twelve men and women who in all likelihood have little or no background in the biological sciences. Please, tell the whole truth and nothing but the truth.

(You might want to review chapter 1 and the case of Mary Jane versus the Nifty Green Yard Care Company as you work on this activity.)

SUMMARY

The successful operation of a living cell depends on its ability to accurately reproduce genes and control chemical reactions. DNA replication results in an exact doubling of the genetic material. The process virtually guarantees that identical strands of DNA will be passed on to the next generation of cells.

Enzymes are responsible for the efficient control of a cell's metabolism. However, the production of protein molecules is under the control of the nucleic acids, the primary control molecules of the cell. The structure of the nucleic acids DNA and RNA determines the structure of the proteins, while the structure of the proteins determines their function in the cell's life cycle. Protein synthesis involves the decoding of the DNA into specific protein molecules and the use of the intermediate molecules, mRNA and tRNA, at the ribosome. Errors in any of the codons of these molecules may produce observable changes in the cell's functioning and lead to the death of the cell.

Methods of manipulating DNA have led to the controlled transfer of genes from one kind of organism to another. This has made it possible for bacteria and other organisms to produce a number of human gene products.

CHAPTER GLOSSARY

adenine (a′den-ēn″) A double-ring nitrogen-containing molecule in DNA and RNA. It is the complementary base of thymine or uracil.

amino acid (ah-mēn′o ă′sid) A subunit of protein.

anticodon (an″′te-ko′don) A sequence of three nitrogen-containing bases on a tRNA molecule capable of forming bonds with three complementary bases on an mRNA codon during translation.

catalyst (cat′uh-list) A chemical that speeds up a reaction but is not used up in the reaction.

chromatin fibers (kro′mah-tin fi′bers) The double DNA strands with attached proteins; also called *nucleoproteins.*

codon (ko′don) A sequence of three nucleotides of an mRNA molecule that directs the placement of a particular amino acid during translation.

cytosine (si′to-sēn) A single-ring nitrogen-containing molecule in DNA and RNA. It is complementary to guanine.

denature (de-nā′chur) To permanently change the protein structure of an enzyme so that it loses its ability to function.

deoxyribonucleic acid (DNA) (de-ok″se-ri-bo-nu-kle′ik as′id) A polymer of nucleotides that serves as genetic information. In prokaryotic cells, it is a double-stranded loop. In eukaryotic cells, it is found in strands with attached proteins. When tightly coiled, it is known as a *chromosome.*

DNA replication (rep″lĭ-ka′shun) The process by which the genetic material (DNA) of the cell reproduces itself prior to its distribution to the next generation of cells.

enzyme (en′zīm) A specific protein that acts as a catalyst to change the rate of a reaction.

gene (jēn) Any molecule, usually a segment of DNA, that is able to (1) replicate by directing the manufacture of copies of itself; (2) mutate, or chemically change, and transmit these changes to future generations; (3) store information that determines the characteristics of cells and organisms; and (4) use this information to direct the synthesis of structural and regulatory proteins.

genetic engineering (jĕ-net′ik en-je-nēr′ing) The science of gene manipulation.

genetic medicine (jĕ-net′ik med′ĭ-sin) The art and science of manipulating genes for the diagnosis and treatment of disease and the maintenance of health.

guanine (gwah′nēn) A double-ring nitrogen-containing molecule in DNA and RNA. It is the complementary base of cytosine.

messenger RNA (mRNA) (mes′-en-jer) A molecule composed of ribonucleotides that functions as a copy of the gene and is used in the cytoplasm of the cell during protein synthesis.

mutation (miu-ta′shun) Any change in the genetic information of a cell.

nucleic acids (nu′kle-ik as′ids) Complex molecules that store and transfer information within a cell. They are constructed of subunits known as nucleotides.

nucleoproteins (nu-kle-o-pro′tēnz) The double DNA strands with attached proteins; also called *chromatin fibers.*

nucleotide (nu′kle-o-tīd) The building block of the nucleic acids. Each is composed of a 5-carbon sugar, a phosphate, and a nitrogen-containing portion.

protein (pro′tēn) Macromolecules made up of amino acid subunits attached to each other; groups of polypeptides.

protein synthesis (pro′tēn sin′thĕ-sis) The process whereby the tRNA utilizes the mRNA as a guide to arrange the amino acids in their proper sequence according to the genetic information in the chemical code of DNA.

regulator proteins (reg′yu-la-tor pro′tēnz) Proteins that influence the activities that occur in an organism; for example, enzymes and some hormones.

ribonucleic acid (RNA) (ri-bo-nu-kle′ik as′id) A polymer of nucleotides formed on the template surface of DNA by transcription. Three forms that have been identified are mRNA, rRNA, and tRNA.

ribosomal RNA (rRNA) (ri-bo-sōm′al) A globular form of RNA; a part of ribosomes.

structural proteins (struk′chu-ral pro′tēnz) Proteins that are important for holding cells and organisms together, such as the proteins that make up the cell membrane, muscles, tendons, and blood.

substrate (sub′strāt) A reactant molecule with which the enzyme combines.

thymine (thi′mēn) A single-ring nitrogen-containing molecule in DNA but not in RNA. It is complementary to adenine.

transcription (tran-skrip′shun) The process of manufacturing RNA from the template surface of DNA.

transfer RNA (tRNA) (trans′fur) A molecule composed of ribonucleic acid. It is responsible for transporting a specific amino acid into a ribosome for assembly into a protein.

translation (trans-la′shun) The assembly of individual amino acids into a protein.

uracil (yu′rah-sil) A single-ring nitrogen-containing molecule in RNA but not in DNA. It is complementary to adenine.

CONCEPT MAP TERMINOLOGY

Construct a concept map to represent the relationships among the following concepts.

allele
central dogma
DNA
enzymes

gene
genetic fingerprints
genotype
locus

phenotype
polymerase
replication

RNA
transcription
translation

LABEL • DIAGRAM • EXPLAIN

Create a single strand of DNA nucleotides about nine nucleotides long. Now construct the complementary strand so that you show a duplex molecule. Did you use the base-pairing rule? Next diagram the steps needed to replicate this double-stranded DNA molecule. Use the first strand contructed to show transcription. Label the product mRNA. Now show the anticodons of the tRNA, which are complementary to the codons of the mRNA. Which amino acid is each of the tRNAs attached to? Show the sequence of amino acids of the newly forming "protein."

Multiple Choice Questions

1. Mutations can occur:
   a. because two codons code for the same amino acid
   b. when only one base is changed
   c. only in the "lower organisms"
   d. to proteins
2. The site of protein synthesis is:
   a. at the nuclear membrane
   b. at ribosomes
   c. in the nucleus
   d. outside the cell
3. While one strand of DNA is being transcribed to RNA:
   a. the complementary strand makes tRNA
   b. the complementary strand is unused until it reattaches to its complement
   c. the complementary strand at this point is replicating
   d. mutations are impossible during this short period
4. One way to introduce new DNA into an organism is:
   a. genetic engineering
   b. adding enzymes
   c. removing portions of the mRNA
   d. transcription
5. Removing only one base in a DNA sequence:
   a. usually has no effect on the organism
   b. is unlikely to change any genes
   c. cannot occur without extremes of heat and pressure
   d. can result in a significant change in the information about a protein
6. A major difference between the genetic data of prokaryotic (bacterial) and eukaryotic cells is that in prokaryotes the:
   a. genes are RNA, not DNA
   b. bacterial DNA lacks thymine
   c. DNA is circular in bacteria
   d. DNA is absent in bacteria

Questions with Short Answers

1. What is the difference between a nucleotide and a codon?

2. What are the differences between DNA and RNA?

3. List the sequence of events that takes place when a DNA message is translated into protein.

4. Write a statement that describes the central dogma.

5. What is an enzyme? How does it work?

6. What are DNA polymerase and RNA polymerase and how do they function?

7. How does DNA replication differ from the manufacture of an RNA molecule?

8. If a DNA nucleotide sequence is CATAAAGCA, what is the mRNA nucleotide sequence that would base-pair with it?

9. What are amino acids and how do they relate to proteins?

10. How do tRNA, rRNA, and mRNA differ in function?

part 5

The Human Organism

More and more, maintaining good health has become the responsibility of individuals rather than the duty of health care professionals. Part V covers some of the major systems of the human organism and how to provide for their health and the well-being of the whole organism.

Chapter 17 encourages readers to look at nutrition. What we put into our bodies determines how well our various parts will function. Digestion and processing of food materials to release energy are covered here. This chapter also provides information that will allow you to evaluate the conflicting claims made for various foods and nutritional supplements.

We are all interested in sexual function and dysfunction, but we often get information from the media that is confusing and, in many cases, sensationalized falsehoods. Chapter 18 offers a rational discussion of many of the questions students ask about sex and sexuality. Topics such as reproductive health and pregnancy prevention are covered.

Chapter 19 covers muscle structure and function but also explains how exercise fits into maintaining health. How the circulatory system functions with the respiratory system is also explained in this chapter.

The nervous and endocrine systems integrate various systems of the body. Chapter 20 describes the sense organs, how information is interpreted, and responses by muscles and glands, demonstrating the complexity of the human organism's relationship to its environment.

Child with peony flowers.

part. 5

The Human Organism

chapter

17

Nutrition— Food and Diet

Commercial fishing, Bellevue, Iowa.

Nutrition—Food and Diet

learning objectives

- Be familiar with the five basic food groups of the Food Guide Pyramid; know the number of recommended servings of each group, their sources, and their benefits.

- Give examples of psychological eating disorders and deficiency diseases.

- Distinguish between calorie and kcalorie (kilocalorie).

- Understand the concepts of basal metabolic rate, specific dynamic action, and voluntary activity.

- Recognize the functions of the six types of nutrients.

- Be familiar with the concept of Recommended Dietary Allowances (RDAs).

- Be familiar with the unique nutritional requirements of each stage of the human life cycle.

You Are What You Eat

Have you looked at the label on a cereal box lately? Within the past few years federal regulations concerning food labeling have caused some major changes in what you see on the packages of many foods. The labels contain information about various nutrients in the foods. **Nutrients** are any of the kinds of molecules that the body needs for growth, reproduction, or maintenance. Printed on the labels is information concerning how much of your daily levels of carbohydrate, fat, and protein the products will provide. These amounts are based on a 2,000-kilocalorie-per-day diet. One of the first things listed on the label is a serving size. Serving sizes have been standardized so that manufacturers can no longer make a particular product seem better by basing their measurements on a very tiny or very large serving. This is important because now consumers can more easily compare products. The labels tell you the number of Calories and the amount of fat (both total fat and saturated fat), carbohydrate, and protein per serving.

The information on the labels is provided to help you make better decisions about your diet. A person's **diet** is the food and drink he or she consumes from day to day. Your diet must contain the minimal nutrients necessary to manufacture and maintain your body's structure (bones and muscle) and regulatory molecules (enzymes and hormones) and to supply the energy (ATP) needed to run your body's machinery. If your diet is deficient in

CONCEPT CONNECTIONS

The word **diet** is used in two different ways in our society. In this book the term is used to describe in a matter-of-fact way what you eat every day. However, most people use the word *diet* to refer to a way of eating that is intended to either increase or decrease weight. We do not want to think of *diet* in this way. A diet should not be a regimen you follow only to abandon and return to unhealthy eating habits.

FYI

Calories and Kilocalories   The unit used to measure the amount of energy in foods is the **kilocalorie (kcalorie)**. One kilocalorie is the amount of energy needed to raise the temperature of one *kilogram* of water one degree Celsius. Remember that the prefix *kilo* means "one thousand times" the value listed. Therefore, a kilocalorie is one thousand times more heat energy than a calorie. A **calorie** is the amount of heat energy needed to raise the temperature of one *gram* of water one degree Celsius. The energy unit you will be using as you study nutrition is almost always the kilocalorie. In the United States, we seldom use the "kilo" prefix; instead, we use a capital C on the word *calorie* to designate a kilocalorie (Calorie = *kilo*calorie). Since a true calorie (not a Calorie or *kilo*calorie) is 1/1,000th of a Calorie, and an extremely small unit of energy, it is not practical to use when labeling foods.

nutrients, or if your body cannot process nutrients efficiently, a dietary deficiency and ill health may result. The processes involved in assimilating and utilizing nutrients are collectively known as **nutrition.** A good understanding of nutrition can promote good health and help people avoid disease. Table 17.1 and table 17.2 briefly outline the structure and function of organs involved in digesting food and the enzymes responsible for the chemical breakdown of large food molecules to smaller units.

The Food Guide Pyramid and the Five Food Groups

With all of the advertisements and new products on the market, it might seem impossible to decide what a healthy diet should include. What foods should be avoided and what foods should we eat? How can we make the best decisions for our health?

apply your knowledge

Reading Labels

Read the labels of two similar products, such as two brands of cereal, two candy bars, or two other foods. Compare the information about Calories, vitamins, and minerals.

The next time you are in a fast-food restaurant, ask for nutrition information about their products. Most restaurants will have a brochure or pamphlet containing information about the products they sell in a format similar to that used for nutrition labels on other foods.

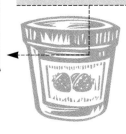

table 17.1

The Human Digestive System

Structure	Function
Mouth	The opening to the gastrointestinal tract.
Oral cavity	Contains the teeth and tongue used to mechanically break down food particles for swallowing.
Salivary glands	Produce *saliva*, which is used to moisten and enzymatically break down food particles.
Pharynx	The back of the throat; nerve endings in the lining of the pharynx are stimulated, causing a reflex contraction of the walls of the esophagus, which transports the food to the stomach.
Esophagus	A tube conducting food particles from the oral cavity to stomach.
Stomach	Gastric juice is added to the food in the stomach. Gastric juice contains enzymes and hydrochloric acid. The major enzyme of the stomach is pepsin, which initiates the breakdown of protein. The pH of gastric juice is very low, generally around 2. Consequently, very few bacteria or protozoa emerge from the stomach alive. Those that do have special protective features that allow them to survive as they pass through the stomach. The food within the stomach may protect the microbes by buffering the pH or by providing a barrier between the microbe and the gastric juice. The entire mixture is churned by the contractions of the three layers of muscle in the stomach wall. The combined activities of enzymatic breakdown, chemical breakdown by hydrochloric acid, and mechanical processing by muscular movement results in a thoroughly mixed liquid called *chyme*. Chyme eventually leaves the stomach through a valve known as the *pyloric sphincter* and enters the small intestine.

Structure	Function

Small intestine	The first part of the small intestine is known as the *duodenum*. In addition to producing enzymes, the duodenum secretes several kinds of hormones that regulate the release of food from the stomach and the release of secretions from the pancreas and liver.
	
	  Emulsification is a physical breakdown in the size of fat globules. It is important because fats are not soluble in water, yet the reactions of digestion must take place in a water-based solution. The activity of bile from the liver is important for the digestion of fats in the intestine. Bile causes large globules of fat to be broken into much smaller units, much as soap breaks up fat particles into smaller units that are suspended in water and washed away. Although the physical processing of food by mechanical forces, pH changes, and emulsification is important in reducing the size of the food particles, food cannot enter the circulatory system unless it is broken down into fundamental chemical units by enzymes and other chemical processes.
	
	  Digested nutrients are transported through the lining of the small intestine and enter the circulatory system for distribution throughout the body.

Pancreas	An organ located near the stomach and liver that produces several kinds of enzymes. The enzyme mixture is capable of breaking down most foods to smaller molecules. Also produces sodium bicarbonate, which neutralizes the acid from the stomach.
Liver	A large organ in the upper abdomen that performs several functions. One of its functions is the secretion of *bile*. When bile leaves the liver, it is stored in the *gallbladder* prior to being released into the duodenum.
Large intestine	Primarily involved in reabsorbing the water that has been added to the food tube as part of the saliva, gastric juice, bile, pancreatic secretions, and intestinal juices.
Rectum	The final portion of the gastrointestinal tract; water is reabsorbed through its surface and the undigestible fecal material is held before elimination from the body.
Anus	The opening through which the undigested fecal material is eliminated from the body.

table 17.2

Digestive Enzymes and Their Functions

Enzyme	Site of Production	Molecules Altered	Molecules Produced
Salivary amylase	Salivary glands	Starch	Smaller polysaccharides (many sugar molecules attached together)
Pepsin	Stomach lining	Proteins	Peptides (several amino acids)
Gastric lipase	Stomach lining	Fats	Fatty acids and glycerol
Chymotrypsin	Pancreas	Polypeptides (long chains of amino acids)	Peptides
Trypsin	Pancreas	Polypeptides	Peptides
Carboxypeptidase	Pancreas	Peptides	Smaller peptides and amino acids
Pancreatic amylase	Pancreas	Polysaccharides	Disaccharides
Pancreatic lipase	Pancreas	Fats	Fatty acids and glycerol
Nuclease	Pancreas	Nucleic acids	Nucleotides
Aminopeptidase	Intestinal lining	Peptides	Smaller peptides and amino acids
Dipeptidase	Intestinal lining	Dipeptides	Amino acids
Lactase	Intestinal lining	Lactose	Glucose and galactose
Maltase	Intestinal lining	Maltose	Glucose
Sucrase	Intestinal lining	Sucrose	Glucose and fructose
Nuclease	Intestinal lining	Nucleic acids	Nucleotides

Planning a diet around basic food groups is generally easy. The four basic food groups first developed and introduced in 1953 (figure 17.1) to serve as guidelines in maintaining a balanced diet have been modified and updated several times. In May 1992, the U.S. Department of Agriculture released the results of its most recent study on how best to educate the public about daily nutrition. The federal government adopted the **Food Guide Pyramid** of the Department of Agriculture as one of its primary tools to help the general public plan for good nutrition (figure 17.2). The Food Guide Pyramid describes five basic groups of foods with guidelines for the

(a) Meat group

(b) Fruit group

(c) Vegetable group

(d) Cereal group

(e) Dairy group

(f) Oils, sugars, and alcohol

Figure 17.1 Serving Sizes of Basic Foods These photos show typical serving sizes for many types of foods. Each item in each photo is equivalent to one serving. On a daily basis, an average adult should eat six to eleven servings of cereals, two to three servings from the meat group, three to five servings from the vegetable group, two to four servings from the fruit group, and two to three servings from the dairy group. Fats, sugars, and alcohol should be used sparingly.

Part 5   The Human Organism

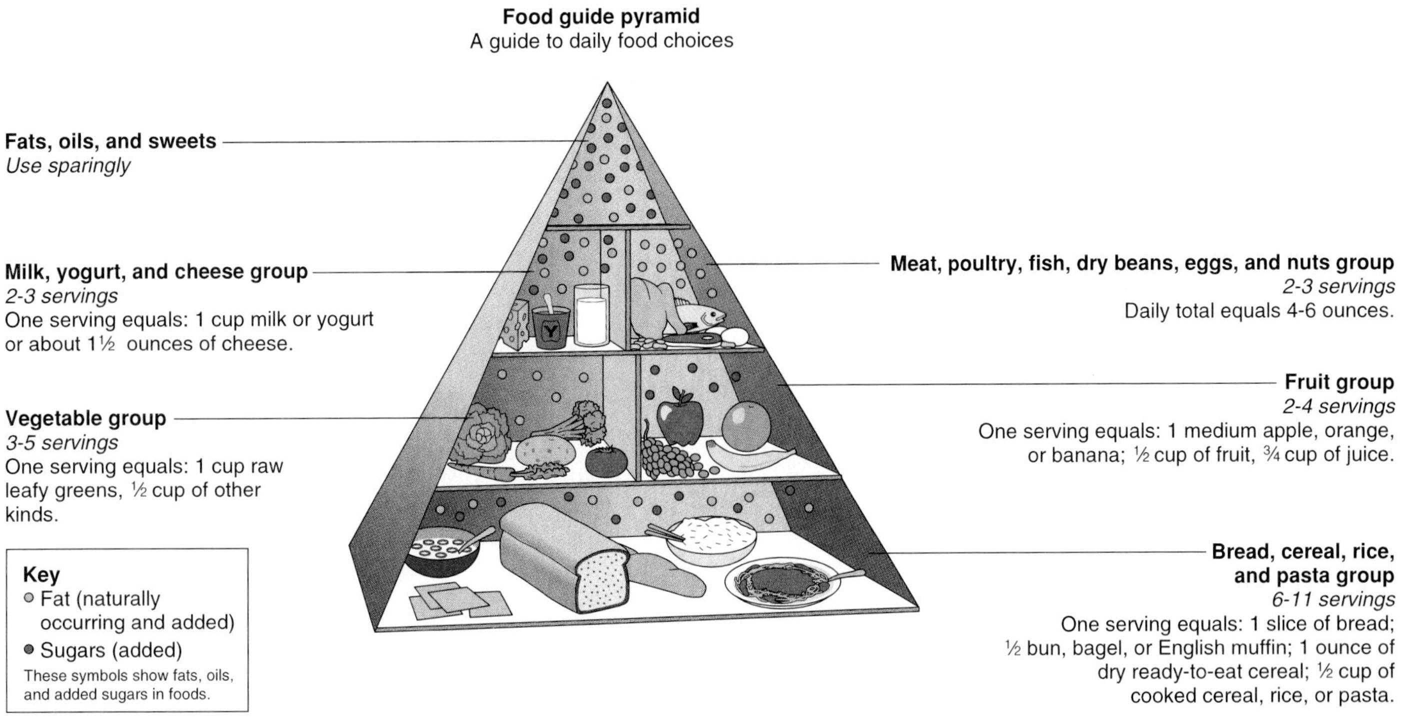

Food guide pyramid
A guide to daily food choices

Fats, oils, and sweets
Use sparingly

Milk, yogurt, and cheese group
2-3 servings
One serving equals: 1 cup milk or yogurt or about 1½ ounces of cheese.

Vegetable group
3-5 servings
One serving equals: 1 cup raw leafy greens, ½ cup of other kinds.

Key
○ Fat (naturally occurring and added)
● Sugars (added)
These symbols show fats, oils, and added sugars in foods.

Meat, poultry, fish, dry beans, eggs, and nuts group
2-3 servings
Daily total equals 4-6 ounces.

Fruit group
2-4 servings
One serving equals: 1 medium apple, orange, or banana; ½ cup of fruit, ¾ cup of juice.

Bread, cereal, rice, and pasta group
6-11 servings
One serving equals: 1 slice of bread; ½ bun, bagel, or English muffin; 1 ounce of dry ready-to-eat cereal; ½ cup of cooked cereal, rice, or pasta.

Figure 17.2 The Food Guide Pyramid In May 1992, the Department of Agriculture released a new guide to good eating. This Food Guide Pyramid suggests that we eat particular amounts of five different food groups while decreasing our intake of fats and sugars. This guide should simplify our menu planning while helping to ensure that we get all of the recommended amounts of basic nutrients.

amounts needed daily. Some of the important aspects of the Food Guide Pyramid that differ from previous information provided by the federal government include decreasing our intake of fats and sugars while increasing our daily servings of fruits and vegetables. In addition, the new guidelines suggest significantly increasing the amount of grain products we eat each day.

Group 1: Meat, Poultry, Fish, and Dry Beans

This group contains most of the things we eat as a source of protein; meat, fish, nuts, peas, tofu, and eggs are considered a part of this group (figure 17.1a). It is recommended that we include four to six ounces (100–170 grams) of these items in our daily diet. For example, you could have two ounces (56 grams) of meat and two ounces of one of the other alternatives each day to meet this guideline. This means that one double cheeseburger greatly exceeds

the recommended daily intake. Eating excessive amounts of protein can stress the kidneys by causing higher concentrations of calcium in the urine, increasing the demand for water to remove toxic keto acids, and leading to weight gain. It should be noted that vegetarians must pay particular attention to acquiring adequate sources of protein because they have eliminated a major source from their diet. (See Bio feature: Which Kind of Vegetarian Are You? on page 399.)

Group 2: Dairy Products

All of the cheeses, ice cream, yogurt, and milk are in this group (figure 17.1e). Two to three servings of dairy products are recommended each day. Servings of milk or yogurt should be about 1 cup (about 250 ml) or contain 100 calories. The serving size for cheese is about 1.5 ounces (40 grams). Using product labels will help you determine the appropriate serving size of individual items. This group provides min-

erals such as calcium but also provides water, vitamins, carbohydrates, and protein. Keep in mind, however, that cheese contains large amounts of cholesterol and fat in each serving.

Group 3: Vegetables

The Food Guide Pyramid suggests three to five servings of vegetables each day. Items in this group include nonsweet plant materials such as broccoli, carrots, cabbage, corn, green beans, potatoes, lettuce, and spinach (figure 17.1c). A serving is considered to be one cup of raw leafy vegetables or a half cup (125 ml) of other types. It is wise to include as much variety as possible in this group. If you eat only carrots, several cups each day can become very boring. Also, a variety of vegetables will provide a greater range of vitamins. Increasing evidence indicates that cabbage, broccoli, and cauliflower can provide some protection from certain types of cancers. This is a good reason to include these foods in your diet.

Proteins

Proteins are composed of amino acids linked together by peptide bonds; however, not all proteins contain the same amino acids. Proteins can be divided into two main groups, the *complete proteins* and the *incomplete proteins*. Complete proteins contain all the amino acids necessary for good health, while incomplete proteins lack certain amino acids that the body must have to function efficiently. Table 17.3 lists the *essential amino acids,* those that cannot be produced by the human body. Without adequate amounts of these amino acids in the diet, a person may develop protein-deficiency diseases. Proteins are essential components of hemoglobin and cell membranes, as well as of antibodies, enzymes, some hormones, hair, muscle, and collagen, a connective tissue fiber. Plasma proteins are important because they can serve as buffers and maintain appropriate water balance. Proteins also provide a last-ditch source of energy when carbohydrate and fat consumption falls below protective levels.

The body's need for a mixture of proteins and carbohydrates is vitally important. Proteins are present in the structures of the human body, but they cannot be stored to offset times of protein deficiency. If you are on a high-protein diet, the amino acids are not stored; they are either used in protein synthesis or used for energy release. A unique relationship exists between carbohydrates and proteins, called **protein-sparing.** When adequate amounts of carbohydrates are present in the diet, the body's proteins do not have to be tapped as a source of energy. *They are spared from being used.* Only when carbohydrate consumption falls below an adequate level will the body use proteins as an energy source. Most people in developed countries have a misconception with regard to the amount of protein necessary in their diets. The total amount necessary is actually quite small and can be easily met.

table 17.3

Sources of Essential Amino Acids

Essential Amino Acids	Food Sources
Threonine	Dairy products, nuts, soybeans, turkey
Lysine	Dairy products, nuts, soybeans, green peas, beef, turkey
Methionine	Dairy products, fish, oatmeal, wheat
Arginine (essential to infants only)	Dairy products, beef, peanuts, ham, shredded wheat, poultry
Valine	Dairy products, liverwurst, peanuts, oats
Phenylalanine	Dairy products, peanuts, calves' liver
Histidine (essential to infants only)	Human and cow's milk and standard infant formulas
Leucine	Dairy products, beef, poultry, fish, soybeans, peanuts
Tryptophan	Dairy products, sesame seeds, sunflower seeds, lamb, poultry, peanuts
Isoleucine	Dairy products, fish, peanuts, oats, macaroni, lima beans

The essential amino acids are required in the diet for protein building and, along with the nonessential amino acids, allow the body to metabolize all nutrients at an optimum rate. Combinations of different plant foods can provide essential amino acids even if complete protein foods (*e.g., meat, fish, and milk*) are not in the diet.

Minerals

All **minerals** are inorganic elements found in nature; they cannot be built by the body. Because they are elements, they cannot be broken down or changed by metabolism or cooking. They commonly occur in many foods and in water. Minerals retain their characteristics whether they are in foods or in the body, and each mineral plays a different role in metabolism. Minerals can function as regulators, activators, transmitters, and controllers of various enzymatic reactions. For example, sodium ions (Na^+) and potassium ions (K^+) are important in the transmission of nerve impulses, while magnesium ions (Mg^{++}) facilitate energy release during reactions involving ATP. Without iron, not enough hemoglobin would be formed to transport oxygen, a condition called *anemia,* and a lack of calcium may result in *osteoporosis,* a condition that results from calcium loss and leads to painful, weakened bones. Many minerals are important in your diet. In addition to those just mentioned, you need chlorine, cobalt, copper, iodine, phosphorus, sulfur, and zinc to remain healthy.

Vitamins

Vitamins are a class of nutrients that cannot be manufactured by the body but are essential in minute amounts to the body's metabolism. Vitamins do not serve as a source of energy, but they help in many enzymatically controlled reactions. They function with specific enzymes to speed the rate of certain chemical reactions. Some enzymes do not function alone but require the attachment of a vitamin to complete their structure. For this reason such vitamins are called *coenzymes.* For example, a B-complex vitamin (niacin) helps enzymes in the respiration of carbohydrates.

Most vitamins are acquired from food; however, vitamin D may be formed when ultraviolet light strikes a molecule already in your skin, converting this molecule to vitamin D. This means that vitamin D is not really a vitamin at all. It came to be known as a vitamin because of the mistaken idea that it is only acquired through food rather than being formed in the skin on exposure to sunshine. It would be more correct to call vitamin D a hormone, but most people do not. Because vitamin D is required for the uptake of calcium from the small intestine, milk fortified with vitamin D is commonly sold. The milk is the source of calcium and the vitamin D required for its uptake.

Foods in this group provide vitamins A and C as well as water and minerals. They also provide fiber, which assists in the proper functioning of the digestive tract.

Group 4: Fruits

This group includes such sweet plant products as melons, berries, apples, oranges, and bananas (figure 17.1b). Since these foods tend to be high in natural sugars, it is recommended that you eat only two to four servings daily. A small apple, half a grapefruit, a half cup of grapes, or six ounces of fruit juice is considered a serving.

Group 5: Grain Products

Group 5 includes vitamin-enriched or whole-grain cereals and grain products such as breads, bagels, buns, crackers, dry and cooked cereals, pancakes, pasta, and tortillas (figure 17.1d). Items in this group are typically dry and seldom need refrigeration. The grain products provide most of your kcalorie requirements. You should have six to eleven servings from this group each day. This is a major change from previous recommendations of four servings each day. A serving is considered about a half cup, or one ounce (125 ml), or containing about 100 kcalories. Using product labels will help to determine the appropriate serving size.

Cereals and grains provide fiber and are a rich source of the B vitamins in your diet. As you decrease the intake of proteins in the meat and poultry group, you should increase your intake of items from this group. These foods give you a full feeling and many of them are very low in fat.

Fats, Oils, and Sweets

Fats, oils, refined sugars, and alcohol should be used moderately (figure 17.1f). Sometimes alcohol and refined sugars (sweets) are called "empty calories" because they are high in energy content but provide minimal nutrients. Consuming these in excess can easily lead to weight problems. Fats present a different problem. Although some fats do have some essential nutrients, they are also very high in energy content. The diet of North Americans includes excessive amounts of fat. Fats include cooking oils, spreads (mayonnaise and margarine), and oils in salad dressings. Foods prepared by frying in fat or oil are extremely high in fat content.

Amounts and Sources of Nutrients

In order to give people some guidelines for planning a diet that provides adequate amounts of the six classes of nutrients, nutritional scientists in the United States and many other countries have developed nutrient standards. In the United States, these guidelines are known as the **Recommended Dietary Allowances,** or **RDAs.** RDAs are dietary recommendations, not requirements or minimum standards. They are based on the needs of a healthy person already eating an adequate diet. RDAs do not apply to a person with medical problems who is under stress or suffering from malnutrition. The amount of each nutrient specified by the RDAs has been set relatively high so that most of the population eating those quantities will be meeting their nutritional needs. Keep in mind that since everybody is different, eating the RDA amounts may not meet your personal needs. You may have a special need for additional amounts of a particular nutrient if you have an unusual metabolic condition.

General sets of RDAs have been developed for four groups of people: infants, children, adults, and pregnant and breast-feeding women. The U.S. RDAs are used when food manufacturers prepare product labels. The federal government requires by law that labels list ingredients from the greatest to the least in quantity. The volume in the package must be stated along with the weight and the name of the manufacturer or distributor. If any nutritional claim is made, it must be supported by factual information.

Carbohydrates

When the word *carbohydrate* is mentioned, many people think of table sugar, but carbohydrates also include more complex molecules such as starch, glycogen, and cellulose. Each of these molecules has a different structural formula and different chemical properties, and plays a different role in the body. Many carbohydrates taste sweet and stimulate the appetite. Complex carbohydrates like starch or glycogen are digested to simpler molecules that can be utilized in aerobic cellular respiration, and the energy of carbohydrates is used to manufacture ATP. Carbohydrates are also converted by the body into molecules that can be used to manufacture necessary components of structures such as nucleic acids. Carbohydrates can also be a source of fibers that slow the absorption of nutrients and stimulate peristalsis in the intestinal tract.

A diet deficient in carbohydrates results in a condition in which primarily fats are converted to ATP. In this situation, fats are metabolized to keto acids, resulting in a potentially dangerous change in the body's pH. A carbohydrate deficiency may also result in the body's use of proteins as a source of energy. In extreme cases this can be fatal, since the oxidation of protein results in an increase in toxic, nitrogen-containing compounds. And keep this in mind: As with other nutrients, if there is an excess of carbohydrates in the diet, they are converted to lipids and stored by the body in fat cells—and you gain weight.

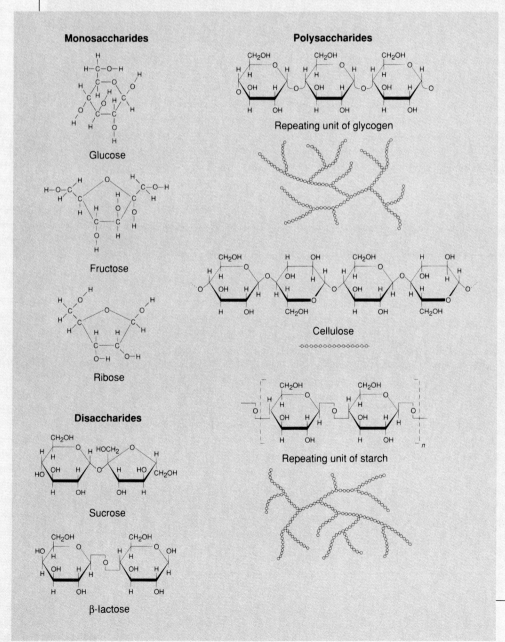

Monosaccharides

Glucose

Fructose

Ribose

Disaccharides

Sucrose

β-lactose

Polysaccharides

Repeating unit of glycogen

Cellulose

Repeating unit of starch

The Structure and Role of Various Carbohydrates The diet includes a wide variety of carbohydrates. Some are monosaccharides (simple sugars) while others are more complex disaccharides, trisaccharides, and polysaccharides. The complex carbohydrates differ from one another depending upon the type of monosaccharides that are linked together. Notice that the complex carbohydrates shown are primarily from plants. With the exception of milk, animal products are not a good nutritional source of carbohydrates because animals do not store them in great quantities or use large amounts of them as structural materials.

BIO feature

Fiber in Your Diet

Fiber is also called *bulk* and *roughage*. Most fibers are a variety of different kinds of indigestible polysaccharides. The five types of fiber are described in table 17.A.

Most fiber is indigestible by human enzymes, but a small amount can be digested by beneficial bacteria normally found in the intestinal tract. Since only a negligible amount of kcalories are obtained through dietary fiber, it is not considered to be a source of energy; however, it does serve other important functions, as shown in table 17.B.

While dietary fiber is usually of great benefit, increasing your fiber intake (or increasing it too rapidly) may result in diarrhea, loss of vitamins and minerals, and damage to the lining of the intestinal tract.

table 17.A

The Five Types of Fiber

Type	Dietary Source	Nature
Cellulose and hemicellulose	Fruits, vegetables, beans, oat and wheat bran, nuts, seeds, whole-grain flour	Wood fiber; plant-cell glue; polysaccharide of glucose
Gums, mucilages, and pectins	Vegetables, fruits, seeds, beans, oats, barley	Plant secretions; galactose-containing polysaccharide
Lignin	Seeds, vegetables, whole grains	Stiffens plant-cell walls; complex of alcohols and organic acids
Algal polysaccharides	Colloids used in chocolate milk, puddings, pie fillings	Carrageenan; polysaccharide extracted from marine algae and seaweeds
Methyl cellulose	Synthetic products	Polysaccharides

table 17.B

Important Functions of Fiber

Benefit	Positive Effect of Dietary Fiber
1. Relieves constipation/diarrhea	Fiber holds water; softens stool to prevent constipation; forms gels to thicken stool to prevent diarrhea.
2. Hemorrhoid control	Softer stools ease elimination to prevent weakening of rectal muscles and protruding of swollen veins.
3. Weight control	Creates feeling of fullness that promotes weight loss; can be used as replacement for fats and sweets.
4. Reduces colon cancer	Speeds movement through intestinal tract reducing exposure time of cancer-causing agents; does same for bile (associated with cancer risk).
5. Reduces blood lipids/ cardiovascular disease	Binds bile, cholesterol, and other lipids to carry them out of the body.
6. Benefits blood glucose/ controls diabetes	Mildly stimulates insulin production and causes a gradual increase in blood glucose.
7. Controls appendicitis	Loosens stool to prevent packing in the appendix and possible infection.
8. Controls diverticulosis	Exercises digestive tract muscles so that they retain their tone; resists bulging of the wall of the intestinal tract into pouches, called diverticula, that could become infected.

Lipids

An important class of nutrients is technically known as *lipids,* but many people use the term *fats.* This is unfortunate and may lead to some confusion because fat is only one of three subclasses of lipids. The subclasses—phospholipids, steroids, and true fats—play important roles in human nutrition. Phospholipids are essential components of all cell membranes. Many steroids are hormones that help to regulate a variety of body processes. The lipids sometimes referred to as *true fats* (also called *triglycerides*) are an excellent source of energy. They are able to release 9 kcalories of energy per gram compared to the 4 kcalories per gram released by carbohydrate or protein.

When broken down to glycerol and fatty acids, some true fats provide the body with an *essential fatty acid* called linoleic acid. Linoleic acid is called an essential fatty acid because it cannot be produced by the human body and, therefore, must be a part of the diet. This essential fatty acid is required by the body for such things as normal growth, blood clotting, and healthy skin. A diet high in linoleic acid has also been shown to help reduce the amount of the steroid cholesterol in the blood. Some dietary fats also are a source of fat-soluble vitamins, such as vitamins A, D, E, and K.

Fat is an insulator against low temperatures and internal heat loss and is an excellent shock absorber. Deposits in back of the eyes serve as cushions when the head suffers a severe blow. During starvation, these deposits are lost, and the eyes become deeply set in the eye sockets, giving the person a ghostly appearance. The pleasant taste and "mouth feel" of many foods is the result of fats. Their ingestion provides that full feeling after a meal because they lengthen the time the food is held in the stomach. You may have heard people say, "When you eat Chinese food, you're hungry a half hour later." Since Chinese foods contain very little animal fat, it's understandable that after such a meal the stomach will empty soon, and people will not have that full feeling very long.

Water

Water is crucial to all life and plays many essential roles. You may be able to survive weeks without food, but you would die in a matter of days without water. It is known as the universal solvent because so many types of molecules are soluble in it. The human body is about 65 percent water. Even dense bone tissue consists of 33 percent water. All the chemical reactions in living things take place in water. It is the primary component of blood, lymph, and body-tissue fluids. Inorganic and organic nutrients and waste molecules are also dissolved in water. Dissolved inorganic ions, such as sodium (Na^+), potassium (K^+), and chloride (Cl^-), are called *electrolytes* because they form a solution capable of conducting electricity. The concentration of these ions in the body's water must be regulated in order to prevent electrolyte imbalances.

Excesses of many types of wastes are eliminated from the body dissolved in water; that is, they are excreted from the kidneys as urine or in small amounts from the lungs or skin through evaporation. In a similar manner, water acts as a conveyor of heat. Water molecules are also essential reactants in the various reactions of metabolism. Without it, the breakdown of molecules such as starch, proteins, and lipids would be impossible. With all these important roles played by water, it's no wonder that nutritionists recommend that you drink the equivalent of at least eight glasses each day. This amount of water can be obtained from tap water, soft drinks, juices, and numerous foods such as lettuce, cucumbers, tomatoes, and applesauce.

Dietary Goals and Guidelines for the United States

Dietary Goals for the United States

The Select Committee on Nutrition and Human Needs of the U.S. Senate (1977) has established seven dietary goals for United States citizens.

1. To avoid becoming overweight, consume only as much energy (kcalories) as is expended; if overweight, decrease energy intake and increase energy expenditure.
2. Increase the consumption of complex carbohydrates and "naturally occurring" sugars from about 22% of energy intake to about 48% of energy intake.
3. Reduce the consumption of refined and other processed sugars by about 45% to account for about 10% of the total energy intake.
4. Reduce overall fat consumption from approximately 42% to about 30% of energy intake.
5. Reduce saturated fat consumption to account for about 10% of total energy intake; balance that with polyunsaturated and monounsaturated fats, which should account for about 10% of the energy intake each.
6. Reduce cholesterol consumption to about 300 milligrams a day.
7. Limit the intake of sodium by reducing the intake of salt (sodium chloride) to about 5 grams a day.

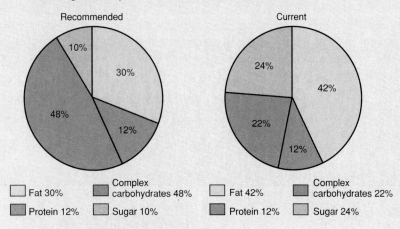

Recommended

Fat 30%
Complex carbohydrates 48%
Protein 12%
Sugar 10%

Current

Fat 42%
Complex carbohydrates 22%
Protein 12%
Sugar 24%

Dietary Guidelines for the United States

The U.S. Department of Health and Human Services recommends the following guidelines.

1. Eat a variety of foods.
2. Maintain desirable weight.
3. Avoid too much fat, saturated fat, and cholesterol.
4. Eat foods with adequate starch and fiber.
5. Avoid too much sugar.
6. Avoid too much sodium.
7. If you drink alcoholic beverages, do so in moderation.

The dietary goals differ from the dietary guidelines; the goals state nutritional objectives in terms of nutrients, while the guidelines translate them into types of food.

Consider, for example, a product label that claims that a serving of cereal provides 25 percent of the RDA for vitamin A. This means that you are getting at least one-fourth of your RDA of vitamin A from a single serving of that cereal. To figure your total RDA of vitamin A, consult a published RDA table for adults. It will tell you that an adult male requires 1,000 and an adult female 800 international units of vitamin A per day. Twenty-five percent of this is 250 and 200 international units, respectively—the amount supplied by a serving of that cereal. You will need to get the additional amounts (750 for men and 600 for women) by having more of that cereal or by eating other foods that contain vitamin A. If a product claims to have 100 percent of the RDA of a particular nutrient, that amount must be present in the product. However, restricting yourself to that one product will surely deprive you of many other nutrients necessary for good health. Ideally, you should eat a variety of complex foods containing a variety of nutrients to ensure that all your health requirements are met.

Basal Metabolism and Weight Control

Sometimes we are surprised at how few kcalories we are using when we exercise. Table 17.4 lists some physical activities and the kcalories per hour they require. The majority of our

table 17.4

Energy Requirements	
Kinds of Activity	**Kilocalories (per Hour)**
Walking up stairs	1,100
Running (a jog)	570
Swimming	500
Vigorous exercise	450
Slow walking	200
Dressing and undressing	118
Sitting at rest	100

This list of activities shows the amount of energy expended (measured in kilocalories) if the activity is performed for an hour.

energy expenditure occurs through muscular activities, but everyone requires a certain amount of energy to maintain basic body functions while at rest. Much of the energy required is used to keep your body temperature constant. The **basal metabolic rate (BMR)** is an estimate of this amount of energy and is usually measured in kcalories per hour per square meter of body surface.

Basal metabolic rate represents the amount of energy your body requires while lying still. Since few of us rest twenty-four hours a day, we require more than the energy needed for basal metabolism. Besides age, sex, weight, and height, basal metabolism depends upon a number of other factors, such as climate, altitude, physical condition, genetic factors, previous diet, and time of the year. A good general indicator of the number of kcalories needed above basal metabolism is the type of occupation a

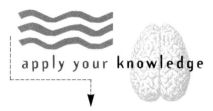

apply your **knowledge**

Estimating Your BMR

To estimate your basal metabolism, calculate your skin-surface area from table 17.5. To use the table, locate your height in the left column and your weight in the right column. Place a straightedge between these two points. The straightedge will cross the middle column and show your skin-surface area in square meters. For example, a person 160 centimeters tall (5'3") who weighs 50 kilograms (110 pounds) has 1.5 square meters of skin-surface area. The heat production in kcalories released per square meter of skin varies with a person's age and sex (see table 17.6). A twenty-year-old, 160-centimeter, 50-kilogram female uses 886 kcalories per day for each square meter of skin. Therefore, she needs 1,329 kcalories per day (1.5 × 886 = 1,329 kcalories) just to maintain her body. To calculate your basal metabolism, determine your skin-surface area using table 17.5 and multiply this figure by the kcalorie figure determined from table 17.6.

Skin surface area × kilocalories per square meter of skin = BMR

table 17.5

Table for Determining Body-Surface Area

Height		Square Meters of Skin Surface Area	Weight	
cm	ft		lb	kg

Height scale (cm / ft): 191 / 3; 183 / 6 ft; 175 / 9; 168 / 6; 160 / 3; 152 / 5 ft; 145 / 9; 137 / 6; 130 / 3; 122 / 4 ft; 114 / 9; 108 / 6

Square Meters of Skin Surface Area scale: 2.3, 2.2, 2.1, 2.0, 1.9, 1.8, 1.7, 1.6, 1.5, 1.4, 1.3, 1.2, 1.1, 1.0, 0.9, 0.8, 0.7, 0.6

Weight scale (lb / kg): 220 / 101; 210 / 96; 200 / 92; 190 / 86; 180 / 82; 170 / 77; 160 / 73; 150 / 68; 140 / 64; 130 / 59; 120 / 54; 110 / 50; 100 / 45; 90 / 41; 80 / 36; 70 / 32; 60 / 27; 50 / 23; 40 / 18; 30 / 14; 20 / 9

From *Dynamic Anatomy and Physiology*, 4th ed., by Langley et al., McGraw-Hill, 1974.

table 17.6

Kilocalories per Day per Square Meter of Skin

Age	Male	Female	Age	Male	Female
6	1,265	1,217	20–24	984	886
8	1,229	1,154	25–29	967	878
10	1,188	1,099	30–40	948	876
12	1,147	1,042	40–50	912	847
14	1,109	984	50–60	886	826
16	1,073	924	60–70	859	806
18	1,030	895	70–80	828	782

table 17.7

Additional Kilocalories as Determined by Occupation

Occupation	Kilocalories Needed above Basal Metabolism
Sedentary (student)	500–700
Light work (business person)	750–1,200
Moderate work (laborer)	1,250–1,500
Heavy work (professional athlete)	1,550–5,000 and up

These are general figures and will vary from person to person depending on the specific activities performed in the job.

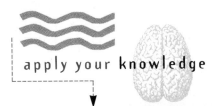

apply your knowledge

Figuring Your Energy Requirements

How do you determine how many kcalories you should have in your diet? After you have determined your basal metabolism, add to that kcalorie value the kcalorie energy you expend on a daily basis. That includes the energy you use in voluntary muscular activities such as work and sports (see table 17.7) plus the amount of energy required to digest and assimilate your food. This latter kcalorie amount is called your **specific dynamic action (SDA)** and is equal to approximately 10 percent of your total daily kcalorie intake.

You must also keep track of your kcalorie input. That means keeping an accurate diet record for at least a week. Record everything you eat and drink and determine the number of kcalories in those nutrients. This can be done by estimating the amounts of protein, fat, and carbohydrate (including alcohol) in your foods. Roughly speaking, 1 gram of carbohydrate is the equivalent of 4 kcalories, 1 gram of fat is the equivalent of 9 kcalories, 1 gram of protein is the equivalent of 4 kcalories, and 1 gram of alcohol, 7 kcalories. Most nutrition books have food-composition tables that tell you how much protein, fat, and carbohydrate is in a particular food. Do the arithmetic and determine your total kcalorie intake for the week. If your intake (your diet) in kcalories equals your output (basal metabolism + voluntary activity + SDA), you should not have gained any weight! You can double-check this by weighing yourself before and after your week of record keeping. To lose two pounds each week, reduce your kcalorie intake by 1,000 kcalories per day. Be careful not to reduce your total daily intake below 1,200 kcalories unless you are under the care of a physician. Below that level, you may not be getting all the vitamins required for efficient metabolism and you could cause yourself harm. To gain two pounds, increase your intake by 1,000 kcalories per day.

person has (table 17.7). If a twenty-year-old, 160-centimeter, 50-kilogram female were a bank teller, she would need between 750 and 1,200 kcalories per day above her basal metabolism of 1,329 kcalories. Therefore, her total daily need would be somewhere between 2,079 and 2,529 kcalories per day. Using tables 17.5, 17.6, and 17.7, calculate your daily caloric requirements. (The label on the cereal box that was described at the beginning of the chapter used 2,000 calories as a typical energy requirement.)

The body's metabolism is designed to convert excess carbohydrates (glucose) or proteins to fat, which serves as a source of stored energy. When needed, fat is converted to glucose to be used as a ready source of energy. Although energy doesn't weigh anything, the nutrients that contain the energy do. Weight control is a matter of balancing dietary intake with output measured in kcalories. There is a limit to the rate at which a moderately active human body can use fat as an energy source. *At the most, an average person can lose one or two pounds of fat per week when dieting.* This loss represents true weight loss and not a decrease in weight from water loss. Many diets promise large and rapid weight loss but in fact only result in temporary water loss. Decreasing your kcalorie intake by 500 to 1,000 kcalories per day while maintaining a balanced diet, including proteins, carbohydrates, and fats, will result in a meaningful loss. Decreasing caloric intake *and* increasing activity is the most effective way of reducing weight. For those who need to gain weight, increasing kcalorie intake by 500 to 1,000 kcalories per day will result in an increase of one or two pounds per week, provided the low weight is not the result of a health problem.

It is difficult to keep track of your caloric and nutrient intake on a daily basis. You probably will not write down every morsel of food you put in your mouth. This is precisely the reason the new pyramid guide was established. It can help you develop good eating habits with much less effort.

If, like millions of others, you feel that you are overweight, you have probably tried numerous diet (eating) plans. Not all of these plans are the same, and not all are suitable to your particular situation. If an eating plan is to be valuable in promoting good health, it must satisfy your needs in several ways. It must provide you with needed kcalories, protein, lipids, and carbohydrates. It should also contain readily available foods from all the basic food groups, and it should provide enough variety to prevent you from becoming bored with the plan and going off the diet. A diet should not be something you follow only to abandon and return to unhealthy eating habits. You need to be aware that there are some biological reasons a person is a particular weight. Not all of us can be model thin. Look at your parents, grandparents, brothers, and sisters for a clue about the way in which your family genetics and lifestyle might be influencing your weight.

Psychological Eating Disorders

To healthy individuals it may seem impossible that someone would behave in ways that could permanently damage their health and well-being. Yet in the United States many people have trouble eating a sound diet. Sometimes we do not use the information we have about what to eat. This can lead to eating disorders.

Eating disorders are grouped into three categories—obesity, bulimia, and anorexia nervosa. All three disorders are founded in psychological problems of one kind or another and are strongly influenced by the culture in which we live.

Obesity

As a person engages in the activities of everyday life, the body constantly uses energy acquired from food. At no time are the food molecules entering the body permanently locked into specific parts of cells. They are constantly being changed and exchanged; fat molecules, for example, are completely exchanged about every four weeks. This exchange is easily seen when a person either gains or loses weight. In either case, the molecules of the body are rearranged. The nature of the rearrangement depends on the amount of incoming food and the amount

Which Kind of Vegetarian Are You?

A vegetarian is a person who, to one degree or another, eliminates animal products from his or her diet. Vegetarians face a special problem: they must obtain the essential nutrients from fewer food groups. In order to do this they must learn which nonanimal foods contain these nutrients and plan their diets accordingly.

Nutritionists usually describe two basic types of vegetarians, with many variations in each group. The first type, the *lacto-ovo* vegetarians, use milk (*lacto* = milk) and eggs (*ovo* =egg), which are considered to be animal products, but avoid animal flesh; that is, meat, fish, and poultry. The *pure* or *strict* vegetarians are also known as *vegans*; they avoid all animal products and flesh and use only plant foods in planning their diets. Such people need to take vitamin B_{12} supplements or drink soy milk fortified with this vitamin in order to receive sufficient amounts to

prevent the vitamin-deficiency disease called *pernicious anemia*. Be assured that an adequate vegetarian diet is healthful because of the minimal amount of saturated fats in these foods and the fact that vegetarian diets are high in insoluble dietary fiber.

Dietary planning for vegetarians can be tricky. In order to obtain complete protein (RDA = 44 grams for an adult), they must learn complementary protein relationships. Since the only source of complete protein in a single food group is from animal products or flesh, strict vegetarians must choose plant products that will complement one another and provide all the essential amino acids. They need to do so in order to avoid abnormal fetal development and poor adult health. The following table should help in planning a vegetarian diet that will provide the essential amino acids and the total amount of recommended protein.

Protein Source 1	Protein Source 2	Total Grams
2 cups rice	1/2 cup beans	15 g
2 1/2 cups rice	2 oz tofu	12 g
1 cup rice	1 cup milk	10 g
6–7 tortillas	1/2 cup beans	14 g
2 slices whole wheat bread	2 tablespoons peanut butter	5.5 g
1 medium potato	1 cup milk	9 g
1 cup macaroni	2 oz cheddar cheese	13 g
2 tablespoons peanut butter	2 tablespoons sunflower seeds	7 g

Surgical Treatments for Obesity

A number of surgical operations to control obesity have been tried and discarded as ineffective remedies or because they resulted in unacceptable complications. Two procedures in use in the early 1990s have advanced beyond the experimental stage.

Vertical banded gastroplasty (figure A) and related techniques consist of constructing a small pouch with a restricted outlet near the entrance to the stomach. This reduces the amount of food the person can eat at one time. The outlet may be externally reinforced to prevent disruption or widening.

Gastric bypass procedures (figure B) involve constructing a pouch in the stomach whose outlet is connected to the small intestine. Food bypasses the stomach and so is not as completely digested. Thus less food is absorbed from the small intestine.

Choosing between these procedures involves the surgeon's preference and consideration of the patient's eating habits. The somewhat greater weight loss after the gastric bypass procedure must be balanced against its risk of nutritional deficiencies, especially of vitamins.

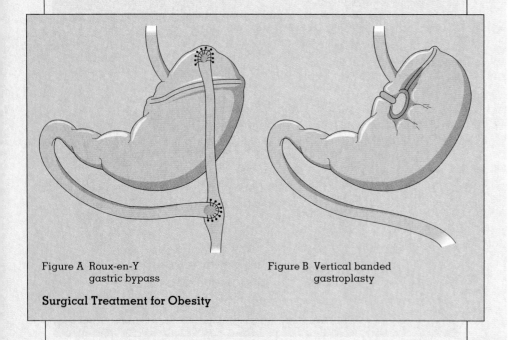

Figure A Roux-en-Y
gastric bypass

Figure B Vertical banded
gastroplasty

Surgical Treatment for Obesity

of activity. If a person's activity level increases and eating habits remain the same, there will be a loss of weight. If the activity level drops and eating habits remain the same, the person will gain weight. Of course, there are extremes of both processes. **Obesity** occurs when people consistently take in more food energy than is necessary to meet their daily requirements.

Obesity is probably the most familiar eating disorder and the one that is most publicized. It is the condition of being overweight to the extent that a person's health and life span are adversely affected. Many people suffering from this condition use food as a psychological crutch. They attempt to cope with the problems they face by overeating. Overeating to solve problems is encouraged by our culture. In addition, Americans celebrate almost all social occasions with food. Gatherings of almost every type are considered incomplete without some sort of food and drink. If snacks (usually high-calorie foods) are not made available by the host, many people feel uneasy or even unwelcome. It is also true that Americans and people of other cultures show love and friendship by sharing a meal. Many photographs in family albums have been taken at mealtime. Controlling obesity can be very difficult, since it requires basic changes in a person's eating habits, lifestyle, and value system. For some, these changes may require professional help.

While many cases of obesity are psychologically based, some have been demonstrated to have a strong biological component. It appears that some obese individuals have a chemical imbalance of the nervous system that prevents them from feeling "full" until they have eaten an excessive amount of food. This imbalance prevents the brain from "turning off" the desire to eat after a reasonable meal. Research into the nature and action of this brain chemical indicates that if obese people

CONCEPT CONNECTIONS

People who gain a great deal of unnecessary weight and are 15 percent to 20 percent above their ideal weight are defined as **obese.**

Individuals who binge on food and then empty their stomachs by vomiting have an eating disorder known as **bulimia.** People who have this disorder are also likely to use laxatives and diuretics even when they are medically unnecessary.

Anorexia nervosa is an eating disorder characterized by a fear of becoming overweight. These individuals are unable to eat healthy quantities of food and suffer severe and prolonged weight loss.

lacking this chemical receive it in pill form, they can feel full even when their food intake is decreased by 25 percent.

Bulimia

Bulimia ("hunger of an ox" in Greek) is a disorder that involves a cycle of eating binges and purges. It is sometimes called the silent killer because it is difficult to detect. Bulimics are usually of normal size or overweight. The cause of the disorder is thought to be psychological, stemming from depression, low self-esteem, displaced anger, a need to be in control of one's body, or a personality disorder. The cycle usually begins with an episode of overeating followed by elimination of the food by induced vomiting or excessive use of laxatives and diuretics, or both. Vomiting may be induced physically or by the use of some nonprescription drugs. Case studies have shown that bulimics may take forty to sixty laxative tablets a day to rid themselves of food. For some, the laxative becomes addictive. Diuretics are also used by bulimics to increase water loss through urination. The binge–purge cycle results in a variety of symptoms that can be deadly. The following is a list of the major symptoms observed in many bulimics:

Excessive water loss
Diminished blood volume
Extreme potassium, calcium, and sodium
   deficiencies
Kidney malfunction
Increase in heart rate
Loss of rhythmic heartbeat
Lethargy
Diarrhea
Severe stomach cramps
Damage to teeth and gums
Loss of body proteins
Migraine headaches
Fainting spells
Increased susceptibility to infections

Anorexia Nervosa

Anorexia nervosa (figure 17.3) is a nutritional deficiency disease characterized by severe, prolonged weight loss. An anorexic person's fear of becoming overweight is so intense that even though weight loss occurs, it does not lessen the fear of obesity, and the person continues to diet, often even refusing to maintain the optimum body weight for his or her age, sex, and height. Anorexic individuals starve themselves to death. This nutritional deficiency disease is thought to stem from sociocultural factors. Our society's preoccupation with weight loss

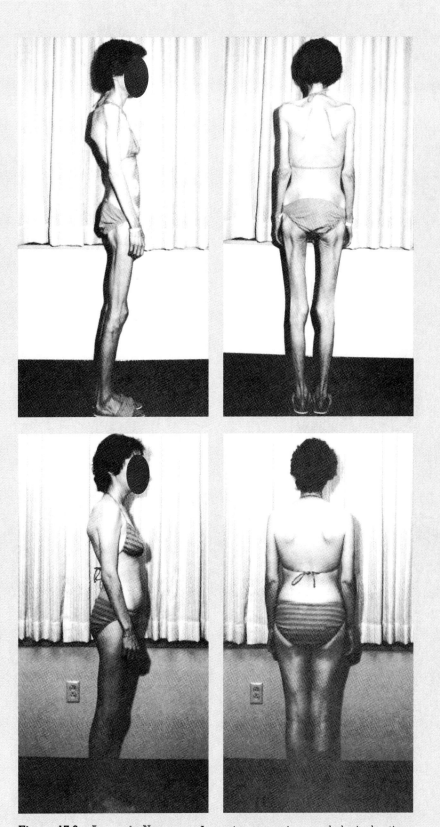

Figure 17.3 Anorexia Nervosa Anorexia nervosa is a psychological eating disorder afflicting many Americans. These photographs were taken of an individual before and after treatment. Restoring a person with this disorder requires both medical and psychological efforts.

and the desirability of being thin strongly influences this disorder. Individuals with anorexia are mostly adolescent and preadolescent females, although the disease does occur in males and older women. Just turn on your television or radio, or look at newspapers, magazines, or billboards, and you can see how our culture encourages people to be thin. Male and female models are thin. Muscle protein is considered to be healthy and fat to be unhealthy. Unless you are thin, so the advertisements imply, you will never be popular, get a date, or even marry. Here are some of the symptoms of anorexia nervosa:

Thin, dry, brittle hair
Degradation of fingernails
Constipation
Amenorrhea (lack of menstrual periods)
Decreased heart rate
Loss of body proteins
Weaker-than-normal heartbeat
Calcium deficiency
Osteoporosis
Hypothermia (low body temperature)
Hypotension (low blood pressure)
Increased skin pigmentation
Reduction in size of uterus
Inflammatory bowel disease
Slowed reflexes
Fainting
Weakened muscles

Deficiency Diseases

In the United States, a wide variety of foods are available. Living in an economically developed country, most of us are likely to have funds to purchase healthy foods. It is difficult for many people in the United States to recognize that some people in other countries are unable to get adequate amounts of food or the variety of foods necessary to provide for all of their dietary needs. In fact, it has been estimated that even in the United States, a full 25 percent of our children are not eating a balanced diet. Dietary problems include deficiencies in protein, vitamins, and minerals.

Without minimal levels of the essential amino acids in the diet, a person may develop health problems that can ultimately lead to death. In many parts of the world, large populations of people live on diets that are very high in carbohydrates and fats but low in complete protein. This is easy to understand, since carbohydrates and fats are inexpensive to grow

Figure 17.4 Kwashiorkor This starving child shows the symptoms of kwashiorkor, a protein-deficiency disease. If treated with a proper diet containing all amino acids, the disease can be cured.

and process in comparison to proteins. For example, corn, rice, wheat, and barley are all high-carbohydrate foods. Corn and its products (meal, flour) contain protein, but it is an incomplete protein that lacks the amino acid tryptophan. Without this amino acid, many necessary enzymes cannot be made in sufficient amounts to keep a person healthy. One protein-deficiency disease is called *kwashiorkor*, and the symptoms are easily seen (figure 17.4). A person with this deficiency has a distended belly, slow growth, and slow movement and is emotionally depressed. If the disease is caught in time, brain damage may be prevented and death averted. This requires a change in diet that includes expensive protein, such as poultry, fish, beef, shrimp, or milk. As the world food problem increases, these expensive foods will be in even shorter supply and will become more and more costly.

Very little carbohydrate is stored in the body. If you starve yourself, this small amount will last as a stored form of energy only for about two days. After the stored carbohydrate has been used, your body begins to use its stored fat deposits as a source of energy; the proteins will be used last. During the early

stages of starvation, the amount of fat in the body will steadily decrease, but the amount of protein will drop only slightly (figure 17.5). This can continue only up to a certain point. For about the first six weeks of this starvation period, the fat acts as a protein protector. Metabolizing proteins as a last resort is important. You would not want to metabolize proteins early during starvation since they play vital functions as enzymes, hormones, muscles, and structural material in the body. After about six weeks, however, so much fat has been lost from the body that proteins are no longer protected, and cells begin to use them as a source of energy. This results in a loss of proteins from the cells that prevents them from carrying out their normal functions. When not enough enzymes are available to do the necessary cellular jobs, the cells die.

The lack of a particular vitamin in the diet can result in a **vitamin deficiency disease.** A great deal has been said in the popular press about the need for vitamin and mineral supplements in diets. Some people claim that supplements are essential, while others claim that a well-balanced diet provides adequate amounts of vitamins and minerals. Supporters of vitamin supplements have even claimed that extremely high doses of certain vitamins can prevent ill health or create "supermen." It is very difficult to evaluate many of these claims, since the functioning of vitamins and minerals and their regulation in the body is not completely understood. In fact, the minimum daily requirements for a number of vitamins have not been determined.

Nutrition Through the Life Cycle

Nutritional needs vary throughout life and involve many factors, including age, sex, reproductive status, and level of physical activity. Infants, children, adolescents, adults, and the elderly all require slightly different amounts and kinds of nutrients.

Infancy

A person's total energy requirements are highest during the first twelve months of life: 100 kcalories per kilogram of body weight per day. Fifty percent of this energy is required for an infant's BMR. Infants (birth to twelve months) triple their weight and increase their length by 50 percent during the first year; this is their so-called first growth spurt. They require nutrients that are high in easy-to-use kcalories, proteins, vitamins, minerals, and water because they are growing and producing new cells and tissues. For many reasons, the food that most easily meets these needs is human breast milk (table 17.9). Even with breast milk's many nutrients, many physicians strongly recommend multivitamin supplements as a part of an infant's diet.

Childhood

As infants reach childhood, their dietary needs change. The rate of growth generally slows between one year of age and puberty, and girls increase in height and weight slightly faster than boys do. The body becomes more lean, bones elongate, and the brain reaches 100 percent of its adult size between the ages of 6 and 10. To adequately meet growth and energy needs during childhood, protein intake should be high enough to maintain the development of new tissues. Minerals, such as calcium, zinc, and iron, and vitamins are also necessary to support growth and prevent anemia. While many parents continue to provide their children with multivitamin supplements, such supplements should be given only after a careful evaluation of their children's diets. Four groups of children are at particular risk and should receive such supplements:

> children from deprived families and those suffering from neglect or abuse;
> children who have anorexia or poor eating habits, or who are obese;
> pregnant teens; and
> children who are strict vegetarians.

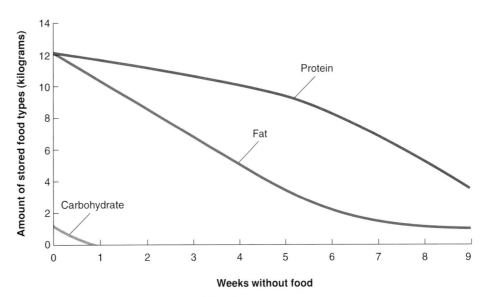

Figure 17.5 Starvation and Stored Foods   Starving yourself results in a very selective loss of the kinds of nutrients stored in the body. Notice how the protein level in the body has the slowest decrease of the three nutrients. This protein-conservation mechanism enables the body to preserve essential amounts of enzymes and other vital proteins.

table 17.9

A Comparison of Human Breast Milk and Cow's Milk

Nutrient	Human Milk	Cow's Milk
Energy (kcalories/1,000g)	690	660 (whole milk)
Protein (grams per liter)	9	35
Fat (grams per liter)	40	38
Lactose (grams per liter)	68	49
Vitamins		
A (International Units)	1,898	1,025
D (activity units)	40	14
E (International Units)	3.2	0.4
K (micrograms)	34	170
Thiamine (B_1) (micrograms)	150	370
Riboflavin (B_2) (micrograms)	380	1,700
Niacin (B_3) (milligrams)	1.7	0.9
Pyridoxine (B_6) (micrograms)	130	460
Cyanocobalamine (B_{12}) (micrograms)	0.5	4
Folic acid (micrograms)	41–84.6	2.9–68
Ascorbic acid (C) (micrograms)	44	17
Minerals (all in milligrams)		
Calcium	241–340	1,200
Phosphorus	150	920
Sodium (depends on distributor)	160	560
Potassium	530	1,570
Iron	0.3–0.56	0.5
Iodine	200	80

All milks are not alike. Each milk is unique to the species that produces it for its young, and each infant has its own special growth rate. Humans have one of the slowest infant growth rates, and human milk contains the least amount of protein. Since cow's milk is so different, many pediatricians recommend that human infants be fed either human breast milk or formulas developed to be comparable to breast milk during the first twelve months of life. The use of cow's milk is discouraged. This table lists the relative amounts of different nutrients in human breast milk and cow's milk.

During childhood, eating habits are erratic and often cause parental concern. Children often limit their intake of milk, meat, and vegetables while increasing their intake of sweets. To get around these problems, parents can provide calcium by serving cheeses, yogurt, and cream soups as alternatives to milk. Meats can be made more acceptable if they are in easy-to-chew, bite-size pieces, and vegetables may be more readily accepted if smaller portions are offered more frequently. Steering children away from sugar by offering sweets in the form of fruits can help reduce dental caries. You can better meet the dietary needs of children by making food available on a more frequent basis, such as every three to four hours.

Adolescence

The nutrition of an adolescent is extremely important because during this period the body changes from nonreproductive to reproductive. Puberty is usually considered to last between five and seven years. Before puberty, males and females tend to have similar proportions of body fat and muscle, both of which make up between 15 percent and 19 percent of the body's weight. Lean body mass, primarily muscle, is about equal in males and females. However, this soon changes. Female body fat increases to about 23 percent, and in males it decreases to about 12 percent. Males double their muscle mass in comparison to females.

The changes in body form that take place during puberty constitute the second growth spurt. Because of their more rapid rate of growth and unique growth patterns, males require more protein, iron, zinc, and calcium than do females. During adolescence, youngsters will gain as much as 20 percent of their adult height and 50 percent of their adult weight, and many body organs will double in size. Nutritionists have taken these growth patterns and spurts into account by establishing RDAs for males and females who are 10 to 20 years old, including requirements at the peak of their growth spurt. RDAs at the peak of the growth spurt are about twice what they are for adults and children.

Adulthood

People who have completed the changes associated with adolescence are considered to have entered adulthood. Most of the nutritional information available to the public through the press, television, and radio focuses on this stage in the life cycle. During adulthood, the body has

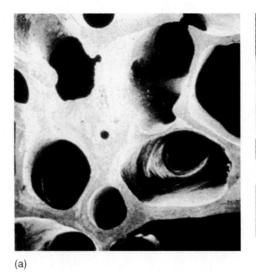 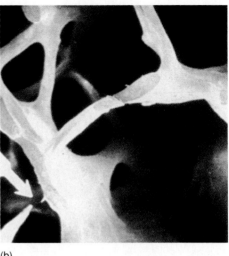

(a)   (b)

Figure 17.6 Osteoporosis   These photographs are of a healthy bone (a) and a section of bone from a person with osteoporosis (b). This nutritional deficiency disease results in a change in the density of the bones as a result of the loss of bone mass. Bones that have undergone this change look "lacy" or like Swiss cheese, with larger than normal holes. A few risk factors found to be associated with this disease are being female and fair skinned; having a sedentary lifestyle; using alcohol, caffeine, and tobacco; and having reached menopause.

entered a plateau phase, and diet and nutrition focus on maintenance and disease prevention. Nutrients are used primarily for tissue replacement and repair. Weight gain occurs easily, while weight loss is difficult. Since the BMR slows, as does physical activity, the need for food energy decreases from about 2,700 kcalories in average young adult males (ages 20 to 40) to about 2,050 for elderly men. For women, the corresponding numbers decrease from 2,000 to 1,600 kcalories. Protein intake for most U.S. citizens is in excess of the recommended amount. The RDA standard for protein is about 56 grams for men and 44 grams for women. About 25 to 50 percent should come from animal foods to ensure intake of the essential amino acids. The rest should be from plant-protein foods such as whole grains, legumes, nuts, and vegetables.

An adult who follows a well-balanced diet should have no need for vitamin supplements; however, improper diet, disease, or other conditions might require that supplements be added. The two minerals that demand special attention are calcium and iron, especially for women. A daily intake of 1,200 milligrams of each should prevent osteoporosis, a condition of calcium loss from bones (figure 17.6), and allow adequate amounts of hemoglobin to be manufactured to prevent anemia in women over 50 and men over 60. In order to reduce the risk of chronic diseases such as heart attack and stroke, adults should be sure to eat a balanced diet, participate in regular exercise programs, control their weight, avoid cigarettes and alcohol, and practice stress management.

Pregnant and Lactating Women

Risk-management practices that help avoid chronic adult diseases become even more important when planning pregnancy. Studies have shown that an inadequate supply of the essential nutrients can result in infertility, spontaneous abortion, and abnormal fetal development. The period of pregnancy and milk production (lactation) requires special attention to the diet to ensure proper fetal development, a safe delivery, and a healthy milk supply. Recent studies have shown that a lack of folic acid in the diet of a pregnant female can result in abnormal development of the spinal cord of the fetus.

The daily amount of essential nutrients must be increased during pregnancy and lactation, as should the kcaloric intake. Kilocalories must be increased by 300 per day to meet the needs of an increased BMR, the development of the uterus, breasts, and placenta, and the work required for fetal growth. Some of these kcalories can be obtained by drinking milk, which simultaneously supplies calcium needed for fetal bone development. Intake of protein should be increased by 75 percent. (Such an increase in protein will probably be sufficient to provide the increase in calories.)

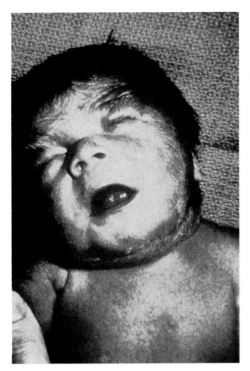
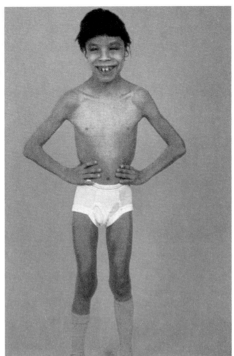

Figure 17.7 Fetal Alcohol Syndrome   The one-day-old infant on the left displays fetal alcohol syndrome while the child on the right with the syndrome is eight years old. The eight-year-old was diagnosed at birth and has spent all of his life in a foster home where the quality of care has been excellent. His IQ has remained stable at 40 to 45.

Two essential nutrients, folic acid and iron, should be obtained through prenatal supplements since they are so essential to cell division and development of the fetal blood supply.

The mother's nutritional status affects the developing baby in several ways. If she is under 15 years of age or has had three or more pregnancies in a two-year period, her nutritional stores may be inadequate to support a successful pregnancy. The use of drugs such as alcohol, caffeine, nicotine, and "hard" drugs (e.g., heroin) can result in decreased nutrient exchange between the mother and fetus. In particular, heavy smoking can result in low birth weights, while alcohol abuse is responsible for fetal alcohol syndrome (figure 17.7).

Seniors

As people move into their sixties and seventies, digestion and absorption of all nutrients through the intestinal tract is not impaired but does slow. The number of cells undergoing mitosis is reduced, resulting in an overall loss in the number of body cells. With age, complex organs such as the kidneys and brain function less efficiently, and protein synthesis becomes inefficient. Energy requirements for the elderly decrease as the BMR slows, physical activity decreases, and eating habits also change.

The change in eating habits is particularly significant, since it can result in dietary deficiencies. For example, linoleic acid, the essential fatty acid, may fall below required levels as an older person reduces the amount of food eaten. The same is true for some vitamins and minerals. Therefore, it may be necessary to supplement the diet daily with one tablespoon of vegetable oil. Vitamin E, multiple vitamins, or a mineral supplement may also be necessary. The loss of body protein means that people must be sure to meet their daily RDA for protein. As with all stages of the life cycle, regular exercise is important in maintaining a healthy, efficiently functioning body. Recent studies have shown that regular exercise in the form of weight training can significantly improve older people's ability to coordinate their bodies. Weight training decreases the incidence of injury due to falls, increases the ability to rise from a sitting or laying position, and greatly improves the ability to control grasping.

Nutrition for Fitness and Sports

In the past few years there has been heightened interest in fitness and sports. Along with this, more people have developed an appreciation for the role nutrition plays in fueling activities, controlling weight, and building muscle. The cell-respiration process is the source of the energy needed to take a leisurely walk or run a marathon. However, just which molecules are respired and when depends on whether or not you warm up before you exercise and how much effort you exert during exercise. The molecules respired by muscle cells to produce ATP depends on the kind of exercise. During a long, brisk walk, the heart and lungs of most people should be able to keep up with the muscle cells' requirement for oxygen. Studies have shown that under these moderate exercise conditions, fats stored in cells and fatty acids from the circulatory system are used as the prime sources of energy. Glycogen stored in muscle cells and glucose in the blood are not utilized. This is why such moderate, short-term exercise is most beneficial in weight loss.

But what about vigorous exercise for longer periods? Without a warm-up period, your muscle cells will have to begin respiring muscle-stored glycogen. At first the heart and lungs are not able to supply all the oxygen needed. Muscle cells use anaerobic respiration to provide ATP for muscle contraction, but there will be an increase in the amount of acid by-products. About five minutes into vigorous exercise, 20 percent of muscle glycogen is gone and little fat has been respired. By this time, the heart and lungs are "warmed up" and are able to provide the oxygen necessary to respire the fats and fatty acids. ATP output increases dramatically. A person who has not warmed up experiences this metabolic shift as a "second wind."

From this point on, all four sources of energy—glucose, glycogen, fats, and fatty acids—are utilized. If exercise is suddenly stopped, the high concentration of fatty acids will be converted to keto acids and the exerciser could experience kidney problems; a loss of sodium, calcium, and other minerals; and a change in the pH of the blood. For this reason, many exercise physiologists suggest a cooldown period of light exercises that allow the body to slowly shift back to normal metabolism. At approximately 50 minutes into exercise the cells shift to using glycogen. If the exerciser continues full-out, the next metabolic shift takes place at about 80 minutes. It is known as "hitting the wall." This is the sudden onset of disabling fatigue that occurs when the limited amount of stored glycogen falls below a certain level.

To avoid or postpone hitting the wall, many marathon athletes practice *carbohydrate loading*. This should only be done by those engaged in periods of hard exercise or competition lasting for 90 minutes or more. It requires following a week-long diet and exercise program. A typical program would be as follows: On the first day, the muscles needed for the event are exercised for 90 minutes, and carbohydrates such as fruits, vegetables, or pasta provide a source of dietary kcalories. On the second and third days of carbohydrate loading, the person continues the 50 percent carbohydrate diet, but the period of exercise is reduced to 40 minutes. On the fourth and fifth days, the workout period is reduced to 20 minutes and the carbohydrates are increased to 70 percent of total kcalorie intake. On the sixth day, the day before competition, the person rests and continues the 70 percent carbohydrate diet. Following this program increases muscle glycogen levels and makes it possible to postpone "hitting the wall."

Conditioning in athletes involves changes in several aspects of the body's functioning. The heart muscle becomes stronger and blood is pumped more efficiently. The number of capillaries increases and exchange of gases and nutrients is enhanced. The number of mitochondria in muscle cells increases so that fats are respired more efficiently and for longer periods, and weight control becomes easier.

The amount of protein in an athlete's diet has also been investigated. Understand that increasing dietary protein does not automatically increase strength, endurance, or speed. In fact, most Americans eat as a part of their normal diets the 10 percent additional protein that athletes require. Supplementing your diet with protein-rich foods such as "power" drinks or nutrient bars is unnecessary. Increasing protein intake will not automatically increase muscle size. Only when there is a need will the protein be used to increase muscle mass. That means exercise. Your body will build the muscle it needs in order to meet the demands you place on it. Vitamins and minerals operate in much the same way. No supplements should be required as long as your diet is balanced and complex. Your meals should provide the vitamins and minerals needed to sustain your effort.

Athletes must monitor their water intake because dehydration can cripple an athlete very quickly. A water loss of only 5 percent of the body weight can decrease muscular activity by as much as 30 percent. One way to replace water is to drink 1 to 1½ cups (250–375 ml) of water 15 minutes before exercising and a half cup during each exercise session. In addition, drinking 16 ounces (480.00 ml) of cool tap water for each pound of body weight lost during exercise is enough to prevent dehydration. Another method is to use diluted orange juice (1 part juice to 5 parts water). This is an excellent way to replace water and resupply a small amount of lost glucose and salt. Salt pills (so-called electrolyte pills) are not recommended.

SUMMARY

To maintain good health, people must receive nutrient molecules that can enter cells and function in metabolic processes. The federal government has developed a Food Guide Pyramid of five groups: Group 1—Meat, Poultry, Fish, and Dry Beans; Group 2—Dairy Products; Group 3—Vegetables; Group 4—Fruits; Group 5—Grain Products. Fats, oils, and alcohol should be limited in the diet.

The proper quantity and quality of nutrients are essential to good health. Nutritionists have classified nutrients into six groups: carbohydrates, proteins, lipids, minerals, vitamins, and water. Energy for metabolic processes may be obtained from carbohydrates, lipids, and proteins, and is measured in kilocalories. An important measure of the amount of energy required to sustain a human at rest is the basal metabolic rate. To meet this and all additional requirements, the United States has established the RDAs, Recommended Daily Allowances, for each nutrient.

Should there be chemical or psychological problems associated with a person's normal metabolism, a variety of disorders may occur, including obesity, anorexia nervosa, bulimia, kwashiorkor, and vitamin-deficiency diseases. As people move through the life cycle, their nutritional needs change, requiring a reexamination of their eating habits in order to maintain good health. Nutrition and diet are also vital to peak athletic performance.

CHAPTER GLOSSARY

anorexia nervosa (an″o-rek′se-ah ner-vo′ sah) A nutritional deficiency disease characterized by severe, prolonged weight loss due to fear of becoming obese. This eating disorder is thought to stem from sociocultural factors.

basal metabolic rate (BMR) (ba′sal mĕt-a-bol′ik rāt) The amount of energy required to maintain normal body activity while at rest.

bulimia (bu-lim′e-ah) A nutritional deficiency disease characterized by a binge-and-purge cycle of eating. It is thought to stem from psychological disorders.

calorie (kal′o-re) The amount of heat energy necessary to raise the temperature of one gram of water one degree Celsius.

diet (di′et) The food and drink consumed by a person from day to day.

fiber (fi′ber) Natural (plant) or industrially produced polysaccharides that are resistant to digestion by human enzymes in the intestinal tract.

Food Guide Pyramid (food gīd pi′ra-mĭd) A tool developed by the U.S. Department of Agriculture to help the general public plan for good nutrition. It contains guidelines for required daily intake from each of the five food groups.

kilocalorie (kil″o-kal′o-re) A measure of heat energy one thousand times larger than a calorie; also called a *kcalorie*.

minerals (mĭn′er-alz) Inorganic elements essential to metabolism that cannot be manufactured by the body but are required in low concentrations.

nutrients (nu′tre-ents) Molecules required by the body for growth, reproduction, and repair.

nutrition (nu-trī′shun) Collectively, the processes involved in taking in, assimilating, and utilizing nutrients.

obese (o-bēs′) A term describing a person who gains a great deal of unnecessary weight and is 15 percent to 20 percent above ideal weight.

protein-sparing (pro'tēn spe'ring) The conservation of proteins by first using carbohydrates and fats as a source of energy.

Recommended Dietary Allowance (RDA) (re-ko-men'ded di'e-te-re a-lao'ans) U.S. dietary guidelines for a healthy person that focus on the six classes of nutrients.

specific dynamic action (SDA) (spe-si'fik di-na'mik ak'shun) The amount of energy required to digest and assimilate food. SDA is equal to approximately 10 percent of your total daily kcalorie intake.

vitamin deficiency disease (vi'tah-min de-fish'en-se di-zēs') Poor health caused by the lack of a certain vitamin in the diet, such as scurvy for lack of vitamin C.

vitamins (vi'tah-minz) Organic molecules that cannot be manufactured by the body but are required in very low concentrations.

 # EXPLORATIONS Interactive Software

Life Span and Lifestyle

This interactive exercise allows students to explore how their lifestyle decisions influence their life expectancy. The exercise presents a diagram of a typical human life span, shown as an age distribution of U.S. life expectancies (how long a U.S. citizen of a certain age can expect to live). The student explores the change in probability of survival when certain activities are initiated. The student may begin or stop smoking at any age (cigarettes induce cancer as well as lung and heart disease); eat or stop eating a diet high in animal fat (which induces heart disease and stroke, and is highly correlated with breast cancer in women); eat or stop eating a high-protein diet (which seems to induce colon cancer); and initiate or cease regular exercise (which counteracts cardiovascular disease). By varying patterns of behavior, the student soon learns that survival reflects lifestyle.

1. Can exercise counteract the lowered life expectancy due to smoking?
2. Is it more beneficial to give up smoking or a red-meat diet?
3. Is smoking more harmful at one age than another?
4. Is there any age when diet doesn't matter?
5. What lifestyle is associated with the longest life expectancy? The shortest?

Diet and Weight Loss

This interactive exercise allows students to explore how diet and exercise interact to determine whether we gain or lose weight by altering both the kinds of food eaten and the amounts of each consumed, as well as the amount of exercise. The Exploration presents a diagram of central metabolism, tracing the path of food to ATP production and/or fat.

Students will learn the power of the basic equation "food minus energy used equals fat." By varying the proportions of carbohydrate, protein, and fat, students can investigate how different diets influence the "burn fat/make fat" decision, learning that the key is the level of blood glucose, not ATP. For example, a high-calorie diet can actually cause you to lose weight by burning fat if that diet is very low in fat.

1. Can you invent a diet that reduces weight without reducing overall consumption of calories?
2. In what ways can you lose weight without exercise?
3. If you consume large but not unreasonable portions of food that is high in calories (rich in fats) three times a day, how much exercise is required to *lose* weight?

CONCEPT MAP TERMINOLOGY

Construct a concept map to represent the relationships among the following concepts.

anorexia nervosa	fiber	nutrients
basal metabolism	Food Guide Pyramid	nutrition
bulimia	kilocalorie (kcalorie)	obese
calorie	minerals	vitamin
diet		

LABEL•DIAGRAM•EXPLAIN

Make a list of everything you eat for 24 hours. Be sure to include the amounts of each item. Keep a diary of the foods you eat as you eat them, rather than trying to remember what you ate, how much, and when. Then modify that list of foods to form an ideal daily diet for you. Be sure to use the Food Guide Pyramid and include sufficient amounts of water.

Multiple Choice Questions

1. One of the most significant changes in the new Food Guide Pyramid is that it recommends that you eat _____ servings of complex carbohydrates.
   a. only one
   b. up to 3
   c. as many as 5
   d. up to 11 servings per day
2. One of the major benefits of eating fruits and vegetables is that they provide:
   a. much of the kcalories needed
   b. needed vitamins and minerals
   c. an alternative source of fats
   d. the essential amino acids
3. Which of the following is a psychological eating disorder?
   a. anorexia nervosa
   b. beri beri
   c. colon cancer
   d. appendicitis
4. The metric unit kcalorie is equal to:
   a. a British thermal unit
   b. 1,000 calories
   c. 0.001 of a calorie
   d. the same as a calorie
5. The energy needed to maintain an individual at rest is termed:
   a. voluntary activity
   b. basal metabolic rate
   c. specific dynamic action
   d. caloric intake
6. Which of these is not one of the six classes of nutrients?
   a. water
   b. minerals
   c. sugars
   d. alcohol
7. The RDA or recommended dietary allowance is:
   a. based on the needs of a healthy person already eating an adequate diet
   b. the amount of caloric intake you should have
   c. determined by the kinds of foods you eat
   d. computed based on your basal metabolic rate, your level of activity, and your specific dynamic action
8. Different sets of RDAs have been developed for which of the following?
   a. the amount of weight you wish to lose
   b. how much a person weighs
   c. age and reproductive condition
   d. limited amount of food available
9. According to the Food Guide Pyramid, intake of which of the following should be limited?
   a. fat
   b. protein
   c. tobacco products
   d. minerals and vitamins
10. A good way to get adequate vitamins is to:
   a. eat lots of fruits
   b. eat large amounts of protein
   c. eat many different kinds of food from each of the five food groups
   d. reduce the amount of fat in the diet

Questions with Short Answers

1. List the six classes of nutrients and give an example of each.

2. Name the five basic food groups and give two examples of each.

3. What are basal metabolic rate, specific dynamic action, and voluntary muscular activity?

4. During which phase of the life cycle is a person's demand for kcalories per unit of body weight the highest?

5. Why are some nutrients referred to as essential? Name them.

6. What do the initials RDA stand for?

7. List four of the dietary guidelines.

8. Americans are currently consuming 42 percent of their kcalories in fat. According to the dietary goals, what should that percentage be?

Human Reproduction,
Sex, and Sexuality

Grandfather with grandson, Pacific Islands, Micronesia, Yap.

Human Reproduction,
Sex, and Sexuality

Biologists have long considered the function of sex and sexuality in light of their value to the population or species. Without sexual exchange of genetic material, variation within a population is reduced, which makes changes in the gene pool less likely. We must recognize that human sexual behavior has an evolutionary basis but also includes strong cultural influences. Our sexual nature is partly genetic and partly learned.

Certainly the behaviors of courtship, mating, and child-rearing and the division of labor within groups are more complex in social animals, including humans. This complexity is demonstrated in the elaborate social behaviors surrounding mate selection and the establishment of families. It is difficult to draw the line between the biological development of sexuality and the social establishment of customs related to the sexual aspects of human life.

Male or Female

The biological mechanism that determines female or male gender in humans has been well documented. When a human egg or sperm cell is produced, it contains twenty-three chromosomes, one of which is a sex-determining chromosome.

There are two kinds of sex-determining chromosomes: the X chromosome and the Y chromosome (figure 18.1). The two sex-determining chromosomes, X and Y, do not carry equivalent amounts of information, nor do they have equal functions. In addition to their function in determining sex, the X chromosome carries typical information about the production of specific proteins. The Y chromosome carries information for determining maleness and few other genes.

When a human sperm cell is produced, it carries twenty-three chromosomes, one of which is a sex-determining chromosome. Unlike eggs, which always carry an X chromosome, half of the sperm cells carry an X chromosome and the other half carry a Y chromosome. If an X-carrying sperm cell fertilizes the X-containing egg cell, the resulting embryo will develop into a female. It is the absence of the Y chromosome that determines femaleness. Femaleness appears to be the "pre-set" condition. A typical human

learning objectives

- Describe the sex-determining chromosomes and how they function in determining gender.

- Describe the maturation processes involved in the development of a fetus.

- Know the different hormones that regulate the maturation of the female and male human reproductive systems at puberty and describe their involvement in sexual function in an adult.

- List the structures of the male and female human reproductive systems.

- Describe how hormones function in the processes of ovulation and pregnancy.

- Understand the differences between the production and development of sperm cells and egg cells.

- List the events necessary for fertilization and pregnancy to occur in humans.

Sexuality from Different Points of View

Probably nothing interests us more than sex and sexuality. By **sexuality,** we mean all the factors that contribute to one's female or male nature. These include the structure and function of the sex organs and the behaviors that involve these structures. We have an intense interest in facts about our own sexual nature and the sexual behavior of others. We question why there are differences between males and females in behavior, attitude, sexual structures, and functions.

Professionals in different fields view sex and sexuality in slightly different ways. Psychologists consider sex to be a strong *drive*. They describe the sex drive as a basic impulse to satisfy a biological, social, or psychological need. The sex drive does not have life-sustaining implications for the individual, as do the drives for food or water. Other social scientists classify sexuality as an *appetite*, strong desire, or *urge* that is not as strong as a drive. Whether we call this interest in sexual matters a drive, an appetite, or an urge, it provides a great deal of motivation for many activities from birth to death.

Figure 18.1 Chromosomal Determination of Sex Human males and females are differentiated by the chromosomes they have. Males have an X and a Y chromosome (a) while females have two X chromosomes (b).

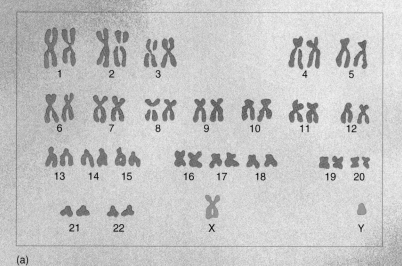

(a)

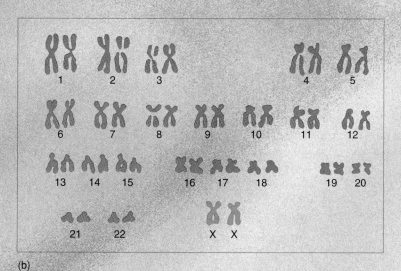

(b)

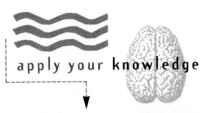

apply your knowledge

Acting Your Sex

How do you think the following facts may be related to our evolutionary past? How are these characteristics expressed in our behavior today?

1. Males do not know when ovulation is occurring in females.
2. Females have a different distribution of fat and muscle than males do.
3. Most of the time males consider contraception a female responsibility.
4. Females consider sharing and communication important in a relationship.
5. Males are strongly influenced by the physical appearance of a potential sexual partner.

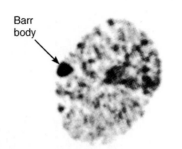

Barr body

Figure 18.2 Barr Body The extra dark body in this white blood cell from a woman is the nonfunctioning X chromosome.

The one X chromosome of the male functions as expected, and the Y chromosome directs the expression of traits that occur only in males. The tightly coiled structure in the cells of female mammals is called a *Barr body* (figure 18.2).

Sexual Development before Birth

When a baby is born, it already has been programmed to be a particular sex. Development of embryonic gonads begins very early during fetal growth. First, a group of cells begins to differentiate into primitive gonads. Within a matter of weeks, these gonads will become testes if a Y

female has an X chromosome from each parent. If a Y-carrying sperm cell fertilizes the egg, a male embryo develops.

The embryo becomes male or female based on the sex-determining chromosomes that control the differentiation of the sex organs, the testes and ovaries. If the Y chromosome is present, the embryonic gonads begin to differentiate into testes about seven weeks after conception. If the embryo does not have a Y chromosome, the gonads differentiate into female sex organs later.

Researchers are interested in how females, with two X chromosomes, handle the double dose of genetic material in comparison to males, who have only one X chromosome. M. L. Barr discovered that a darkly staining body was generally present in female cells but was not present in male cells. It was postulated, and has since been confirmed, that this structure is an X chromosome that is largely nonfunctional. Therefore, although female cells have two X chromosomes, only one is functional; the other X chromosome coils up tightly and does not direct the manufacture of proteins.

FYI

In some Olympic events it would be unfair for males and females to compete against each other. When the gender of an individual is questioned, eligibility for female events can be determined by the presence of the Barr body in the contestant's cells.

chromosome is present. As soon as the gonad has differentiated into an embryonic testis, it begins to produce hormones (figure 18.3) that influence the further development of the embryo, causing it to complete its sexual differentiation.

If not influenced by the presence of the Y chromosome, the embryo will begin to develop ovaries at about the twelfth week after fertilization. The outer portion of the embryonic gonad develops into an ovary due to the production of the "female" hormones in response to the presence of the X chromosome. The ovaries remain within the abdominal cavity. Production of egg cells begins at this time, but the process is interrupted and

the cells do not complete the development into egg cells until several years after birth. When the young woman reaches sexual maturity the continuation of this process allows the production of mature egg cells.

Following birth, sexuality plays only a small part in human physical development for several years. During **puberty,** increasing production of sex hormones causes major changes as the individual reaches sexual maturity. After puberty, humans are sexually mature and have the capacity to produce offspring.

Sexual Maturation of Males

At about the seventh month of development, the testes move from a position in the abdominal cavity to the external sac, the **scrotum.** This descent occurs through small openings in the abdominal wall that close after the testes descend into the sac. These closed passageways may rupture later in life, causing the formation of a *hernia*. A rupture can happen when strain (e.g., from improperly lifting heavy objects) causes a portion of the intestine to push through this spot into the scrotum. Sometimes the descent of the testes occurs later, during puberty; if not, there is an increased incidence of testicular cancer. Because of this increased risk, undescended testes may be surgically moved to their normal position in the scrotum.

Males typically begin puberty with a change in hormone levels. **Hormones** are molecules produced in one part of the body that alter the activity or growth of distant organs. One hormone involved in the development of the male sex organs is **follicle-stimulating hormone (FSH),** a pituitary gland secretion also produced by females. This is the primary growth stimulator of the testes and ovaries. FSH produced by the male is responsible for the production of sperm cells by the testes. **Interstitial cell-stimulating hormone (ICSH)** stimulates the testes to produce testosterone. **Testosterone** causes the differentiation of internal and

CONCEPT CONNECTIONS

follicle-stimulating hormone Produced by the pituitary, causes the testes to begin the production of the sperm cells.

interstitial cell-stimulating hormone (ICSH) (also known as luteinizing hormone) Produced by the pituitary, stimulates the testes to produce testosterone, which is also important in the maturation and production of sperm.

testosterone Produced by the testes, is responsible for the maturation of the male reproductive organs and the development of the secondary sex characteristics.

Figure 18.3 Differentiation of Sexual Characteristics The early embryo grows without showing any sexual characteristics. The male and female sexual organs eventually develop from common basic structures. (a) Development of the internal anatomy. (b) Development of external anatomy.

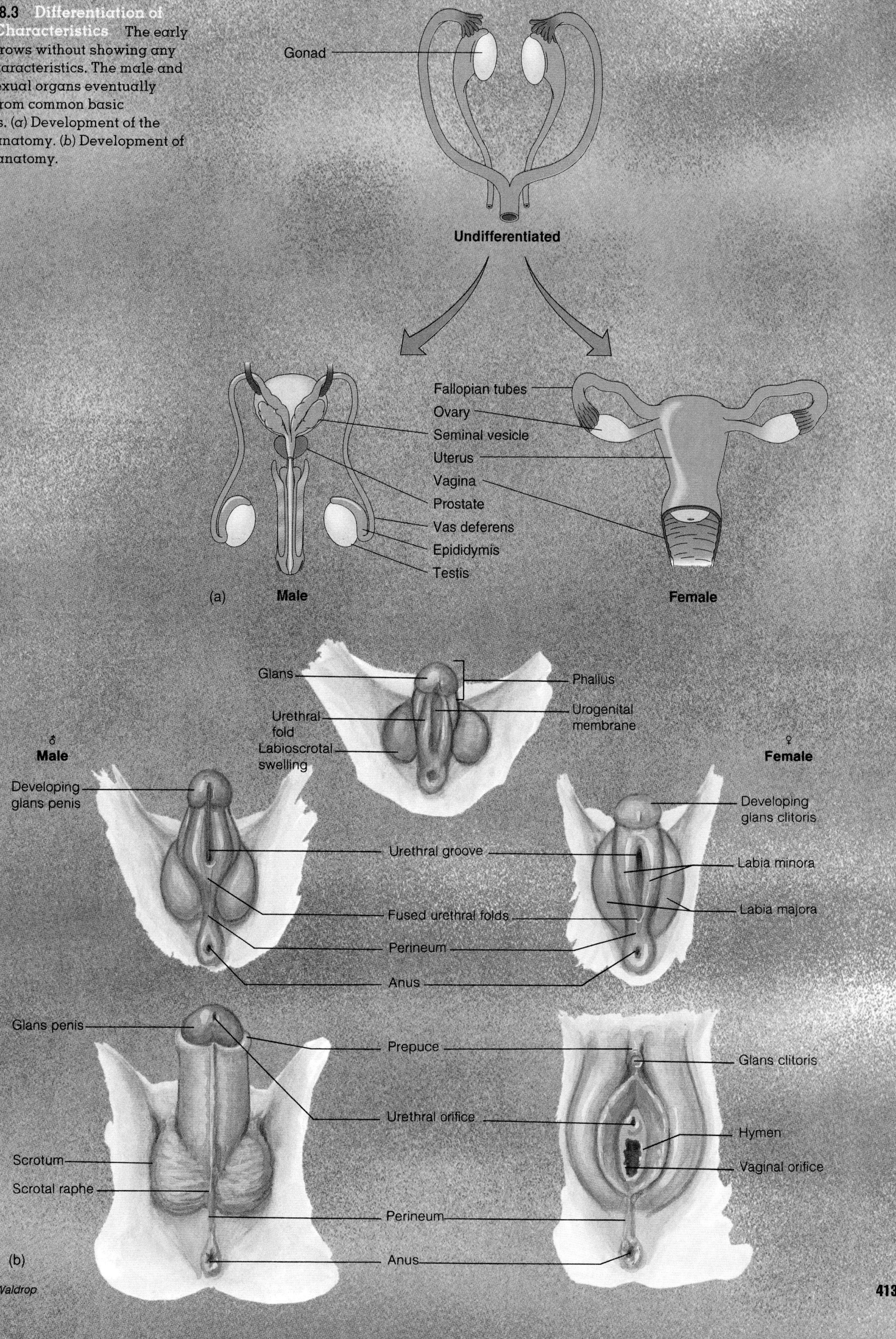

Nondisjunction and Abnormalities

Evidence that the Y chromosome controls male development comes as a result of studying individuals who have an abnormal number of chromosomes. An abnormal meiotic division that results in sex cells with too many or too few chromosomes is called *nondisjunction* (see chapter 14). If nondisjunction affects the X and Y chromosomes, a gamete might be produced that has only twenty-two chromosomes and lacks a sex-determining chromosome, or it might have twenty-four chromosomes, with two sex-determining chromosomes. If a cell with too few or too many sex chromosomes takes part in fertilization, an abnormal embryo develops.

If a normal egg cell is fertilized by a sperm cell with no sex chromosome, the offspring will have only one X chromosome. These people are designated as XO. They develop a collection of characteristics known as *Turner's syndrome*. A *syndrome* is a combination of features or symptoms of a disease. An individual with this condition is female but is generally sterile and matures sexually later than normal. In addition, she may have a thickened neck (termed webbing), hearing impairment, and some abnormalities in the cardiovascular system. While these are typical symptoms of Turner's syndrome, not all individuals with this condition will display the same set of symptoms.

Should one of the gametes involved in fertilization contain an extra sex chromosome while the other contains the normal one, the resulting zygote will be XXX or XXY. An individual who has XXY chromosomes is basically male. This genetic abnormality is termed *Klinefelter's syndrome*, and the symptoms include mental deficiencies and sexual dysfunction due to small testes that do not usually produce viable sperm. Breast and hip development is similar to that of females. Since both of these conditions involve abnormal numbers of X or Y chromosomes, this is strong evidence that these chromosomes are involved in determining sexual development.

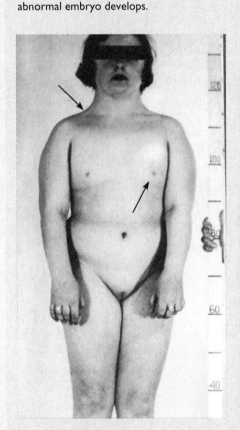

Turner's Syndrome

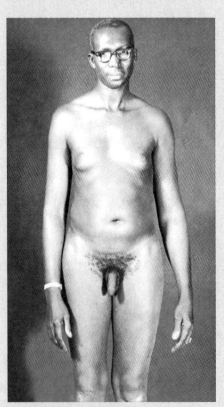

Klinefelter's Syndrome
(before hormone treatment)

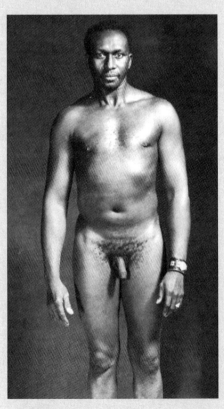

Klinefelter's Syndrome
(after hormone treatment)

BIO *feature*

Hermaphrodites, Gender-Confusion, and Transvestites

In some organisms, the reproductive structures of both sexes are formed as a normal part of their reproductive strategy. These organisms are classified as *hermaphrodites*. In humans, such instances of partial development of the genitalia (sex organs) of both sexes in one individual are rare. These people are referred to as pseudohermaphrodites since they do not have complete sets of male and female organs. Sometimes this abnormal development occurs because the hormone levels are out of balance at critical times in the development of the embryo. This hormonal imbalance may be related to an abnormal number of sex-determining chromosomes, or it may be the result of abnormal functioning of the endocrine glands. Corrective surgery and appropriate counseling sometimes help the individual to live a more normal life. Some hermaphrodites are primarily female with a slightly enlarged clitoris, while others are primarily male with an underdeveloped penis and normal labia and vagina. Such people must be assessed by a physician in consultation with the parents to determine which sexual structures should be retained or surgically reconstructed. The physician may also decide that hormone therapy might be a more successful treatment. These decisions are not easily made because they involve children who have not fully developed their sexual nature.

More and more frequently, we are becoming aware of individuals whose physical gender does not match their psychological gender. These people are sometimes referred to as being *gender-confused*. A male with normal external male genitals may "feel" like a female to one degree or another. The same situation may occur with structurally female individuals. This may mean that, in private, such people dress as the other sex. However, in social situations, they would behave normally for their sex. Others have completely changed their public and private behavior to reflect their inner desire to function as the other sex. A male may dress as a female, work in a traditionally female occupation, and make social contacts as a female. Tremendous psychological and emotional pressures develop from this condition. Frequently, these individuals would like to have gender reassignment surgery: a sex-change operation. This surgery and the follow-up hormonal treatment can cost tens of thousands of dollars and take several years.

Because the most frequent behavior of gender-confused individuals is dressing as a member of the opposite sex, we label them *cross-dressers,* or *transvestites.* Other psychological conditions can cause this same symptom of cross-dressing, so it is not accurate to say that all transvestites are gender-confused individuals. The way we dress is not biological but is related to cultural customs.

The sexual orientation of a person seems to be unrelated to whether they are transvestites. Some males who are attracted to males ("gay") may also be gender-confused, while others are not. The reverse is also true. Some females who are attracted to females ("lesbians") are gender-confused, while others are not. Homosexuality is a complex behavioral pattern that appears to be separate and distinct from gender-confusion.

Hermaphrodite
This individual shows the partial development of the female labia as well as the male scrotum.

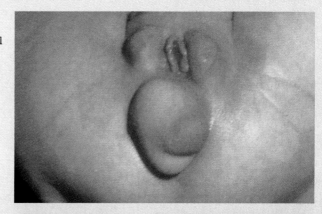

Transvestites

Is Homosexuality Genetic?

According to evidence collected in the past few years, homosexuality seems to be at least in part a biological phenomenon. Research with a small group of individuals suggests that in homosexual men, the INAH-3 "nuclei" of the anterior hypothalamus—a section of the brain that governs sexual behavior— have the anatomical form usually found in females rather than the form typical of heterosexual males. (The term *nuclei* here refers to a group or cluster of cells in a particular region of the brain.) The INAH-3 nuclei of heterosexual males are twice as large as those of homosexual males or heterosexual females. This study depended on autopsies of individuals who died of AIDS. There is some question whether their disease was a factor in the abnormal nuclei in their brains.

The difference in size may be dependent on levels of testosterone just before birth and immediately after birth. Unusual levels of sex hormones at any given time may switch the development of susceptible brain areas such as the hypothalamus from one sex to the other. While some studies show that size differences do exist between some homosexual and some heterosexual male brains, it is important to note that this does not rule out the possibility that childhood or adolescent events could also alter INAH-3 nuclei.

In 1993 and again in 1995, researchers reported that homosexual men share a DNA sequence on their X chromosomes that heterosexual men do not have. This further suggests that homosexual behavior in males has a hereditary component. Perhaps this genetic factor is related to the development of brain structure.

external genital anatomy in a male embryo, and is also important in the production and maturation of sperm. Testosterone and similar hormones are also called **androgens.**

Features that are not directly necessary for sexual reproduction but are characteristic of a sex are called **secondary sex characteristics.** These begin to become apparent at age 13 or 14. Facial hair, underarm and pubic hair, and chest hair are some of the most obvious. The male voice changes as the larynx (voice box) begins to change shape. Body contours also change, and a growth spurt increases height. In addition, the proportion of body muscle increases, while the proportion of body fat decreases. At this time, a boy's body begins to take on the characteristic adult male shape, with broader shoulders and heavier muscles.

In addition to these external changes, the FSH released from the pituitary causes the production of seminal fluid, called **semen,** by the seminal vesicles, the prostate gland, and the bulbourethral glands. Later, FSH stimulates the production of sperm cells. The release of sperm cells and seminal fluid also begins during puberty and is termed **ejaculation.** This release is

CONCEPT CONNECTIONS

Curiosity about the changing body at puberty leads to self-investigation. Studies have shown that sexual activity such as manipulation of the clitoris or penis, which causes pleasurable sensations, is performed by a large percentage of the population. Self-stimulation, frequently to orgasm, is a common result. This self-stimulation is termed **masturbation.** It is a normal part of sexual development during puberty and is also common in adults.

Orgasm is a complex response to mental and physical stimulation that causes rhythmic contractions of the muscles of the reproductive organs and an intense frenzy of excitement.

BIO *feature*

Planning the Sex of Your Child

Frequently, due to cultural or personal desires, a couple may want to choose the sex of their child. In the past, the sex of offspring was determined strictly by chance. If the father produced typical sperm cells, half would be X-containing and therefore produce female children, and half would be Y-containing and produce male children. Recently, it has become possible to change the probability in order to favor production of the desired sex. Reproductive specialists claim a success rate of from 65 percent to 85 percent in obtaining the desired sex of a child, as opposed to a 50 percent chance if no special practices are involved. It is believed that this is done by taking advantage of what we know about the mobility of sperm cells, their viability and pH requirements, and the relationship between the age of the egg cell and its ability to be fertilized.

Generally speaking, a sperm cell carrying an X chromosome is stronger than the Y-carrying sperm cell. It is larger and has an oval head and a shorter tail. It is speculated that anything that favors the stronger sperm cell favors the production of a female child. An acid douche (vinegar and water) creates a slightly acidic environment, which favors sperm cells carrying an X chromosome. Any condition that makes it difficult for sperm to reach the egg works in favor of the stronger X-carrying sperm. Shallow penetration during intercourse, lack of female orgasm, and refraining from intercourse several days before ovulation are all thought to achieve this purpose. The opposite conditions are thought to favor the production of a male offspring. An alkaline douche (baking soda and water), deeper penetration, female orgasm, and frequent intercourse centered around the time of ovulation will favor fertilization by a sperm cell containing a Y chromosome.

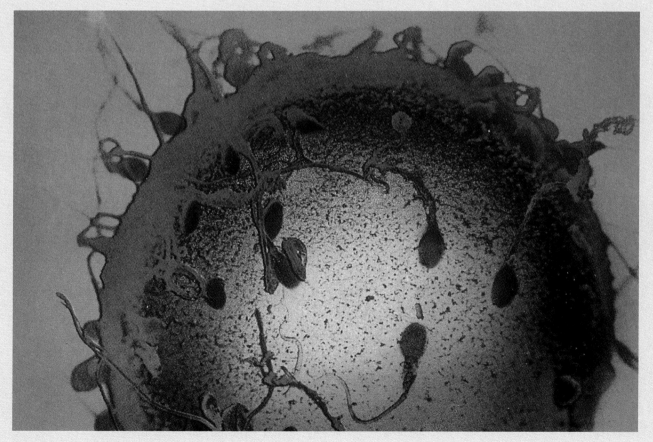

Fertilization

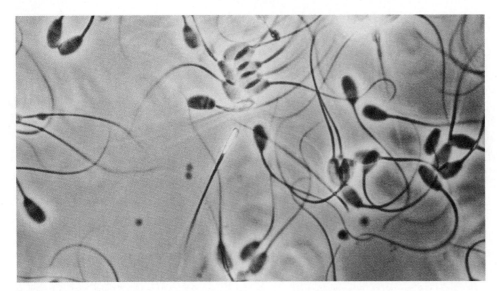

Figure 18.4 **Human Sperm Cells** These cells are primarily DNA-containing packages produced by the male.

cells. Beginning about age 11, some of them specialize and begin the process of *meiosis,* while others continue to divide by mitosis, assuring a constant and continuous supply of sperm-producing cells. Once sperm production begins, the tubules become hollow and sperm can flow into the sperm duct.

Mature sperm have only a small amount of food reserves. Therefore, once they are released and become active swimmers, they live no more than 72 hours (figure 18.6). However, if frozen, the sperm become deactivated, and can live for years outside the testes. This technique is widely used in the breeding of domesticated animals and is even used to store human sperm. Human sperm banks have been established to store the sperm of selected men. These sperm can be used to impregnate women whose husbands are sterile.

Sperm production takes place continuously throughout a male's reproductive life, although the number of sperm produced decreases as a man ages. For reasons not totally understood, a man must be able to release at least 100 million sperm at one ejaculation to be fertile. A healthy male probably releases about 300 million sperm during each act of

generally accompanied by the pleasurable sensations of **orgasm.** The sensations associated with ejaculation may lead to self-stimulation (**masturbation**). Studies of sexual behavior have shown that masturbation is a normal part of developing sexuality and that nearly all men masturbate at some time during their lives.

The Production of Sperm Cells

The primary biological reason for sexual activity is the production of genetically diverse offspring. The process of producing gametes includes meiosis and maturation of **sperm** cells (figure 18.4). The process of producing sperm cells in males takes place in the testes. The two bean-shaped testes are composed of many small sperm-producing tubes and collecting ducts that store the sperm. These tubes are held together by a thin membrane (figure 18.5). The ducts join together and eventually form a long, narrow tube in which sperm cells are stored and mature before ejaculation.

Leading from the testis is the sperm duct, which empties into the urethra, which transports the sperm out of the body through the **penis.** Before puberty, the tubes in the testes are packed with diploid cells. These cells undergo *mitosis* and generate sperm-producing

FYI

The production of each human sperm cell takes approximately 74 days. Typically, 300 million sperm mature in the adult human male each day. After ejaculation, sperm can survive only about 72 hours in the female reproductive tract.

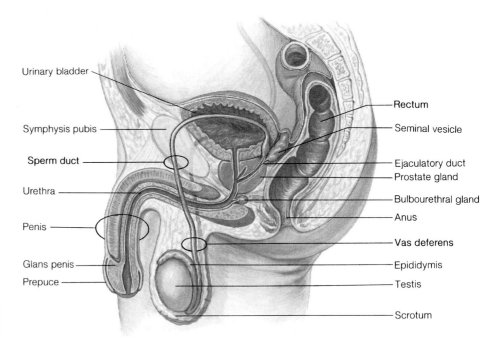

Figure 18.5 **The Human Male Reproductive System** The male reproductive system consists of two testes that produce sperm, ducts that carry the sperm, and various glands. Muscular contractions propel the sperm through the vas deferens and past the seminal vesicles, prostate gland, and bulbourethral glands, where most of the liquid of the semen is added. The semen passes through the urethra of the penis to the outside of the body.

Figure 18.6 Frozen Sperm Sperm can be frozen and preserved for years.

sexual intercourse. Sperm counts are often taken to determine if infertility is a result of inadequate numbers of sperm.

The Maturation of Females

Female children at 8 to 12 years of age typically begin to produce quantities of sex hormones from a portion of the brain called the hypothalamus, as well as from the pituitary gland, the ovaries, and the adrenal glands. This marks the onset of puberty. The **hypothalamus** controls the functioning of many glands throughout the body, including the pituitary. At puberty, the **pituitary gland** begins to produce follicle-stimulating hormone (FSH). The increasing amount of this hormone circulating in the blood of girls causes the ovaries to begin producing larger quantities of **estrogen.** The increasing supply of estrogen is responsible for the many changes in sexual development that can be noted at this time. These changes include: (1) breast growth; (2) changes in the walls of the uterus and vagina; (3) increased blood supply to the **clitoris,** the small erectile structure containing many nerve endings that cause pleasurable sensations; and (4) changes in the pelvic bone structure. Secondary sex characteristics such as the distribution of body hair and the deposition of fat at the hips and breasts also occur during puberty (figure 18.7).

The Production of Eggs

The production of an egg cell starts during prenatal development of the ovary, when diploid cells cease dividing by *mitosis* and enlarge to form potential egg cells. Egg cell development halts at this point, and the cells remain just under the surface of the ovary.

These cells will continue *meiosis* in the normal manner at puberty. At puberty and on a regular basis thereafter until menopause, the sex hormones stimulate a potential egg cell to continue its maturation process, and it goes through the first meiotic division. But in telophase I, the two cells receive unequal portions of cytoplasm. You might think of it as a lopsided division (figure 18.8). The smaller of

CONCEPT CONNECTIONS

follicle-stimulating hormone
Produced by the pituitary; causes the ovaries to begin producing the female sex hormone, estrogen.

estrogen Hormone produced by the ovaries that is responsible for breast growth, changes in the walls of the uterus and vagina, increased blood supply to the clitoris, and changes in the pelvic bone structure. Estrogen also stimulates the female adrenal gland to produce androgens.

androgens Male sex hormones; responsible for the production of pubic hair; they influence the female sex drive and may be involved in the development of acne.

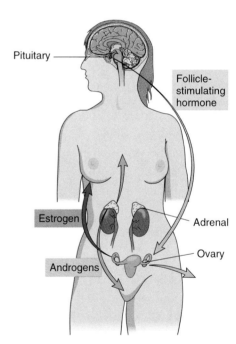

Figure 18.7 Sexual Development in Females Both estrogen and androgens are important in the sexual development of females. Estrogen stimulates breast tissue development, body fat distribution, and many other sexual characteristics. Androgens from the adrenal glands are important in stimulating the growth of body hair and determining the sex drive.

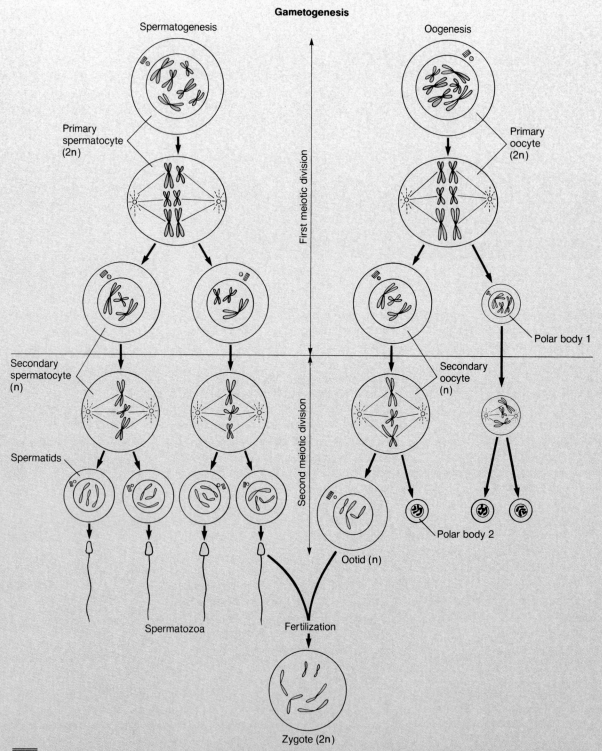

Gametogenesis

Spermatogenesis

Oogenesis

Primary spermatocyte (2n)

Primary oocyte (2n)

First meiotic division

Secondary spermatocyte (n)

Secondary oocyte (n)

Polar body 1

Second meiotic division

Spermatids

Spermatozoa

Ootid (n)

Polar body 2

Fertilization

Zygote (2n)

Figure 18.8 Gametogenesis This diagram illustrates the process of gametogenesis in human males and females. Not all of the 46 chromosomes are shown. Carefully follow the chromosomes as they segregate, recalling the details of the process of meiosis explained previously.

the two cells (polar body) disintegrates later, and the larger haploid cell is released from the ovary in a process called **ovulation.** Before ovulation, the developing **egg** becomes surrounded by a fluid-filled, saclike structure (follicle). As the follicle grows, it pushes against the surface of the ovary (figure 18.9). When this maturation is complete, the developing egg is released. It is swept into the **oviduct** (fallopian tube) by ciliated cells and travels toward the **uterus** (figure 18.10). The empty, ruptured follicle develops into a glandlike structure, which produces hormones (progesterone and estrogen) that prevent the release of other eggs.

If the egg is fertilized, it completes meiosis by proceeding through meiosis II with the sperm DNA inside. During the second meiotic division, the egg again divides unevenly, so that a second polar body forms. None of the polar bodies survive; therefore, only one large egg cell is produced.

As the young female matures, the **menstrual cycle** is established. This cycle involves the periodic growth and shedding of the lining of the uterus. These changes are under the control of a number of hormones produced by the pituitary gland and the ovaries. The ovaries are stimulated to release their hormones by the pituitary gland, which is in turn influenced by the ovarian hormones. Follicle-stimulating hormone (FSH) and luteinizing hormone (LH) are both produced by the pituitary gland. FSH causes the maturation and development of the ovaries, and LH is important in causing ovulation and in maintaining the menstrual cycle (table 18.1). Initially, these two cycles, menstruation and ovulation, may be irregular, which is normal during puberty. Eventually, hormone production becomes regulated so that ovulation and menstruation take place every month in most women, although the length of the cycle varies from one woman to another and may be as short as 21 days or as long as 45 days. If the egg cell is not fertilized, it passes through the **vagina** to the outside during menstruation. During her lifetime, a female releases about three hundred to five hundred eggs.

Hormones control the cycle of changes in breast tissue, in the ovaries, and in the uterus. In particular, estrogen and progesterone stimulate milk production by the breasts and cause the lining of the uterus to become thicker and more vascularized prior to the release of the egg cell. This ensures that if the egg cell becomes fertilized, the resulting embryo will be able to attach itself to the wall of the uterus and receive nourishment. If the cell is not fertilized, the lining of the uterus is shed. Once the wall of the uterus has been shed, it begins to build again.

Fertilization and Pregnancy

In most women, an egg cell is released from the ovary about 14 days before the start of the next menstrual cycle. The menstrual cycle is usually said to begin on the first day of menstrual flow. Therefore, if a woman has a 28-day cycle, the cell is released approximately on day 14 (figure 18.11). Some women, however, have irregular menstrual cycles, and it is difficult to determine just when the egg cell will be released to become available for fertilization. Once the cell is released, it is swept into the oviduct and moved

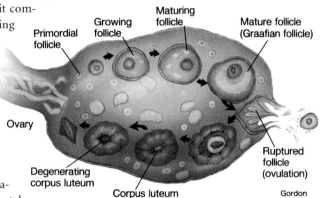

Figure 18.9 Ovulation In the ovary, the egg begins development inside a sac of cells known as a follicle. Each month, one of these follicles develops and releases its product. This release through the wall of the ovary is known as ovulation.

table 18.1

Human Reproductive Hormones

Hormone	Production Site	Target Organ	Function
1. Prolactin (lactogenic or luteotropic hormone)	Pituitary gland	Breasts, ovary	Stimulates milk production; also helps maintain normal ovarian cycle
2. Follicle-stimulating hormone	Pituitary gland	Ovary, testis	Stimulates ovary and testis development; stimulates egg production in females and sperm production in males
3. Luteinizing hormone (interstitial cell stimulating hormone)	Pituitary gland	Ovary, testis	Stimulates ovulation in females and sex-hormone (estrogen and testosterone) production in both males and females
4. Estrogen	Follicle of the ovary	Entire body	Stimulates development of female reproductive tract and secondary sexual characteristics
5. Testosterone	Testes	Entire body	Stimulates development of male reproductive tract and secondary sexual characteristics
6. Progesterone	Ovaries	Uterus, breasts	Causes uterine thickening and maturation; maintains pregnancy
7. Oxytocin	Pituitary gland	Breasts, uterus	Causes uterus to contract and breasts to release milk

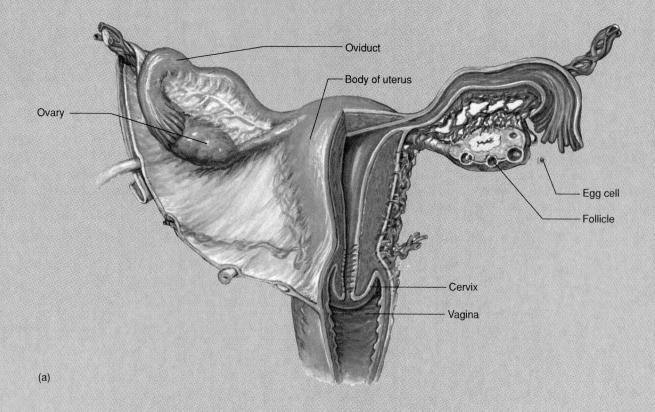

Oviduct

Body of uterus

Ovary

Egg cell

Follicle

Cervix

Vagina

(a)

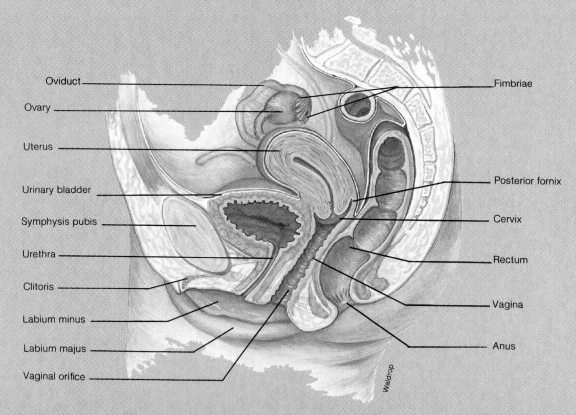

Oviduct

Ovary

Uterus

Urinary bladder

Symphysis pubis

Urethra

Clitoris

Labium minus

Labium majus

Vaginal orifice

Fimbriae

Posterior fornix

Cervix

Rectum

Vagina

Anus

(b)

Figure 18.10 The Human Female Reproductive System (a) After ovulation, the cell travels down the oviduct to the uterus. If it is not fertilized, it is shed when the uterine lining is lost during menstruation. (b) The human female reproductive system, side view.

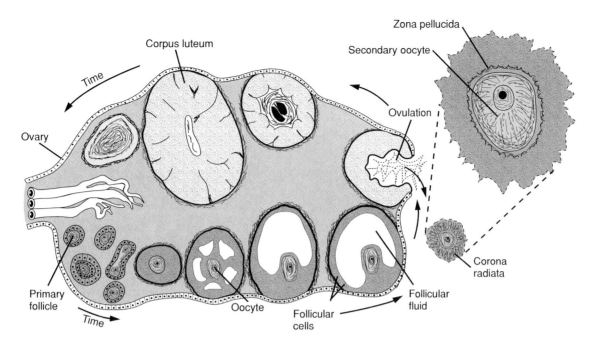

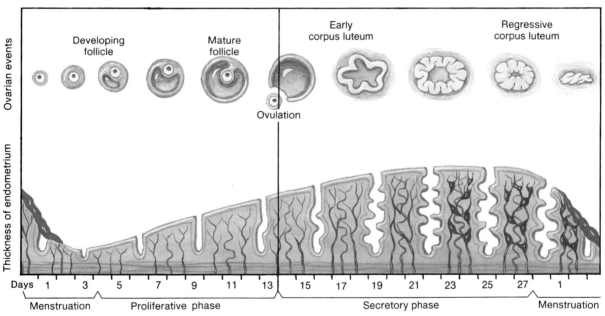

Figure 18.11 The Ovarian Cycle in Human Females The release of a secondary oocyte (ovulation) is timed to coincide with the thickening of the lining of the uterus. The uterine cycle in humans involves the preparation of the uterine wall to receive the embryo if fertilization occurs. Knowing how these two cycles compare, it is possible to determine when pregnancy is most likely to occur.

toward the uterus. If sperm are present, they swarm around as the egg passes down the oviduct, but only one sperm penetrates the outer layer to fertilize it and cause it to complete meiosis II.

The chromosomes from the sperm intermingle with those of the egg, forming a **zygote,** or fertilized egg. As the zygote continues to travel down the oviduct, it begins dividing by mitosis into smaller and smaller cells (figure 18.12). Eventually, a solid ball of cells is produced that becomes hollow. When the embryo is about six days old it becomes embedded, or implanted, in the lining of the uterus. The embryo secretes enzymes that allow it to digest its way into the uterine lining where it becomes surrounded with uterine cells.

The embryo develops a tube that later becomes the brain and nerve cord. Then it produces structures that will become the gut. The formation of the primitive nervous system and the gastrointestinal system are just some of the changes that eventually result in an embryo that is recognizable as human. Most of the time during its development, the embryo is enclosed in a water-filled membrane that protects it and keeps it moist. This membrane, along with other embryonic sacs, fuses with

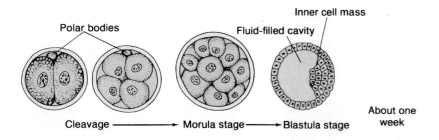

Polar bodies Fluid-filled cavity Inner cell mass

Cleavage ⟶ Morula stage ⟶ Blastula stage

About one week

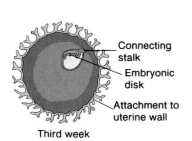

Connecting stalk
Embryonic disk
Attachment to uterine wall

Third week

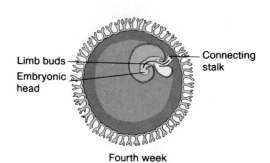

Limb buds
Embryonic head
Connecting stalk

Fourth week

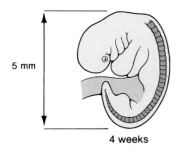

5 mm

4 weeks

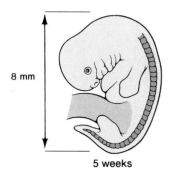

8 mm

5 weeks

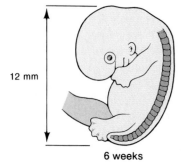

12 mm

6 weeks

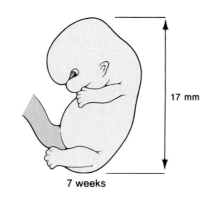

17 mm

7 weeks

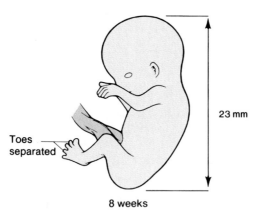

Toes separated

23 mm

8 weeks

Figure 18.12 Human Embryonic Development During the period of time between fertilization and birth, many changes take place in the embryo. Here we see some of the changes that take place during the first eight weeks.

the lining of the uterus to form the **placenta** (figure 18.13). The placenta produces hormones that prevent menstruation and ovulation during the time when an embryo is developing within the uterus. Preventing menstruation means the pregnancy is more likely to continue for the full term. Preventing ovulation means there is no opportunity for an additional fertilization and implantation while the first embryo is developing.

The embryo is contained within the mother's body and relies upon her body to provide all of its needs. Nourishment for the developing embryo is provided by the mother through the placenta. The materials crossing the placenta include oxygen, carbon dioxide, nutrients, and a variety of waste products. Blood from the mother is pumped to the placenta

Figure 18.13 Placental Structure
The embryonic blood vessels that supply the developing child with nutrients and remove metabolic wastes are separate from the blood vessels of the mother. Because of this separation, the placenta can selectively filter many types of incoming materials and microorganisms.

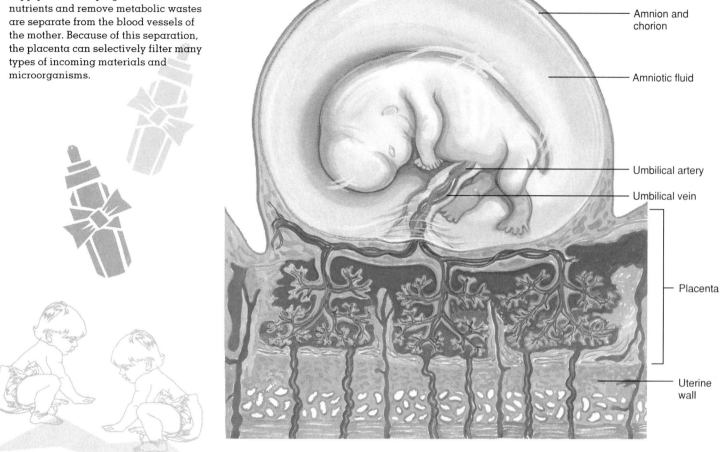

Amnion and chorion

Amniotic fluid

Umbilical artery

Umbilical vein

Placenta

Uterine wall

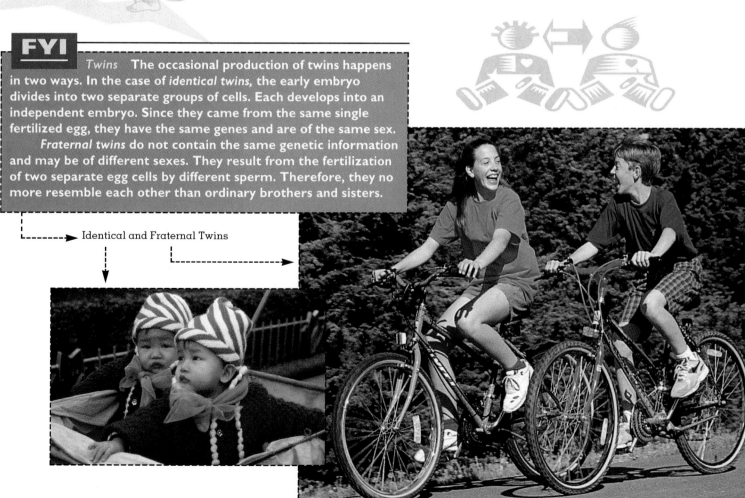

FYI

Twins The occasional production of twins happens in two ways. In the case of *identical twins*, the early embryo divides into two separate groups of cells. Each develops into an independent embryo. Since they came from the same single fertilized egg, they have the same genes and are of the same sex.

Fraternal twins do not contain the same genetic information and may be of different sexes. They result from the fertilization of two separate egg cells by different sperm. Therefore, they no more resemble each other than ordinary brothers and sisters.

Identical and Fraternal Twins

where blood vessels containing blood from the embryo are present. Although the two blood systems are not directly connected, exchange of nutrients and wastes occurs. The materials entering the embryo travel through blood vessels in the umbilical cord. The major parts of the body develop by the tenth week of pregnancy. After this time, the embryo increases in size, and the structure of the body is refined.

Birth

At the end of about nine months, hormone changes in the mother's body stimulate contractions of the muscles of the uterus during a period prior to birth called *labor*. These contractions are stimulated by the hormone oxytocin, which is released from the posterior pituitary. The contractions normally move the baby headfirst through the vagina, or birth canal. One of the first effects of these contractions may be bursting of the membrane (bag of water) surrounding the baby. Following this, the uterine contractions become stronger, and shortly thereafter the baby is born (figure 18.14).

Following the birth of the baby, the placenta, also called the *afterbirth,* is expelled. Once born, the baby begins to function on its own. The umbilical cord collapses and the baby's lungs, kidneys, and digestive system must now support all bodily needs. This change is quite a shock, but the baby's loud protests fill the lungs with air and stimulate breathing.

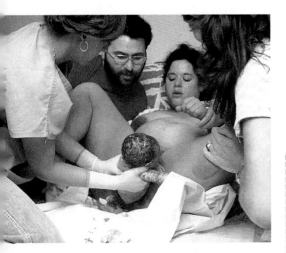

Figure 18.14 Childbirth

After the birth, the mother's breasts, which have undergone changes during the period of pregnancy, are ready to produce milk to feed the baby. Progesterone stimulates the production of milk, and oxytocin stimulates its release. If the baby is breast-fed, the stimulus of the baby's sucking will prolong the time during which milk is produced.

In some cultures, breast-feeding continues for two to three years, and the continued production of milk-producing hormones often delays the reestablishment of the normal cycles of ovulation and menstruation. Many people believe that a woman cannot become pregnant while she is nursing a baby. However, because there is so much variation among women, breast-feeding is not a reliable method of conception control. Many women have been surprised to find themselves pregnant again a few months after delivery.

FYI
Nearly 23 percent of all live births in the United States are cesarean deliveries.

Contraception

Throughout history people have tried various methods of conception control (figure 18.15). In ancient times, conception control was encouraged during times of food shortage or when tribes were on the move from one area to another in search of a new home. Writings as early as 1500 B.C. indicate that the Egyptians used a form of tampon medicated with the ground powder of a shrub to prevent fertilization. This may sound primitive, but we use the same basic principle today to destroy sperm in the vagina.

While the only completely effective method of preventing conception is to abstain from sexual intercourse, many other methods dramatically reduce the chance of fertilization. Contraceptive jellies and foams make the environment of the vagina more acidic, which diminishes the sperm's chances of survival. The spermicidal (sperm-killing) foam or jelly is placed in the vagina before sexual intercourse. When the sperm make contact with the acidic environment, they stop and soon die.

Aerosol foams are an effective method of conception control, but manipulating hormones

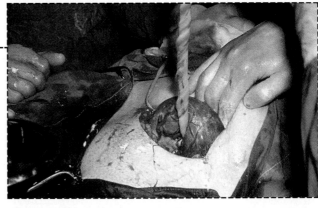

Cesarean Birth

FYI
Complications at Birth In some childbirth cases, the baby is not head down in the uterus before labor. If this occurs, the feet or buttocks appear first. Such a birth is called a *breech birth.* This can be a dangerous situation since the baby's source of oxygen is being cut off as the placenta begins to separate from the mother's body.

If for any reason the baby does not begin breathing on its own, it will not be receiving enough oxygen to prevent the death of nerve cells; thus, brain damage or death can result. A common procedure to resolve this problem is the surgical removal of the baby through the mother's abdomen. This procedure is known as a *cesarean,* or *C-section.*

Figure 18.15 Contraceptive Methods These are the primary methods of conception control used today: (a) oral contraception (pills), (b) contraceptive implants, (c) diaphragm and spermicidal jelly, (d) intrauterine device, (e) spermicidal vaginal foam, (f) depo-provera injection, (g) male condom, and (h) female condom.

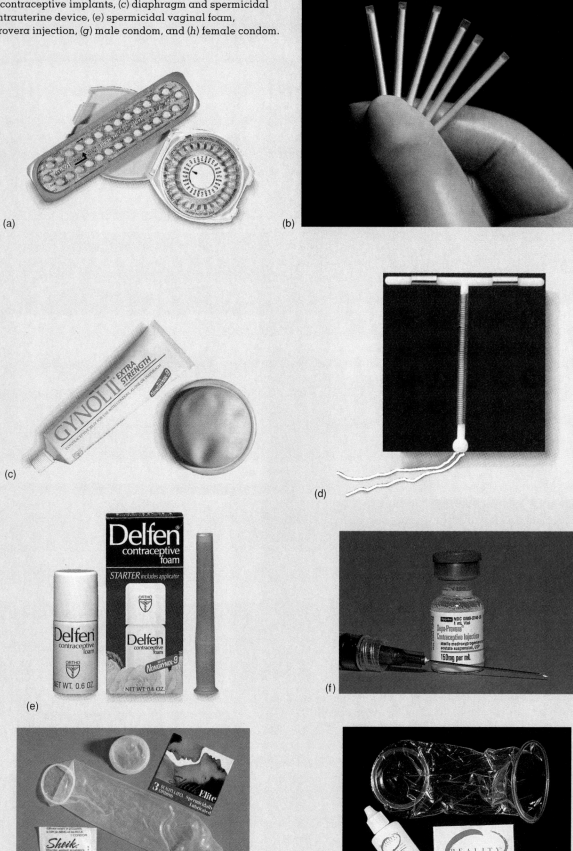

Abortion **A medical procedure often associated with birth control is abortion. Abortion involves various procedures that cause the death and removal of the developing embryo. Although abortion has been used throughout history, it is not a method of conception control; rather, it prevents the normal development of the embryo and causes its death. Abortion is an emotionally charged subject. Some people feel that abortion should be prohibited by law in all cases. Others feel that abortion should be allowed in certain situations, such as in pregnancies that endanger the mother's life or in pregnancies that are the result of rape or incest. Still others feel that abortion should be available to any woman under any circumstances. Regardless of the moral and ethical issues that surround abortion, it is still a common method of terminating unwanted pregnancies.**

The abortion techniques used in the United States today all involve the possibility of infections, particularly if done by poorly trained personnel. The three most common techniques are scraping the inside of the uterus with special instruments (called a *D and C* or *dilation and curettage*), injecting a saline solution into the uterine cavity, and using a suction device to remove the embryo from the uterus. The drug RU 486 is also available in many countries throughout the world and is used in about 15 percent of the elective abortions in France. The medication is administered orally under the direction of a physician, and several days later, a hormone is administered. This usually results in the onset of contractions that expel the embryo. A follow-up examination of the woman is made after several weeks to ensure that there are no serious side effects from the medication.

so that ovulation does not occur is more effective. The first successful method of hormonal control was "the pill." One of the newest methods of conception control also involves hormones. The hormones are contained within small rods or capsules (Norplant), which are placed under a woman's skin. These rods, when properly implanted, slowly release hormones and prevent the maturation and release of eggs. The major advantage of the implant is its convenience. Once the implant has been inserted, the woman can forget about contraceptive protection for several years. If properly inserted, these rods can be removed with a minimal amount of trauma so that the ovarian cycle can be reestablished and fertilization can occur.

Recently made available in the United States is a contraceptive injection. This injection of a modified form of the hormone progesterone (depoprovera) prevents ovulation. The shot is given in the buttocks or arm every three months. When a woman decides that she would like to become pregnant, she discontinues the injections and fertility will return after 6 to 18

months. This form of conception control is proclaimed to be convenient, safe, highly effective, and long-lasting. It also protects women from cancer of the lining of the uterus. Its drawbacks include a long lag between the time a woman stops the injections until she becomes fertile again. Depending on the individual, some women may experience spotting, irregular monthly periods, bloating and weight gain, headaches, depression, loss of interest in sex, and hair loss.

Killing sperm or preventing ovulation are not the only methods of preventing conception. Any method that prevents the sperm from reaching the egg prevents conception. One method is to avoid intercourse during those times of the month when the egg cell may be present. This is known as the *rhythm method* of conception control. While at first glance it appears to be the simplest and least expensive, determining just when an egg cell is likely to be present can be very difficult. If a woman has an irregular menstrual cycle, there may be only a few days each month for intercourse without the chance of fertilization. In addition to calculating

safe days based on the length of the menstrual cycle, a woman can better estimate the time of ovulation by keeping a record of changes in her body temperature and vaginal pH. Both of these changes are tied to the menstrual cycle and can, therefore, help a woman predict ovulation. In particular, at about the time of ovulation, a woman has a slight rise in body temperature. The rise is less than 1°C, so the woman should use an extremely sensitive thermometer. (There is even a digital-readout thermometer on the market that spells out the words *yes* or *no*.)

Other methods of conception control that prevent the sperm from reaching the egg include the diaphragm, cap, sponge, and condom. The diaphragm is a specially fitted membranous shield that is inserted into the vagina before intercourse and positioned so that it covers the opening of the uterus. Because of anatomical differences among females, diaphragms must be fitted by a physician. The effectiveness of the diaphragm is increased if spermicidal foam or jelly is also used. The vaginal cap functions in a similar way.

The male condom is probably the most popular contraceptive device. It is a thin sheath that is placed over the penis at the time of erection. In addition to preventing sperm from reaching the egg, condoms also help to prevent the spread of sexually transmitted diseases (STDs). There are more than 20 recognized STDs, including syphilis, gonorrhea, hepatitis, and AIDS. The most desirable condoms are made of a thin layer of latex that does not reduce the sensitivity of the penis. The condom is most effective if it is prelubricated with a spermicidal material such as nonoxynol-9. This lubricant also has the advantage of providing some protection against the spread of the HIV virus. Recently developed condoms for women are now available for use. One, called the Femidom, is a polyurethane sheath that, once inserted, lines the contours of the woman's vagina. It has an inner ring that sits over the cervix and an outer ring that lies flat against the labia. Research shows that this device protects against STDs and is as effective a contraceptive as the condom used by men.

The intrauterine device (IUD) is not a physical barrier that prevents the gametes from uniting. How this device works is not completely known. It may in some way interfere with the implantation of the embryo. The IUD must be fitted and inserted into the uterus by a physician, who can also remove it if pregnancy is desired.

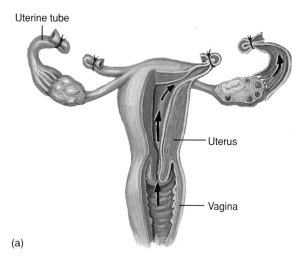

(a)

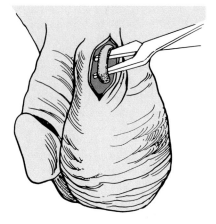

(b)

Figure 18.16   Tubal Ligation and Vasectomy   Two very effective contraceptive methods require surgery. Tubal ligation (a) involves severing the oviducts and suturing or sealing the cut ends. This prevents the sperm cells and the egg from meeting. This procedure is generally considered outpatient surgery, or at most requires a short hospitalization period. Vasectomy (b) requires minor surgery, usually in a clinic under local anesthesia. Following the procedure, minor discomfort may be experienced for several days. The severing and sealing of the vas deferens prevents the release of sperm cells from the body by ejaculation.

table 18.2

The Effectiveness of Contraceptive Methods

Method	Pregnancies per 100 Women per Year[a]	
	High[b]	Low
No contraceptive[c]	80	40
Coitus interruptus	23	15
Condom[d]	17	8
Douche	61	34
Chemicals (spermicides)[e]	40	9
Diaphragm and jelly	28	11
Rhythm[f]	58	14
Pill	2	0.03
IUD	8	3
Sterilization	0.003	0

[a]Data describe the number of women per 100 who will become pregnant in a one-year period while using a given method.

[b]High and low values represent best and worst estimates from various demographic and clinical studies.

[c]In the complete absence of contraceptive practice, 8 out of 10 women can expect to become pregnant within one year.

[d]Effectiveness increases if spermicidal jelly or cream is used in addition.

[e]Aerosol foam is considered to be the best of the chemical barriers.

[f]Use of a clinical thermometer to record daily temperatures increases effectiveness.

From E. Peter Volpe, *Biology and Human Concerns*, 3d ed. Copyright © 1983 Wm. C. Brown Communications, Inc., Reprinted by permission of Times Mirror Higher Education Group, Inc. Dubuque, Iowa. All Rights Reserved.

Two contraceptive methods that require surgery are tubal ligation and vasectomy (figure 18.16). Tubal ligation involves the cutting and tying off of the oviducts and in most cases can be done on an outpatient basis with the use of laser surgery. Only a small incision is required. Ovulation continues as usual, but the sperm and egg cannot unite. Vasectomy can be performed in a physician's office and does not require hospitalization. A small opening is made above the scrotum, and the vas deferens is cut and tied. This prevents sperm from moving through the ducts to the outside. Because most of the sperm-carrying fluid, or semen, is produced by glands other than the testes, a vasectomy does not interfere with normal ejacu-

lation. The sperm that are still being produced die and are reabsorbed in the testes. Neither tubal ligation nor vasectomy interferes with normal sex drives. However, these medical procedures are generally not reversible and should not be considered by those who may want to have children at a future date. The effectiveness of various contraceptive methods is summarized in table 18.2.

Sexual Function in Seniors

At the onset of **menopause** (about the age of 50), a woman's body begins to shift into the postreproductive stage of life. Her hormonal balance slowly begins to change in such a way

that she will no longer release eggs or experience menstruation. At this time, the menstrual cycle becomes less regular and ovulation is often unpredictable. The changes in hormone levels cause many women to experience mood swings and physical symptoms, including cramps and hot flashes.

Occasionally the physical discomfort of menopause reaches a point where it interferes with normal life and the enjoyment of sexual activity. A physician might recommend hormonal treatment to augment the natural production of hormones and relieve the discomfort. This hormonal treatment also helps to control other bodily changes that are occurring at this time. Estrogen supplements, for example, slow the

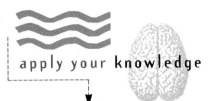

Testing Your STD Knowledge

Many states have instituted an STD HIV/AIDS education requirement prior to obtaining a marriage license. This brief introduction to the subject must take place within two months of the marriage. During the presentation, couples are encouraged to take a voluntary HIV blood test, but no test for any STDs are required of those attending this mandatory class. Knowing what you do about sexuality and the sexual behavior of people in the United States,

(1) does this class seem valuable considering the growth of HIV cases in the United States?
(2) is this an appropriate time in a person's life to learn about STDs?
(3) would you continue such a program?
(4) what alternatives would you suggest?
(5) what would be the content of your class?

loss of calcium from the bones and help prevent osteoporosis. The supplements also help maintain normal vaginal secretions, which makes sexual intercourse more enjoyable.

Human males do not experience a dramatic change in their reproductive lives. Rather, their sexual desires tend to wane slowly as hormone production decreases with age. They produce fewer sperm cells and less seminal fluid. Healthy men can experience a satisfying sex life during aging. The sex lives of seniors show the same variation as those of younger people. The whole range of responses to sexual partners continues but generally in a diminished form. It is reasonable to state that one's sexuality continues from before birth until death.

CONCEPT CONNECTIONS

menarche The time when a female begins to establish her menstrual cycle with the release of egg cells.

menopause The time when a female stops having a regular menstrual cycle. Both menarche and menopause are due to changes in the concentration of hormones in the blood.

Sexually Transmitted Diseases

Diseases currently referred to as *sexually transmitted diseases* (STDs) were formerly called *venereal diseases* (*VDs*). The term *venereal* is derived from the name of the Roman goddess for love, Venus. Although these kinds of illnesses are most frequently transmitted by sexual activity, many can also be spread by other methods of direct contact such as hypodermic needles, blood transfusions, and blood-contaminated materials.

Some of the most important STDs are described here because of their high incidence in the population and our inability to bring some of them under control (table 18.3). For example, there is no known cure for infection with human immunodeficiency virus (HIV), which is responsible for AIDS. There has also been a sharp rise in the number of gonorrhea cases in the United States caused by a form of the bacterium *Neisseria gonorrhoeae* that has become resistant to the drug penicillin. The bacterium produces an enzyme that actually destroys the antibiotic. However, most of the infectious agents can be controlled if diagnosis occurs early and treatment programs are carefully followed by the patient. The spread of STDs during sexual intercourse is significantly diminished by the use of condoms. Other types of sexual contact (i.e., hand, oral, anal) and congenital transmission (i.e., from the mother to the fetus during pregnancy) help to maintain some of these diseases in the population at levels high enough to warrant attention by the United States Public Health Service, the Centers for Disease Control and Prevention, and state and local public health agencies. All of these agencies are involved in attempts to raise the general health of the public. Their investigations have resulted in the successful control of many diseases and the identification of special problems, such as those associated with STDs.

Members of all public health agencies are responsible for warning the public about things that may be dangerous to them. In order to meet these obligations when dealing with sexually transmitted diseases, such as AIDS and syphilis, they encourage the use of one of their most potent weapons, sex education. Individuals must know about their own sexuality if they are to understand the transmission and nature of STDs. Then it will be possible for them to alter their behavior in ways that will prevent the spread of these diseases. The intent is to present people with biological facts, not to scare them. Public health officials do not have the luxury of advancing their personal opinions when it comes to their jobs. The biological nature of sexual behavior is not a moral issue, but biological facts are needed if people are to make intelligent decisions relating to their sexual behavior. It is hoped that through education, people will alter their high-risk sexual behaviors and avoid situations where they could become infected with one of the STDs. As one health official stated, we should be knowledgeable enough about our own sexuality and the STDs to answer the question, Is what I'm about to do worth dying for?

table 18.3

Sexually Transmitted Diseases

Disease	Other Names	Cause	Signs and Symptoms
1. Genital herpes	HSV: type II	Virus	Symptoms apply equally to both ♂ and ♀: Painful, fluid-filled blisters. Fever, tingling sensation, and tenderness at site of infection. Itching sensation, swollen lymph glands, nausea, aches, and fatigue. Infants can be infected at time of birth. Infected children may develop brain damage or other neurological complications.
2. Gonorrhea	Clap, drip	Bacterium	♂ = May infect sex organs, mouth, throat, and rectum; discharge from penis and severe burning during urination. ♀ = May infect sex organs, mouth, throat, and rectum; 80% of women show no visible symptoms during early stages of disease; vaginal discharge and recurring minor symptoms.
3. Syphilis	Syph, chancre (SHAN-ker)	Bacterium	Symptoms apply equally to both ♂ and ♀: Primary stage: painless sore called chancre with gelatinous center appears where bacteria entered body; vagina, penis, mouth, anus, skin. Chancre will disappear on its own with no scaring. Latent period will follow and may last months; no symptoms during this period. Secondary stage: weakness, rash, fever; infection spreads to internal organs and may result in blindness, heart disease, arthritis, insanity, stroke, miscarriages, transmission of bacteria to fetus, or death. If person survives, second latent period will follow that may last years. Tertiary (third) stage: degeneration of tissues will result in death.
4. AIDS/HIV		Virus	Symptoms apply equally to both ♂ and ♀: Initial stages include swollen glands, fever, night sweats, dementia, memory loss, rashes, diarrhea. Later stages (full-blown AIDS) include skin cancers, pneumonia, tuberculosis, herpes infections, cytomegalovirus infection, and death.
5. Candidiasis	Vaginitis, yeast infection	Yeast	♂ = Rarely have symptoms. ♀ = Thick, odorless, white discharge with cottage cheese consistency; intense itching, burning, and reddening of vagina.
6. Chancroid	Soft chancre, *Haemophilus ducreyi*	Bacterium	Symptoms apply equally to both ♂ and ♀: Flat sore on the surface of the skin where bacteria enter; begins as small, red pimples that develop into deep, pus-producing sore; painful with foul odor; may be found on penis or foreskin, thighs, vulva, clitoris, anus, mouth, lips, tongue and swollen lymph nodes.
7. Condyloma acuminatum	Venereal warts, papilloma virus infection, verruca	Virus	♂ = Internal warts: discomfort during urination, mild irritation and itching sensation, pain during sexual intercourse; soft, red/pink fleshy growths. External warts: cauliflower-like cancerous fleshy skin growths; firm and dark in color. ♀ = Internal warts: discomfort during urination, mild irritation and itching sensation, pain during sexual intercourse; soft, red/pink fleshy growths. External warts: cauliflower-like cancerous fleshy skin growths; firm and dark in color. Viruses can be transmitted to infant at birth. Infection is correlated with cervical cancer.

Continued

Sexually Transmitted Diseases (continued)

Disease	Other Names	Cause	Signs and Symptoms
8. *Chlamydia sp.*	Lymphogranuloma venereum, LGV, NGU, urethritis in men	Bacterium	♂ = 50% show no symptoms or may be very mild; prostate infections; urethritis (infection of the urethra); pain or burning during urination; watery or milky discharge. ♀ = May result in sterility due to scarring of fallopian tubes (oviducts); 50% show no symptoms; diagnosis often occurs after unsuccessful attempts to become pregnant; heavier than normal, yellow vaginal discharge; pain or burning during urination; frequent urge to urinate; painful intercourse; fever; lower abdominal pain during sexual intercourse.
9. Cytomegalovirus infection	CMV	Virus	Symptoms apply equally to both ♂ and ♀: Swollen lymph nodes, mild yellowing of tissue (jaundice); fever, headache, muscle pain, tiredness. In newborns: enlarged spleen and liver, mental retardation.
10. Nongonococcal urethritis	NGU	Bacterium	Symptoms apply equally to both ♂ and ♀: Thick discharge (typically seen in the morning), pain or burning sensation during urination; frequent urination; 25% show no symptoms; may cause infertility.
11. Pelvic inflammatory disease	PID, *Chlamydia*, *Neisseria gonorrhoea*	Bacterium	♂ = Does not apply. ♀ = Sterility due to destruction of fallopian tubes (oviducts); severe abdominal pain, fever, and tenderness of uterus and ovaries; abnormal mucus discharge; more severe menstrual periods; tiredness, weakness, nausea, vomiting, painful sexual intercourse; death.
12. Reiter's syndrome	Infectious arthritis	Bacterium	Symptoms apply equally to both ♂ and ♀: Joint pain, eye inflammation, and sores on the skin.
13. Crabs	Pubic lice, *Pthirus pubis*	Body lice	Symptoms apply equally to both ♂ and ♀: Intense itching; bright blue spots on the skin.
14. Trichomoniasis	Trich, vaginitis	Protozoan	♂ = Rarely show symptoms. ♀ = Irritating yellow or green frothy, foul-smelling discharge; burning and itching, especially during urination; vaginal redness and swelling.
15. Viral hepatitis	HBV, HCV	Virus	Symptoms apply equally to both ♂ and ♀: Yellowing of the skin and eyes (jaundice), dark orange urine, gray fecal material, fever, pain in upper right portion of the abdomen, constipation, tiredness, weakness, loss of appetite. May progress to cirrhosis of the liver, liver cancer, and death.
16. Bacterial vaginosis	Nonspecific bacterial infection of vagina, *Gardnerella vaginalis*	Bacterium	♂ = While this disease is called a "vaginosis" and is generally free of symptoms in men, they can carry the microbes responsible for the disease and, therefore, can reinfect women. ♀ = Thin, watery, gray-white, fishy-smelling vaginal discharge.

SUMMARY

The human sex drive is a powerful motivator for many activities in our lives. While it provides for reproduction and modification of the gene pool, it also has a social dimension. Sexuality begins before birth. A person's sex is controlled by sex-determining chromosomes. Human females receive two X chromosomes. Only one of these remains functional; the other remains tightly coiled as a Barr body. A human male receives one X and one Y chromosome. It is the presence of the Y chromosome that causes male development.

At puberty, hormones influence the development of secondary sex characteristics and the functioning of gonads. Mature eggs and sperm begin to be produced, and fertilization is possible.

Sexual reproduction involves the production of gametes by meiosis in the ovaries and testes. The production and release of these gametes is controlled by the interaction of hormones. Human females have specialized structures for the support of the developing embryo, and many factors influence its development in the uterus. Successful sexual reproduction depends on hormone balance, meiotic division, fertilization, placenta formation, diet of the mother, birth, and other health factors. Hormones regulate ovulation and menstruation and may also be used to encourage or discourage ovulation. Fertility drugs and birth-control pills, for example, involve hormonal control. In addition to the pill, a number of other contraceptive methods have been developed, including the diaphragm, condom, IUD, spermicidal jellies and foams, tubal ligation, and vasectomy. In addition, some of these conception control methods also help prevent the spread of sexually transmitted diseases. Hormones continue to direct our sexuality throughout our lives.

CHAPTER GLOSSARY

androgens (an′dro-jenz) Male sex hormones produced by the testes and adrenal glands that cause the differentiation of the internal and external genital anatomy; also important in sex drive of males and females.

clitoris (kli′to-ris) Small erectile structure of the female genitals containing many nerve endings that cause pleasurable sensations.

egg (eg) The haploid sex cell produced by the ovary.

ejaculation (e-jak″u-la′shun) The release of sperm cells and seminal fluid through the penis of a male.

estrogens (es′tro-jens) Female sex hormones that cause the differentiation in the female embryo of the internal and external genital anatomy; responsible for the changes in breasts, vagina, uterus, clitoris, and pelvic bone structure at puberty.

follicle-stimulating hormone (FSH) (fol′ĭ-kul stim′yu-lā-ting hōr′mōn) The pituitary secretion that causes the ovaries to begin to produce larger quantities of estrogen and to develop the follicle and prepare the egg for ovulation.

hormone (hōr′mōn) A chemical substance that is released from glands in the body to regulate other parts of the body.

hypothalamus (hi″po-thal′ă-mus) The region of the brain that causes the production of several kinds of hormones involved in sexual development and function.

interstitial cell-stimulating hormone (ICSH) (in″ter-stĭ′shal sel stim′yu-lā-ting hōr′mōn) The chemical messenger molecule released from the pituitary that causes the testes to produce testosterone.

masturbation (măs″tur-ba′shun) Stimulation of one's own sex organs.

menopause (měn′o-pawz) The period beginning at about age 50 when the ovaries stop producing eggs and estrogen production drops.

menstrual cycle (men′stru-al si′kul) The repeated building up and shedding of the lining of the uterus.

orgasm (or′gaz-um) A complex series of responses to sexual stimulation that result in an intense frenzy of sexual excitement.

oviduct (o′vĭ-dukt) The tube that carries the oocyte to the uterus; also called the *fallopian tube.*

ovulation (ov-yu-la′shun) The release of an egg from the surface of the ovary.

penis (pe′nis) The portion of the male reproductive system that deposits sperm in the female reproductive tract.

pituitary gland (pĭ-tu′ĭ-ta-re gland) The structure in the brain that controls the functioning of other glands throughout the organism.

placenta (plah-sen′tah) An organ made up of tissues from the embryo and the uterus of the mother that allows for the exchange of materials between the mother's bloodstream and the embryo's bloodstream. It also produces hormones.

puberty (pu′ber-te) A time in the life of a developing individual characterized by the increasing production of sex hormones, which cause it to reach sexual maturity.

scrotum (skrō′tum) The sac in which the testes are located.

secondary sex characteristics (sěk′on-dăr-e seks kăr-ak-tě-ris′tiks) Characteristics of the adult male or female, including the typical shape that develops at puberty.

semen (se′men) The sperm-carrying fluid produced by the seminal vesicles, prostate gland, and bulbourethral glands of males.

sexuality (sek″shoo-al′ĭ-te) A term used in reference to the totality of the aspects—physical, psychological, and cultural—of our sexual nature.

sperm (spurm) Haploid male gametes.

testosterone (tes-tos′tur-ōn) The male sex hormone produced in the testes that controls the secondary sex characteristics.

uterus (yu′tur-us) The organ in female mammals in which the embryo develops.

vagina (vuh-ji′nah) The passageway between the uterus and outside of the body; the birth canal.

zygote (zi′gōt) The fertilized egg.

AIDS

This interactive exercise allows students to explore the risk factors associated with the AIDS epidemic by varying key aspects of behavior and policy. Students can investigate the past history of the epidemic and explore the potential future consequences of sexual behavior and drug use, use of safe sex, and level of public education. The potential for explosive growth of the epidemic in Asia will become evident as projections are made for different world regions.

1. Why do so many more people have HIV than AIDS?
2. Can you think of a way to have HIV and never get AIDS?
3. What would be the effect of a mutation in gp120 that allowed it to recognize a mosquito blood cell surface protein as well as CD4?
4. What is the role of public education in the AIDS epidemic?
5. When is the AIDS epidemic going to be over?
6. What do you imagine a successful HIV vaccine might be like?

CONCEPT MAP TERMINOLOGY

Construct a concept map to represent the relationships among the following concepts.

androgens	interstitial cell-stimulating	scrotum
clitoris	hormone (ICSH)	semen
egg	menstrual cycle	sperm
estrogens	oviduct	testosterone
follicle-stimulating	ovulation	uterus
hormone (FSH)	penis	vagina
hormone	pituitary gland	zygote
hypothalamus	placenta	

LABEL•DIAGRAM•EXPLAIN

Write several paragraphs explaining in as much detail as possible what happens when a human female ovulates, the egg is fertilized, and the embryo develops. Include as much information about hormones as you can as you write about these processes.

Multiple Choice Questions

1. In humans, if an embryo at age 7 weeks has both an X and a Y chromosome:
   a. it can develop into either a male or a female
   b. it will develop into a male
   c. it should have already developed female sex organs
   d. because of this early stage in development, it is not possible to predict the sex

2. If an individual has only 45 chromosomes and is female it probably has:
   a. only one X chromosome
   b. only one Y chromosome
   c. only one sex-determining chromosome, which could be either X or Y
   d. resulted from an unfertilized egg cell that abnormally duplicated all of the chromosomes except the sex-determining chromosome

3. If an embryo is not influenced by a sex-determining chromosome, it will:
   a. develop into a male
   b. develop into a female
   c. develop into a transvestite
   d. develop into a hermaphrodite

4. Androgens are the hormones produced by the:
   a. ovary, which causes females to be attracted to members of the opposite sex
   b. adrenal glands, which result in the production of body hair in women
   c. pituitary, which control the production of egg cells
   d. egg, which causes the breasts to enlarge

5. Which of the following structures is part of the male reproductive organs?
   a. placenta
   b. follicle
   c. clitoris
   d. penis

6. Which of the following structures is part of the female reproductive organs?
   a. scrotum
   b. clitoris
   c. penis
   d. testis

7. Which of the following hormones is thought to have the greatest effect on timing ovulation?
   a. androgen
   b. follicle-stimulating hormone
   c. estrogen
   d. oxytocin

8. Unequal division of cells is part of the production of:
   a. egg cells and polar bodies
   b. sperm cells
   c. cells resulting from nondisjunction
   d. all body cells but no sex cells

9. Which of the following will prevent fertilization because of lack of viable sperm cells?
   a. rhythm method of conception control
   b. tubal ligation
   c. spermicidal creams or foams
   d. insertion of intrauterine device

10. Which of the following conception control methods works by influencing ovulation?
    a. oral contraceptive pills
    b. douches
    c. vasectomy
    d. condom used with spermicidal jelly

Questions with Short Answers

1. What variations occur among people with regard to sexuality and gender?

2. What are the effects of the secretions of the pituitary, the gonads, and the adrenal glands at puberty?

3. What structures are associated with the human female reproductive system? What are their functions?

4. What structures are associated with the human male reproductive system? What are their functions?

5. Identify three practices that could reduce the spread of sexually transmitted diseases.

6. How are ovulation and the menstrual cycle related to each other?

7. What changes occur in ovulation and menstruation during pregnancy?

8. What are the functions of the placenta?

9. Describe the methods of conception control.

10. List the events that occur as an embryo matures.

chapter

19

Muscles, *Hearts,* & Lungs

Color-enhanced bronchogram of right human lung.

- Describe how a muscle cell contracts from a molecular point of view.

- Be able to explain such common muscular events as muscle cramps, strains, and increase in muscle size.

- Identify ways in which drugs and nutrient supplements are misused by athletes.

- Know that there are three types of muscle—skeletal, smooth, and cardiac—and how they differ.

- Understand how the circulatory and respiratory systems function together to facilitate the exchange and distribution of gases throughout the body.

- Know the cellular and fluid composition of the blood, and explain the roles played by these components.

- Be able to describe a buffer system and its function, and provide a human example.

- Outline the ABO blood typing system and be familiar with the common blood tests performed by a medical lab.

- Describe the ways in which carbon dioxide is carried in the blood.

- Be able to trace the flow of blood through the human circulatory system.

- Recognize that the circulatory system transports molecules, cells, and heat.

- Understand how carbon dioxide levels, blood pH, and breathing rate are interrelated.

- Be able to explain various common problems related to the health of the heart and circulatory system (e.g., stroke, heart attack, dizziness).

- Be able to trace the flow of air through the human respiratory system.

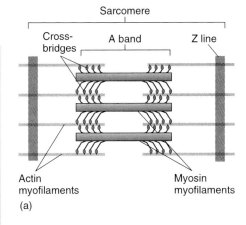

(a)

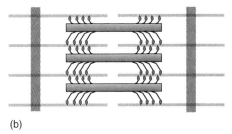

(b)

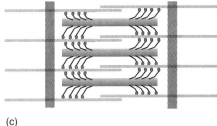

(c)

Figure 19.1 The Subcellular Structure of Muscle The actin-, tropomyosin-, and troponin-containing myofilaments and the myosin-containing myofilaments are arranged in a regular fashion into units called sarcomeres (a). Each sarcomere consists of two sets of actin-containing myofilaments inserted into either end of the bundles of myosin-containing myofilaments (b). The actin-containing myofilaments slide by the myosin-containing myofilaments, shortening the sarcomere (c).

Fitness and the Muscular System

The ability to move is one of the fundamental characteristics of healthy individuals. Through the coordinated contraction of many muscles, the intricate, precise movements of a dancer, basketball player, or writer are accomplished. And yet, muscles are responsible for more than just these major movements of the human body. Unseen, unfelt contractions of extremely small muscles are responsible for activities such as blood flow, food movement through the digestive system, eye focusing, and facial expressions. The size, shape, and distribution of muscles partially determines a person's overall form, which may contribute to one's self-esteem.

We have all seen the kind of development the human body can achieve with regular workouts. Regular exercise, whether it involves walking, swimming, jogging, step exercise, or dance, is important in maintaining our health and physical condition. Some types of physical conditioning involve working out to increase and develop muscles to their maximum. A serious bodybuilder works on each muscle and tries to change its shape, size, and proportion to a form perceived as ideal. The numbers of

FYI

Muscles comprise 40 percent to 50 percent of total body weight in a healthy human.

muscle cells cannot increase, but the amount of protein in the muscle fibers and the number of fibers in each cell certainly can. For this reason, weight lifters pay a great deal of attention to the amount of protein in their diet and the availability of all of the amino acids needed to construct muscle proteins.

If you have a reasonably good diet, consisting of appropriate amounts of carbohydrates, fats, and proteins, you will be able to increase the size of a particular muscle by increasing the amount of the proteins in the individual muscle cells. Dietary proteins (meat, poultry, fish, and dry beans) are digested in the gastrointestinal tract. The component amino acids are absorbed through the wall of the digestive tract and circulated to the cells, where the individual amino acids are reassembled by the process of protein synthesis into new human muscle protein.

A muscle cell responds to stimulation by contracting. Before stimulation, the protein fil-

aments within the individual muscle cells are arranged lengthwise from one end of the cell to the other and spaced at a distance from one another. As the muscle cell contracts, thin filaments consisting of **actin,** *tropomyosin,* and *troponin* "slide" past thick **myosin** filaments so that the distance from end to end is reduced (figure 19.1). This shortens the total length, resulting in a bulging of the muscle cell. When contraction occurs in the thousands of individual muscle cells that make up a muscle, it too

BIO *feature*

Protein Supplements for Bodybuilders and Weight Lifters

Many health food stores sell products composed of protein that are advertised as having the ability to increase muscle size. Package labels carefully list the amino acids contained in the product. Amino acids are the building blocks of proteins. The total weight (measured in grams) of the protein available from the product is also listed. Are these kinds of products a valuable source of protein, and do they promote the type of muscle growth desired by the athletes who use them? Is there a difference between the protein in a health food product and the protein in foods such as meat, fish, poultry, cottage cheese, skim milk, and low-fat yogurt?

There are two major proteins in muscle cells, actin and myosin. These proteins are constructed from the amino acids in our daily diets. An amino acid from a normal diet and the same amino acid from a packaged "health and fitness" food are identical in structure. They are used to produce muscle protein in exactly the same way. Therefore, a normal diet will provide all the amino acids needed to produce muscle. However, these products may also include carbohydrates, fats, vitamins, minerals, and other nutrients.

If there are additional materials in a dietary supplement, such as vitamins or minerals, you should question their value separately. It is also wise to compare the amount of fat, cholesterol, and other materials in both the dietary supplement and regular foods such as milk, cheese, and meat.

How does the cost of a special protein supplement compare to that of a normal diet? It is easy to determine the dollar value of the protein in foods. New federal labeling laws enable you to determine the cost per gram of protein in food products. Just divide the cost of the product by the weight in grams of the protein in the product to discover the cost per gram of protein. If there is a great difference between the two sources in the cost of a gram of protein, you might want to choose the less expensive of the two.

BIO *feature*

How Muscles Contract

The myosin molecules have a shape similar to a golf club. The head of the club-shaped molecule sticks out from the thick myosin-containing filament and can combine with the actin of the thin filament. Two other proteins, troponin and tropomyosin, are associated with the actin and cover the actin in such a way that myosin cannot bind with it. When actin is uncovered and myosin can bind to it, contraction of a muscle can occur when ATP (a source of energy) is utilized.

The process of muscle-cell contraction involves several steps. When a nerve impulse arrives at a muscle cell, it causes the cell to change the distribution of its electrical charges, and it depolarizes. When muscle cells depolarize, calcium ions (Ca^{++}) contained within membranes are released among the actin and myosin filaments. The calcium (Ca^{++}) combines with the troponin molecules, causing the troponin-tropomyosin complex to expose actin so that it can attach with myosin. While the actin and myosin molecules are attached, the head of the myosin molecule can bend as ATP is used and the actin molecule is pulled past the myosin molecule. Thus, a tiny section of the muscle cell shortens. When one of our muscles contracts, thousands of such interactions take place within a tiny portion of a muscle cell, and many cells within a muscle all contract at the same time.

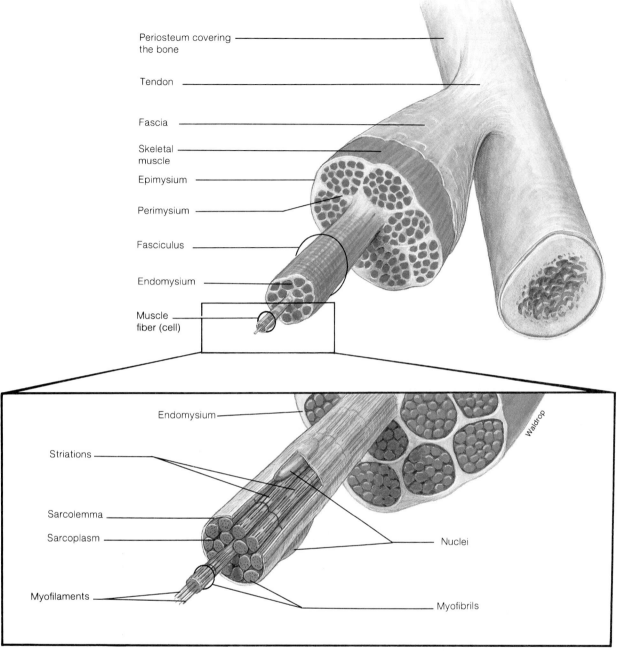

Periosteum covering the bone
Tendon
Fascia
Skeletal muscle
Epimysium
Perimysium
Fasciculus
Endomysium
Muscle fiber (cell)

Endomysium
Striations
Sarcolemma
Sarcoplasm
Myofilaments
Nuclei
Myofibrils

Waldrop

Figure 19.2 Microanatomy of a Muscle Muscles are made up of cells that contain bundles known as myofibrils. The myofibrils are composed of myofilaments of two different kinds, thick myofilaments composed of myosin, and thin myofilaments composed of actin, tropomyosin, and troponin.

shortens and thickens (figure 19.2). When this happens frequently, the muscle cells respond by producing additional protein filaments. If we add weight or resistance (through exercise or weight lifting) to the muscle, protein synthesis is stimulated and additional actin- and myosin-containing filaments are produced.

It is important to recognize that when a muscle is stimulated it can only contract; it cannot actively lengthen. Lengthening occurs passively when the muscle relaxes after the stimulation and contraction. This return can be speeded up by applying a force that will stretch the muscle. In the body, this is accomplished by the use of *antagonistic* muscles. Antagonistic muscles are groups whose actions oppose one another. As a muscle contracts, it pulls on its antagonistic partner to help stretch it out and get it back into the lengthened, ready-to-contract state. For every muscle's action there is an antagonistic muscle that has the opposite action. For example, the biceps muscle causes the arm to flex (bend) as the muscle shortens. The contraction of its antagonist, the triceps muscle on the back of the arm, causes the arm to extend (straighten) and at the same time stretches the unstimulated biceps muscle (figure 19.3).

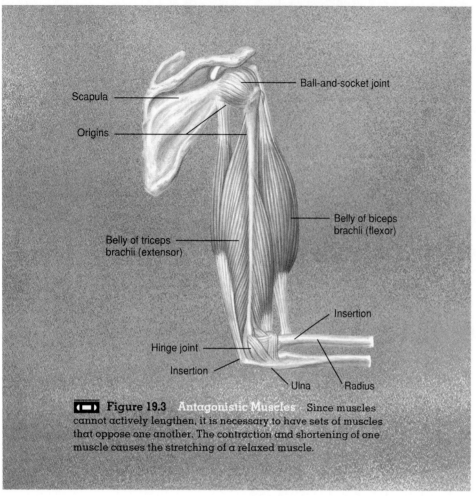

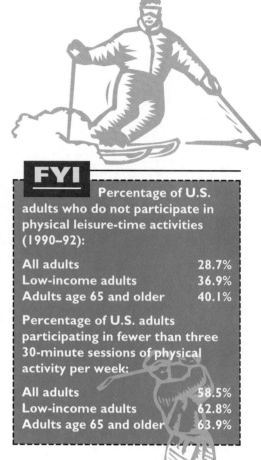

Figure 19.3 Antagonistic Muscles Since muscles cannot actively lengthen, it is necessary to have sets of muscles that oppose one another. The contraction and shortening of one muscle causes the stretching of a relaxed muscle.

Labels in figure: Scapula, Origins, Ball-and-socket joint, Belly of biceps brachii (flexor), Belly of triceps brachii (extensor), Insertion, Hinge joint, Insertion, Ulna, Radius

Skeletal Muscle

There are three major types of muscle: skeletal, smooth, and cardiac. These types differ from one another in several ways. *Skeletal muscle* is voluntary muscle: It is under the control of the nervous system. The nervous system sends a message to skeletal muscles that tells them to contract, moving the legs, fingers, and other parts of the body.

This does not mean that you must make a conscious decision every time you want to move a muscle. Many of the movements we make are learned initially but become automatic as a result of practice. For example, walking, swimming, or riding a bicycle required a great amount of practice originally, but now you perform these movements without thinking about them. You may not be aware or conscious of the fact that your nervous system has stimulated these contractions, but the impulse has been sent from your brain to these muscles. Since the contractions have originated from your nervous system they are, however, still considered to be voluntary actions.

In addition, some voluntary skeletal muscle contractions are the result of reflex reactions and are not learned. When your eyes blink or you shiver, your muscles are being stimulated through a complex nervous system response beginning with the reception of a stimulus. A nerve impulse is generated that travels to your central nervous system and returns to the muscles, which are stimulated to contract. Even though these kinds of contractions are not deliberate, they are still considered voluntary because they are initiated by the nervous system.

Skeletal muscles are constantly bombarded with nerve impulses that result in repeated contractions of differing strength. Many nerve cells run to each muscle, and each

nerve stimulates a specific set of muscle cells. This set of muscle cells and its stimulating nerve cell is called a **motor unit.** Since each muscle consists of many motor units, it is possible to stimulate a few, many, or nearly all at any one time. The number of motor units stimulated determines how forcefully the muscle contracts. This allows a single set of muscles to serve a wide variety of functions. For example, the same muscles of the arms and shoulders that are used to gracefully play a piano can also be used to lift the piano into a moving van.

Individual skeletal muscles are able to contract quickly, but their cells cannot remain contracted for long periods. When you contract a muscle mass for a minute or so, the nerve impulses arriving at the thousands of motor units constantly shift from one motor unit to another. This enables the entire muscle to remain contracted even though individual cells cannot stay in a contracted state.

If the nerves going to a muscle are destroyed, the muscle becomes paralyzed and begins to shrink. Regular nervous stimulation of

skeletal muscle is necessary for muscle to maintain size and strength. Any kind of prolonged inactivity leads to the degeneration of muscles. Muscle maintenance is one of the primary functions of physical therapy and a benefit of regular exercise.

Voluntary Muscles and Bladder Control

Infants have no control over when they empty their bladders. There are two rings of muscles that close the opening of the urinary bladder. When urine accumulates in the urinary bladder and pressure is exerted on the rings of muscles, the muscles relax and the baby wets its diaper. When parents begin "potty training," they begin a period during which the infant learns to gain conscious control over one of the two sets of muscles that control urination.

The two rings or sets of muscles are of two types. One is composed of smooth muscle fibers and contracts on an involuntary basis while the second ring is skeletal muscle and is under voluntary control. The reason the baby does not wet its diaper continuously is because the smooth muscle ring contracts after the bladder is emptied. It stretches when enough urine accumulates in the bladder to stimulate the muscle to relax and the bladder empties. During potty training, the infant is taught to gain voluntary control over the skeletal muscle ring. It is this second ring of muscle that becomes the "controlling" ring. It is important for parents to recognize that until the child matures sufficiently, he or she will not be able to control this ring of skeletal muscle.

Maintaining this voluntary control is not always easy or possible. In some people, emotional stress, certain foods (e.g., aspartame in diet foods and drinks), surgery (e.g., for prostate cancer), traumatic baby deliveries, and allergies (e.g., milk) can interfere with this control mechanism and result in bed wetting or "accidents." Even pregnancy, laughing, and coughing can disrupt this nervous system control and lead to embarrassing situations. The leading cause of *incontinence,* or lack of control over the urinary bladder, is the nervous system deterioration associated with aging.

Control of Urinary Bladder

At the point where the urethra attaches to the urinary bladder, there are two rings of muscles. One is composed of smooth muscle fibers and contracts on an involuntary basis while the second ring is voluntary.

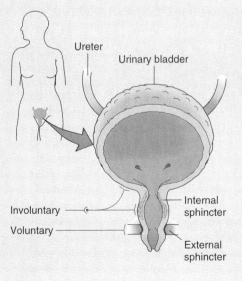

Ureter

Urinary bladder

Involuntary

Voluntary

Internal sphincter

External sphincter

Movement of your body by your muscles requires energy. This energy is made available to you from the food you eat by a process known as *cellular respiration.* When you are working normally, your blood carries oxygen to the muscle cells, and the mitochondria in these cells release the energy from the foods you eat. The waste products of this respiration are carbon dioxide and water. When you are working harder your lungs may not be able to get sufficient oxygen, or your blood may not be able to carry enough oxygen to the muscles. When this happens, lactic acid is produced as a waste product. When lactic acid builds up in your muscle cells, fatigue and pain result.

Athletes have an interest in increasing the strength and endurance of their muscles. To do this, some athletes have resorted to drugs, such as anabolic steroids, which have recently received a great deal of attention. Anabolic steroids are synthetic androgens that stimulate the production of additional muscle. The male sex hormones, testosterone and other androgens, are naturally occurring anabolic steroids that are partly responsible for the differences in body shapes of males and females. Because males produce more of these hormones, they have a lower proportion of fat and they have more muscle protein. This results in males typically having greater strength and greater physical endurance than females. Athletes such as bodybuilders and weight lifters are interested in just these characteristics, so they have been tempted to use artificial hormonelike drugs.

Smooth Muscle

Smooth muscles make up the walls of muscular internal organs, such as the intestinal tract, bladder, blood vessels, and reproductive organs. These are involuntary muscles; they can contract without receiving direct messages from the nervous system, although nervous stimulation can modify their contraction. They have the property of contracting slowly in response to hormones, calcium levels, special cells called pacemaker cells, certain chemicals, or as the result of being stretched. This ability can be demonstrated when researchers remove portions of the digestive system or uterus from experimental animals. If these muscular organs are kept moist, they will go through cycles of contraction without any stimulation from nerve cells. Such contractions help move food along the digestive tract from one portion to the next. Food moving through the digestive

Drug Misuse and Abuse in Sports

There are several types of drugs that are likely to be misused or abused by individuals who compete in sporting events. The table on the right lists some of them with their positive and negative effects.

Anabolic steroids	Increased strength and endurance. Increased proportion of protein and decreased relative amounts of fat in the body. Increased production of hemoglobin, which increases oxygenation of the cells.

	Liver cancer.
	Acne.
	Kidney damage.
	Increased risk of heart disease.
	Muscle spasms.
	Increased cholesterol.
	Stunted growth in children and adolescents.
	Behavior changes: irritability and aggression, hallucinations, manic episodes, major depression, and mood swings.
	Sterility.
	May be addictive.
	Females only:
	    Development of facial hair.
	    Deepening of the voice.
	    Atrophy of the breasts and uterus.
	    Enlargement of the clitoris.
	    Irregularities of menstruation.
	Males only:
	    Shrivelled testes.
	    Diminished hormone secretion and sperm production.
	    Baldness.
	    Excessive development of breast tissue.

Psychomotor stimulants	Amphetamines and related substances (speed) reduce fatigue, reportedly increase alertness and athletic performance.
	Increased aggression, elevated heart and respiratory rate, interference with sleep and possible psychological trauma.

Opium-type drugs
    Morphine
    Codeine
    Heroin
    Demerol®
	Mask pain and thus allow injured athletes to continue to compete.

	Illegal and addictive.
Analgesics and anti-inflammatory drugs Aspirin	Mask pain and thus allow injured athletes to continue to compete.
	Gastrointestinal upset, headache, drowsiness.
Adrenocorticosteroids Prednisone	Mask pain and thus allow injured athletes to continue to compete.
	Gastrointestinal upset, headache, drowsiness.
Local anesthetics Novocain®	Reduce pain due to an isolated injury.
	Further damage to the injured structure can result with continued athletic activity.
Sedative tranquilizers	Reduce tension and stress.
	Addiction, slow breathing rate, interference with reaction time and thus increased risk of injury.

Muscle Pain and Exercise

When we exercise we sometimes experience soreness of a muscle. This might be the result of increased metabolism within the muscle due to the increase in the amount of work we are asking of it. Some experts think that the soreness is due to the increase in waste products, such as lactic acid, which accumulate as a result of the muscle cell functioning with inadequate oxygen. This soreness is likely to go away if you stop the strenuous exercise. However, moderate exercise of the muscle encourages the blood to supply the muscle with oxygen and to remove the waste products that have accumulated. Small tears or other damage to muscles also cause pain and soreness.

Pumping Up

After a short period of exercise, you may notice an increase in the size of the muscles that have been exercised. This temporary increase in size is called being "pumped up." The exercise results in an increase in muscle fluids. In time the blood carries these fluids away, and the muscle returns to normal size. This is the reason you don't stay pumped up.

Muscle Cramps

A cramp is a painful contraction of a muscle. The cramping may be caused by a variety of things, such as low blood sugar levels or the loss of calcium, potassium, or sodium from the muscle cells. When this occurs, the muscle is stimulated to contract as if it had received a nerve impulse. During a cramp the muscle is likely to stay in a highly contracted state for several minutes. To relieve the cramping you can gently massage the muscle. This helps it relax, as does the application of mild heat and resting the affected muscle. However, be careful not to damage the muscle or connective tissue by forcefully extending the muscle, that is, do not press hard on the muscle. To control cramping, you can eat foods containing sodium, potassium, and calcium. Sports drinks are available that can replace the sodium, potassium, calcium, and sugar as well as other ions lost during exercise.

Strains

A strain (muscle pull) is the result of damage to the muscle itself or to the connective tissue (tendon) associated with the muscle. A strain could be the result of over-extending or damaging the muscle during exercise. The muscle becomes painfully inflamed. Strains should be treated with cold packs immediately following the injury.

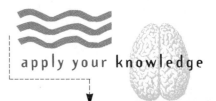

apply your knowledge

Which Type of Exercise Do You Do?

Aerobic exercise occurs when the muscles being contracted are supplied with sufficient oxygen to continue aerobic cellular respiration (see chapter 3):

$$C_6H_{12}O_6 + 6O_2 \rightarrow 6CO_2 + 6H_2O + Energy$$

This type of exercise involves long periods of activity with elevated breathing and heart rate. It results in strengthened chest muscles, which enable more complete exchange of air during breathing. It also improves the strength of the heart, enabling it to pump more efficiently. Flow of blood to the muscles also improves. All these changes increase endurance.

Anaerobic exercise takes place when insufficient amounts of oxygen reach the contracting muscle cells and they shift to anaerobic cellular respiration:

$$C_6H_{12}O_6 \rightarrow lactic\ acid + less\ Energy$$

The buildup of lactic acid results in muscle pain and eventually prevents further contraction. Anaerobic exercise involves explosive bouts of activity, as in sprints or jumping. This kind of exercise increases muscle strength but does little to improve endurance.

Resistance exercise occurs when muscles contract against an object that does not allow the muscles to move that object. This type of exercise does not improve the ability of your body to deliver oxygen to your muscles, nor does it increase your endurance. However, it does stimulate your muscle cells to manufacture more contractile protein fibers.

Make a list of your activities that would be considered aerobic, anaerobic, or resistance exercise. Which activities would result in weight reduction? Weight gain? Improved cardiovascular fitness?

system constantly stretches the smooth muscles in the walls of the intestinal tract. This stretching causes the muscles to contract sequentially. Hormones produced by the portion of the small intestine known as the duodenum also influence the smooth muscles of the digestive system to contract.

Smooth muscle also has the ability to stay contracted for long periods without becoming fatigued. Many kinds of smooth muscle, such as the muscle of the uterus, also respond to the presence of hormones, specifically the hormone **oxytocin,** which is released from the posterior pituitary and causes strong contractions of the uterus during labor and childbirth.

Cardiac Muscle

Cardiac muscle is the muscle that makes up the heart. It has the ability to contract rapidly like skeletal muscle, but does not require nervous stimulation to do so. Pacemaker cells are responsible for stimulating cardiac muscle. Nervous stimulation can, however, cause the heart to speed or slow its rate of contraction. Hormones, such as epinephrine (adrenalin) and norepinephrine, also influence the heart by increasing its rate and strength of contraction. Cardiac muscle is unable to stay contracted. It will contract quickly but must have a short period of relaxation before it will be able to contract again. This is seen in the continuous rhythmic pumping function of the heart. Table 19.1 summarizes the differences between skeletal, smooth, and cardiac muscle.

Healthy Hearts and the Circulatory System

Muscles remain healthy with continual use, the proper nutrients, and removal of waste products as they accumulate (figure 19.4). Transporting nutrients and waste products is the responsibility of the circulatory system.

FYI
The human body contains 100,000 km (60,000 miles) of blood vessels.

table 19.1

Characteristics of Different Kinds of Muscle

Kind of Muscle	Stimulus	Length of Contraction	Rapidity of Response
Skeletal	Nervous	Short, tires quickly	Most rapid
Smooth	1. Self-stimulated	Long, doesn't tire quickly	Slow
	2. Also responds to nervous and endocrine systems		
Cardiac	1. Self-stimulated	Short, cannot stay contracted	Rapid
	2. Also responds to nervous and endocrine systems		

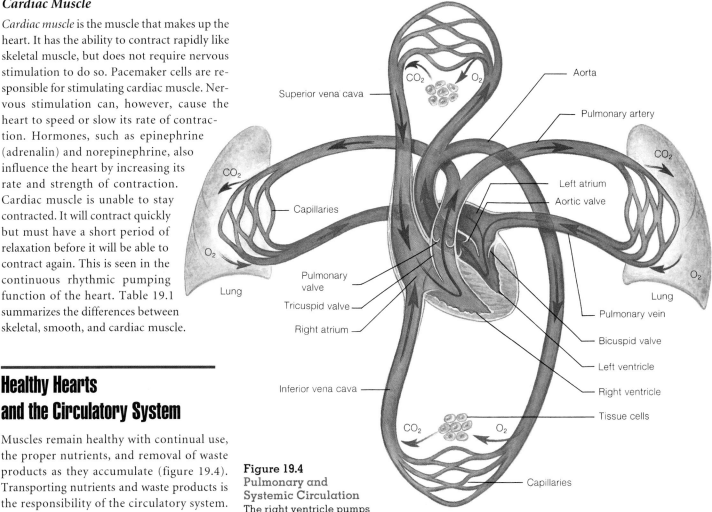

Figure 19.4
Pulmonary and Systemic Circulation
The right ventricle pumps blood that is poor in oxygen to the two lungs where the blood receives oxygen and turns bright red. The blood is then returned to the left atrium by way of four pulmonary veins. This part of the circulatory system is known as the pulmonary circulation. The left ventricle pumps oxygen-rich blood to all parts of the body except the lungs. This blood returns to the right atrium, depleted of its oxygen, by way of the superior vena cava from the head region and the inferior vena cava from the rest of the body. This portion of the circulatory system is known as the systemic circulation.

Hemoglobin and Carbon Monoxide Poisoning

Hemoglobin is capable of picking up the molecule carbon monoxide (CO), a waste product of many types of burning that can be found in car exhaust, fireplace exhaust, tobacco smoke, and other sources. When CO is inhaled, it becomes attached to the hemoglobin molecules in red blood cells in much the same way that oxygen (O_2) combines with hemoglobin. However, the hemoglobin does not readily release the CO as it would the O_2. As more and more red blood cells pick up CO, the number of red blood cells capable of transporting O_2 decreases and may reach such low levels that it is fatal. This is what occurs when a person is inside a building with inadequate ventilation to remove exhaust gases (CO). As the oxygen level in the brain decreases, the person "falls asleep" (i.e., brain functions slow and eventually stop) and dies.

Your blood carries materials through a series of blood vessels and distributes oxygen and food to individual cells. At the same time, it picks up waste products and carries them away. The heart is primarily responsible for the movement of blood through the blood vessels (table 19.2).

The Nature of Blood

Blood is a fluid that consists in part of a watery medium called **plasma.** This fluid plasma contains many kinds of dissolved molecules and larger suspended cellular components. (You have probably seen plasma if you have had a scrape or burn that released clear fluid but did not bleed.) The primary function of the blood is to transport molecules, cells, and heat from one place to another. The major kinds of molecules that are distributed by the blood are respiratory gases (oxygen and carbon dioxide), nutrients of various kinds, waste products, and chemical messengers (hormones). Blood has special characteristics that allow it to distribute respiratory gases efficiently. Although little oxygen is carried as free, dissolved oxygen in the plasma, *red blood cells* (*RBCs*) contain **hemoglobin.** Hemoglobin is an iron-containing molecule that can pick up and transport oxygen (O_2).

Table 19.3 lists the variety of cells found in blood. While the red, hemoglobin-containing red blood cells serve in the transport of oxygen, the *white blood cells* (*WBCs*) carried in the blood defend against harmful agents. These cells act like a police force that helps the body resist many diseases. They constitute the core

table 19.2

Structures in the Circulatory System

Heart	A pump that transports the blood through the body.
Arteries	The largest blood vessels that distribute blood to the body tissues. They flow into the smaller arterioles.
Arterioles	Smaller vessels carrying blood from the arteries into the capillaries.
Capillaries	Smallest thin-walled tubes that allow gas and nutrient exchange with individual cells throughout the body. Capillaries are located between arterioles and venules.
Venules	Small collecting tubes into which the capillaries empty their contents to be returned to the heart. Venule blood flows into the veins.
Veins	Large collecting tubes with one-way valves. These valves prevent back flow and encourage return of the blood to the heart.
Blood	Tissue composed of fluid and various types of cells that is pumped through the circulatory system.

table 19.3

The Composition of Blood

Component	Quantity Present	
Plasma	55%	
Water		91.5%
Protein		7.0%
Other materials		1.5%
Cellular material	45%	
Red blood cells (erythrocytes)		4.3–5.8 million/mm³
White blood cells (leukocytes)		5–9 thousand/mm³
Lymphocytes		25%–30% of white cells present
Monocytes		3%–7% of white cells present
Neutrophils		57%–67% of white cells present
Eosinophils		1%–3% of white cells present
Basophils		less than 1% of white cells present
Platelets		130–360 thousand/mm³

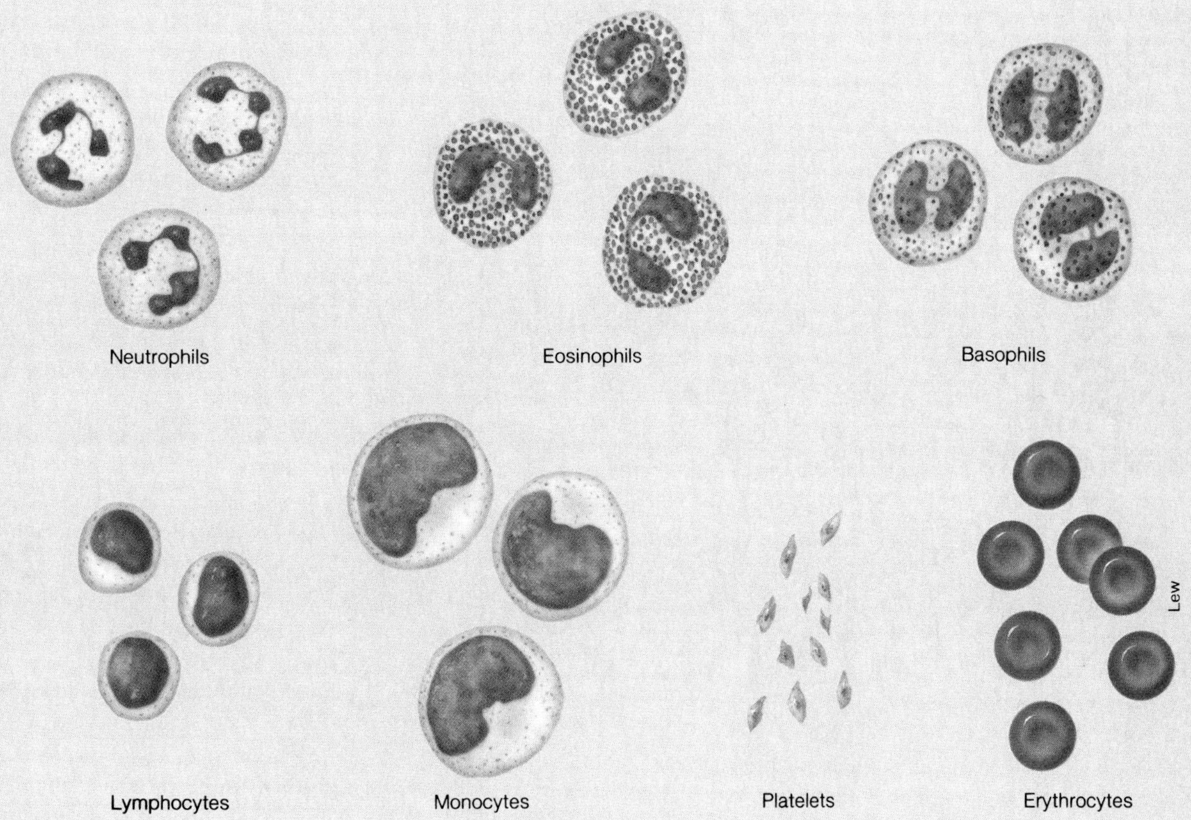

Neutrophils

Eosinophils

Basophils

Lymphocytes

Monocytes

Platelets

Erythrocytes

BIO *feature*

Blood Tests

A blood test can tell a physician many things about how a patient's body is functioning. Some of the most frequent tests performed on a patient's blood are listed here.

ABO typing

This test identifies the blood type of the person as either A, B, O, or AB. Blood typing is useful should the patient require a blood transfusion, since the recipient must receive a matching type in order for the transplantation of blood to be successful. If the wrong type is transfused, the blood cells could clump in the circulatory system, clogging the vessels, or the body could destroy the transfused blood through a rejection response.

CBC (complete blood count)

White blood cell count (WBC)
Abnormal white blood cell numbers may indicate that the patient's body is fighting infection.

Differential white cell count (Diff)
This more accurately indicates which types of white blood cells are circulating. Elevation in the count of a particular type indicates certain types of infection or abnormalities.

Red blood count (RBC)
Low RBC counts could indicate anemia, while elevated counts could indicate abnormalities in the bone marrow, the blood forming organ.

Hematocrit (HCT) (packed cell volume)
A decreased amount may indicate such conditions as anemia, leukemia, iron-deficiency anemia, or cirrhosis of the liver. An increase may indicate severe dehydration or shock.

Hemoglobin (Hgb)
A decrease could indicate anemia, hyperthyroidism, cirrhosis of the liver, or severe hemorrhage. An increase might indicate chronic obstructive pulmonary disease, congestive heart failure, or severe burns.

Red blood cell indices
These tests focus on the size and hemoglobin content of the red blood cells. Changes in these values could indicate different kinds of anemias.

Stained red cell examination (blood smear)
This enables the lab technician to investigate the size, shape, and other characteristics of red blood cells. Diseases such as sickle-cell anemia may be diagnosed in this manner.

Platelet count
Platelets are involved in blood coagulation. Changes in the amount of platelets could indicate an inability to form blood clots or an abnormally high rate of clot formation.

General biochemistry profile (12 channel test)

Abnormal amounts of any of the following materials in the blood, along with other symptoms, may be helpful in making a diagnosis.

total protein (TP)
albumin
calcium (Ca^{++})
inorganic phosphorus
cholesterol
glucose
blood urea nitrogen (BUN)
uric acid
creatinine
total bilirubin
alkaline phosphatase
aspartate transaminase (AST)

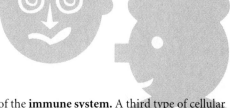
of the **immune system.** A third type of cellular material in the blood are the *platelets.* The platelets are partially responsible for the formation of blood clots.

The plasma also carries nutrient molecules from the gut to other locations where they are modified, metabolized, or incorporated into cell structures. Amino acids and simple sugars are dissolved in the blood and carried through the body. Lipids, which are not water-soluble, are carried as suspended particles called lipoproteins. Most lipids do not enter the bloodstream directly from the gut but are carried to the bloodstream by the lymphatic system. Other organs, like the liver, manufacture or modify molecules for use elsewhere; therefore, they must constantly receive raw materials and distribute their products to the cells that need them through the flow of blood.

Heat is also transported by the blood. Just as your car engine generates heat when it runs, metabolic activities generate heat that must be lost from the body. To handle excess heat, your car has a radiator filled with fluid. The radiator has folds that provide a large surface over which cool air can pass. Your body eliminates heat in a similar way! Blood is shunted to the surface of the body through many small capillaries, where heat can be removed when cooler air passes over the skin. In addition, humans have the ability to sweat. The evaporation of sweat from the body surface removes heat and helps to maintain constant body temperature.

Vigorous exercise produces a large amount of heat. To eliminate this excess energy, the body increases the diameter of the capillaries, resulting in more blood being pumped closer to the surface of the skin. When this happens, the heat energy is lost, helping to maintain the inner, core body temperature in the normal range. You can see this change in circulation in the red face of a person who is working out. If the body is losing heat too rapidly, the diameter of the capillaries is reduced and blood flow is shunted away from the skin, conserving the heat energy and preventing its loss.

The Heart

In order for a fluid to flow through a tube, there must be a pressure difference between the two ends of the tube—high pressure at one end and low pressure at the other end. Water flows through pipes because it is under pressure. Because the pressure is higher behind a faucet than at the spout, water flows from the spout when the faucet is opened. The circulatory system can be analyzed from the same point of view. The heart is a muscular pump that provides the high pressure necessary to push the blood throughout the body. It must continue its cycle of contraction and relaxation, or blood stops flowing and body cells are unable to satisfy their immediate needs. Some

How White Blood Cells Protect

Some white blood cells (WBCs) are capable of a process called **phagocytosis,** in which cells wrap around a particle and engulf it. When phagocytosis occurs, the material to be engulfed touches the surface of the cell and causes a portion of its outer membrane to be indented. The indented cell membrane is pinched off inside the cell to form a sac containing the engulfed material. This sac, composed of a single membrane, is called a *vacuole*. Once inside the cell, the material inside the vacuole can be digested, and when the membrane of the vacuole is broken down, its contents will be released into the cytoplasm of the cell.

When harmful microorganisms (e.g., bacteria, viruses, fungi), cancer cells, or toxic molecules enter the body, WBCs (1) recognize, (2) boost their abilities to engulf, (3) move toward, (4) engulf, and (5) destroy the problem causers. While most phagocytes can move from the bloodstream into the surrounding tissue, *monocytes* undergo such a striking increase in size that they are given a different name following their differentiation— *macrophages*. Macrophages can be found throughout the body and are the most active of the phagocytes.

The other white cells, *lymphocytes,* work with phagocytes to provide

protection. Two major types are *T-lymphocytes* (*T-cells*) and *B-lymphocytes* (*B-cells*). T-cells are central to a complex set of events called the *cell-mediated immune response*. This highly complex response involves the release of chemical messengers that coordinate the immune response by increasing the population of T- and B-cells. It also stimulates B-cell and macrophage activities. Some T-cells are capable of directly killing dangerous cells by destroying their cell membranes.

B-cells are the source of protein molecules known as *antibodies* or *immunoglobulins*. They are central to a series of events called *antibody-mediated immunity*. Antibodies are released into the body by B-cells when the B-cells are stimulated by the presence of dangerous agents. The antibodies are specifically constructed so that they will bind to particular agents. Agents that stimulate the production of antibodies and then combine with the antibodies are called *antigens* or *immunogens*. Examples include viruses, bacteria, fungi, poisons, transplanted tissues, and cancer cells. The antibody-antigen combination renders the antigen harmless by destroying its poisonous properties, preventing it from infecting human cells, or stimulating phagocytosis.

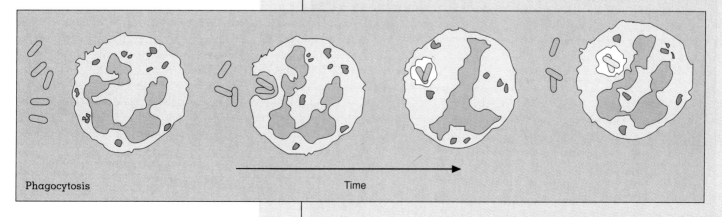

Phagocytosis   Time

BIO feature

Blood Typing

When you get a transfusion or give blood to a blood bank for transfusions to other people, it is important to know what kinds of proteins are located on the surface of the red blood cell membranes, because antibodies in the plasma will attack cells that have different proteins.

Type A blood has the "A" protein on the red blood cell (RBC) surface and antibodies against "B" proteins in its plasma.

Type B blood has the "B" protein on the RBC surface and antibodies against "A" proteins in its plasma.

Type AB blood has both the "A" and the "B" proteins on the RBC surface and does not have any antibodies against "A" or "B" proteins in its plasma.

Type O blood has neither "A" nor "B" proteins on the RBC surface but has antibodies against both "A" and "B" proteins in its plasma.

People who have type O blood are often called universal donors. Their red blood cells do not have A or B proteins, so the cells are not attacked by plasma antibodies. Furthermore, they are donating such small amounts of antibody against the recipients' red blood cells that even if some of the recipients' red blood cells are destroyed, there will be enough left unharmed for a successful transfusion.

If you have type A blood you should not get a transfusion of type B or AB blood. When a type A person receives B or AB blood, the antibodies in that person's plasma will damage the donor RBCs. Similarly, type B people should not receive type A or AB blood. Type AB people can receive blood from anyone; thus they are called universal recipients.

cells, such as brain cells, are extremely sensitive to having their flow of blood interrupted because they require a constant supply of glucose and oxygen. Others, such as muscle cells or skin cells, are better able to withstand temporary interruptions of blood flow.

The **heart** is composed of four chambers. Two of these chambers, the **right** and **left atria,** are relatively thin-walled, weak structures that collect blood from the major veins and empty it into the larger, more muscular ventricles (figure 19.5). Most of the flow of blood from the atria to the ventricles is caused by the lowered pressure produced within the ventricles as they relax. The contraction of the thin-walled atria assists in emptying them more completely.

The walls of the **right** and **left ventricles** are composed of powerful muscles whose contraction forces blood to flow through the arteries to all parts of the body. The valves between the atria and ventricles, known as *atrioventricu-*

lar valves, are important one-way valves that allow the blood to flow from the atria to the ventricles but prevent flow in the opposite direction. The valve between the left atrium and left ventricle is also known as the mitral or bicuspid valve. Similarly, the valve between the right atrium and right ventricle is known as the tricuspid valve. There are valves in the aorta and pulmonary artery, known as *semilunar valves.*

The **aorta** is the large artery that carries blood from the left ventricle to the body, and the **pulmonary artery** carries blood from the right ventricle to the lungs. The semilunar valves prevent blood from flowing back into the ventricles. If the atrioventricular or semilunar valves are damaged or function improperly, the efficiency of the heart as a pump is diminished, and the person may develop an enlarged heart or other symptoms. Malfunctioning heart valves are often diagnosed because they cause abnormal sounds as the blood passes through them. These sounds are referred to as heart murmurs. Similarly, if the ventricles are weakened because of infection, damage from a heart attack, or lack of exercise, the pumping efficiency of the heart is reduced and the person develops symptoms that may include chest pain, shortness of breath, or fatigue. The pain is caused by the heart muscle not getting sufficient blood to satisfy its needs. If this persists, the portion of the heart muscle not receiving an adequate supply of blood will die.

The Lymphatic System **Considerable amounts of water and dissolved materials leak through the walls of the capillaries. This liquid is known as lymph.** Lymph bathes the cells, but it must eventually be returned to the circulatory system by lymph vessels or tissue swelling will occur. Return is accomplished by the **lymphatic system,** a collection of thin-walled tubes that branch throughout the body. These tubes collect lymph that is filtered from the circulatory system and ultimately empty it into major blood vessels near the heart.

Some of this leakage through the capillary walls is normal, but the flow is subject to changes in pressure inside the capillaries as well as changes in pressure in the surrounding tissues. Changes in the permeability of the capillary wall will also affect flow through the wall. If pressure inside the capillary increases, more fluid may leak from the capillaries into the tissues and cause swelling. This swelling is called edema, and it is common in circulatory disorders. Another cause of edema is an increase in the permeability of capillaries. It is commonly associated with injury to a part of the body: a sprained ankle or a smashed thumb are examples.

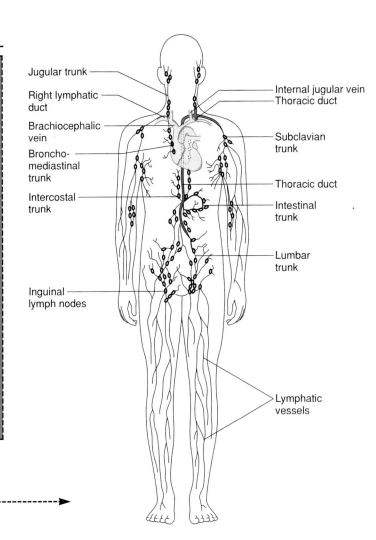

Figure 19.5 Anatomy of the Heart.

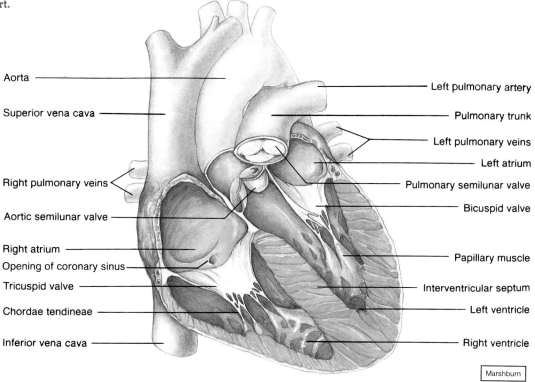

Shortness of breath and fatigue have the same cause, since the heart is not able to pump adequate amounts of blood to the lungs, muscles, and other parts of the body.

The right and left sides of the heart have slightly different jobs, since they pump blood to different parts of the body. The right side of the heart receives blood from the general body and pumps it through the pulmonary arteries to the lungs, where exchange of oxygen and carbon dioxide takes place. This is called **pulmonary circulation.** The larger, more powerful left side of the heart receives blood from the lungs and delivers it through the aorta to all parts of the body. This is known as **systemic circulation.** The systemic circulation is responsible for gas, nutrient, and waste exchange in all parts of the body except the lungs (figure 19.6).

Blood Pressure in the Arteries and Veins

Arteries and veins are the tubes that transport blood from one place to another within the body. Figure 19.7 compares the structure and function of arteries and veins. **Arteries** carry blood away from the heart; the blood is under considerable pressure from the contraction of the ventricles. A typical pressure that would be recorded in a large artery while the heart is contracting would be about 120 millimeters of mercury. The pressure that would be recorded while the heart is not contracting would be about 80 millimeters of mercury. A blood pressure reading includes both of these numbers and would be recorded as 120/80.

The walls of arteries are relatively thick and muscular. Healthy arteries have the ability

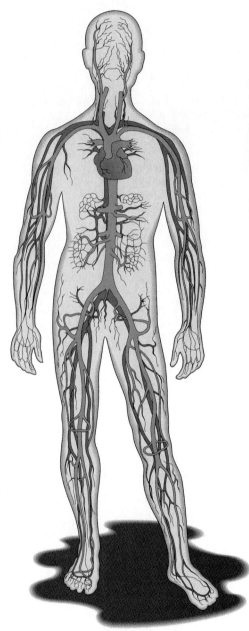

Figure 19.6 Systemic Circulation

to expand as blood is pumped into them and return to normal as the pressure drops. This lessens the peak pressure within the arteries and reduces the likelihood that they will burst. If arteries become hardened and less resilient, the peak blood pressure rises and the arteries are more likely to rupture. The elastic nature of the arteries is also responsible for assisting the flow of blood. When the arteries return to

normal from their stretched condition, they give a little push to the blood that is flowing through them.

Arteries branch into smaller and smaller blood vessels as the blood is distributed from the large aorta to millions of tiny capillaries. **Capillaries** are thin-walled tubes that allow gas and nutrient exchange with individual cells. Some of the smaller arteries may contract or relax to regulate the flow of blood to specific parts of the body. Major parts of the body that receive differing amounts of blood, depending on need, are the digestive system, muscles, and skin.

Veins collect blood from the capillaries and return it to the heart. The pressure in these blood vessels is very low. Some of the largest veins may have a blood pressure of zero. The walls of veins are not as muscular as those of arteries. Because of the low pressure, veins must have valves that prevent the blood from flowing backward, away from the heart.

Since pressure in veins is so low, muscular movements of the body are important in helping to return blood to the heart. When muscles of the body contract, they compress nearby veins, and this pressure pushes blood along in the veins. Because the valves allow blood to flow only toward the heart, this activity acts as an additional pump to help return blood to the heart. People who sit or stand for long periods without using their leg muscles tend to have a considerable amount of blood pool in the veins of their legs and lower body. Thus, less blood may be available to go to the brain and the person may faint.

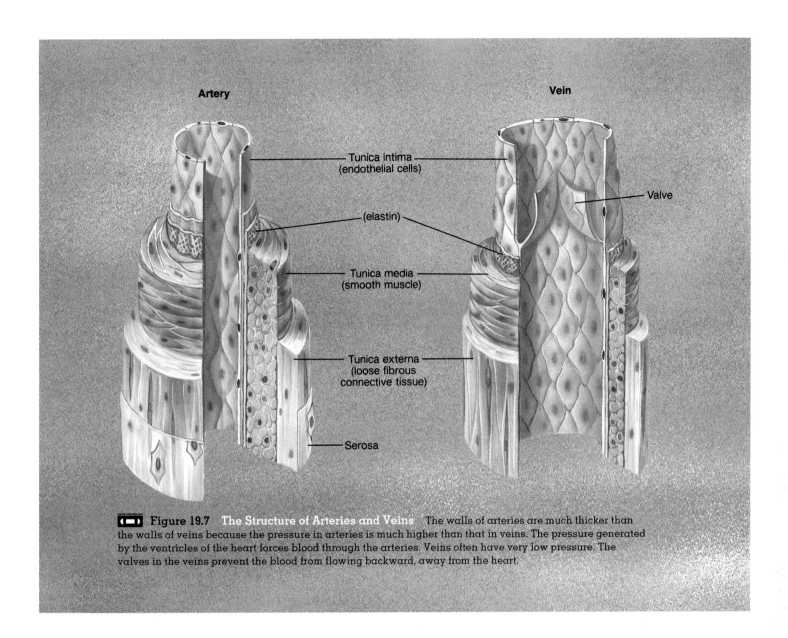

Figure 19.7 **The Structure of Arteries and Veins** The walls of arteries are much thicker than the walls of veins because the pressure in arteries is much higher than that in veins. The pressure generated by the ventricles of the heart forces blood through the arteries. Veins often have very low pressure. The valves in the veins prevent the blood from flowing backward, away from the heart.

Labels in figure:
- Artery
- Vein
- Tunica intima (endothelial cells)
- (elastin)
- Valve
- Tunica media (smooth muscle)
- Tunica externa (loose fibrous connective tissue)
- Serosa

Gas Exchange and the Respiratory System

One of the most important jobs of the circulatory system is to carry oxygen to the cells so that they can release energy from food. The cells then use the energy to do their jobs, and the circulatory system is responsible for removing the waste products generated. This gas exchange occurs twice in an organism. As the blood flows past a cell such as a muscle cell, oxygen diffuses from the blood into the muscle; at the same time carbon dioxide diffuses from the muscle cell into the blood. This is the first exchange. As circulation continues and the blood flows through the lung, the second exchange takes place. This time the carbon dioxide from the blood diffuses into the lung spaces, and oxygen diffuses from the lungs into the blood. Red blood cells are involved in the transport of both oxygen and carbon dioxide. Hemoglobin molecules in red blood cells combine with oxygen to allow for large amounts of oxygen to be carried. When the blood reaches the tissues, the oxygen is released from the hemoglobin.

Red blood cells are involved in the transport of carbon dioxide (CO_2) in two ways. Some is carried on hemoglobin molecules. However, most of the carbon dioxide is carried through the body in the form of bicarbonate ions (HCO_3^-) in the blood plasma. Red blood cells contain an enzyme that speeds the conversion of CO_2 to HCO_3^-.

The amount of carbon dioxide in the blood is the primary mechanism that causes changes in the rate and depth of a person's breathing. Carbon dioxide is a waste product of aerobic cellular respiration that becomes

FYI

The air that you inhale typically contains 21% oxygen (O_2) and .04% carbon dioxide (CO_2). Exhaled air contains 16% oxygen and 4.5% carbon dioxide.

Stroke and Heart Attack

It is estimated that in the United States, by the age of 60, one out of every five individuals will have suffered a heart attack. A heart attack is the result of an interruption in the blood flow to the cells in the heart. When the arteries that supply the heart (coronary arteries) become clogged with fatty deposits or a blood clot, blood flow is temporarily interrupted. Severe chest pain, dizziness, and nausea are symptoms of a temporary blockage. Much more severe is a longer lasting blockage that could result in the death of some of the heart muscle cells.

Treatment of this blockage might include dissolving a blood clot using drugs, or surgical replacement of parts of the coronary arteries, known as a heart bypass. Heart disease is responsible for 700,000 deaths annually in the United States.

A stroke is similar to a heart attack, but blockage occurs in the arteries that service parts of the brain. If the blockage is temporary the symptoms are dizziness, double vision, or inability to speak, but if brain cells die due to lack of oxygen, permanent loss of function or death may result.

Coronary Artery Bypass

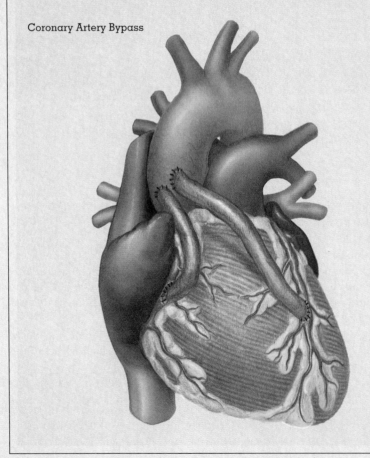

Nelson

toxic in high quantities because it combines with water to form carbonic acid:

$$CO_2 + H_2O \rightarrow H_2CO_3$$

Exercising causes an increase in the amount of carbon dioxide in the blood because muscles are burning glucose more rapidly. This makes the blood more acidic (lowers the pH). Certain brain cells are sensitive to changes in blood pH. When they sense a lower blood pH, nerve impulses are sent more frequently to the muscles responsible for breathing. These muscles contract more rapidly and more forcefully, resulting in more rapid, deeper breathing. Since more air is being exchanged per minute, carbon dioxide is lost from the lungs more rapidly. When exercise stops, carbon dioxide levels fall, blood pH rises, and breathing eventually returns to normal (figure 19.8). Bear in mind, however, that moving air in and out of the lungs is of no value unless oxygen is diffusing into the blood and carbon dioxide is diffusing out.

The **lungs** are organs of the body that allow gas exchange to take place between the air and blood (table 19.4). Breathing is the process of pumping air in and out of the lungs. Figure 19.9 illustrates the various parts of the respiratory system. Inhalation and exhalation are accomplished by the movement of a muscular organ known as the **diaphragm,** which separates the lung cavity from the abdominal cavity. In addition, muscles located between the ribs are attached to the ribs in such a way that their contraction causes the chest wall to move outward and upward, which increases the size of the chest cavity. During inhalation, the diaphragm moves downward and the external muscles of the chest wall contract, causing the volume of the chest cavity to increase. This results in a lower pressure in the chest cavity compared to the outside air pressure. Consequently, air flows from the outside high pressure area through the trachea to the lungs. During normal relaxed breathing, exhalation is accomplished by the chest wall and diaphragm simply returning to their normal position. Muscular

Figure 19.8 The Control of Breathing Rate The rate of breathing is controlled by specific cells in the brain that sense the pH of the blood. When the CO_2 increases, the pH drops (becomes more acid), and the brain sends more frequent messages to the diaphragm and intercostal muscles, causing the breathing rate to increase. More rapid breathing increases the rate at which CO_2 is lost from the blood; thus the blood pH rises (becomes less acid) and the breathing rate decreases.

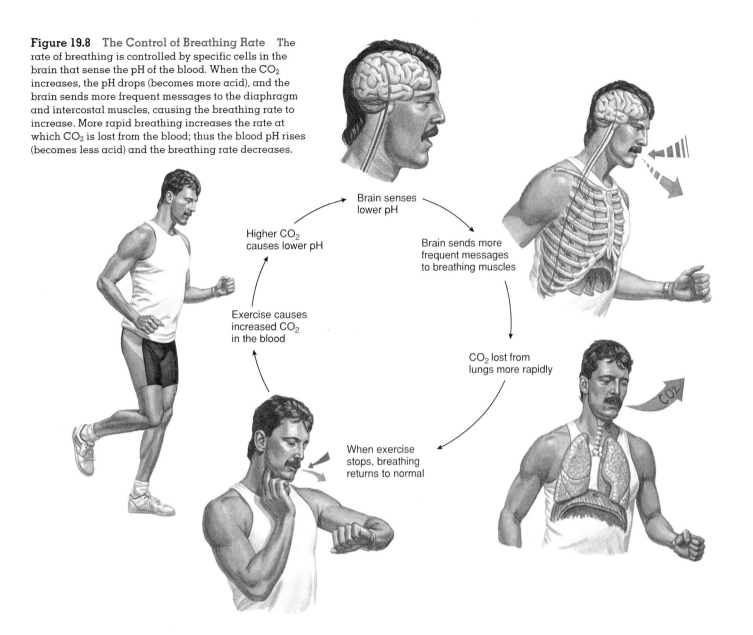

Brain senses lower pH

Higher CO_2 causes lower pH

Brain sends more frequent messages to breathing muscles

Exercise causes increased CO_2 in the blood

CO_2 lost from lungs more rapidly

When exercise stops, breathing returns to normal

table 19.4

Respiratory System Anatomy

Lungs	Two saclike structures that allow gas exchange to take place between the air and blood.
Nose, mouth, and throat	Parts of the air-transport pathway that change the humidity and temperature of the air and clean the air as it passes.
Trachea	Tube supported by rings of cartilage that prevent its collapse. This tube connects the throat with two branches, the bronchial tubes.
Bronchial tubes	Two major tubes that branch from the trachea and deliver air to smaller and smaller branches. These tubes are supported by cartilage.
Bronchioles	Smallest tubes for conducting air. They contain smooth muscle and are therefore capable of constricting.
Alveoli	Clusters of tiny sacs, where the exchange of gases takes place between the air and blood.

contraction is not involved (figure 19.10). During exercise, the body's demand for oxygen increases. The only way that the breathing system can respond is by exchanging the gases in the lungs more rapidly. This can be accomplished both by increasing the breathing rate and by increasing the volume of air exchanged with each breath. Increase in volume exchanged per breath is accomplished in two ways. First, the muscles of inhalation can contract more forcefully, resulting in a greater change in the volume of the chest cavity. In addition, the lungs can be emptied more completely by contracting the muscles of the abdomen, which force the abdominal contents upward and compress the lungs. A set of internal muscles also helps to compress the chest. You are familiar with both of these mechanisms.

Buffers

Since many kinds of chemical activities, such as enzyme-controlled reactions, are sensitive to changes in the pH of the blood and other body fluids, it is important to regulate their pH within very narrow ranges. Remember that pH is the scale used to describe the amount of acidity in a system. Lower pH numbers indicate more acid and larger numbers indicate more base or alkalinity. The normal human blood pH is about 7.4.

Although the respiratory system and kidneys are involved in regulating the pH of the blood, there are several buffer systems in the blood that prevent wide fluctuations in pH. **Buffers** are mixtures of weak acids and the ions they become (the salt of the weak acid). A buffer system tends to maintain constant pH because it can either accept or release hydrogen ions (H^+). The weak acid can release hydrogen ions (H^+) if a base is added to the solution, and the negatively charged ion of the salt can accept hydrogen ions (H^+) if an acid is added to the solution (figure A).

One example of a buffer system frequently encountered in the body is the carbonic acid/bicarbonate ion buffer system. In this system, carbonic acid (H_2CO_3) is a weak acid and bicarbonate ion (HCO_3^-) is the weak base.

$$H_2CO_3 \rightarrow HCO_3^- + H^+$$

Addition of an acid to the mixture causes the additional hydrogen ions to attach to HCO_3^- to form H_2CO_3. This removes the additional hydrogen ions from solution and ties them up in the H_2CO_3, so that the amount of free hydrogen ions remains constant (the pH remains the same).

$$H_2CO_3 \overset{\text{Added } H^+}{\underset{\longrightarrow}{\longleftarrow}} HCO_3^- + H^+$$

If a base is added to the mixture, hydrogen ions are released to tie up the hydroxyl ions; the pH remains unchanged.

$$H_2CO_3 \overset{\text{Added } OH^-}{\underset{\longrightarrow}{\longleftarrow}} H^+ + HCO_3^- + HOH$$

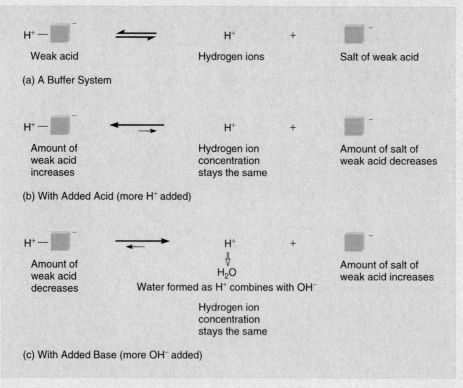

(a) A Buffer System

(b) With Added Acid (more H^+ added)

(c) With Added Base (more OH^- added)

Figure A

Figure 19.9 **Respiratory Anatomy**
Although the alveoli of the lungs are
where gas exchange takes place, there
are many other important parts of the
respiratory system. The nasal cavity
cleans, warms, and humidifies the air
entering the lungs. The trachea is also
important in cleaning the air going to
the lungs.

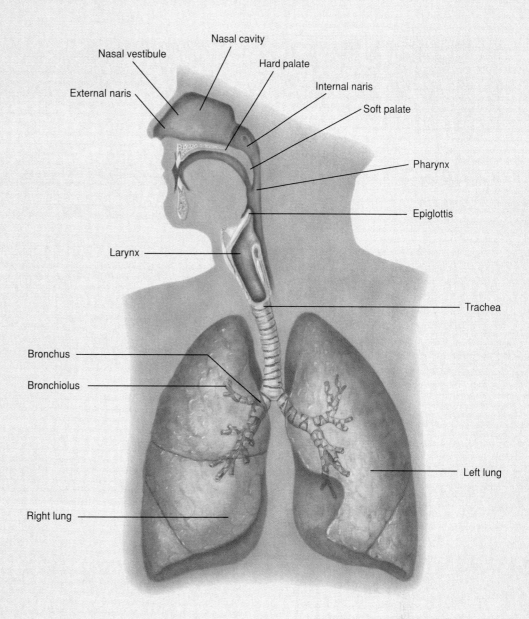

Figure 19.10 **Breathing Movement**
During inhalation, the diaphragm and
external intercostal muscles between
the ribs contract, causing the volume of
the chest cavity to increase. During a
normal exhalation, these muscles relax,
and the chest volume returns to normal.

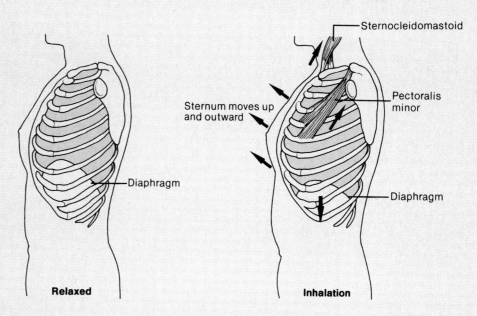

BIO *feature*

Cigarette Smoking and Your Health

Cigarette smoking is becoming less and less acceptable in our society. The banning of smoking in public buildings and on domestic air flights attests to this fact. Yet in spite of social pressure to quit smoking, research linking smoking with lung and heart disease, and evidence that even secondhand smoke can be harmful, over one-fifth of American adults are smokers (a smoker in this case is defined as someone who has smoked 100 or more cigarettes in his or her lifetime). Among Americans with a high school education or less, smoking is even more prevalent.

Hazards of Cigarette Smoking

Bronchitis

Cigarette smoking is the leading cause of chronic bronchitis, which involves the inflammation of the bronchi. A common symptom of bronchitis is a harsh cough that expels a greenish-yellow mucus.

Emphysema

Emphysema is a progressive disease in which some of the alveoli are lost. People afflicted with this disease have less and less respiratory surface area and experience greater difficulty getting adequate oxygen, even though they may be breathing more rapidly. It may be caused by cigarette smoke and other air pollutants that damage alveoli. This damage reduces the capacity of the lungs to exchange gases with the bloodstream. A common symptom of emphysema is difficulty exhaling. Several years of an emphysema sufferer's forced breathing can increase the size of the chest and give it a barrel appearance.

Asthma

Cigarette smoke is one of many environmental factors that may trigger an asthma attack. Asthma is an allergic reaction that results in the narrowing of the lungs' air passages and the excess production of fluids that limit the amount of air that can enter the lungs. Symptoms of asthma include coughing, wheezing, and difficulty breathing.

Lung Cancer

Lung cancer develops twenty times more frequently in heavy smokers than in nonsmokers. Typically, lung cancer starts in the bronchi. Cigarette smoke and other pollutants cause cells below the surface of the bronchi to divide at an abnormally high rate. This malignant growth may spread through the lung and move into other parts of the body.

Occurrence of cancers of the mouth, larynx, esophagus, pancreas, and bladder are also significantly greater in smokers than in nonsmokers.

Pneumonia

Cigarette smokers have an increased risk of developing pneumonia. Pneumonia involves the infection or inflammation of alveoli, which leads to fluid filling the alveolar sacs. Pneumonia is typically caused by the bacterium *Streptococcus pneumoniae* but in some cases is caused by other bacteria, fungi, protists, or viruses.

SURGEON GENERAL'S WARNING: Smoking Causes Lung Cancer, Heart Disease, Emphysema, And May Complicate Pregnancy.

Smoking during Pregnancy

Cigarette smoking during pregnancy has been linked to low birth weight and higher rates of fetal and infant death. Children of mothers who smoke during pregnancy have a higher incidence of heart abnormalities, cleft palate, and sudden infant death syndrome (SIDS). Nursing infants of smoking mothers have higher than normal rates of intestinal problems, and infants exposed to secondhand smoke have an increased incidence of respiratory disorders.

Heart Disease

Smoking is a major contributor to heart disease. The action of nicotine from cigarette smoke results in constriction of blood vessels and the reduction of blood flow.

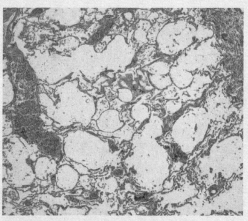

(a) Normal healthy lung

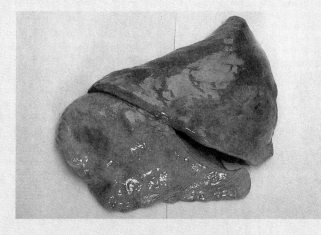

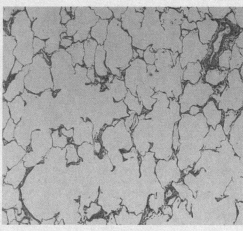

(b) Lung damage from smoking

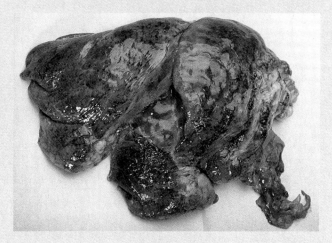

SUMMARY

There are three types of muscles found in the human body. Skeletal muscle is voluntary, serviced by the brain and spinal cord. The contractions involve different types of protein fibers. These muscle cells mass together to form whole muscles. The strength and duration of their contraction is influenced by the number and type of nerve impulses sent to them. Exercise can increase muscle mass and in some cases result in muscle injury. Many types of drugs are known to interfere with normal muscular activities. Smooth muscles differ from skeletal muscles in that they are stimulated to contract by factors other than the central nervous system. Smooth muscles are involuntary. Smooth muscle is associated with blood vessels, the intestinal tract, the bladder, and other internal organs. Cardiac muscle is found exclusively in the heart. It contracts when stimulated by special internal pacemaker cells.

The circulatory system consists of a pump, the heart, and blood vessels that distribute the blood to all parts of the body. The blood is a carrier fluid that transports molecules, cells, and heat. The exchange of materials between the blood and body cells takes place through the walls of the capillaries. Since the flow of blood can be regulated by the contraction of blood vessels, blood can be sent to different parts of the body at different times. Hemoglobin in red blood cells is very important in the transport of oxygen. Blood pH is maintained at relatively constant levels by a bicarbonate buffer system.

The respiratory system consists of the lungs and associated tubes that allow air to enter and leave the lungs. The diaphragm and muscles of the chest wall are important in the process of breathing. In the lungs, tiny sacs called alveoli, in association with capillaries, provide a large surface, which allows for rapid exchange of oxygen and carbon dioxide.

CHAPTER GLOSSARY

actin (ak′tin) A protein found in the thin filaments of muscle fibers that binds to myosin.

aorta (a-or′tah) The large blood vessel that carries blood from the left ventricle to the majority of the body.

arteries (ar′te-rēz) The blood vessels that carry blood away from the heart.

atria (ā′trē-ah) Thin-walled sacs of the heart that receive blood from the veins of the body and empty it into the ventricles.

blood (blud) The fluid medium consisting of cells and plasma that assists in the transport of materials and heat.

buffer (bu′fer) A mixture of a weak acid and the salt of a weak acid that operates to maintain a constant pH.

capillaries (cap′i-lair-ēz) Tiny blood vessels through which exchange between cells and the blood takes place.

diaphragm (di′uh-fram) A muscle separating the lung cavity from the abdominal cavity that is involved in exchanging the air in the lungs.

heart (hart) The muscular pump that forces the blood through the blood vessels of the body.

hemoglobin (he′mo-glo-bin) An iron-containing molecule found in red blood cells, to which oxygen molecules bind.

immune system (i-mūn′ sis′tem) A system of white blood cells specialized to provide the body with resistance to disease. There are two types of protection: antibody-mediated immunity and cell-mediated immunity.

lung (lung) A respiratory organ in which air and blood are brought close to one another and gas exchange occurs.

lymph (limf) Liquid material that leaves the circulatory system to surround cells.

lymphatic system (lim-fa′tik sis′tem) A collection of thin-walled tubes that collects, filters, and returns lymph from the body to the circulatory system.

motor unit (mo′tur yoo′nit) All of the muscle cells stimulated by a single nerve cell.

myosin (mi′o-sin) A protein molecule found in the thick filaments of muscle fibers that attaches to, flexes, and moves along actin molecules.

oxytocin (ok″si-to′sin) A hormone released from the posterior pituitary that causes contraction of the uterus.

phagocytosis (fa″jo-si-to′sis) The process by which a cell wraps around a particle and engulfs it.

plasma (plaz′mah) The watery matrix that contains the molecules and cells of the blood.

pulmonary artery (pul′muh-na-rē ar′tuh-rē) The major blood vessel that carries blood from the right ventricle to the lungs.

pulmonary circulation (pul′muh-na-rē ser-ku-la′shun) The flow of blood through certain chambers of the heart and blood vessels to the lungs and back to the heart.

systemic circulation (sis-te′mik ser-ku-la′shun) The flow of blood through certain chambers of the heart and blood vessels to the body and back to the heart.

veins (vānz) The blood vessels that return blood to the heart.

ventricles (ven′tri-klz) The powerful muscular chambers of the heart whose contractions force blood to flow through the arteries to all parts of the body.

 # EXPLORATIONS   Interactive Software

Smoking and Cancer

This interactive exercise allows the student to explore the relationship between smoking and cancer. The exercise presents a diagram of a human chromosome, showing the location of four genes that regulate cell growth. When their activities are disabled, they actively promote growth. In this exercise, all four must be disabled for cancer to be initiated. Students investigate the relationship between smoking and cancer by varying the amount an individual smokes and seeing how long it takes before all four genes have been mutated to a cancer-causing state.

1. What role does dose play in the probability that smoking will lead to cancer?

2. Can you discover a "safe" amount of smoking?

3. Is the twentieth cigarette smoked in a day more or less dangerous than the first?

4. How much does smoking one pack of cigarettes a day increase the likelihood that you will get cancer?

Muscle Contraction

This interactive exercise allows students to explore how the proteins in muscle cells interact with one another to produce muscle contraction. The exercise presents a diagram of a myofilament, with myosin "walking" along actin. Students can explore how changing the ATP concentration affects the speed of muscle contraction, learning that diminished ATP slows rather than weakens the force of an individual myofibril's contraction. Students can then investigate a muscle fiber made of many myofilaments, and explore how the force of the fiber's contraction depends upon changes in calcium ion levels, and why a more forceful muscle contraction results from repeated firing of the motor neuron.

1. Why are skeletal muscles able to pull but not push?
2. How can the force exerted by an individual sarcomere be changed?
3. What would be the expected effect of punching a tiny hole in the sarcoplasmic reticulum?
4. What factor limits how rapidly a tetanus can be achieved in a skeletal muscle?
5. What is the effect of calcium deficiency upon muscle contraction?

Evolution of the Human Heart

This interactive exercise allows students to learn how the human heart works by exploring the consequences of the changes that took place during its evolution. The exercise presents a cross-sectional diagram of a heart, as it would appear in a fish, at the beginning of the vertebrate heart's evolution. The fish heart has four sequential chambers and no internal septum. The students can explore the consequences of the evolutionary changes that have taken place by forming a septum and extending it through the heart while watching the effects on blood pressure and oxygen delivery to the tissues. In the fish, blood pressure is low while oxygenation is high. In land vertebrates, the evolutionary goal is to maximize both.

1. Why do lungs decrease blood pressure?
2. What is the advantage of high blood pressure?
3. Why does a fish drown in air?
4. Why does a bird's heart work "better" than a frog's?

Immune Response

This interactive exercise allows students to explore the way the human body defends itself from cancer and against invasion by viruses and microbes by varying the effectiveness of the body's defenses. The exercise presents a cross section of the human body, with a variety of white blood cells circulating, whose numbers change in response to an infection. The "infection" may be a virus (HIV), a microbe (*Pneumocystis carinii*), or a cancer (Kaposi's sarcoma), each of which elicits a different response from the immune system. Students can investigate how the defense works by altering the numbers of particular cell types, raising or lowering the number of macrophages, helper T cells, memory T cells, etc. When changes are made that mimic those produced in an HIV infection, there is drastic damage to the immune system.

1. Why are there *two* immune responses?
2. What happens if helper T cells are removed?
3. What is the effect of reducing the macrophage population?
4. Is long-term immunity to past infections, like mumps, destroyed by HIV infection?

CONCEPT MAP TERMINOLOGY

Construct a concept map to represent relationships among the following concepts.

aorta
arteries
atria
blood
capillaries

heart
hemoglobin
plasma
pulmonary artery

pulmonary circulation
systemic circulation
veins
ventricles

LABEL • DIAGRAM • EXPLAIN

Using the names of the various structures of the circulatory system and the respiratory system, trace a drop of blood containing plasma and red blood cells through the body. Start with the blood as it is returning to the heart from the muscles in the lower part of the body. Explain what is happening to the blood and blood cells as you pass through the heart and lungs.

Multiple Choice Questions

1. Which of the following best describes how a muscle contracts?
   a. muscle protein filaments of one type slide past protein filaments of another type
   b. the protein filaments within a muscle shorten as they coil into a helix shape
   c. part of the muscle loops past itself, forming thicker proteins that are shorter
   d. muscle proteins called actin change into the muscle protein myosin, which is shorter and thicker

2. A muscle increases in size in which of the following ways?
   a. increasing the number of muscle cells in a particular muscle
   b. changing the proportion of actin to myosin within the cells
   c. increasing the amount of protein filaments within the component cells
   d. decreasing the distance from one end of the muscle to the other

3. Which of the following may frequently be misused by an athlete who wishes to increase muscle mass?
   a. health food supplements
   b. steroid hormones
   c. fluids containing ions and salts, such as Gatorade
   d. diuretics to reduce the amount of water in the body

4. Muscle cells that are able to sustain a contraction for long periods of time best describes which type of muscle?
   a. cardiac muscles
   b. skeletal muscles
   c. smooth muscles
   d. none of the above fit the description

5. The major function of the respiratory system is to:
   a. provide motion in the lower abdomen
   b. control the exchange of gases between air and blood
   c. move air into and out of the lungs
   d. constantly change the atmospheric pressure within the chest cavity

6. Which part of the circulatory system has major responsibility for transport of carbon dioxide?
   a. red blood cells
   b. white blood cells
   c. platelets
   d. plasma

7. Maintaining a constant pH is a function of:
   a. buffers
   b. water
   c. platelets
   d. immune system

8. One thing that a typical blood test determines is:
   a. relative number of red blood cells
   b. the temperature of the blood
   c. size of the heart
   d. diameter of the capillaries

9. Which of the following is commonly carried by the blood?
   a. excess acid
   b. heat from the muscle cells
   c. nerve impulses
   d. toxic materials

10. Your breathing rate is controlled by:
    a. levels of carbon dioxide in the blood
    b. how strong your diaphragm muscle is
    c. amount of oxygen in the atmosphere
    d. size of the lung

Questions with Short Answers

1. What are the functions of the arteries, veins, blood, heart, and capillaries?

2. How do red blood cells assist in the transportation of oxygen and carbon dioxide?

3. Describe the mechanics of breathing.

4. How are blood pH and breathing interrelated?

5. How do skeletal, smooth, and cardiac muscle differ in (1) speed of contraction, (2) ability to stay contracted, and (3) cause of contraction?

6. Describe the mechanics of muscle contraction.

7. List behaviors that could lead to good health based on what you have learned from this chapter.

8. How are the circulatory and respiratory systems related?

9. Describe several factors that may lead to malfunctions of the muscular system, circulatory system, and respiratory system.

10. Describe the functioning and give examples of voluntary and involuntary muscles.

chapter

20

The Human Body's Coordinating Mechanisms:

The Nervous and Endocrine Systems

Aitutaki spear fishers in sea, Cook Islands, South Pacific

The Human Body's Coordinating Mechanisms: The Nervous and Endocrine Systems

learning objectives

- Describe the ionic events of a nerve impulse.

- Describe the molecular events at the synapse.

- List the structural differences between the endocrine system and the nervous system.

- Describe how information is sent by both the nervous system and the endocrine system.

- Recognize that the endocrine system is under negative-feedback control.

- Understand that the endocrine system is able to regulate growth.

- Describe how chemicals, light, and sound are detected.

Integration of Input: Nervous and Endocrine Systems

A large, multicellular organism that consists of many different kinds of systems must have some way of integrating various functions so that it can survive. If the organism does not respond appropriately to stimuli, it may die. A *stimulus* is any change in the environment that the organism can detect. Some stimuli, like light or sound, are typically external to the organism; others, like the pain generated by an infection, are internal. The reaction of the organism to a stimulus is known as a *response* (figure 20.1).

The nervous and endocrine systems are the major systems of the body that integrate stimuli and generate appropriate responses. The **nervous system** consists of a network of cells with fibrous extensions that carry information throughout the body. The **endocrine system** consists of a number of glands that communicate with one another and with other tissues through chemicals distributed throughout the organism.

Although the functions of the nervous and endocrine systems can overlap and be interrelated, these two systems have quite different methods of action. The nervous system functions very much like a telephone system. A message is sent along established pathways from a specific initiating point to a specific end point, and the transmission is very rapid. The endocrine system functions in a manner analogous to a radio broadcast system. Messenger molecules are distributed throughout the body by the circulatory system so that all cells receive them. However, only those cells that have the proper receptor sites can respond to the molecules. In the same way, a radio signal will go to all the radios in a particular area, but only those that are tuned to the correct frequency can receive the message.

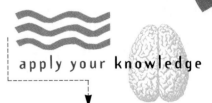

apply your knowledge

Testing the Waters

Obtain three containers large enough to place both of your hands inside. Pour cold water into the first container, warm water into the second, and water at room temperature into the third. Place your right hand in the container with cold water and your left hand in the container with warm water. After 60 seconds, place both hands in the container with room temperature water. Describe the temperature sensation of the room temperature water.

Sensory Input

The activities of the nervous and endocrine systems are often responses to some kind of input received from the sense organs. Sense organs of various types are located throughout the body. Many of them are located on the surface, where environmental changes can be easily detected. Hearing, sight, and touch are good examples of such senses. Other sense organs are located within the body and indicate to the organism how its various parts are changing. For example, pain and pressure are

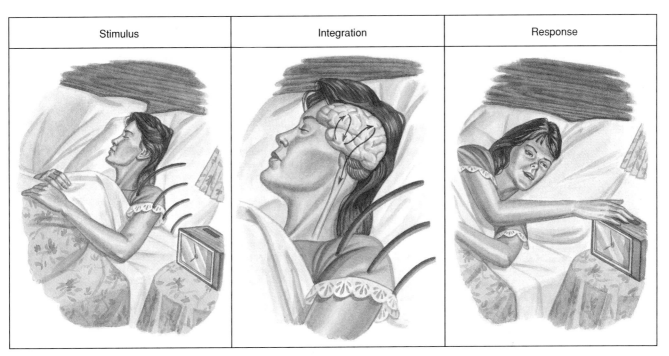

Stimulus	Integration	Response

Figure 20.1 **Stimulus-Response** A stimulus is any detectable change in the surroundings of an organism. When an organism receives a stimulus, it processes the information and may ignore the stimulus or generate a response to it.

often used to monitor internal conditions. The sense organs detect changes, but the brain is responsible for **perception**—the recognition that a stimulus has been received. Sensory abilities involve many different kinds of mechanisms, including chemical recognition, the detection of energy changes, and the monitoring of forces.

Chemical Detection

All cells have receptors on their surfaces that can bind selectively to molecules they encounter. Binding certain molecules to receptors on the cell surface can cause messages to be sent to the central nervous system, informing it of some change in the surroundings. In other cases, a molecule binding to the cell surface may cause a particular gene to be expressed, and the cell responds by changing the molecules it produces. This is typical of the way the endocrine system receives and delivers messages.

Many kinds of cells have specific binding sites that are limited to a few kinds of detectable molecules. Others, such as the taste buds on the tongue, appear to respond to classes of molecules. The taste buds are able to distinguish four kinds of tastes: sweet, sour, salt, and bitter. These different kinds of taste buds are found at specific locations on the surface of the tongue (figure 20.2). The taste buds that give us the sensation of sour appear to respond to the presence of hydrogen ions (H^+), since acids taste sour. Many kinds of ionic compounds, including sodium chloride, can stimulate the taste buds that give us the sensation of a salty taste. The sensation of sweetness can be stimulated by many kinds of organic molecules, including sugars, artificial sweeteners, and lead salts.

It is also important to understand that much of what we often refer to as **taste** involves such inputs as temperature, texture, and smell. Cold coffee has a different taste than hot coffee, even though they are chemically the same. Lumpy cooked cereal and smooth cereal

have different tastes. If you are unable to smell food, it doesn't taste as it should, which is why you sometimes lose your appetite when you have a stuffy nose. We still have much to learn about how the tongue detects chemicals and the role that other associated senses play in modifying taste.

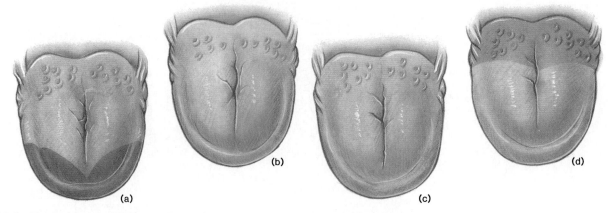

Figure 20.2 **The Location of Different Taste Sensors** The four primary tastes are sweet, sour, salt, and bitter. These different kinds of taste buds are located on specific regions of the tongue. (*a*) Sweet receptors are located at the tip of the tongue; (*b*) sour receptors are located on the sides of the tongue; (*c*) salt receptors are located at the tip and on the sides of the tongue; (*d*) bitter receptors are located on the back of the tongue.

The other major chemical sense, the sense of smell, is much more versatile; it can detect thousands of different molecules at very low concentrations. The cells that make up the **olfactory epithelium,** the part of the nasal cavity that responds to smells, apparently bind molecules to receptors on their surface. Exactly how this can account for the large number of recognizably different odors is unknown, but the receptor cells are extremely sensitive. In some cases a single molecule of a substance is sufficient to cause a receptor cell to send a message to the brain, where the sensation of odor is perceived. These sensory cells also fatigue rapidly. You have probably noticed that when you first walk into a room, specific odors are readily detected, but after a few minutes you are unable to detect them. Most perfumes and aftershaves are undetectable after 15 minutes of continuous stimulation by them.

Many internal sense organs also respond to specific molecules. For example, the brain and aorta contain cells that respond to concentrations of hydrogen ions, carbon dioxide, and oxygen in the blood. Remember, too, that the endocrine system relies on the detection of specific messenger molecules to trigger its activities.

Light Detection

The eyes primarily respond to changes in the flow of light energy. The structure of the eye is designed to focus light on a light-sensitive layer of the back of the eye known as the **retina** (figure 20.3). There are two kinds of receptors in the retina of the eye. The cells called

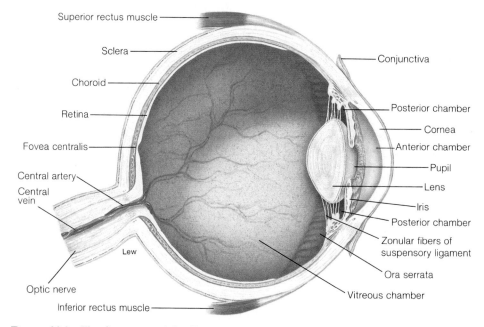

Labels: Superior rectus muscle, Sclera, Choroid, Retina, Fovea centralis, Central artery, Central vein, Optic nerve, Inferior rectus muscle, Lew, Conjunctiva, Posterior chamber, Cornea, Anterior chamber, Pupil, Lens, Iris, Posterior chamber, Zonular fibers of suspensory ligament, Ora serrata, Vitreous chamber

Figure 20.3 **The Structure of the Eye** The eye contains a cornea and lens that focus the light on the retina of the eye. The light causes pigments in the rods and cones of the retina to decompose. This leads to the depolarization of these cells and the stimulation of neurons that send messages to the brain.

rods respond to a broad range of wavelengths of light and are responsible for black-and-white vision. Since rods are very sensitive to light, they are particularly useful in dim light. Rods are located over most of the retinal surface except for the area of most acute vision known as the **fovea centralis.** The other receptor cells, called **cones,** are found throughout the retina but are particularly concentrated in the fovea centralis. Cones are not as sensitive to light, but they can detect

Vision Problems

Many structures associated with the eye may be faulty.

1. If the muscles that move the right and left eyeballs do not coordinate the movement of both eyes, they may cause crossed eyes.

2. If the cornea of the eye becomes clouded (usually associated with age), the light will not be able to pass into the sensitive areas of the eye. This condition can be corrected by transplanting a cornea from someone else.

3. If the lens becomes cloudy, a condition known as *cataract,* the eye will not be able to transmit and focus the light on the light-sensitive retina.

4. Lenses that are misshapen will not be able to focus the light on the retina. Auxiliary lenses (glasses or contact lenses) can help to bend and focus the light.

5. If the retina, the light-sensitive area of the eye, is not closely associated with the back of the eye, accumulating fluid may cause it to pull away so that distortion of vision results. The retina can be reattached by surgery.

6. If the retina does not have the proper assortment of light-sensitive pigments, an individual may not be able to distinguish red and green colors of light. This condition is hereditary.

7. Night-blindness is a condition that results from lack of vitamin A. In order to be able to see well, people with night-blindness need to have a great deal of light, more than is normally available at night. Treatment is a dietary increase in foods rich in vitamin A.

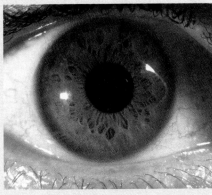

(a) Normal eye

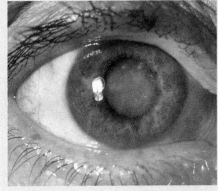

(b) Clouded eye

(c) Comparison of view through normal and clouded eyes

different wavelengths of light. This combination of receptors gives us the ability to detect color when light levels are high, but we rely on black-and-white vision at night. There are three different varieties of cones: one type responds best to red light, another responds best to green light, and the third responds best to blue light. Stimulation of various combinations of these three kinds of cones allows us to detect different shades of color (figure 20.4).

Rods and the three different kinds of cones each contain a pigment that decomposes when struck by light of the proper wavelength and sufficient strength. The pigment found in rods is called **rhodopsin.** This change in the structure of rhodopsin causes the rod to be stimulated. Cone cells have a similar mechanism of action, and each of the three kinds of cones has a different pigment. Since rods and cones can transfer their information to neurons, they stimulate a neuron and cause a message to be sent to the brain. Thus, the pattern of color and light intensity recorded on the retina is detected by rods and cones and converted into a series of nerve impulses that are received and interpreted by the brain.

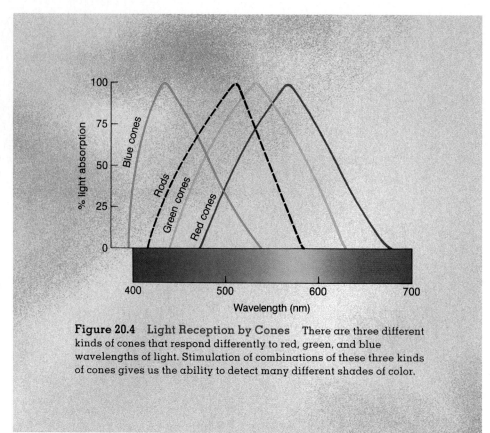

Figure 20.4 **Light Reception by Cones** There are three different kinds of cones that respond differently to red, green, and blue wavelengths of light. Stimulation of combinations of these three kinds of cones gives us the ability to detect many different shades of color.

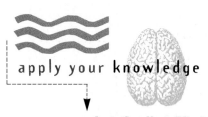

apply your knowledge

Locating Your Blind Spot

Look at the sketch of the structure of the eye on page 466. Notice the place where the nerve cells all come together to form the optic nerve. At this point on the retina there are no rods or cones. An image that falls on this point is not able to be seen. We call this the blind spot. While viewing the figure with the • and the +, close the left eye and look at the •. Move the page closer to and farther away from your face (7 to 30 cm, or 3 to 12 inches). Notice that at some distances you can see the + even when you are not looking at it, and at other distances it is not visible. When the + is not visible, its image is falling on the blind spot on your retina where there are no receptor cells.

Check with other students to see if everyone has a certain distance where the + becomes invisible. Does the distance between the • and the + have anything to do with the distance from your face where the + disappears?

Sound Detection

Ears respond to changes in sound waves. Sound is produced by the vibration of molecules. Consequently, the ears are detecting changes in the quantity of energy and the quality of sound waves. Sound has several characteristics. Loudness, or volume, is a measure of the intensity of sound energy that arrives at the ear. Very loud sounds will literally vibrate your body, and can cause hearing loss if they are too intense. Pitch is a quality of sound that is determined by the frequency of the sound vibrations. High-pitched sounds have short wavelengths; low-pitched sounds have long wavelengths.

Figure 20.5 shows the anatomy of the ear. The sound that arrives at the ear is first funneled by the external ear to the **tympanum,** also known as the *eardrum.* The cone-shaped nature of the external ear focuses sound on the tympanum and causes it to vibrate at the same frequency as the sound waves reaching it. Attached to the tympanum are three tiny bones known as the **malleus** (hammer), **incus** (anvil), and **stapes** (stirrup). The malleus is attached to the tympanum, the incus is attached to the malleus and stapes, and the stapes is attached to a small, membrane-covered opening called the **oval window** in a snail-shaped structure known as

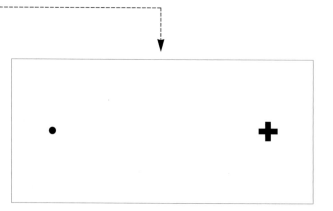

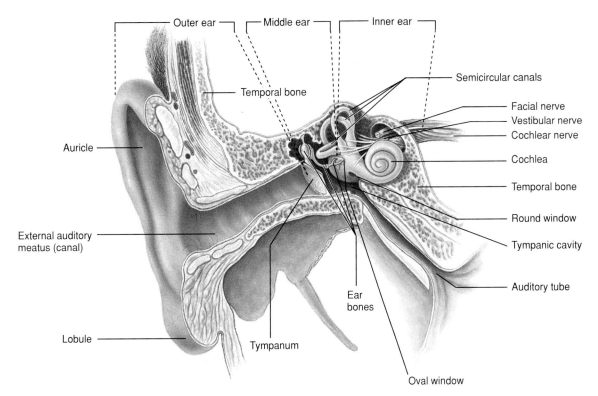

Figure 20.5 **The Anatomy of the Ear** The ear consists of an external cone that directs sound waves to the tympanum. Vibrations of the tympanum move the ear bones and vibrate the oval window of the cochlea, where the sound is detected. The semicircular canals monitor changes in the position of the head, helping us to maintain balance.

the **cochlea.** The vibration of the tympanum causes the tiny bones (malleus, incus, and stapes) to vibrate, and they in turn cause a corresponding vibration in the membrane of the oval window.

The cochlea of the ear is the structure that detects sound. It consists of a snail-shaped set of fluid-filled tubes. When the oval window vibrates, sound waves are transferred to the fluid in the cochlea, causing a membrane in the cochlea, called the **basilar membrane,** to vibrate (figure 20.6). High-pitched, short-wavelength sounds cause the basilar membrane to vibrate at the base of the cochlea near the oval window. Low-pitched, long-wavelength sounds vibrate the basilar membrane far from the oval window. Loud sounds cause the basilar membrane to vibrate more vigorously than do faint sounds. Cells on this membrane are stimulated by its vibrations. Since they transmit their information to neurons, messages can be sent to the brain.

Because sounds of different wavelengths stimulate different portions of the cochlea, the brain is able to determine the pitch of a sound.

Most sounds consist of a mixture of pitches. Louder sounds stimulate the membrane more forcefully, causing the sensory cells in the cochlea to send more nerve impulses per second. Thus, the brain is able to perceive the loudness of various sounds, as well as the pitch.

Associated with the cochlea is a set of fluid-filled tubes called the **semicircular canals.** In the walls of these canals and chambers are cells similar to those found on the basilar membrane. These cells are stimulated by movements of the head and by the position of the head with respect to the force of gravity. The constantly changing position of the head results in sensory input that is important in maintaining balance.

Touch

What we normally call the sense of *touch* consists of a variety of different kinds of input. Some receptors respond to pressure, others to temperature, and others, which we call *pain receptors,* usually respond to cell damage. When these receptors are appropriately stimulated, they send a message to the brain. Since recep-

tors are stimulated in particular parts of the body, the brain is able to localize the sensation. Not all parts of the body are equally supplied with these receptors. The tips of the fingers, lips, and external genitals have the highest density of these nerve endings, whereas the back, legs, and arms have far fewer receptors.

Some internal receptors, such as pain and pressure receptors, are important in allowing us to monitor our internal activities. Many pains generated by the internal organs are often

BIO feature

Noise Pollution

Noise is referred to as unwanted sound. However, noise can be more than just unpleasant sound. Research has shown that exposure to noise can cause physical, as well as mental, harm. The loudness of the noise is measured by decibels (db). Decibel scales are logarithmic rather than linear. Thus, the change from 40 db (a quiet library) to 80 db (a dishwasher or garbage disposal) represents a ten thousandfold increase in sound loudness.

The frequency or pitch of a sound is also a factor in determining its degree of harm. The most common sound pressure scale for high-pitched sounds is the A scale, whose units are written dbA. Hearing loss begins with prolonged exposure (eight hours or more) to 80 to 90 dbA levels of sound

pressure. Sound pressure becomes painful at around 140 dbA and can kill at 180 dbA (table 20.1).

In addition to hearing loss, noise pollution is linked to a variety of human ailments, ranging from nervous tension headaches to neuroses. Research has also shown that noise may cause blood vessels to constrict (which reduces the blood flow to key body parts), disturbs unborn children, and sometimes causes seizures in epileptics. The U.S. Environmental Protection Agency has estimated that noise causes about 40 million U.S. citizens to suffer hearing damage or other mental or physical effects. Up to 64 million people are estimated to live in homes affected by aircraft, traffic, or construction noise.

table 20.1

Intensity of Noise

Source of Sound	Intensity in Decibels (db)
Jet aircraft at takeoff	145
Pain occurs	**140**
Hydraulic press	130
Jet airplane overhead (160 meters/523.20 feet)	120
Unmuffled motorcycle	110
Subway train	100
Farm tractor	98
Gasoline lawn mower	96
Food blender	93
Heavy truck (15 meters/yards away)	90
Heavy city traffic	90
Vacuum cleaner	85
Hearing loss after long exposure	**85**
Garbage disposal unit	80
Dishwasher	65
Normal speech	**60**
Window air conditioner	60

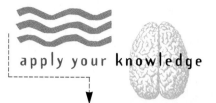
Sensory Input and Maintaining Your Balance

Using removable tape, attach a flashlight to your back so that it will shine on the ceiling. Turn on the flashlight and turn off the lights in a darkened room. Focus on an object in front of you. Stand quietly for a minute or two. Have a partner describe the pattern of the light from the flashlight on the ceiling. Describe what you did consciously to remain standing quietly.

Repeat the exercise, but this time blindfold or close your eyes. Explain any differences in the way the light from the flashlight moves on the ceiling.

perceived as if they were somewhere else. For example, the pain associated with heart attack is often perceived to be in the left arm. Pressure receptors in joints and muscles are important in providing information about the degree of stress being placed on a portion of the body. This is also important information to feed back to the brain so that adjustments can be made in movements to maintain posture.

The Structure of the Nervous System

Information concerning the surroundings of the organism is accumulated by the sense organs. These senses send messages through nerves to the spinal cord and brain, where the information is integrated with experiences of the past. The central nervous system then responds with messages to certain muscles or glands. The basic unit of the nervous system is a specialized cell called a **neuron,** or **nerve cell.** A typical neuron consists

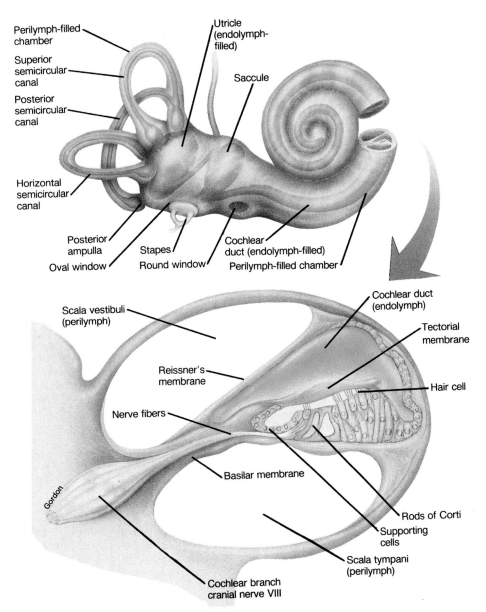

Figure 20.6 The Basilar Membrane The cells that respond to vibrations and stimulate neurons are located in the cochlea. Vibrations of the oval window cause the fluid in the cochlea to vibrate, and the basilar membrane moves also. This movement causes the receptor cells to depolarize and send a message to the brain.

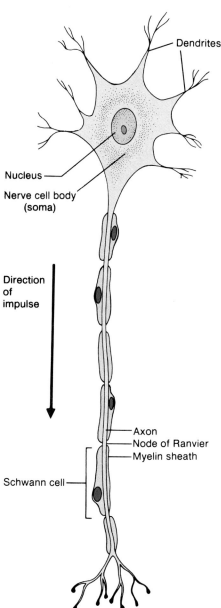

Figure 20.7 The Structure of a Nerve Cell Nerve cells consist of a nerve cell body that contains the nucleus and several fibrous extensions. The shorter, more numerous fibers that carry impulses to the nerve cell body are dendrites. The long fiber that carries the impulse away from the cell body is the axon.

FYI The brain of the average adult human contains about 1,000 billion neurons and weighs about 1.3 kg (3 lbs).

of a central body that contains the nucleus and is called the **soma,** or **cell body,** and several long, cytoplasmic extensions called **nerve fibers.** There are two kinds of fibers: **axons,** which carry information away from the cell body, and **dendrites,** which carry information toward the cell body (figure 20.7). Typically, a cell has one axon and several dendrites.

Neurons are arranged into two major systems. The **central nervous system,** which consists of the brain and spinal cord, is surrounded by the skull and the vertebrae of the spinal column. It receives input from sense organs, interprets information, and generates responses. The **peripheral nervous system** is located outside the skull and spinal column and consists of bundles of long fibers called **nerves.** There are two different sets of neurons in the peripheral nervous system. **Motor neurons** carry information from the central nervous

system to muscles and glands, and **sensory neurons** carry input from sense organs to the central nervous system. Motor neurons typically have one long axon that runs from the spinal cord to a muscle or gland, while sensory neurons have long dendrites that carry input from the sense organs to the central nervous system.

Activities at the Synapse

Between the fibers of adjacent neurons in a chain is a space called the **synapse.** When a neuron is stimulated, an impulse passes along its length from one end to the other. When the impulse reaches a synapse, a molecule called a **neurotransmitter** is released into the synapse from the axon. It diffuses across the synapse and binds to specific receptor sites on the dendrite of the next neuron. When enough neurotransmitter molecules have bound to the second neuron, an impulse is initiated in it as well. Several kinds of neurotransmitters are produced by specific neurons. These include dopamine, epinephrine, acetylcholine, and several other molecules. The first neurotransmitter identified was **acetylcholine.** Acetylcholine

The Nature of the Nerve Impulse

The message that travels along a neuron is known as a **nerve impulse.** This transmission of information involves a series of chemical events. An impulse occurs because of several characteristics of the cell membrane. The cell membrane is *differentially permeable*; that is, it allows only certain kinds of ions to diffuse through it. There are also proteins in the membrane that can actively transport specific ions from one side of the membrane to the other. One of the ions that is actively transported from cells is the sodium ion (Na^+). At the same time sodium ions (Na^+) are being transported out of cells, potassium ions (K^+) are being transported into the normal resting cells. However, there are more sodium ions (Na^+) transported out than potassium ions (K^+) transported in.

Because a normal resting cell has more positively charged ions on the outside of the cell than on the inside, a small but measurable voltage exists across the membrane of the cell. **Voltage** is a measure of the electrical charge difference that exists between two points or objects. The voltage difference between the inside and outside of a cell membrane is about 70 millivolts (0.07 volt). The cell membrane is therefore polarized in the same sense that a battery is polarized, with a positive and a negative pole. A resting neuron has its positive pole on the outside of the cell membrane and its negative pole on the inside of the membrane (figure A).

When a cell is stimulated, the cell membrane changes its permeability and lets sodium ions (Na^+) pass through it. The membrane is thus **depolarized**; it loses its difference in charge as sodium ions (Na^+) diffuse into the cell from the outside. They do so because they are in greater concentration outside the cell than inside. When the membrane becomes more permeable, the sodium ions are able to diffuse into the cell, toward the area of lower concentration. The depolarization of one point on the cell membrane causes the adjacent portion of the cell membrane to change its permeability, and it also depolarizes. Thus, a wave of depolarization passes along the length of the neuron from one end to the other (figure B). The passage of an impulse along any portion of the neuron is a momentary event, since sodium ions (Na^+) begin to be actively pumped out of the cell just as soon as they enter. This reestablishes the original polarized state, and the membrane is said to be *repolarized*. When the nerve impulse reaches the end of the axon, it induces the release of a chemical that stimulates depolarization of the next neuron in the chain or contraction of a muscle.

Figure A The Polarization of Cell Membranes All cells, including nerve cells, have an active transport mechanism that pumps Na⁺ out of cells and simultaneously pumps K⁺ into them. The end result is that there are more Na⁺ ions outside the cell and more K⁺ ions inside the cell. In addition, negative ions such as Cl⁻ are more numerous inside the cell. Consequently, the outside of the cell is more positive (+) compared to the inside, which is negative (−).

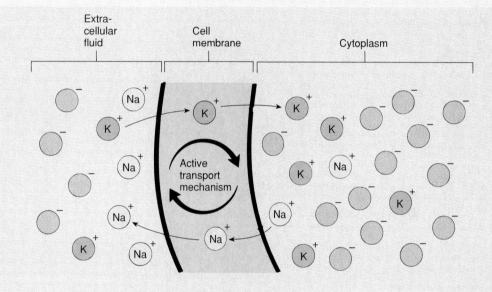

Figure B A Nerve Impulse
When a nerve cell is stimulated, a small portion of the cell membrane depolarizes as Na⁺ flows into the cell through the membrane. This encourages the depolarization of an adjacent portion of the membrane, and it depolarizes a short time later. In this way a wave of depolarization passes down the length of the nerve cell. Shortly after a portion of the membrane is depolarized, the ionic balance is reestablished. It is repolarized and ready to be stimulated again.

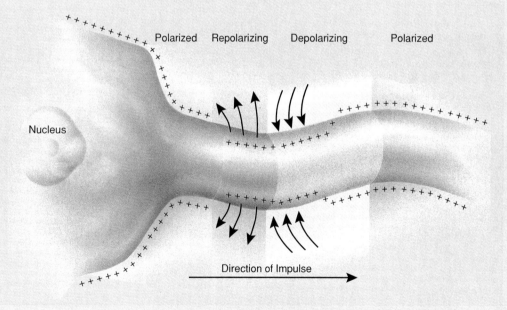

BIO *feature*

Things That Can Go Wrong with Nerve Cells

Paralysis is the inability to contract voluntary muscles. It can have many causes such as

- a tumor of the brain or spinal cord,
- dislocation of part of the backbone, which pinches nerves,
- injury that severs a nerve, and
- infection by viruses such as the poliovirus.

Inflammation of the nerves, sometimes called *neuritis,* is characterized by pain in areas served by the nerve.

Infection may be by viruses such as herpes zoster, which causes chicken pox and shingles. In this infectious disease, the herpes zoster viruses, which may first cause chicken pox, can travel into nerve cells and lay dormant for many years. When the environment surrounding these cells changes, dormant viruses may become reactivated, causing the disease known as shingles.

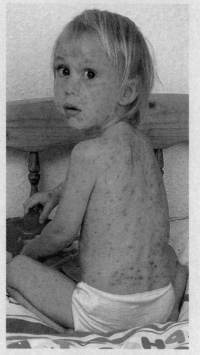

(a) Child with chicken pox

(b) Shingles infection

molecules are manufactured in the soma and migrate down the axon where they are stored until needed (figure 20.8).

If acetylcholine continues to occupy receptors, however, the neuron continues to be stimulated again and again. An enzyme called **acetylcholinesterase** destroys acetylcholine and prevents this from happening. The destruction of acetylcholine allows the second neuron in the chain to return to normal and be ready to accept another burst of acetylcholine from the first neuron when it arrives. Neurons must also constantly manufacture new acetylcholine molecules or they will exhaust their supply and be unable to conduct an impulse across a synapse.

Because of the way the synapse works, impulses can go in only one direction: only axons secrete acetylcholine, and only dendrites have acetylcholine receptors. This explains why there are sensory and motor neurons to carry messages to and from the central nervous system.

The nervous system is organized in a fashion similar to a computer. Information enters the computer by way of input devices such as the keyboard, optical scanner, or voice activation (stimuli) and is interpreted in the computer's central processing unit (brain). Messages can be sent by way of cables (motor and sensory neurons) to be displayed on a screen, printed, or cause other mechanical responses. This concept allows us to understand how the functions of various portions of the nervous system have been identified. It is possible to electrically stimulate or damage specific portions of the nervous system in experimental animals. With this kind of experimentation it is possible to determine the functions of different parts of the brain and other parts of the nervous system. For example, since peripheral nerves carry bundles of both sensory and motor fibers, damage to a nerve may result in both a lack of feeling and an inability to move.

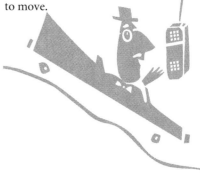

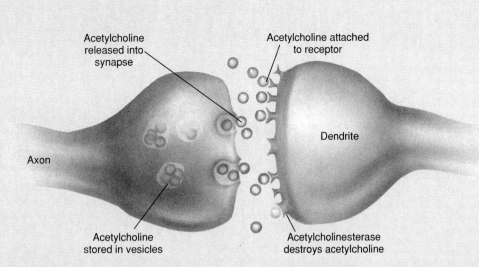

Acetylcholine released into synapse

Acetylcholine attached to receptor

Dendrite

Axon

Acetylcholine stored in vesicles

Acetylcholinesterase destroys acetylcholine

Figure 20.8 **Events at the Synapse** When a nerve impulse reaches the end of an axon, it releases a neurotransmitter into the synapse. In this illustration, the neurotransmitter is acetylcholine. When acetylcholine is released into the synapse, acetylcholine molecules diffuse across the synapse and bind to receptors on the dendrite, initiating an impulse in the next neuron. Acetylcholinesterase is an enzyme that destroys acetylcholine, preventing continuous stimulation of the dendrite.

FYI

Interfering with the Synapse Certain drugs, such as curare, botulism toxin, cocaine, and nicotine, interfere with activities at the synapse. They may cause paralysis or overstimulation of the nervous system.

Curare is a muscle-relaxing drug that was originally used as an arrowhead poison by South American Indians. This chemical functions by binding to acetylcholine receptor sites without stimulating the neuron. As a result, motor nerves function and release acetylcholine, but the acetylcholine is blocked from receptor sites by the curare and the message to contract is never received by the muscle. Curarelike drugs are used today as muscle relaxants during surgery.

In one type of food poisoning, a toxin released by the bacterium *Clostridium botulinum* inhibits the release of acetylcholine from nerve cells. This prevents muscle contraction. This bacterium is commonly found in the soil and may enter the body through foods that have not been preserved in a way that prevents these endospore-forming bacteria from growing. As they reproduce in the anaerobic environment of canned tuna, green beans, or other foods, they release botulism toxin. It has been said that a single teaspoon of this toxin, distributed equally among all people of the world, would destroy the human population. Keep in mind, however, that this toxin is easily destroyed by heating. Simply cooking your canned green beans will make them safe.

Caffeine and nicotine are considered stimulants because they make it easier for impulses to pass through the synapse. Drinking coffee, taking caffeine pills, or smoking will increase sensory input to the brain.

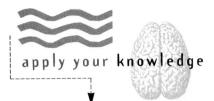

Testing Your Memory

Memory is the ability to recall a thought or piece of information. It can be *short-term*, which means the information can be recalled only for a few minutes or hours, or it may be *long-term*, which means that you can recall the information or thought for days or even years. We can enhance our memories by a process known as memory *consolidation* when we recall information frequently. Just how memory causes changes in the structure of the brain is not completely understood. We do know that memories seem to be more closely associated with the cortex or outer parts of the cerebrum.

To test your memory, try to memorize these shapes in only 10 seconds!

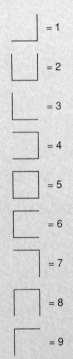

This might seem to be an impossible task to accomplish in such a short time. However, your ability to remember can be made more rapid and accurate by association with already known information. Once you have made the association, it becomes easy to recall very complex facts. For an easy way to memorize the shapes and numbers listed, see the solution diagrammed at the end of the chapter on page 487.

This Is Your Brain on Drugs

The human body's response to crack cocaine stems from the drug's ability to block neurotransmitter inactivation. Under normal conditions, a neurotransmitter will be inactivated once it has bonded to a receptor site and stimulated a neuron. Crack blocks the inactivation of dopamine, leading to a sense of euphoria. Continual use of crack can deplete the supply of neurotransmitter. When this occurs, the sense of euphoria changes to depression and anxiety. Other effects of crack use include:

(1) convulsions
(2) abnormal heart rate
(3) constriction of blood vessels and high blood pressure
(4) weight loss
(5) insomnia
(6) vulnerability to disease

Action of Drugs on the Nervous System

Substance	How it Acts	Addictive	Controlled
Alcohol	Toxic to cells. Interferes with release of acetylcholine.	Not thought to be, though emotional dependency is common.	Restricted sales. Illegal to sell to minors.
Sedatives such as valium and librium	Slows transmission to motor units.	Yes.	Prescription required.
Barbiturates	Anesthetic; dissolves in lipid part of membrane. Interferes with the production of synapse chemicals.	Yes.	Prescription required.
Solvents such as hexene	Toxic to nerve cells generally. Interferes with the production of synapse chemicals.	No.	Available in many products not sold primarily for their analgesic or anesthetic use. Glue and spray paints are common sources.
Lead and other heavy metals	Transmission speed is slowed; generally toxic to nerve cells.	No.	No longer sold as additives to products such as paint or gasoline.
Nicotine	Acts on the synapse; makes passage of impulses through synapses easier; mimics the effects of adrenalin.	Yes, possibly the most addictive drug readily available.	Restricted sales. Illegal to sell to minors.
THC from marijuana	Affects the synapse chemicals. Rapidly transmitted throughout tissues. Can remain in tissues for long period of time. Causes sleepiness and/or euphoria.	Yes, in light of symptoms of withdrawal after discontinued use.	Illegal to possess or sell.
LSD	Can cause long-term "high" of 6 hours or more and state of arousal.	Not thought to be addictive, since withdrawal symptoms are slight.	Illegal to possess or sell.
PCP	A rapid anesthetic, which binds to receptors in the brain.	Chronic users report memory loss, anxiety, and personality changes as withdrawal symptoms.	Illegal to possess or sell.
Opiates such as morphine, codeine, and heroin	Inhibits release of synapse chemicals.	Yes.	Prescriptive control for most opiates. Heroin is not used therapeutically and is illegal to possess or sell.
Cocaine and amphetamines	Blocks production of synapse chemical dopamine. Activates reward pathways in brain. Produces euphoric state resulting from the interference of dopamine uptake.	Craving symptoms reported but not necessarily associated with withdrawal symptoms.	Illegal to possess or sell.
Caffeine	Heightens alertness.	Not strongly addictive, but minor withdrawal symptoms, including headache and fatigue, may last for several days.	Not controlled, widely available.

Alcohol Consumption

Alcohol can have a variety of effects on the central nervous system, resulting in poor coordination, delayed reflexes, behavioral changes, and blackouts. The degree of these effects is determined by

- body size,
- rate of consumption,
- amount of food in the stomach,
- type and amount of beverage consumed, and
- tolerance to alcohol.

Signs of intoxication occur when blood alcohol levels reach 2,000 mg/l. Alcohol poisoning appears at 4,000 to 5,000 mg/l, and death results at 6,000 to 8,000 mg/l. Alcohol appears to act on the lipids of cell membranes and at high concentrations depolarizes nerve membranes.

The annual alcohol consumption among Americans 16 years of age and over is:

	Total (gallons)	Per capita (gallons)
Beer	5.7 billion	30.2
Wine	517 million	2.7
Spirits	371 million	1.9

The percentage of Americans aged 18 and over who drink five or more alcoholic beverages on one or more occasions per month, also called binge drinkers, is:

All adults:	15.2%
Males 18–34:	35.2%

When a small amount of ethyl alcohol enters the body, it is broken down into CO_2 and H_2O. However, as consumption increases or continues over a long period, it is broken down into a chemical (acetaldehyde) that directly poisons the liver. This liver damage is called *cirrhosis*. While many things can cause cirrhosis of the liver, alcoholism is the most common cause. Symptoms of cirrhosis include abdominal pain, constipation or diarrhea, and weight loss. Alcohol consumption affects a person's mood, judgment, behavior, concentration, and consciousness. It causes a person to become restless, nervous, and irritated. Alcohol has adverse effects on sexual response, disrupts sleep patterns, and lowers resistance to infection. With increased dependence on alcohol, the body's level of the vitamin folic acid decreases, resulting in chronic hemolytic anemia.

When alcoholics stop drinking alcohol, they experience *delirium tremens,* or alcohol withdrawal. The symptoms experienced during delirium tremens are caused by the excessive loss of magnesium from the body. This results in a variety of symptoms, including excessive excitability, weakness, tremors, slow involuntary twisting and writhing movements, tetani, seizures, abnormal heartbeats, mood alterations (depression, apathy, apprehension), vertigo, confusion, and delirium.

The Brain

The functions of specific portions of the brain have been identified. Certain parts of the brain are involved in controlling fundamental functions such as breathing and heart rate. Others are involved in decoding sensory input, and still others are involved in coordinating motor activity. The human brain also has considerable capacity to store information and create new responses to environmental stimuli. Figure 20.9 shows a diagram of the brain and some of the major locations of specific functions. The brain has many specialized regions where neurons produce specific neurotransmitter molecules that are used only to stimulate specific cells that have the proper receptor sites. As we learn more about the functioning of the brain, we are finding more kinds of specialized neurotransmitter molecules.

Glands of the Body

Glands are organs that manufacture molecules that are secreted either through ducts or into surrounding tissue, where they are picked up by the circulatory system. The **exocrine glands** are those that secrete to the surface of the body or into one of the tubular organs of the body, such as the gut or reproductive tract. Examples are the salivary glands, intestinal mucous glands, and sweat glands. Some of these glands, such as salivary glands and sweat glands, are under nervous control. When stimulated by the nervous system, they secrete their contents. **Endocrine glands** have no ducts and secrete their products into the circulatory system. We have already talked about several of these: the pituitary, thyroid, ovary, and testis are examples.

Many exocrine glands are under hormonal control. For example, many of the digestive

CONCEPT CONNECTION

exocrine glands organs that manufacture molecules that are secreted through ducts.

endocrine glands organs that manufacture molecules (hormones) that are secreted into surrounding tissues.

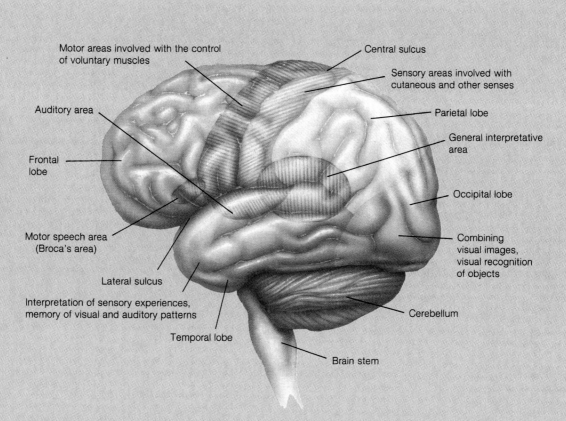

Figure 20.9 Specialized Areas of the Brain Each portion of the brain has particular functions. Although we do not know all of the brain's functions and where they are located, general regions and their functions have been identified.

Labels: Motor areas involved with the control of voluntary muscles; Central sulcus; Sensory areas involved with cutaneous and other senses; Auditory area; Parietal lobe; General interpretative area; Frontal lobe; Occipital lobe; Motor speech area (Broca's area); Combining visual images, visual recognition of objects; Lateral sulcus; Interpretation of sensory experiences, memory of visual and auditory patterns; Cerebellum; Temporal lobe; Brain stem

enzymes of the stomach and intestine are secreted in response to local hormones produced in the gut. These are circulated through the blood to the digestive glands, which respond by secreting the appropriate digestive juice.

Endocrine System Function

As mentioned previously, the nervous system functions much like a telephone system. By contrast, the endocrine system is basically a broadcasting system in which glands secrete messenger molecules, called **hormones,** that are distributed throughout the body by the circulatory system. However, each kind of hormone attaches only to appropriate receptor molecules on the surfaces of certain cells. The cells that receive the message typically

Left Brain vs. Right Brain

The right and left hemispheres of the brain appear very similar, but they possess several functional differences. Most notably, the right hemisphere controls the left side of the body and the left hemisphere controls the right side of the body. There are also several intellectual attributes associated with each hemisphere. The left hemisphere appears to be more important in analytical reasoning and the right hemisphere in creative thinking. Sometimes people are referred to as "right brain" or "left brain" dominant because they excel intellectually in fields associated with a particular hemisphere of the brain.

Left hemisphere	Right hemisphere
language	arts, aesthetics
analytical skills	spacial relationships
reasoning	imagination

Left Brain Activities

Fill in the next number in the series

1, 4, 7, 10, __?__

Recall from chapter 5 that bacteria are prokaryotic

If Ann is John's mother but John is not Ann's son, how are they related?

Right Brain Activities

Locate the continent just east of North America

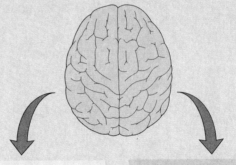

Draw what the 3-dimensional object would look like if you folded on the dotted lines

Hum this familiar melody.

Left Brain and Right Brain Activities.

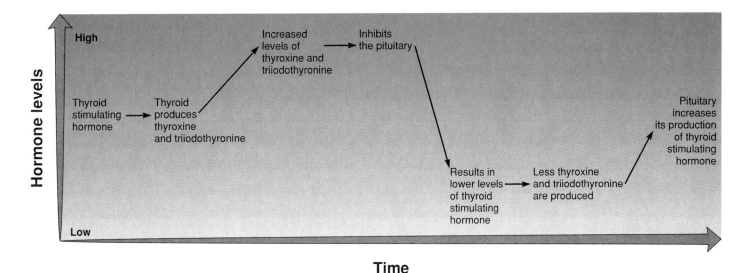

Figure 20.10 Control of Thyroid Hormone Levels The levels of thyroxine and triiodothyronine increase and decrease in response to the amount of thyroid-stimulating hormone present. Increased levels of the thyroid hormones cause the pituitary to stop producing thyroid-stimulating hormone so that eventually the thyroid hormone levels decrease. This is an example of negative-feedback control.

respond in one of three ways. Some cells release products that have been previously manufactured, other cells are stimulated to synthesize molecules or to begin metabolic activities, and some are stimulated to divide and grow.

These different kinds of responses mean that some endocrine responses are relatively rapid, while others are very slow. For example, the release of the hormones **epinephrine** and **norepinephrine** from the adrenal medulla causes a rapid change in the behavior of an organism. The heart rate increases, blood pressure rises, blood is shunted to muscles, and the breathing rate increases. You have certainly experienced this reaction many times in your lifetime.

Another hormone, called **antidiuretic hormone,** acts more slowly. It is released from the posterior pituitary gland and regulates the rate at which the body loses water through the kidneys by encouraging the reabsorption of water from their collecting ducts. The effects of this hormone can be noticed in a matter of minutes to hours. Insulin is another hormone whose effects are quite rapid. **Insulin** is produced by the pancreas and stimulates cells—particularly muscle, liver, and fat cells—to take up glucose from the blood. After a meal that is high in carbohydrates, the level of glucose in the blood begins to rise, stimulating the pancreas

to release insulin. The increased insulin causes glucose levels to fall as the sugar is taken up by cells. People with diabetes have insufficient or improperly acting insulin and therefore have difficulty regulating glucose levels in their blood.

The responses that result from the growth of cells may take weeks or years to occur. For example, **growth-stimulating hormone** is produced by the anterior pituitary gland over a period of years and results in typical human growth. After sexual maturity, the amount of this hormone generally drops, and body growth stops. Sexual development is also largely the result of the growth of specific tissues and organs. The male sex hormone **testosterone,** produced by the testes, causes the growth of male sex organs and a change to the adult body form. The female counterpart, **estrogen,** results in the development of female sex organs and body form. In all of these cases, it is the release of hormones over long periods, continually stimulating the growth of sensitive tissues, that results in the normal developmental pattern. The absence or inhibition of any of these hormones early in life changes the normal growth process.

Glands within the endocrine system typically interact with one another and control production of hormones. One common control mechanism is called **negative-feedback**

control. In any negative feedback system, an increase in the stimulus causes a reduction of the response. One of the most common negative-feedback control mechanisms is a household furnace thermostat. When the temperature of the room (stimulus) has increased to the set temperature, the thermostat sends a signal to the furnace to shut off (response). Negative feedback systems work in your body in a similar manner. In negative-feedback control the increased amount of one hormone interferes with the production of a different hormone in the chain of events. The production of **thyroxine** and **triiodothyronine** by the thyroid gland exemplifies this kind of control. The production of these two hormones is stimulated by increased production of a hormone from the anterior pituitary called **thyroid-stimulating hormone.** The control lies in the quantity of the hormone produced. When the anterior pituitary produces high levels of thyroid-stimulating hormone, the thyroid is indeed stimulated. But when increased amounts of thyroxine and triiodothyronine are produced, these hormones have a negative effect on the pituitary so that it decreases its production of thyroid-stimulating hormone, leading to reduced production of thyroxine and triiodothyronine. As a result of the interaction of these hormones, their concentrations are maintained within certain limits (figure 20.10).

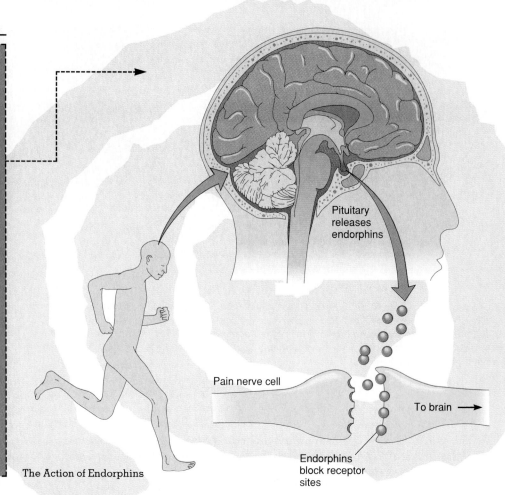

The Action of Endorphins

It is possible for the nervous and endocrine systems to interact. The pituitary gland is located at the base of the brain and is divided into two parts. The posterior pituitary is directly connected to the brain and develops from nerve tissue. The other part, the anterior pituitary, is produced from the lining of the roof of the mouth in early fetal development. Certain pituitary hormones are produced in the brain and transported down axons to the posterior pituitary where they are stored before being released. The anterior pituitary also receives a continuous input of messenger molecules from the brain, but these are delivered by way of a special set of blood vessels that picks up hormones produced by the hypothalamus and delivers them to the anterior pituitary.

The pituitary gland produces a variety of hormones that are responsible for causing other endocrine glands, such as the thyroid, ovaries and testes, and adrenals, to secrete their hormones. Pituitary hormones also influence milk production, skin color, body growth, mineral regulation, and blood glucose levels (figure 20.11).

Because the pituitary is constantly receiving information from the brain, many kinds of sensory stimuli to the body can affect the functioning of the endocrine system. One example is the way in which the nervous system and endocrine system interact to influence the menstrual cycle. At least three different hormones are involved in the cycle of changes that affect the ovary and the lining of the uterus (see chapter 18). It is well documented that stress caused by tension or worry can interfere with the normal cycle of hormones and delay or stop menstrual cycles. In addition, young women living in groups, such as in college dormitories, often have their menstrual cycles synchronized. Although the exact mechanism involved in this phenomenon is unknown, it is suspected that input from the nervous system causes this synchronization.

It has been known for centuries that changes in the levels of sex hormones cause changes in the behavior of animals. Castration (removal of the testes) of male domesticated animals, such as cattle, horses, and pigs, is sometimes done in part to reduce their aggressive behavior and make them easier to control. In humans, the use of anabolic steroids to increase muscle mass is known to cause behavioral changes and "moodiness."

Although we still tend to think about the nervous and endocrine systems as being separate and different, it is becoming clear that they are interconnected. These two systems cooperate to bring about appropriate responses to environmental challenges. The nervous system is specialized for receiving and sending short-term messages, while activities that require long-term, growth-related actions are handled by the endocrine system.

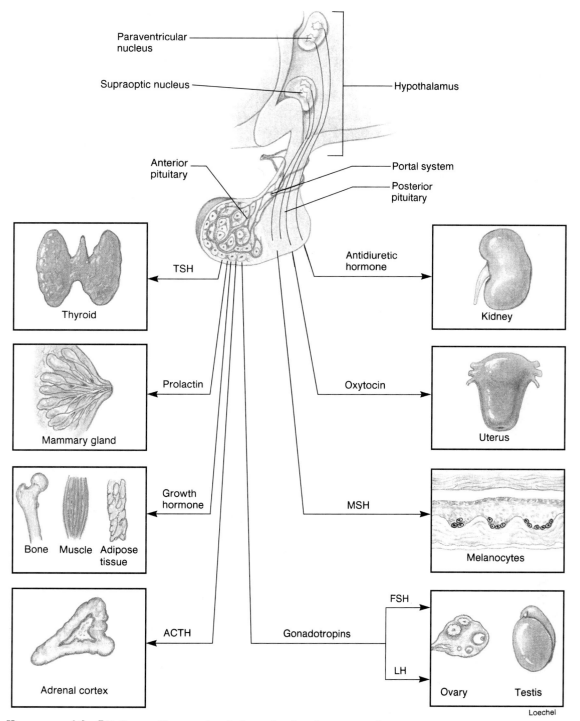

Figure 20.11 Hormones of the Pituitary The anterior pituitary gland produces several hormones that regulate growth and the secretions of target tissues. The posterior pituitary produces hormones that change the behavior of the kidney and uterus but do not influence the growth of these organs.

FYI

A Young Bird's Fancy   In many animals, the changing length of the day causes hormonal changes related to reproduction. In the spring, birds respond to lengthening days and begin to produce hormones that gear up their reproductive systems for the summer breeding season. The pineal body, a portion of the brain, serves as the receiver of light stimuli and changes the amounts of hormones secreted by the pituitary, resulting in changes in the levels of reproductive hormones. These hormonal changes modify the behavior of birds. Courtship, mating, and nest-building behaviors increase in intensity. Therefore, it appears that a change in hormone level is affecting the behavior of the animal; the endocrine system is influencing the nervous system.

Light

Changing length of day stimulates growth of reproductive organs

Reproductive organs secrete hormones that cause changed behavior

Light Controls Breeding

Growth Responses The hormones produced by the endocrine system can have a variety of effects. As mentioned earlier, hormones can stimulate smooth muscle to contract and can influence the contraction of cardiac muscle as well. Many kinds of glands, both endocrine and exocrine, are caused to secrete as a result of a hormonal stimulus. However, the endocrine system has one major effect that is not equaled by the nervous system: hormones regulate growth. Several examples of the many kinds of long-term growth changes that are caused by the endocrine system were given earlier in the chapter. Growth-stimulating hormone is produced over a period of years to bring about the increase in size of most of the structures of the body. The absence of this hormone results in a person with small body size. It is important to recognize that the amount of growth-stimulating hormone present varies from time to time. It is present in fairly high amounts throughout childhood and results in steady growth. It also appears to be present at higher levels at other times, resulting in growth spurts. Finally, as adulthood is reached, the level of this hormone falls, and growth stops, but these growth hormones are required for normal healing processes. Some research indicates that there is a relationship between decreasing quantities of growth hormone and aging processes.

Testosterone produced during adolescence influences the growth of bone and muscle to provide men with larger, more muscular bodies. In addition, there is growth of the penis, growth of the larynx, and increased growth of hair on the face and body. The primary female hormone, estrogen, causes growth of reproductive organs and development of breast tissue. It is also involved, along with other hormones, in the cyclic growth and sloughing off of the wall of the uterus.

Pituitary Giantism This woman received more growth-stimulating hormone than do people of normal height.

SUMMARY

Throughout this chapter we have been comparing the functions of the nervous and endocrine systems, the kinds of effects they have, and their characteristics. Table 20.2 summarizes these differences.

A nerve impulse is caused by sodium ions entering the cell as a result of a change in the permeability of the cell membrane. Thus, a wave of depolarization passes down the length of a neuron to the synapse. The axon of a neuron secretes a neurotransmitter, such as acetylcholine, into the synapse, where these molecules bind to the dendrite of the next cell in the chain, resulting in an impulse in it as well. The acetylcholinesterase present in the synapse destroys acetylcholine so that it does not repeatedly stimulate the dendrite.

Several kinds of sensory inputs are possible. Many kinds of chemicals can bind to cell surfaces and be recognized. This is probably how the sense of taste and the sense of smell function. Light energy can be detected because light causes certain molecules in the retina of the eye to decompose and stimulate neurons. Sound can be detected because fluid in the cochlea of the ear is caused to vibrate, and special cells detect

table 20.2

Comparison of the Nervous and Endocrine Systems

System	Method of Action	Effects
Nervous	1. Nerve impulse travels along established routes. 2. Neurotransmitters allow impulse to cross synapses. 3. Rapid action.	1. Causes skeletal-muscle contraction. 2. Modifies contraction of smooth and cardiac muscle. 3. Causes gland secretion.
Endocrine	1. Hormones released into bloodstream. 2. Receptors bind hormones to their target organs. 3. Often slow to act.	1. Stimulates smooth-muscle contraction. 2. Stimulates gland secretion. 3. Regulates growth.

this movement and stimulate neurons. The sense of touch consists of a variety of receptors that respond to pressure, cell damage, and temperature.

Glands are of two types: exocrine glands, which secrete through ducts into the cavity of an organ or to the surface of the skin, and endocrine glands, which release their secretions into the circulatory system. Digestive glands and sweat glands are examples of exocrine glands. Endocrine glands such as the ovaries, testes, and pituitary gland change the activities of cells and often cause responses that result in growth over a period of time. It is becoming clear that the endocrine system and the nervous system are interrelated. Actions of the endocrine system can change how the nervous system functions, and the reverse is also true. Much of this interrelation takes place in the association between the brain and the pituitary gland.

CHAPTER GLOSSARY

aetylcholine (a-sēt″l-kō′lēn) A neurotransmitter secreted into the synapse by many axons and received by dendrites that initiates a nerve impulse.

acetylcholinesterase (a-sēt″l-kō″li-nes′te-rās) An enzyme present in the synapse that destroys acetylcholine.

antidiuretic hormone (an-ti-di″u-re′tik hōr′mōn) A hormone produced by the pituitary gland that stimulates the kidney to reabsorb water.

axon (ak′sahn) A neuronal fiber that carries information away from the nerve cell body.

basilar membrane (ba′si-lar mem′brān) A membrane in the cochlea containing sensory cells that are stimulated by the vibrations caused by sound waves.

central nervous system (sen′trul ner′vus sis′tem) The portion of the nervous system consisting of the brain and spinal cord.

cochlea (kok′lē-ah) The part of the ear that converts sound into nerve impulses.

cones (kōnz) Light-sensitive cells in the retina of the eye that respond to different colors of light.

dendrites (den′drīts) Neuronal fibers that receive information from axons and carry it toward the nerve cell body.

depolarized (de-po′la-rīzd) Having lost the electrical difference existing between two points or objects.

endocrine glands (en′do-krin glandz) Glands that secrete into the circulatory system.

endocrine system (en′do-krin sis′tem) A number of glands that communicate with one another and other tissues through chemical messengers transported throughout the body by the circulatory system.

epinephrine (e″pi-nef′rin) A hormone produced by the adrenal medulla that increases heart rate, blood pressure, and breathing rate.

estrogen (es′tro-jen) The female sex hormone responsible for the development of female anatomy.

exocrine glands (ek′sa-krin glandz) Glands that secrete through ducts to the surface of the body or into hollow organs of the body.

fovea centralis (fo′ve-ah sen-tral′is) The area of sharpest vision on the retina, where light is normally focused.

gland (gland) An organ that manufactures and secretes a material either through ducts or directly into the circulatory system.

growth-stimulating hormone (grōth sti′mu-la-ting hōr′mōn) A hormone produced by the anterior pituitary gland that stimulates tissues to grow.

hormones (hōr′mōnz) Chemical messengers secreted by endocrine glands.

incus (in′kus) The ear bone that is located between the malleus and the stapes.

insulin (in′se-len) A hormone produced by the pancreas that regulates the amount of glucose in the blood.

malleus (ma′le-us) The ear bone that is attached to the tympanum.

motor neurons (mo′tur noor′onz) Neurons that carry information from the central nervous system to muscles or glands.

negative-feedback control (ne′ga-tiv feed′bak con-trōl′) A kind of control mechanism in which the product of one activity inhibits an earlier step in the chain of events.

nerve fibers (nerv fi′berz) Cytoplasmic extensions of nerve cells; axons and dendrites.

nerve impulse (nerv im′puls) A series of changes that take place in the neuron, resulting in a wave of depolarization that passes from one end of the neuron to the other.

nerves (nervz) Bundles of neuronal fibers.

nervous system (ner′vus sis′tem) A network of neurons that carry information from sense organs to the central nervous system and from the central nervous system to muscles and glands.

neuron (noor′on) The cellular unit, consisting of a cell body and fibers, that makes up the nervous system; also called *nerve cell*.

neurotransmitter (noor″o-trans′mit-er) A molecule released by the axons of neurons that stimulates other cells.

norepinephrine (nor-e″pi-nef′rin) A hormone produced by the adrenal medulla that increases heart rate, blood pressure, and breathing rate.

olfactory epithelium (ol-fak′to-re e″pi-thē′le-um) The cells of the nasal cavity that respond to chemicals.

oval window (o′val win′dō) The membrane-covered opening of the cochlea, to which the stapes is attached.

perception (per-sep′shun) Recognition by the brain that a stimulus has been received.

peripheral nervous system (pu-ri′fe-ral ner′vus sis′tem) The fibers that communicate between the central nervous system and other parts of the body.

retina (re′ti-nah) The light-sensitive region of the eye.

rhodopsin (ro-dop′sin) A light-sensitive, purple-red pigment found in the retinal rods that is important for vision in dim light.

rods (rahdz) Light-sensitive cells in the retina of the eye that respond to low-intensity light but do not respond to different colors of light.

semicircular canals (se-mi-ser′ku-lar ca-nalz′) A set of tubular organs associated with the cochlea that senses changes in the movement or position of the head.

sensory neurons (sen′so-re noor′onz) Neurons that send information from sense organs to the central nervous system.

soma (so′mah) The cell body of a neuron, which contains the nucleus.

stapes (sta′pēz) The ear bone that is attached to the oval window.

synapse (si′naps) The space between the axon of one neuron and the dendrite of the next, where chemicals are secreted to cause an impulse to be initiated in the second neuron.

testosterone (tes-tos′tur-ōn) The male sex hormone.

thyroid-stimulating hormone (thi′roid sti′mu-la-ting hōr′mōn) A hormone secreted by the pituitary gland that stimulates the thyroid to secrete thyroxine.

thyroxine (thi-rok′sin) A hormone produced by the thyroid gland that speeds up the metabolic rate.

triiodothyronine (tri″ī-ō″do-thī′row-nen) A hormone produced by the thyroid gland that speeds up the metabolic rate; similar to thyroxine but more potent.

tympanum (tim′pa-num) The eardrum.

voltage (vōl′tij) A measure of the electrical difference that exists between two different points or objects.

Answer to problem on p. 476.

1	2	3
4	5	6
7	8	9

EXPLORATIONS Interactive Software

Drug Addiction

This interactive exercise allows students to learn about the physical basis of drug addiction by exploring the direct consequences of the addictive drug cocaine on a nerve. The exercise presents an animated diagram of a single nerve synapse within the human brain. The neurotransmitter dopamine crosses this synapse to produce feelings of pleasure in the central nervous system. The student can explore the consequences of introducing cocaine into the synapse, watching it bind up the transporter that normally removes dopamine from the synapse. The result is that dopamine levels stay high and fire the synapse repeatedly, producing euphoria. Students can then watch the synapse as it adjusts to this higher neurotransmitter level by lowering the number of its postsynaptic receptor channels. Now there is no pleasure without the drug, which is how addiction starts. By varying the amount and frequency of cocaine use, the student can explore how patterns of drug use reinforce addiction.

1. How do neurotransmitters pass a nerve impulse across a synapse?
2. How are neurotransmitters removed from a synapse after a nerve impulse has passed?
3. What would happen if neurotransmitters were *not* removed from a synapse?
4. How do nerves respond to prolonged high levels of neurotransmitter?
5. How can addiction be reversed?
6. Can you discover a way to use cocaine and avoid addiction?

Nerve Conduction

This interactive exercise allows students to explore how voltage-gated channels cause a nerve impulse to pass down a motor axon by enabling them to alter the architecture of the neuron. The exercise presents a diagram of a motor neuron axon, showing the series of voltage-gated channels in the membrane. Students can investigate the consequences of extending or reducing the zones covered by the myelin sheath, measuring the speed of conduction along the axon. By altering the diameter of the axon, students can explore the surprisingly great influence of axon diameter upon the speed with which the impulse travels down the axon.

1. If you were to stimulate the tip of an axon, would a nerve impulse move back toward the cell body?
2. Could you generate an action potential by opening voltage-gated potassium channels?
3. Can you increase the speed a nerve impulse travels by increasing the magnitude of the action potential?
4. Can you increase the speed of conduction by shortening the intervals between nodes of Ranvier?

Synaptic Transmission

This interactive exercise allows students to explore the ways neurons employ chemicals to pass nerve impulses from one cell to another by enabling them to vary the nature of the chemicals. The exercise presents a diagram of a junction between a neuron and a target cell, containing a variety of chemically-gated and voltage-gated channels. Students can investigate the difference between excitatory and inhibitory synapses by varying the nature of the chemical released into the synapse and seeing what kind of channels are opened in response. Students will also observe the consequences of the channel opening to transmission of the nerve impulse.

1. What happens if a neuron releases both stimulatory and inhibitory neurotransmitters into a synapse simultaneously?

2. Does increasing the amount of neurotransmitter alter the speed of transmission across the synapse?
3. Can a neurotransmitter be excitatory in one synapse and inhibitory in another?
4. What is the advantage of a synapse over a direct physical connection between two nerves?
5. Why can't the nerve impulse leap across the synaptic cleft electrically, like it does when jumping from one node of Ranvier to another in saltatory conduction?

Hormone Action

This interactive exercise allows students to explore how the hormone insulin regulates levels of sugar in the blood. The exercise presents a diagram of a human body with liver and muscles highlighted and levels of circulating blood glucose indicated. An insert shows a section of plasma membrane with insulin receptors in cross section. By varying diet and amount of exercise, students can investigate how the interaction of insulin and glucagon keeps levels of blood glucose constant despite wide fluctuations in dietary intake of calories and utilization of calories for exercise.

1. How does insulin know what cells to affect?
2. Why does the body use *two* hormones to regulate blood glucose levels?
3. Can you envision a way that obese individuals could avoid contracting type II diabetes?
4. Why cannot a diabetic person simply eat more to compensate for the lack of insulin?

CONCEPT MAP TERMINOLOGY

Construct a concept map to represent the relationships among the following concepts.

antidiuretic hormone	exocrine glands	norepinephrine
endocrine glands	gland	testosterone
endocrine system	growth-stimulating hormone	thyroid-stimulating hormone
epinephrine	hormones	thyroxine
estrogen	negative-feedback control	triiodothyronine

LABEL•DIAGRAM•EXPLAIN

Describe how the endocrine system and the nervous system would interact and exactly what might happen with each if you were to be suddenly awakened from a dream by the sound of a smoke detector.

Multiple Choice Questions

1. When a nerve impulse is propagated along a nerve:
   a. positive and negative charges alternate like a current
   b. the message weakens as it moves along the cells
   c. positive sodium ions pass through the cell membrane more readily and balance the charge inside and out
   d. the cell has a positive and a negative end
2. Acetylcholine is released at:
   a. a synapse
   b. the time of nerve stimulation
   c. the point where the response is to happen
   d. each end of the cell
3. A major way in which the endocrine system differs from the nervous system is that:
   a. the message goes both directions in a nerve
   b. a hormone is broadcast and only receptive structures are able to respond
   c. endocrine messages only deal with things inside the organism
   d. the nervous system response time is much longer than the endocrine system response time
4. The message from the endocrine system is:
   a. an electrical impulse
   b. units of energy
   c. changes in the polarity of cells
   d. chemicals released by glands
5. Negative feedback control requires that:
   a. information concerning quantity is used to control production
   b. the only things that can be regulated are toxic or negative things
   c. the food eaten controls how much positive value it will have on the organism
   d. too much of a material will result in the formation of more of that material
6. Which one of the following is *not* involved in regulating the growth of an organism?
   a. insulin
   b. estrogen
   c. androgens produced by the testis
   d. growth-stimulating hormone from the anterior pituitary

7. Chemicals such as odors can be detected:
   a. only when in high concentrations
   b. when associated with particular tastes
   c. even if there are very few molecules present
   d. for long periods of time
8. Taste is more limited than smell because:
   a. there are only four separate taste bud types
   b. taste fatigues the receptor cells more rapidly
   c. smell requires shorter exposure than taste
   d. taste is the first sense organ to age
9. The cells in the eye that are receptive to light are located:
   a. in the lens
   b. behind the optic nerve
   c. lining the brain cavity
   d. on the retina
10. Sounds are caused by stimulations of:
    a. pitch receptors within the earlobe
    b. vibrations of membranes in the cochlea of the ear
    c. stringlike cells similar to strings on a violin
    d. a hammer cell inside of the ear

Questions with Short Answers

1. Describe how changing the permeability of the cell membrane and the movement of sodium ions cause a nerve impulse.

2. What is the role of acetylcholine in a synapse? What is the role of acetylcholinesterase?

3. List three ways in which the nervous system differs from the endocrine system.

4. Give an example of the interaction between the endocrine system and the nervous system.

5. Give an example of negative-feedback control in the endocrine system.

6. Identify three chemicals that interfere with normal synapse function.

7. What is actually detected by the nasal epithelium, taste buds, cochlea of the ear, and retina of the eye?

8. How do exocrine and endocrine glands differ?

9. List three hormones and give their functions.

10. List the differences between the following:
    a. central and peripheral nervous systems

    b. motor and sensory nervous systems

    c. anterior and posterior pituitary

appendix

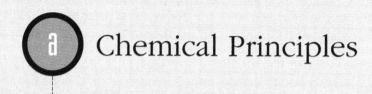 

a Chemical Principles

Basic Structures

Everything on earth is part of what we call *matter*. **Matter** is anything that has weight (mass) and also takes up space (volume). Both of these characteristics depend on the amount of matter you are dealing with; the greater the amount, the greater its mass and volume.

Characteristics that are independent of the amount of matter include *density* and *activity*. **Density** is the weight of a certain volume of material; it is frequently expressed as grams per cubic centimeter. For example, a cubic centimeter of lead is very heavy and a cubic centimeter of aluminum is very light. Lead has a higher density than aluminum. The activity of matter depends almost entirely on its composition. All matter is composed of one or more types of substances called *elements*. **Elements** are the basic building blocks from which all things are made. You already know the names of some of these elements: oxygen, iron, aluminum, silver, carbon, and gold. The sidewalk, water, air, and your body are all composed of various types of elements.

The Atomic Nucleus

In order to understand the way elements act, we need to understand what they are composed of. The smallest part of an element that still acts like that element is called an **atom.** When we use a **chemical symbol** such as Al for aluminum or C for carbon, it represents one atom of that element. The atom is constructed of three major particles; two of them are in a central region called the **atomic nucleus.** The third type of particle is in the region surrounding the nucleus (figure A.1). The weight, or mass, of the atom is concentrated in the nucleus. One major group of particles located in

the nucleus is the **neutrons;** they were named *neutrons* to reflect their lack of electrical charge. **Protons,** the second type of particle in the nucleus, have a positive electrical charge. **Electrons,** found in the area surrounding the nucleus, have a negative charge.

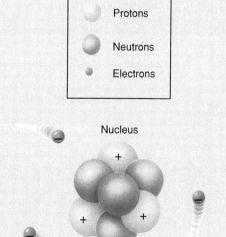

Figure A.1 Atomic Structure The nucleus of the atom contains the protons and the neutrons, which are the massive particles of the atom. The electrons, much less massive, are in constant motion around the nucleus.

An atom is neutral in charge because the number of positively charged protons is balanced by the number of negatively charged electrons. You can determine the number of either of these two particles in a balanced atom if you know the number of the other particle. For instance, carbon with six protons has six electrons, oxygen with eight electrons has eight protons, and hydrogen with one proton has one electron.

The atoms of each kind of element have a specific number of protons. For example, oxygen always has eight protons and no other element has that number. Carbon always has six protons. The **atomic number** of an element is the number of protons in an atom of that element; therefore, each element has a unique atomic number. Since oxygen has eight protons, its atomic number is eight. The mass of a proton is 1.67×10^{-24} grams. Since this is an extremely small mass and is awkward to express, it is said to be equal to one **atomic mass unit,** abbreviated **AMU.** One AMU is actually 1/12 of the mass of a particular carbon atom, but is very close to the mass of each proton (table A.1).

table A.1

Comparison of Atomic Particles

	Protons	Electrons	Neutrons
Location	Nucleus	Outside nucleus	Nucleus
Charge	Positive (+)	Negative (−)	None (neutral)
Number present	Identical to the atomic number	Equal to number of protons	Mass number minus atomic number
Mass	1 AMU	1/1,836 AMU	1 AMU

Although all atoms of the same element have the same number of protons, they do not always have the same number of neutrons. In the case of oxygen, over 99 percent of the atoms have eight neutrons, but there are others with more or fewer neutrons. Each atom of an element with a particular number of neutrons is called an **isotope.**

The most common isotope of oxygen has eight neutrons, but another isotope of oxygen has nine neutrons. We can determine the number of neutrons by comparing the masses of the isotopes. The **mass number** of an atom is the number of protons plus the number of neutrons in the nucleus. The mass number is customarily used to compare different isotopes of the same element. An oxygen isotope with a mass number of sixteen AMUs is composed of eight protons and eight neutrons and is identified as ^{16}O. Oxygen 17, or ^{17}O, has a mass of seventeen AMUs.

Eight of these units are due to the eight protons that every oxygen atom has; the rest of the mass is due to nine neutrons ($17 - 8 = 9$). Figure A.2 shows different isotopes of hydrogen.

The **periodic table of the elements** (see the accompanying FYI box) lists all the elements in order of increasing atomic number (number of protons). In addition, this table lists the mass number of each element. You can use these two numbers to determine the number of the three

Periodic Table of the Elements

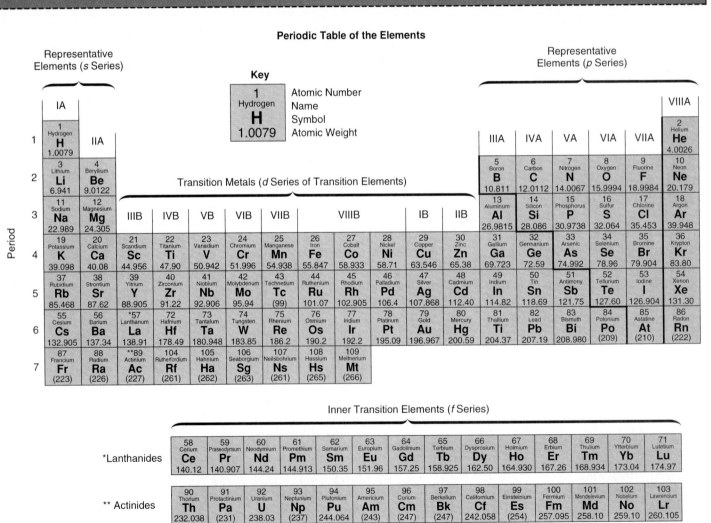

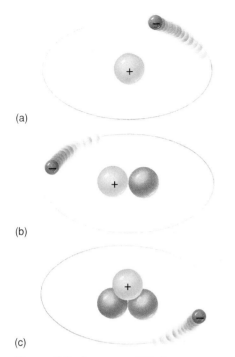

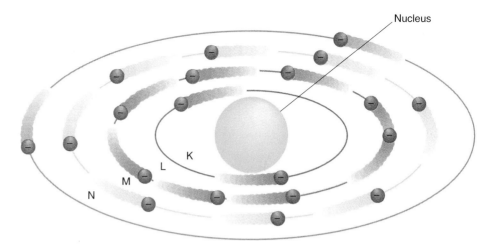

Nucleus

Figure A.3 **The Bohr Atom** Several decades ago we thought that electrons revolved around the nucleus of the atom in particular paths, or tracks. Each track was labeled with a letter: K, L, M, N, and so on. Each track was thought to be able to hold a specific number of electrons moving at a particular speed. These electron tracks were described as quanta of energy.

Figure A.2 Isotopes of Hydrogen The most common form of hydrogen is the isotope that is 1 AMU. (*a*) It is composed of one proton and no neutrons. (*b*) The isotope deuterium is 2 AMU and has one proton and one neutron. (*c*) Tritium, 3 AMU, has two neutrons and one proton. Each of these isotopes of hydrogen also has one electron, but since the mass of an electron is so small, they do not contribute significantly to the mass as measured in AMU.

major particles in an atom—protons, neutrons, and electrons. Look at the periodic table and find helium in the upper right-hand corner (He). Two is its atomic number; thus, every helium atom will have two protons. Since the protons are positively charged, the nucleus will have two positive charges that must be balanced by two negatively charged electrons. The mass of helium is given as 4.003. This is the calculated average mass of a group of helium atoms. Most of them have a mass of four—two protons and two neutrons. Generally, you will need to work only with the most common isotope, so the mass number should be rounded to the nearest whole number. If it is a number like 4.003, use 4 as the most common mass. If the mass number is a number like 39.95, use 40 as the nearest whole number. Look at several atoms in the periodic table. You can easily determine the number of protons and the number of neutrons in the most common isotopes of almost all of these atoms.

Since isotopes differ in the number of neutrons they contain, it is logical to assume that some isotopes have characteristics that are different from those of the most common form of the element. For example, there are many isotopes of iodine. The most common isotope of iodine is ^{127}I; it has a mass number of 127. A different isotope of iodine is ^{131}I; its mass number is 131 and it is **radioactive.** This means that it is not stable and that its nucleus disintegrates, releasing energy and particles from its nucleus. The energy can be detected by using photographic film or a Geiger counter. If a physician suspects that a patient has a thyroid gland that is functioning improperly, ^{131}I may be used to help confirm the diagnosis. The thyroid normally collects iodine atoms from the blood and uses them in the manufacture of the body-regulating chemical thyroxine. If the thyroid gland is working properly to form thyroxine, the radioactive iodine will collect in the gland, where its presence can be detected. If no iodine has collected there, the physician knows that the gland is not functioning correctly and can take steps to help the patient.

Electron Distribution

Electrons are the negatively charged particles of an atom that balance the positive charges of the protons in the atomic nucleus. Notice in table

A.1 that the mass of an electron is a tiny fraction of the mass of a proton. This mass is so slight that it usually does not influence the AMU of an element. But electrons are important even though they do not have a major effect on the mass of the element. The number and position of the electrons in an atom are responsible for the way atoms interact with each other.

Electrons are constantly moving at great speeds and tend to be found in specific regions some distance from the nucleus. The position of an electron at any instant in time is determined by several factors. First, since protons and electrons are of opposite charge, electrons are attracted to the protons in the nucleus of the atom. Second, counterbalancing this is the force created by the movement of the electrons, which tends to cause them to move away from the nucleus. Third, the electrons repel one another because they have identical charges. The balance of these three forces creates a situation in which the electrons of an atom tend to remain in the neighborhood of the nucleus but are distributed apart from one another. Electron distribution is not random; electrons are likely to be found in certain locations.

When chemists first described the atom, they tried to account for the fact that electrons seemed to be traveling at one of several different speeds about the atomic nucleus. Electrons did not travel at intermediate speeds. Because of this, it was thought that electrons followed a particular path, or orbit, similar to the orbits of the planets about the sun.

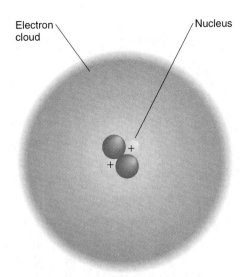

Figure A.4   **The Electron Cloud**   So fast are the electrons moving around the nucleus that they can be thought of as forming a cloud, rather than an orbit or track. You might think of the electron cloud as hundreds of photographs of an atom. Each photograph shows where an electron was at the time the picture was taken. But when the next picture is taken, the electron is somewhere else. Although we are able to determine where an electron is at a given time, we do not know the path it uses to go from there to where it is the next time we determine its position.

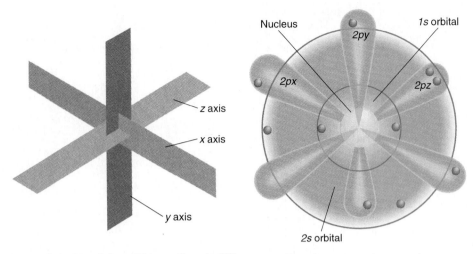

Figure A.5   **The Second Energy Level of Electrons**   The electrons on the second energy level all have about the same amount of energy, but more energy than the electrons in the first energy level. The electrons are most likely to be located in the four regions labeled 2s, 2px, 2py, and 2pz.

The model of an atom shown in figure A.3 is called the Bohr atom because Niels Bohr, a Danish physicist, advanced the theory that electrons move in discrete circular orbits about a nucleus. In the Bohr model, the electrons with the greatest amount of energy are farther from the nucleus. Think of swinging a weight on an elastic strap. As you swing the weight around your head, it makes a path a certain distance from you. If you swing it harder to make it go faster, to give it more energy, the path is a bigger circle a greater distance from your head. It was thought that electrons could follow only certain paths. No electrons were thought to go at intermediate speeds between the particular paths. The speeds or paths were called *quanta* (singular *quantum*), meaning a certain amount of energy. From the collection of early experimental data, it was thought that only two electrons could exist in the first quantum, or *shell*, eight electrons could occupy the second shell, eight (or sometimes eighteen) the third shell, and so on. These shells, also known as *energy levels*, were labeled *K, L, M*, and so forth.

The Modern Model of the Atom

Several decades ago, as more experimental data were gathered and interpreted, we began to think of the *K* shell not as a particular pathway but as a region, or space, within which electrons were likely to be. In this more modern model of the atom, each region, or **orbital,** is able to hold a maximum of two electrons. Each orbital is designated with a number that indicates the major energy level and a letter that indicates the kind of space the electrons occupy. The first orbital is lowest in energy and is designated as *1s*. The *1* indicates it is the first energy level from the nucleus and the *s* is used to help us remember that the space is spherical in shape. (Originally the *s* indicated something entirely different but a lucky happenstance allows us to use it to remember the spherically-shaped space.) Thus, the electrons in a helium atom would be located in the area described as an electron cloud in figure A.4. The area is labeled the *1s* orbital, and that orbital is full with its two electrons.

If an atom has more than two electrons, not all of these have the same amount of energy. Neon, for example, has ten electrons. The first two we would say are located in the first energy level, just as are the two in the helium atom. They are designated as being in the *1s* orbital. The rest of the electrons—the other eight—are in a higher, second energy level. (This second energy level is similar to the Bohr model of atomic structure with eight electrons located on the second orbit or path.) All eight of these electrons, however, do not have exactly the same energy and they are not likely to occupy the same spacial area.

We now think that two of these eight electrons have an amount of energy that makes it likely that they will occupy a special area of the second energy level, designated as the *2s* orbital. The *2* indicates that it is the second energy level and the *s* helps us remember that the shape of the space the electrons occupy is spherical. The other six electrons have slightly more energy and they tend to occupy three areas as far away from each other as possible, but still on the second energy level. You might think of these three areas as propeller-shaped areas on the *x, y,* and *z* axes (figure A.5). Each propeller-shaped area can hold a maximum of two electrons, so the eight electrons of the *L* shell of the Bohr atomic model can now be more accurately described as being located in one spherical area and three propeller-shaped areas at right angles to each other. By convention, we indicate these areas as the *2s,* the *2px,* the *2py,* and the *2pz* orbitals.

The third energy level (formerly called the *M* shell) contains electrons that have a greater amount of energy than those in the second energy level. These electrons are distributed in four different orbitals, which are designated the *3s*, *3px*, *3py*, and *3pz* (figure A.6). You can see how cluttered the graphic representation of the atom in figure A.6 becomes when you try to account for the number and location of all its protons, neutrons, and electrons. This will become even more difficult as we deal with larger and larger atoms. A simpler way to represent the atom is shown in figure A.7. The arrows on the diagram represent the electrons. In order to diagram the structure of an atom and place the electrons in their proper orbitals, you must start filling the spaces at the *1s* level and move outward. Each orbital is filled with two electrons. If the atom contains more than two electrons, proceed to the second energy level. At the second energy level there are four different orbitals (*2s*, *2px*, *2py*, *2pz*). The *2s* is filled with electrons first, before any additional electrons are placed in the *p* orbitals. After you have filled the *2s* orbital, begin adding electrons one at a time to each of the three *p* orbitals. An electron is added to the *2px*, a second is added to the *2py*, and a third to the *2pz*. Additional electrons are then added in this same sequence until each orbital contains two electrons. Then you can continue to the third energy level and beyond using the same pattern.

An atom such as potassium, with nineteen protons and nineteen electrons, would have two electrons in the first energy level (*1s*). In the second energy level, there would be two electrons in the *2s* orbital; two electrons in each of the *2p* orbitals; two electrons in the *3s* orbital; two in each of the *3px*, *3py*, and *3pz* orbitals; and one electron in the *4s* orbital.

Ions

Now that you know the rules for positioning electrons in their proper orbitals, it would be convenient if all atoms always followed these rules. Remember that atoms are electrically neutral when they have equal numbers of protons and electrons. Certain atoms, however, are able to exist with an unbalanced charge. These unbalanced, or charged, atoms are called **ions.** The ion of sodium is formed when one of the eleven electrons of the sodium atom escapes. Let's look at the electron distribution to explain how and why this happens.

The sodium nucleus is composed of eleven positive charges (protons) insulated from each other by twelve neutrons. (The most

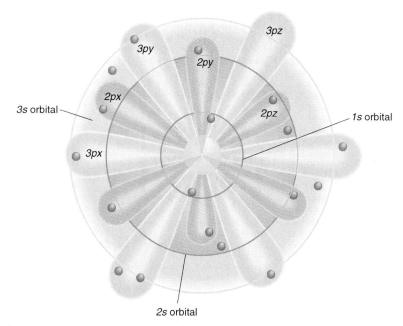

Figure A.6 **The Third Energy Level of Electrons** The electrons in these outer clouds all have about the same amount of energy. The areas where they are located are labeled *3s*, *3px*, *3py*, and *3pz*.

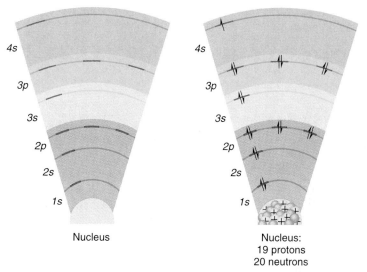

Figure A.7 **Electron Configuration** This chart is like a theater seating chart. The lines represent places where electrons can be. Each line can hold a maximum of two electrons. The electrons are most likely to be in the lower energy levels. They will go in an empty area of the same energy level before they will occupy an area that already has a negative electron in it. The filled-in chart on the right is of the atom potassium, number 19.

common isotope of sodium is sodium 23, which has twelve neutrons.) The eleven electrons that balance the charge are most likely positioned as follows: two electrons in the first energy level, eight in the second energy level, and one in the third energy level. Focus your attention on the outermost electron. It has more

energy than any of the other electrons. But because it is farther from the nucleus than any other electron, it is not as strongly attracted to the positive charges in the nucleus. This is similar to gravitational attraction—the closer to earth an object is, the greater the gravitational pull. Since this electron is the least attracted to the nucleus and has the most kinetic energy, when conditions are right it might escape from the sodium atom. What remains when the electron leaves the atom is the ion. In this case, the sodium ion is composed of the eleven positively charged protons and the twelve neutral neutrons—but only ten electrons. The fact that there are eleven positive and only ten negative charges means that there is an excess of one positive charge. We still use the chemical symbol Na to represent the ion, but we add the $^+$ to indicate that it is no longer a neutral atom but an electrically charged ion (Na^+). It is easy to remember that a positive ion is formed because it loses negative electrons.

The sodium ion is relatively stable because its outermost energy level is full. A sodium atom will lose one electron from its third major energy level so that the second energy level becomes outermost and is full of electrons. Similarly, magnesium loses two electrons from its third major energy level so that the second major energy level, which is full with eight electrons, becomes outermost. When a magnesium atom (Mg) loses two electrons, it becomes a magnesium ion (Mg^{++}). The periodic table of the elements is arranged so that all atoms in the first column become ions in a similar way. That is, when they form ions, they do so by losing one electron. Each becomes a $^+$ ion. Atoms in the second column of the periodic table become $^{++}$ ions when they lose two electrons. Those atoms at the extreme right of the periodic table of the elements do not become ions; they tend to be stable as atoms. These atoms are called *inert* because of their lack of activity. They seldom react because their protons and electrons are equal in number and they have a full outer energy level; therefore, they are not likely to lose electrons.

The column to the left of these gases contains atoms that lack a full outer energy level. They all require an additional electron. Fluorine with its nine electrons would have two in the *K* shell (*1s* orbital) and seven in the *L* shell (two in *2s*, two in *2px*, two in *2py*, and one in *2pz*). The second major energy level can hold a total of eight electrons. You can see that one additional electron could fit into the *2pz* orbital. Whenever the atom of fluorine can, it will accept an extra electron so that its outermost energy level is full. When it does so, it no longer has a balanced charge. When it accepts an extra electron, it has one more negative electron than positive protons; thus, it has become a negative ion (F^-) (figure A.8).

Similarly, chlorine will form a $^-$ ion. Oxygen, in the next column, will accept two electrons and become a negative ion with two extra negative charges (O^{--}). If you know the number and position of the electrons, you are better able to hypothesize whether or not an atom will become an ion and, if it does, whether it will be a positive ion or a negative ion. You can use the periodic table of the elements to help you determine an atom's ability to form ions. This information is useful as we see how ions react to each other.

Chemical Bonds

A variety of physical and chemical forces act on atoms and make them attractive to each other. Each of these results in a particular arrangement of atoms or association of atoms. The

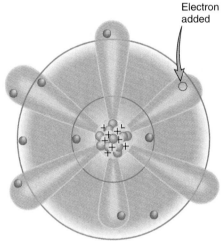

Electron added

Fluoride ion (F^-)

9 protons
10 neutrons
9 electrons
1 acquired electron

Figure A.8 Fluoride Ion When the fluorine atom accepts an additional electron, it becomes a negative ion. Negative ions are indicated with a minus sign and their names often end in -ide.

forces that combine atoms and hold them together are called **chemical bonds.** There are several types of chemical bonds. They differ from each other with respect to the kinds of attractive forces holding the atoms together. The bonding together of atoms results in the formation of a *compound*. This **compound** is composed of a specific number of atoms (or ions) joined to each other in a particular way. We generally use the chemical symbols for each of the component atoms when we designate a compound. Sometimes there will be a small number behind the chemical symbol. This number indicates how many atoms of that particular element are used in the compound. The group of chemical symbols and numbers is termed a **formula;** it will tell you what elements are in a compound and also how many atoms of each element are required. For example, $CaCl_2$ tells us that the compound of calcium chloride is composed of one calcium atom and two chlorine atoms (figure A.9).

The properties of a compound are very different from the properties of the atoms that make up the compound. Table salt is composed of the elements sodium and chlorine bound together. Both sodium and chlorine are very poisonous by themselves. Yet, when they are combined as salt, the compound is a nontoxic substance, essential for living organisms.

Ionic Bonds

When positive and negative ions are near each other, they are mutually attracted because of their opposite charges. This attraction between ions of opposite charge results in the formation of a stable group of ions. This attraction is termed an **ionic bond.** Compounds that form as a result of attractions between ions are called *ionic compounds* (see figure A.9) and are very important in living systems. We can categorize these ionic compounds into three different groups known as acids, bases, and salts.

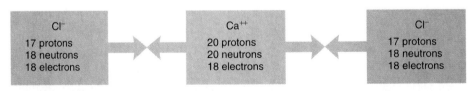

Cl^-	Ca^{++}	Cl^-
17 protons	20 protons	17 protons
18 neutrons	20 neutrons	18 neutrons
18 electrons	18 electrons	18 electrons

Figure A.9 Calcium Chloride This combination of a calcium ion and two chloride ions makes up the compound calcium chloride. The formula of the compound is $CaCl_2$. Notice that the two positive charges on the calcium ion are offset by the two chloride ions, each of which has an overabundance of only one negative charge.

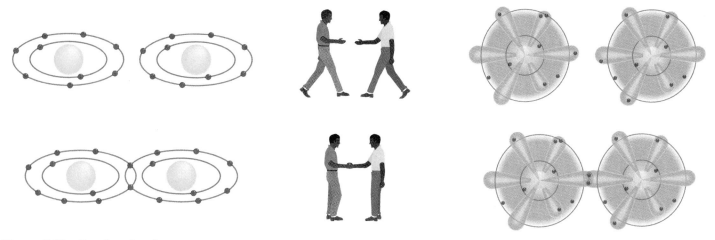

Figure A.10 Covalent Bonds   When two atoms come sufficiently close to each other that the locations of the outermost electrons overlap, an electron from each one can be shared to "fill" that outermost energy-level area. The two men depicted above have to get close enough so that their hands can overlap to form a handshake. At the left, using the Bohr model, the *L*-shells of the two atoms overlap, and so each shell appears to be full. Using the modern model at the right, the propeller-shaped orbitals of the second energy level of each atom overlap, so that each propeller appears to have a full orbital. Notice that just as it takes two hands to form a handclasp, it takes two electrons to form a covalent bond.

Covalent Bonds

In addition to ionic bonds, there is a second strong chemical bond known as a *covalent bond*. A **covalent bond** is formed by two atoms that share a pair of electrons. This sharing can occur when two atoms have orbitals that overlap one another. A covalent bond should be thought of as belonging to each of the atoms involved. You can visualize the bond as people shaking hands: the people are the atoms, the hands are electrons to be shared, and the handshake is the combining force (figure A.10). Generally, this sharing of a pair of electrons is represented by a single straight line between the atoms involved. The reason covalent bonds form relates to the arrangement of electrons within the atoms. Many elements do not tend to form ions. They will not lose electrons, nor will they gain electrons. Instead, these elements get close enough to other atoms that have unfilled outer orbitals and share electrons. If the two elements have orbitals that overlap, the electrons can be shared. By sharing electrons, the unfilled outer energy levels of each atom will be filled. Both atoms become more stable as a result of the formation of this covalent bond.

Molecules are defined as the smallest particles of chemical compounds. They are composed of a specific number of atoms arranged in a particular pattern. For example, a molecule of water is composed of one oxygen atom bonded covalently to two atoms of hydrogen. The shared electrons are in the second energy level of oxygen, and the bonds are almost at right angles to each other. Now that you realize how and why bonds are formed, it makes sense that only certain numbers of certain atoms will bond with each other to form molecules. Chemists also use the term *molecule* to mean the smallest naturally occurring part of an element or compound. Using this definition, one atom of iron is a molecule because one atom is the smallest natural piece of the element. Hydrogen, nitrogen, and oxygen tend to form into groups of two atoms. Molecules of these elements are composed of two atoms of hydrogen, two atoms of nitrogen, and two atoms of oxygen, respectively.

Hydrogen Bonds

Molecules that are composed of several atoms sometimes have an uneven distribution of charge. This may occur because the electrons involved in the formation of bonds may be located on one side of the molecule. This makes that side of the molecule slightly negative and the other side slightly positive. One side of the molecule has possession of the electrons more than the other side. When a molecule is composed of several atoms that have this uneven charge distribution, the whole molecule may show a positive side and a negative side. We sometimes think of such a molecule as a tiny magnet with a positive pole and a negative pole. This polarity of the molecule may influence how the molecule reacts with other molecules.

When several of these polar molecules are together, they orient themselves so that the partially positive end of one is near the partially negative end of another. This attraction between two molecules is called a **hydrogen bond.** Since hydrogen has the least attractive force for electrons when it is combined with other elements, the hydrogen electron tends to spend more of its time encircling the other atom's nucleus than its own. The result is the formation of a polar molecule. When the negative pole of this molecule is attracted to the positive pole of another similar polar molecule, the hydrogen will usually be located between the two molecules. Since the hydrogen serves as a bridge between the two molecules, this weak bond has become known as a *hydrogen bond.*

We usually represent this attraction as three dots between the attracted regions. This weak bond is not responsible for forming molecules, but it is important in determining how groups of molecules are arranged. Water, for example, is composed of polar molecules that form hydrogen bonds (figure A.11). Because of this, individual water molecules are less likely to separate from each other. They need a large input of energy to become separated. This is reflected in the relatively high boiling point of water. In addition, when a very large molecule, such as a protein or DNA (which is long and threadlike), has parts of its structure slightly positive and other parts slightly negative, these two areas will attract each other and result in coiling or folding of the chain of molecules in particular ways (figure A.11, right).

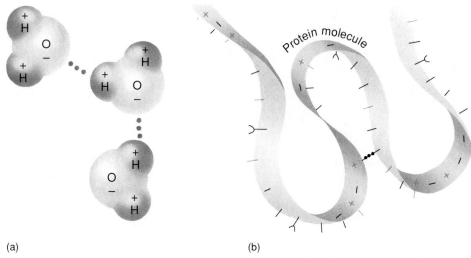

(a)   (b)

Figure A.11 Hydrogen Bonds (a) Water molecules arrange themselves so their positive portions are near the negative portions of other water molecules. The attractions are indicated as three dots. (b) The large protein molecule here also has polar areas. When the molecule is folded so that the partially positive areas are near the partially negative areas, a slight attraction forms that tends to keep it folded.

Chemical Reactions

When molecules interact with each other and rearrange their chemical bonds, we say that they have undergone a **chemical reaction.** A chemical reaction usually involves a change in energy as well as some rearrangement in the molecular structure. We frequently use a chemical shorthand to express what is going on. An arrow ($\rightarrow$) indicates that a chemical reaction is occurring. The arrowhead points to the materials that are produced by the reaction; we call these the **products.** On the other side of the arrow, we generally show the materials that are going to react with each other; we call these the **reactants.** Some of the most fascinating information we have learned recently concerns the way in which living things manipulate chemical reactions to release or store chemical energy. Figure A.12 shows the chemical shorthand used to indicate several reactions. The

chemical shorthand is called an *equation.* Look closely at the equations and identify the reactants and products in each.

Organic Molecules

Chemical principles and concepts apply to all types of matter, nonliving as well as living. Living systems are composed of various types of molecules. The things we just described did not contain carbon atoms and so were classified as **inorganic molecules.** The following is mainly concerned with more complex structures, **organic molecules,** which contain carbon atoms arranged into rings or chains.

The original meaning of the terms *inorganic* and *organic* is related to the fact that organic materials were thought to be either alive or produced only by living things. Therefore, a very strong link exists between organic

chemistry and the chemistry of living things, which is called **biochemistry,** or biological chemistry. Modern chemistry has considerably altered the original meaning of the terms *organic* and *inorganic,* since it is now possible to manufacture unique organic molecules that cannot be produced by living things. Many of the materials we use daily are the result of the organic chemist's art. Nylon, aspirin, polyurethane varnish, silicones, Plexiglas, food wrap, Teflon, and insecticides are just a few of the unique molecules that have been invented by organic chemists.

In other instances, organic chemists have taken their lead from living organisms and have been able to produce organic molecules more efficiently, or in forms that are slightly different from the original natural molecule. Some examples of these are rubber, penicillin, some vitamins, insulin, and alcohol.

Carbon: The Central Atom

All organic molecules, whether they are natural or synthetic, have certain common characteristics. The carbon atom, which is the central atom in all organic molecules, has some unusual properties. Carbon is unique in that it can combine with other carbon atoms to form long chains. In many cases, the ends of these chains may join together to form ring structures (figure A.13). Only a few other atoms have this ability. What is really unusual is that these bonding sites are all located at equal distances from one another. If you were to take a rubber ball and stick four nails into it so that they were equally distributed around the ball, you would have a good idea of the geometry involved. These bonding sites are arranged this way because in the carbon atom there are four electrons in the second energy level. These four electrons in the *L* shell, or the *2s, 2px, 2py,* and *2pz* orbitals, are all as far away from each other as possible (figure A.14). Carbon atoms are usually involved in covalent bonds. Since carbon has four places it can bond, the carbon atom can combine with four other atoms. This is the case with the methane molecule, which has four hydrogen atoms attached to a single carbon atom. Methane is a colorless and odorless gas usually found in natural gas (figure A.15).

$$HCl \quad + \quad NaOH \longrightarrow NaCl \quad + \quad H_2O$$

$$C_6H_{12}O_6 \quad + \quad 6\,O_2 \longrightarrow 6\,H_2O \ + \ 6\,CO_2 \ + \ energy$$

$$C_6H_{12}O_6 \ + \ C_6H_{12}O_6 \longrightarrow C_{12}H_{22}O_{11} \ + \ H_2O$$

Figure A.12 Chemical Equations The three equations here use chemical shorthand to indicate that the chemical bonds in the reactants have been rearranged to form the products. Along with the rearrangement of the chemical bonds, the energy content has changed.

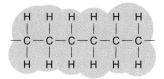

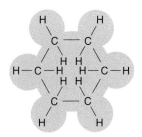

Figure A.13 A Ring or Chain Structure The ring structure shown on the bottom is formed by joining the two ends of a chain of carbon atoms.

Figure A.14 Bonding Sites of a Carbon Atom The arrangement of bonding sites around the carbon is similar to a ball with four equally spaced nails in it. Each of the four bondable electrons inhabits an area as far away from the other three as possible.

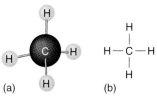

(a)   (b)

Figure A.15 A Methane Molecule A methane molecule is composed of one carbon atom bonded with four hydrogen atoms. These bonds are formed at the four bonding sites of the carbon. (a) For the sake of simplicity, all future diagrams of molecules will be two-dimensional drawings, though in reality they are three-dimensional molecules. (b) Each line in the diagram represents a covalent bond between the two atoms where a pair of electrons is being shared.

$$O = C = O$$

$$H - \overset{\displaystyle H}{\underset{\displaystyle H}{C}} - C = O$$

Figure A.16 Double Bonds These diagrams show several molecules that contain double bonds. A double bond is formed when two atoms share two pairs of electrons with each other.

Some atoms may be bonded to a single atom more than once. This results in a slightly different arrangement of bonds around the carbon atom. An example of this type of bonding occurs when oxygen is attracted to a carbon. Oxygen has two bondable electrons. If it shares one of these with a carbon and then shares the other with the same carbon, it forms a *double bond*. A **double bond** is two covalent bonds formed between two atoms that share two pairs of electrons. Oxygen is not the only atom that can form double bonds, but double bonds are common between it and carbon. The double bond is denoted by two lines between the two atoms:

$$-C = O$$

Two carbon atoms might form double bonds between each other and then bond to other atoms at the remaining bonding sites. Figure A.16 shows several compounds that contain double bonds.

Although most atoms can be involved in the structure of an organic molecule, only a few are commonly found. Hydrogen (H) and oxygen (O) are almost always present. Nitrogen (N), sulfur (S), and phosphorus (P) are also very important in specific types of organic molecules.

An enormous variety of organic molecules is possible because carbon is able to bond at four different sites, form long chains, and combine with many other kinds of atoms. The types of atoms in the molecule are important in determining the properties of the molecule. The three-dimensional arrangement of the atoms within the molecule is also important. Since most inorganic molecules are small and involve few atoms, there is usually only one way in which a group of atoms can be arranged to form a molecule. There is only one arrangement for a single oxygen atom and two hydrogen atoms in a molecule of water. In a molecule of sulfuric acid, there is only one arrangement for the sulfur atom, the two hydrogen atoms, and the four oxygen atoms.

Consider, however, these two organic molecules:

dimethyl ether

and

ethyl alchohol

Both the dimethyl ether and the ethyl alcohol contain two carbon atoms, six hydrogen atoms, and one oxygen atom, but they are quite different in their arrangement of atoms and in the chemical properties of the molecules. While the first is an ether, the second is an alcohol. Since the ether and the alcohol have the same number and kinds of atoms, they are said to have the same **empirical formula,** which in this case is written C_2H_6O. An empirical formula simply indicates the number of each kind of atom within the molecule. When the arrangement of the atoms and their bonding within the molecule is indicated, we call this a **structural formula.** Figure A.17 shows several structural formulas for the empirical formula $C_6H_{12}O_6$. Molecules that have the same empirical formula but different structural formulas are called **isomers.**

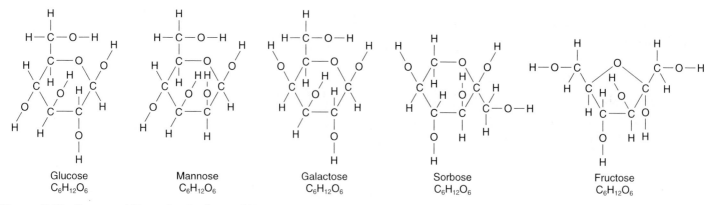

Glucose
$C_6H_{12}O_6$

Mannose
$C_6H_{12}O_6$

Galactose
$C_6H_{12}O_6$

Sorbose
$C_6H_{12}O_6$

Fructose
$C_6H_{12}O_6$

Figure A.17 Structural Formulas for Several Hexoses Several six-carbon sugars are represented here. Each has the same empirical formula, but they each have a different structural formula. They also act differently from one another.

The Carbon Skeleton and Functional Groups

To help us understand organic molecules a little better, let's consider some of their similarities. All organic molecules have a **carbon skeleton,** which is composed of rings or chains of carbons. It is this carbon skeleton in the organic molecule that determines the overall shape of the molecule. The differences between various organic molecules depend on the length and arrangement of the carbon skeleton. In addition, the kinds of atoms that are bonded to this carbon skeleton determine the way the organic compound acts. Attached to the carbon skeleton are specific combinations of atoms called **functional groups.** These functional groups determine specific chemical properties. By learning to recognize some of the functional groups, it is possible to identify an organic molecule and to predict something about its activity. Figure A.18 shows some of the functional groups that are important in biological activity. Remember that a functional group does not exist by itself; it must be a part of an organic molecule.

Common Organic Molecules

One way to make organic chemistry more manageable is to organize different kinds of compounds into groups on the basis of their similarity of structure or the chemical properties of the molecules. Frequently you will find that organic molecules are composed of subunits that are attached to each other. If you

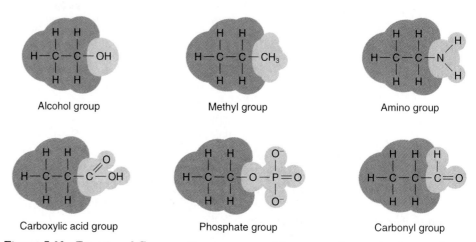

Alcohol group

Methyl group

Amino group

Carboxylic acid group

Phosphate group

Carbonyl group

Figure A.18 Functional Groups These are some of the groups of atoms that frequently attach to a carbon skeleton. Notice that in each case the carbon skeleton is unchanged; just the group attached to it is changed. The functional groups (in color) determine how the molecule will act.

recognize the subunit, then the whole organic molecule is much easier to identify. It is similar to distinguishing between a passenger train and a freight train by recognizing the individual cars unique to each.

When there are several subunits (*monomers*) bonded together, the molecule is referred to as a macromolecule or a *polymer.* The word *monomer* means a single unit, while the term *polymer* means composed of many parts. The plastics industry has polymer chemistry as its foundation. The monomers in a polymer are usually combined by a **dehydration synthesis reaction.** This reaction results in the synthesis or formation of a macromolecule when water is removed from between the two smaller component parts. For example, when a monomer with an "—OH group" attached to its carbon

skeleton approaches another monomer with an available hydrogen, dehydration synthesis can occur. Figure A.19 shows the removal of water from between two such subunits. Notice that in this case, the structural formulas are used to help identify just what is occurring. However, the chemical equation also indicates the removal of the water. You can easily recognize a dehydration synthesis reaction, because the reactant side of the equation shows numerous small molecules, while the product side lists fewer, larger products and water.

The reverse of a dehydration synthesis reaction is known as *hydrolysis.* **Hydrolysis** is the process of splitting a larger organic molecule into two or more component parts by the addition of water. Digestion of food molecules in the stomach is an important example of hydrolysis.

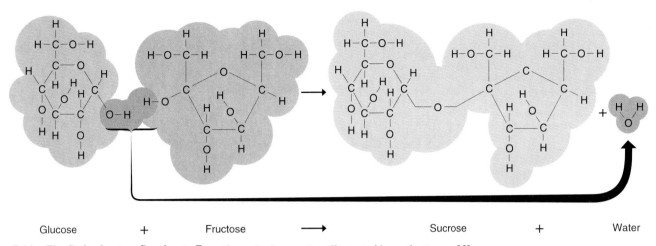

| Glucose | + | Fructose | ⟶ | Sucrose | + | Water |

Figure A.19 **The Dehydration Synthesis Reaction** In the reaction illustrated here, the two —OH groups form water, and the oxygen that remains acts as an attachment site between the two larger sugar molecules. Many structural formulas appear to be complex at first glance, but if you look for the points where subunits are attached and dissect each subunit, they are much simpler to deal with.

Carbohydrates

One class of organic molecules, **carbohydrates,** is composed of carbon, hydrogen, and oxygen atoms linked together to form monomers called *simple sugars* or *monosaccharides.* The empirical formula for a simple sugar is easy to recognize because there are equal numbers of carbons and oxygens and twice as many hydrogens—for example, $C_3H_6O_3$ or $C_5H_{10}O_5$. We usually describe the kinds of simple sugars by the number of carbons in the molecule. The ending *-ose* is a clue that indicates you are dealing with a carbohydrate. A *triose* has three carbons, a *pentose* has five, and a *hexose* has six. If you remember that the number of carbons equals the number of oxygen atoms and that the number of hydrogens is double that number, these names tell you the empirical formula for the simple sugar. Simple sugars, such as glucose, fructose, and galactose, provide the chemical energy necessary to keep organisms alive. These simple sugars combine with each other by dehydration synthesis to form **complex carbohydrates** (figure A.19). When two simple sugars bond to each other, a *disaccharide* is formed; when three bond together, a *trisaccharide* is formed (figure A.20). Generally we call a complex carbohydrate that is larger than this a *polysaccharide* (many sugar units). In all cases, the complex carbohydrates are formed by the removal of water from between the sugars. Some common examples of polysaccharides are starch and glycogen. Cellulose is an important polysaccharide used in constructing the cell walls of plant cells. Humans cannot digest (*hydrolyze*) this complex carbohydrate, so we are not able to use it as an energy source. Plant

cell walls add bulk or fiber to our diet, but no calories. Fiber is an important addition to your diet because it helps to control weight, reduce the risk of colon cancer, and control constipation and diarrhea.

FYI

Chemical Shorthand You have probably noticed that sketching the entire structural formula of a large organic molecule takes a great deal of time. If you know the structure of the major functional groups, you can use several shortcuts to more quickly describe chemical structures.

When multiple carbons with two hydrogens are bonded to each other in a chain, we sometimes write it as follows:

or we might write it this way:

$$-CH_2-CH_2-CH_2-CH_2-CH_2-CH_2-$$

or more simply, we may write it as follows: $(-CH_2-)_6$. If the six carbons were in a ring, we probably would not label the carbons or hydrogens unless we wished to focus upon a particular group or point. We would probably draw the ring with only hydrogen attached as follows:

Or

Don't let these shortcuts throw you. You will soon find that you will be putting an **—OH** group onto a carbon skeleton and neglecting to show the bond between the oxygen and hydrogen, just like a professional.

Simple sugars can be used by the cell as components of other, more complex molecules, such as the molecule adenosine triphosphate (ATP). This molecule is important in energy transfer. It has a simple sugar (ribose)

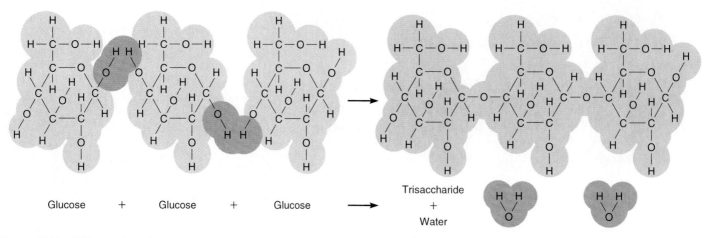

Glucose   +   Glucose   +   Glucose   ⟶   Trisaccharide + Water

Figure A.20   **A Trisaccharide**   Three simple sugars are attached to each other by the removal of two waters from between them. This is an example of a complex carbohydrate.

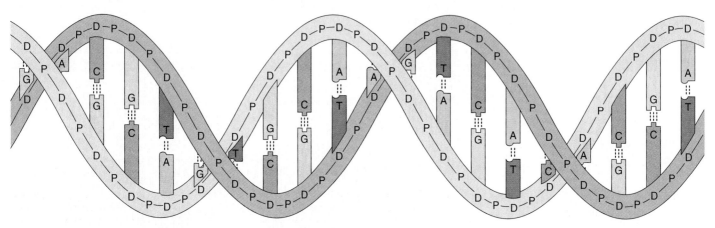

Figure A.21   **Nucleic Acid**   Deoxyribonucleic acid (DNA) is an organic molecule composed of four nucleotides. Each nucleotide is composed of a sugar, deoxyribose (D); an inorganic phosphate (P); and one of four nitrogen-containing bases: adenine (A), cytosine (C), guanine (G), or thymine (T).

as part of its structural makeup. The building blocks of the genetic material (DNA) also have a sugar component (figure A.21).

Lipids

We generally describe **lipids** as large organic molecules that do not easily dissolve in water. Just like carbohydrates, the lipids are composed of carbon, hydrogen, and oxygen. They do not, however, have the same ratio of carbon, hydrogen, and oxygen in their empirical formulas. Lipids generally have very small amounts of oxygen in comparison to the amounts of carbon and hydrogen. Fats, phospholipids, and steroids are all examples of lipids, but they are all quite different in structure.

Fats are important organic molecules that are used to provide energy. The building blocks of a fat are a glycerol molecule and fatty acids. The **glycerol** is a carbon skeleton that has three alcohol groups attached to it. Its chemical formula is $C_3H_5(OH)_3$. A **fatty acid** is a long-chain carbon skeleton that has a carboxylic acid functional group. If the carbon skeleton has as much hydrogen bonded to it as possible, we call it **saturated.** The saturated fatty acid below is stearic acid, a component of

solid meat fats. Notice that at every point in this structure the carbon has as much hydrogen as it can hold. Saturated fats are generally found in animal tissues and tend to be solids at room temperatures. Some examples of saturated fats are butter, whale blubber, suet, lard, and fats associated with such meats as steak or pork chops.

If the carbons are double-bonded to each other at one or more points, the fatty acid is said to be **unsaturated.** The unsaturated

stearic acid

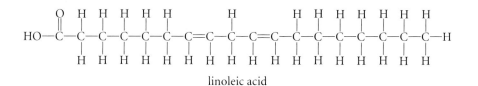

linoleic acid

fatty acid above is linoleic acid, a component of sunflower and safflower oils. Notice that there are several double bonds between the carbons and fewer hydrogens than in the saturated fatty acid. Unsaturated fats are frequently plant fats or oils and are usually liquids at room temperature. Peanut oil, corn oil, and olive oil are considered unsaturated because they have double bonds between the carbons of the carbon skeleton. A polyunsaturated fatty acid is one that has a great number of double bonds in the carbon skeleton. When glycerol and three fatty acids are combined by three dehydration synthesis reactions, a fat is formed.

Fats are important molecules for storing energy. There is twice as much energy in a gram of fat as in a gram of sugar. This is important to an organism because fats can be stored in a relatively small space and still yield a high amount of energy. Fats in animals also provide protection from heat loss. Some animals have a layer of fat under the skin that serves as an insulating layer. The thick layer of blubber in whales, walruses, and seals prevents the loss of internal body heat to the cold, watery environment in which they live. This same layer of fat, together with the fat deposits around some internal organs—such as the kidneys and heart—serves as a cushion that protects these organs from physical damage. If a fat is formed from a glycerol molecule and three attached fatty acids, it is called a *triglyceride;* if two, a *diglyceride;* and if one, a *monoglyceride* (figure A.22).

Phospholipids are a class of water-insoluble molecules that resemble fats but contain a phosphate group (PO_4) in their structure (figure A.23). One of the reasons phospholipids are important is that they are a major component of membranes in cells. Without these lipids in our membranes, the cell contents would not be able to be separated from the exterior environment. Some of the phospholipids are better known as the *lecithins.* Lecithins are found in cell membranes and also help in the emulsification of fats. They help to separate large portions of fat

into smaller units. This allows the fat to mix with other materials. Lecithins are added to many types of food for this purpose (chocolate bars, for example). Some people take lecithin as nutritional supplements because they believe it leads to healthier hair and better reasoning ability. But once inside your intestines, lecithins are destroyed by enzymes, just like any other phospholipid.

Steroids, a third group of lipid molecules, are characterized by their arrangement of interlocking rings of carbon. They often serve as hormones that aid in regulating body processes. One steroid molecule with which you are probably familiar is cholesterol. Serum cholesterol (the kind found in your blood

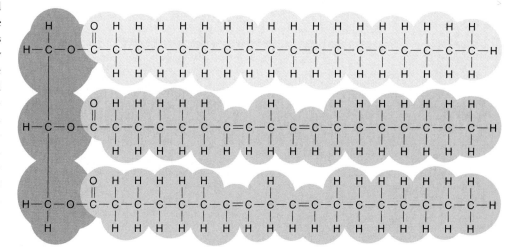

Figure A.22 A Fat Molecule The arrangement of the three fatty acids attached to a glycerol molecule is typical of the formation of a fat. The structural formula of the fat appears to be very cluttered until you dissect the fatty acids from the glycerol; then it becomes much more manageable. This example of a triglyceride contains a glycerol molecule, two unsaturated fatty acids (linoleic acid), and a third saturated fatty acid (stearic acid).

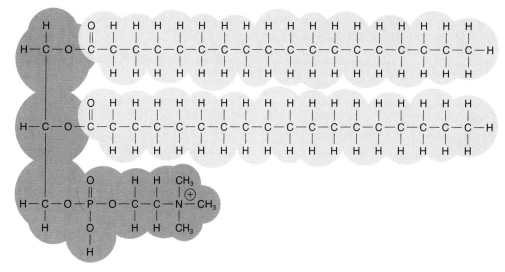

Figure A.23 A Phospholipid Molecule This molecule is similar to a fat but has a phosphate group in its structure. The phosphate group is bonded to the glycerol by a dehydration synthesis reaction. This phospholipid contains glycerol, two fatty acids, and the phosphate-containing portion. Molecules like this are known as the lecithins.

Fat and Your Diet **When triglycerides are eaten in fat-containing foods, digestive enzymes hydrolyze them into monoglycerides and fatty acids. These molecules are absorbed by the intestinal tract and coated with protein to form** *lipoprotein,* **as shown in the accompanying diagram.**

There are four types of lipoproteins in the body: (1) chylomicrons, (2) very-low-density-lipoproteins (VLDL), (3) low-density-lipoproteins (LDL), and (4) high-density-lipoproteins (HDL). Chylomicrons are very large particles formed in the intestine and are between 80 percent and 95 percent triglycerides in composition. As the chylomicrons circulate through the body, the triglycerides are removed by cells in order to make hormones, store energy, and build new cell parts. When most of the triglycerides have been removed, the remaining portions of the chylomicrons are harmlessly destroyed. The VLDLs and LDLs are formed in the liver. VLDLs contain all types of lipid, protein, and 10 percent to 15 percent cholesterol, while the LDLs are about 50 percent cholesterol. As with the chylomicrons, the body taps these molecules for their lipids. However, in some people, high levels of LDLs in the blood are associated with the disease *atherosclerosis* **(hardening of the arteries). While in the blood, LDLs may stick to the insides of the vessels, forming hard deposits that restrict blood flow and contribute to high blood pressure, strokes, and heart attacks. Even though they are 30 percent cholesterol, a high level of HDLs (made in the intestine) in comparison to LDLs is associated with a lower risk of atherosclerosis. One way to reduce the risk of this disease is to lower your intake of LDLs. This can be done by reducing your consumption of saturated fats, since these are most easily converted by your body into LDLs and cholesterol.**

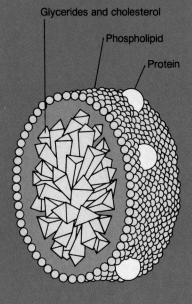

Glycerides and cholesterol

Phospholipid

Protein

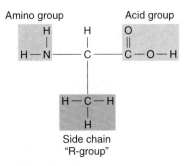

Amino group   Acid group

Side chain "R-group"

Figure A.24   The Structure of an Amino Acid An amino acid is composed of a short carbon skeleton with three functional groups attached: an amino group, a carboxylic acid group (acid group), and an additional variable group ("R-group"). The variable group determines which specific amino acid is constructed.

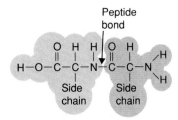

Peptide bond

Side chain   Side chain

Figure A.25   A Peptide Bond The bond that results from a dehydration synthesis reaction between two amino acids is called a peptide bond. This bond forms as a result of the removal of the hydrogen and hydroxyl groups. In the formation of this bond, the nitrogen is bonded directly to the carbon.

associated with lipoproteins) has been implicated in many cases of atherosclerosis. This steroid is made by your body for use as a component of cell membranes. It is also used by your body to make bile acids. These products of your liver are channeled into your intestine to emulsify fats. Cholesterol is also necessary for the manufacture of vitamin D. Cholesterol molecules in the skin react with ultraviolet light to produce vitamin D, which assists in the proper development of bones and teeth.

A large number of steroid molecules are hormones. Some of them regulate reproductive processes such as egg and sperm production, while others regulate such things as salt concentration in the blood. Athletes have been known to use certain hormonelike steroids to increase their muscular bulk. The medical

community is certain that use of these chemicals is potentially harmful, possibly resulting in liver dysfunction, sex-characteristic changes, changes in blood chemistry, and even death.

Proteins

Proteins are polymers made up of monomers known as *amino acids.* An **amino acid** is a short carbon skeleton that contains an amino group (a nitrogen and two hydrogens) on one end of the skeleton and a carboxylic acid group at the other end (figure A.24). In addition, the carbon skeleton may have one of several different side chains on it. There are about twenty amino acids that are important to cells. All are identical except for their side chains.

The amino acids can bond together by dehydration synthesis reactions. When two amino acids form a bond by removal of water, the nitrogen of the amino group of one is bonded to the carbon of the acid group of another. This bond is termed a **peptide bond** (figure A.25).

Any amino acid can form a peptide bond with any other amino acid. They fit together in a specific way, with the amino group of one bonding to the acid group of the next. You can imagine that by using twenty different amino acids as building blocks, you can construct millions of different combinations. Each of these combinations is termed a **polypeptide chain.** A specific polypeptide is composed of a specific sequence of amino acids bonded end to end. This is called its *primary structure.* A listing of the amino acids in their proper order within a particular polypeptide constitutes its primary structure. The specific sequence of

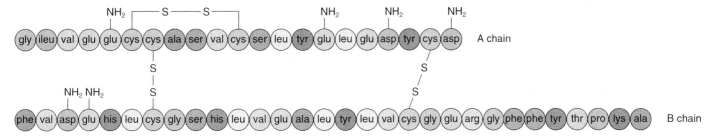

Figure A.26   An Insulin Molecule   The protein insulin is composed of two polypeptide chains bonded together at specific points by reactions between the side chains of particular amino acids. The side chains of one interact with the side chains of the other and form a particular three-dimensional shape. The bonds that form between the polypeptide chains are called disulfide bonds.

amino acids in a polypeptide is controlled by the genetic information of an organism.

The string of amino acids in a polypeptide is likely to twist into a particular shape: a coil or a pleated sheet. These forms are referred to as the *secondary structure* of polypeptides. For example, some proteins (e.g., hair) take the form of a *helix:* a shape like that of a coiled telephone cord. The helical shape is maintained by hydrogen bonds formed between different amino acid side chains at different locations in the polypeptide. Hydrogen bonds do not form molecules but result in the orientation of one part of a molecule to another part of a molecule. Other polypeptides form hydrogen bonds that cause them to make several flat folds that resemble a pleated skirt.

It is also possible for a single polypeptide to contain one or more coils and pleated sheets along its length. As a result, these different portions of the molecule can interact to form an even more complex three-dimensional structure. This occurs when the coils and pleated sheets twist and combine with each other. The complex three-dimensional structure formed in this manner is the polypeptide's *tertiary* (third degree) *structure.* A good example of tertiary structure can be seen when a coiled phone cord becomes so twisted that it folds around and back on itself in several places. The oxygen-holding protein found in muscle cells, myoglobin, displays tertiary structure: it is composed of a single helical molecule containing 153 amino acids folded back and bonded to itself in several places.

Frequently, several different polypeptides, each with its own tertiary structure, twist around each other and chemically combine. The larger, three-dimensional structure formed by these interacting polypeptides is referred to as the protein's *quaternary* (fourth degree) *structure.* The individual polypeptide chains are

bonded to each other by the interactions of certain side chains, which can form disulfide bonds (figure A. 26). Quaternary structure is displayed by the protein molecules called *antibodies,* which are involved in fighting diseases such as mumps and chicken pox.

Individual polypeptide chains or groups of chains forming a particular configuration are proteins. The structure of a protein is closely related to its function. We will consider two aspects of the structure of proteins: the sequence of amino acids within the protein and the overall three-dimensional shape of the molecule. Any changes in the arrangement of amino acids within a protein can have far-reaching effects on its function. For example, normal hemoglobin found in red blood cells consists of two kinds of polypeptide chains called the alpha and beta chains. The beta chain is 146 amino acids long. If just one of these amino acids is replaced by a different one, the hemoglobin molecule may not function properly. A classic example of this results in a condition known as *sickle-cell anemia.* In this case, the sixth amino acid in the beta chain, which is normally glutamic acid, is replaced by valine. This minor change causes the hemoglobin to fold differently, and the red blood cells that contain this altered hemoglobin assume a sickle shape when the body is deprived of an adequate supply of oxygen.

When a particular sequence of amino acids forms a polypeptide, the stage is set for that particular arrangement to bond with another polypeptide in a certain way. Think of a telephone cord that has curled up and formed a helix (its secondary structure). Now imagine that at several irregular intervals along that cord, you have attached magnets. You can see that the magnets at the various points along the cord will attract each other, and the curled cord will form a particular three-dimensional shape.

You can more closely approximate the complex structure of a protein if you imagine several curled cords, each with magnets attached at several points. Now imagine these magnets as bonding the individual cords together. The globs or ropes of telephone cords approximate the quaternary structure of a protein. This shape can be compared to the shape of a key. In order for a key to do its job effectively, it has to have particular bumps and grooves on its surface. Similarly, if a particular protein is to do its job effectively, it must have a particular shape. The protein's shape can be altered by changing the order of the amino acids that causes different cross linkages to form. Changing environmental conditions also influences the shape of the protein.

Energy in the form of heat or light may break the hydrogen bonds within protein molecules. When this occurs, the chemical and physical properties of the protein are changed and the protein is said to be **denatured.** A common example of this occurs when the gelatinous, clear portion of an egg is cooked and the protein changes to a white solid. Some medications are proteins and must be protected from denaturation so as not to lose their effectiveness. Insulin is an example. For protection, such medications may be stored in brown-colored bottles or kept under refrigeration.

The thousands of kinds of proteins can be placed into two categories. Some proteins are important for maintaining the shape of cells and organisms—they are usually referred to as **structural proteins.** The proteins that make up the cell membrane, muscle cells, tendons, and blood cells are examples of structural proteins. The other kinds of proteins, **regulator proteins,** help determine what activities will occur in the organism. These regulator proteins include enzymes and some hormones. These molecules help control the chemical activities of cells and organisms. Some examples of enzymes are the digestive enzymes in the stomach and the mouth. Two hormones that are regulator proteins are insulin and oxytocin. Insulin is produced by the pancreas and controls the amount of glucose in the blood. If insulin production is too low, or if the molecule is improperly constructed, glucose molecules are not removed from the bloodstream at a fast enough rate. The excess sugar is then eliminated in the urine. Other symptoms of excess sugar in the blood include excessive thirst and even loss of consciousness. The disease caused by improperly functional insulin is known as *diabetes*. Oxytocin, a second protein hormone, stimulates the contraction of the uterus during childbirth. It is also an example of an organic molecule that has been produced artificially and is used by physicians to induce labor.

Nucleic Acids

The last group of organic molecules that we will consider are the *nucleic acids*. **Nucleic acids** are complex molecules that store and transfer information within a cell. They are constructed of fundamental monomers known as **nucleotides.** Each nucleic acid is a polymer composed of nucleotides bonded together. There are eight different nucleotides: each is constructed of a phosphate group, a sugar, and an organic nitrogenous base.

The two kinds of sugar that can be part of the nucleotide are ribose and deoxyribose. These are five-carbon simple sugars. The phosphate group is attached to the sugar and the nitrogenous base is attached to another part of the sugar. There are five common organic molecules containing nitrogen that are likely to be part of the nucleotide structure. The nitrogen-containing bases are adenine, guanine, cytosine, thymine, and uracil.

The nucleotides can then connect to form long chains by dehydration synthesis reactions. The long chains of nucleotides are of two types—RNA (ribonucleic acid) and DNA (deoxyribonucleic acid). The RNA forms a single polymer, whereas the DNA is generally composed of two matching polymers twisted together and held by hydrogen bonds. These two types of molecules contain the information needed for the formation of particular sequences of amino acids; they determine what kinds of proteins an organism can manufacture. The mechanisms of storing and using this information are replication and protein synthesis.

appendix

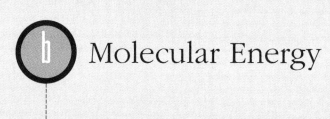

b Molecular Energy

Molecules have a certain amount of energy, and therefore are able to move. While we cannot see the movement of the individual molecules, we can deduce several things about their movement by measuring their activity and noting the results of their movement.

energy. Considering the amount of energy in the molecules of each state of matter helps us explain changes such as freezing and melting. When a liquid becomes a solid, its molecules lose some of their energy; when it becomes a gas, its molecules gain energy.

Molecular Motion

All molecules have a certain amount of **kinetic energy,** the energy of motion. The amount of energy that a molecule has is related to how fast it moves. **Temperature** is a measure of this velocity or energy of motion. The higher the temperature, the faster the molecules are moving. The three **states of matter**—solid, liquid, and gas—can be explained by thinking of the relative amounts of energy possessed by the molecules of each. A **solid** contains molecules packed tightly together. The molecules vibrate in place and are strongly attracted to each other. They are moving rapidly and constantly bump into each other. The amount of kinetic energy in a solid is less than that in a liquid of the same material. A **liquid** has molecules still strongly attracted to each other, but slightly farther apart. Since they are moving more rapidly, they sometimes slide past each other as they move. This gives the flowing property to a liquid. Still more energetic are the molecules of a **gas.** The attraction the gas molecules have for each other is overcome by the speed with which the individual molecules move. Since they are moving the fastest, their collisions tend to push them farther apart, and so a gas expands to fill its container. A common example of a substance that displays the three states of matter is water. Ice, liquid water, and water vapor are all composed of the same chemical—H_2O. The molecules are moving at different speeds in each state because of the difference in kinetic

Diffusion

Because the cell membrane is composed of phospholipid and protein molecules that are in constant motion, temporary openings are formed that allow small molecules to cross from one side of the membrane to the other. Molecules close to the membrane are in constant motion as well. They are able to move into and out of a cell by passing through these openings in the membrane. **Net movement** is the movement of molecules in one direction minus the movement of molecules in the opposite direction. The net movement of a particular kind of molecule from an area of higher concentration to an area of lower concentration is called **diffusion.** If the concentration of a specific kind of molecule is higher on the outside of a cell membrane than on the inside, those molecules will diffuse through the membrane into the cell if the membrane is permeable to the molecules.

The rate of diffusion is related to the kinetic energy and size of the molecules. Since diffusion only occurs when molecules are unevenly distributed, the relative concentration of the molecules is important in determining how fast diffusion occurs. The difference in concentration of the molecules is known as a **concentration gradient** or **diffusion gradient.** When the molecules are equally distributed, no such gradient exists.

Diffusion can take place only as long as there are no barriers to the free movement of molecules. In the case of a cell, the membrane

permits some molecules to pass through, while others are not allowed to pass or only allowed to pass more slowly. This permeability is based on size, ionic charge, and solubility of the molecules involved. The membrane does not, however, distinguish direction of movement of molecules; therefore, the membrane does not influence the direction of diffusion. The direction of diffusion is determined by the relative concentration of specific molecules on the two sides of the membrane, and the energy that causes diffusion to occur is supplied by the kinetic energy of the molecules themselves (figure B.1).

Diffusion is an important means by which materials are exchanged between a cell and its environment. Since the movement of the molecules is random, the cell has little control over the process; thus, diffusion is considered a passive process. For example, animals are constantly using oxygen in various chemical reactions. Consequently, the oxygen concentration in cells always remains low. The cells then contain a lower concentration of oxygen in comparison to the oxygen level outside of the cell. This creates a diffusion gradient, and the oxygen molecules diffuse from the outside of the cell to the inside of the cell.

In large animals, many of the cells are buried deep within the body; if it were not for the animals' circulatory systems, there would be little opportunity for cells to exchange gases directly with their surroundings. The circulatory system is a transportation system within a body composed of blood vessels of various sizes. These vessels carry many different molecules from one place to another. Oxygen may diffuse into blood through the membranes of the lungs,

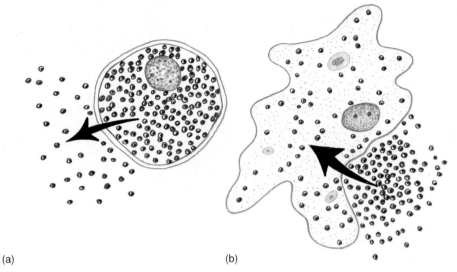

(a)   (b)

Figure B.1 Diffusion As a result of molecular motion, molecules move from areas where they are concentrated to areas where they are less concentrated. This figure shows (a) molecules leaving a cell by diffusion and (b) molecules entering a cell by diffusion. The direction is controlled by concentration, and the energy necessary is supplied by the kinetic energy of the molecules themselves.

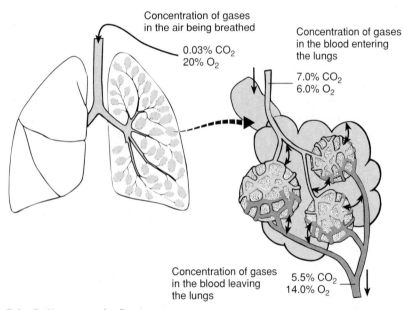

Concentration of gases
in the air being breathed

0.03% CO_2
20% O_2

Concentration of gases
in the blood entering
the lungs

7.0% CO_2
6.0% O_2

Concentration of gases
in the blood leaving
the lungs

5.5% CO_2
14.0% O_2

Figure B.2 Diffusion in the Lungs As blood enters the lungs, it has a higher concentration of carbon dioxide and a lower concentration of oxygen than the air in the lungs. The concentration gradient of oxygen is such that it diffuses from the lungs into the blood, and the concentration gradient of carbon dioxide is such that it diffuses from the blood into the lungs. These two different diffusions happen simultaneously, and the direction of diffusion is controlled by the relative concentrations of each kind of molecule in the blood and in the lungs.

gills, or other moist surfaces of the animal's body. The circulatory system then transports the oxygen-rich blood throughout the body. The oxygen automatically diffuses into cells that are low in oxygen. The opposite is true of carbon dioxide. Animal cells constantly produce carbon dioxide, and so there is always a high concentration of it within the cells. These molecules diffuse from the cells into the blood, where the concentration of carbon dioxide is lower. The blood is pumped to the moist surface (gills, lungs, etc.), and the carbon dioxide diffuses into the surrounding environment, which has a lower concentration of this gas. In a similar manner, many other types of molecules constantly enter and leave cells (figure B.2).

Dialysis and Osmosis

Another characteristic of all membranes is that they are differentially permeable. (The terms *selectively permeable* and *semipermeable* are synonyms.) **Differential permeability** means that a membrane will allow certain molecules to pass across it and will prevent others from doing so. Molecules that are able to dissolve in phospholipids, such as vitamins A and D, can pass through the membrane rather easily; however, many molecules cannot pass through at all. In certain cases, the membrane differentiates on the basis of molecular size; that is, the membrane allows small molecules, such as water, to pass through and prevents the passage of larger molecules. The membrane may also regulate the passage of ions. If a particular portion of the membrane has a large number of positive ions on its surface, positively charged ions in the environment will be repelled and prevented from crossing the membrane.

We make use of diffusion across a differentially permeable membrane when we use a dialysis machine to remove wastes from the blood. If a kidney is unable to function normally, blood from a patient is diverted to a series of tubes composed of differentially permeable membrane. The toxins that have concentrated in the blood diffuse into the surrounding fluids in the dialysis machine, and the cleansed blood is returned to the patient. Thus, the machine functions in place of the kidney.

Water is a molecule that easily diffuses through cell membranes. The net movement of water molecules through a differentially permeable membrane is known as **osmosis.** In any osmotic situation, there must be a differentially permeable membrane separating two solutions. For example, if a solution of 90 percent water and 10 percent sugar is separated by a differentially permeable membrane from a solution of 80 percent water and 20 percent sugar, osmosis will occur. The membrane allows water molecules to pass freely but prevents the larger sugar molecules from crossing. There is a higher concentration of water molecules in one solution (compared to the concentration of water molecules in the other), so more of the water molecules move from the solution with 90 percent water to the solution with 80 percent water. Be sure that you recognize that osmosis is really diffusion in which the diffusing substance is water,

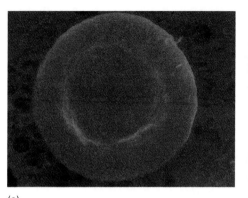

(a)

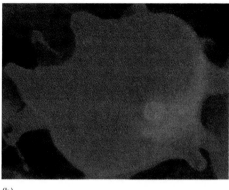

(b)

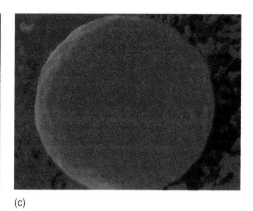

(c)

Figure B.3 Osmotic Influences on Cells The cells in these three photographs were subjected to three different environments. In (a), the cell is isotonic to its surroundings. The water concentration inside the red blood cell and the water concentration in the environment are in balance, so movement of water into the cell equals movement of water out of the cell, and the cell has a normal shape. In (b), the cell is in a hypertonic solution. Water has diffused from the cell to the environment because a higher concentration of water was in the cell, and the cell has shrunk. Photo (c) shows a cell that has accumulated water from the environment because a higher concentration of water was outside the cell than in its protoplasm. The cell is in a hypotonic solution so it has swollen.

and that the regions of different concentrations are separated by a membrane that is permeable to water.

A proper amount of water is required if a cell is to function efficiently. Too much water in a cell may dilute the cell contents and interfere with the chemical reactions necessary to keep the cell alive. Too little water in the cell may result in a buildup of poisonous waste products. As with the diffusion of other molecules, osmosis is a passive process because the cell has no control over the diffusion of water molecules. This means that the cell can remain in balance with an environment only if that environment does not cause the cell to lose or gain too much water.

Many organisms have a concentration of water and dissolved materials within their cells that is equal to that of their surroundings. When the concentration of water is equal on both sides of the cell membrane, the cell is said to be **isotonic** to its surroundings. This is particularly true of simple organisms that live in the ocean. The ocean has many kinds of salts dissolved in it, and such organisms as sponges, jellyfishes, and protozoa can be isotonic because the amount of material dissolved in their cellular water is equal to the amount of salt dissolved in the ocean's water.

However, if an organism is going to survive in an environment that has a different concentration of water than its cells, it must expend energy to maintain this difference. Organisms that live in freshwater have a lower concentration of water (higher concentration of dissolved materials) than their surroundings and tend to gain water by osmosis very

rapidly. They are said to be **hypertonic** to their surroundings, and the surroundings are **hypotonic.** These two terms are always used to compare two different solutions. The hypertonic solution is the one with more dissolved material and less water; the hypotonic solution has less dissolved material and more water. Organisms whose cells gain water by osmosis must expend energy to eliminate any excess if they are to keep from swelling and bursting (figure B.3).

Under normal conditions, when we drink small amounts of water the cells of the brain will swell a little, and signals are sent to the kidneys to rid the body of excess water. By contrast, marathon runners may drink large quantities of water in a very short time following a race. This rapid addition of water to the body may cause abnormal swelling of brain cells because the excess water cannot be gotten rid of rapidly enough. If this happens, the person may lose consciousness or even die because the brain cells have swollen too much.

Plant cells also experience osmosis. If the water concentration outside the plant cell is higher than the water concentration inside, more water molecules enter the cell than leave. This creates internal pressure within the cell. But plant cells do not burst, because they are surrounded by a rigid cell wall. Lettuce cells that are crisp are ones that have gained water so that there is high internal pressure. Wilted lettuce has lost some of its water to its surroundings so that it has only slight internal cellular water pressure. Osmosis occurs when you put salad dressing on a salad. Because the dressing

has a very low water concentration, water from the lettuce diffuses from the cells into the surroundings. Salad that has been "dressed" too long becomes limp and unappetizing.

So far, we have considered only situations in which the cell has no control over the movement of molecules. Cells cannot rely solely on diffusion and osmosis, because many of the molecules they require either cannot pass through the membrane or occur in relatively low concentrations in the cell's surroundings.

Controlled Methods of Transporting Molecules

Some molecules move across the membrane by combining with specific carrier proteins. When the rate of diffusion of a substance is increased in the presence of a carrier, we call this **facilitated diffusion.** Since this is diffusion, the net direction of movement is in accordance with the concentration gradient. Therefore, this is considered a passive transport method, although it can only occur in living organisms with the necessary carrier proteins. One example of facilitated diffusion is the movement of glucose molecules across the membranes of certain cells. In order for the glucose molecules to pass into these cells, specific proteins are required to carry them across the membrane. The action of the carrier does not require an input of energy other than the kinetic energy of the molecules.

When molecules are moved across the membrane from an area of *low* concentration to an area of *high* concentration, the cell must be expending energy. The process of using a carrier protein to move molecules up a concentration gradient is called **active transport** (figure B.4). Active transport is very specific: only certain molecules or ions are able to be moved in this way, and they must be carried by specific proteins in the membrane. The action of the carrier requires an input of energy other than the kinetic energy of the molecules; thus the name *active transport*. For example, some ions, such as sodium and potassium, are actively pumped across cell membranes. Sodium ions are pumped out of cells up a concentration gradient. Potassium ions are pumped into cells up a concentration gradient.

In addition to active transport, two other methods are used to actively move materials into cells. **Phagocytosis** is the process that cells use to wrap membrane around a particle (usually food) and engulf it. This is the process leukocytes (white blood cells) in your body use to surround invading bacteria, viruses, and other foreign materials. Because of this, these kinds of cells are called *phagocytes*. When phagocytosis occurs, the material to be engulfed touches the surface of the phagocyte and causes a portion of the outer cell membrane to be indented. The indented cell membrane is pinched off inside the cell to form a sac containing the engulfed material. This sac, composed of a single membrane, is called a **vacuole.** Once inside the cell, the membrane of the vacuole is broken down, releasing its contents inside the cell.

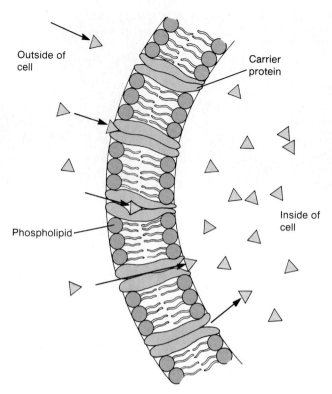

Figure B.4 Active Transport One possible method whereby active transport could cause materials to accumulate in a cell is illustrated here. Notice that the concentration gradient is such that if simple diffusion were operating, the molecules would leave the cell. The action of the carrier protein requires an active input of energy other than the kinetic energy of the molecules. Thus this process is termed active transport.

Phagocytosis is used by many types of cells to acquire large amounts of material from their environment. However, if the cell is not surrounding a large quantity of material, but is merely engulfing some molecules dissolved in water, the process is termed **pinocytosis.** In this process, the sacs that are formed are very small in comparison to those formed during phagocytosis. Because of this size difference, they are called **vesicles.** In fact, an electron microscope is needed in order to see them.

The processes of phagocytosis and pinocytosis differ from active transport in that the cell surrounds large amounts of material with a membrane rather than taking the material in through the membrane, molecule by molecule. The movement of materials into the cell by either phagocytosis or pinocytosis is referred to as *endocytosis*. The transport of material from the cell by the reverse processes is called *exocytosis*.

Cell Size

The size of a cell is directly related to its level of activity and the rates of movements of molecules across cell membranes. In order to stay alive, a cell must have a constant supply of nutrients, oxygen, and other molecules. It must also be able to get rid of carbon dioxide and other waste products that are harmful to it. The larger a cell becomes, the more difficult it is for the cell to satisfy these requirements; consequently most cells are very small. There are a few exceptions to this general rule, but they are easily explained. Egg cells, like the yolk of a hen's egg, are very large cells. However, the only part of an egg cell that is metabolically active is a small spot near its surface. The central portion of the egg is simply inactive stored food called *yolk*. Similarly, some plant cells are very large but consist of a large, centrally located region filled with water. Again, the metabolically active portion of the cell is at the surface, where exchange by diffusion or active transport is possible.

 EXPLORATIONS   **Interactive Software**

Active Transport

This interactive exercise allows students to explore how substances are transported across membranes against a concentration gradient (that is, toward a region of higher concentration). The exercise presents a diagram of a coupled channel within a membrane through which amino acids are pumped into the cell. By altering ATP concentrations, the student can speed or slow the operation of the ATP-driven sodium/potassium pump and explore the consequences for amino acid transport. Similarly, the student can alter the cellular or extracellular levels of amino acid and investigate the effect on cellular expenditure of ATP. Because the amino acid transport channel is coupled to the ATP-driven sodium/potassium pump, students will discover that both ATP and amino acid levels have important influences.

1. How does the level of ATP influence the operation of the sodium/potassium pump?

2. Does a fall in cellular levels of ATP inhibit cellular uptake of amino acids?

3. Does an increase in extracellular levels of amino acid always lead to an increased expenditure of ATP?

4. As long as ATP is readily available, is there any condition under which an increase in extracellular levels of amino acid does not result in increased transport of amino acid into the cell?

glossary

acetylcholine (a-sēt″l-ko′lēn) A neurotransmitter secreted into the synapse by many axons and received by dendrites that initiates a nerve impulse.

acetylcholinesterase (a-sēt″l-ko″li-nes′te-rās) An enzyme present in the synapse that destroys acetylcholine.

acid (as′id) Any compound that releases a hydrogen ion in a solution.

acquired characteristics (a-kwird′ kar″ak-ter-iss′tiks) A characteristic of an organism gained during its lifetime, not determined genetically, and therefore not transmitted to the offspring.

actin (ak′tin) A protein found in the thin filaments of muscle fibers that binds to myosin.

active transport (ak′tive trans′port) Use of a carrier molecule to move molecules across a cell membrane often from an area of lower concentration to an area of higher concentration. The carrier requires an input of energy.

adaptive radiation (uh-dap′tiv rā-de-ā′shun) A specific evolutionary pattern in which there is a rapid increase in the number of kinds of closely related species.

adenine (a′den-ēn″) A double-ring nitrogen-containing molecule in DNA and RNA. It is the complementary base of thymine or uracil.

aerobic cellular respiration (a ro′bik sel′yu-lar res″pi-ra′shun) The conversion of oxygen and food, such as carbohydrates, to carbon dioxide and water. During this conversion, energy is released.

age distribution (aj dis″tri-biu′shun) The number of organisms of each age in a population.

algae (al′je) Protists that have cell walls and chlorophyll and can therefore carry on photosynthesis.

alleles (a-lēlz′) Alternative forms of a gene for a particular characteristic.

amino acid (ah-mēn′o a′sid) A subunit of protein.

anaphase (an′a-faze) The third stage of mitosis, characterized by splitting of the centromeres and movement of the chromosomes to the poles.

androgens (an′dro-jenz) Male sex hormones produced by the testes and adrenal glands that cause the differentiation of the internal and external genital anatomy; also important in the sex drive of males and females.

angiosperms (an′je-o″spurmz) Plants that produce flowers and fruits.

annual (an′yu-uhl) A plant that completes its life cycle in one year.

anorexia nervosa (an″o-rek′se-ah ner-vo′sah) A nutritional deficiency disease characterized by severe, prolonged weight loss due to fear of becoming obese. This eating disorder is thought to stem from sociocultural factors.

anthropomorphism (an-thro-po-mor′fizm) The assigning of human feelings, emotions, or meanings to the behavior of animals.

anticodon (an″te-ko′don) A sequence of three nitrogen-containing bases on a tRNA molecule capable of forming bonds with three complementary bases on an mRNA codon during translation.

antidiuretic hormone (an-ti-di″u-re′tik hor′mōn) A hormone produced by the pituitary gland that stimulates the kidney to reabsorb water.

aorta (a-or′tah) A large blood vessel that carries blood from the left ventricle to the majority of the body.

arteries (ar′te-rēz) The blood vessels that carry blood away from the heart.

assumption (e-sump′shun) A speculation; something you think might be true.

atria (ā′tre-ah) Thin-walled sacs of the heart that receive blood from the veins of the body and empty it into the ventricles.

autosomes (aw′to-somz) Chromosomes that are composed of the genes that determine general body features.

autotrophs (aw-to-trofs) Organisms able to use light energy to produce organic nutrients from organic materials; a self-feeder.

axon (ak′sahn) A neuronal fiber that carries information away from the nerve cell body.

bacteria (bak-tir′e-ah) Unicellular organisms of the kingdom Prokaryotae that have the genetic ability to function in various environments.

basal metabolic rate (BMR) (ba′sal meta-bol′ik rāt) The amount of energy required to maintain normal body activity while at rest.

base (bas) Any compound that releases a hydroxyl group in a solution.

basilar membrane (ba′si-lar mem′brān) A membrane in the cochlea containing sensory cells that are stimulated by the vibrations caused by sound waves.

behavior (be-hav′yur) How an organism acts, what it does, and how it does it.

behavioral isolation (be-hav′yu-ral i-so-la′shun) A genetic isolating mechanism that prevents interbreeding between species because of differences in behavior.

benthic (ben′thik) A term used to describe organisms that live in bodies of water, attached to the bottom or to objects in the water.

biennial (bi-e′ne-al) A plant that requires two years to complete its life cycle.

binomial system of nomenclature (bi-no′mi-al sis′tem ov no′men-kla-ture) A naming system that uses two Latin names, genus and species, for each type of organism.

biogenesis (bi-o-jen′uh-sis) The concept that life originates only from preexisting life.

biological amplification (bi-o-loj′i-cal am″pli-fi-ka′shun) The accumulation of a compound in increasing concentrations in organisms at successively higher trophic levels.

biomass (bi′o-mas) The dry weight of a collection of designated organisms.

biomes (bi′omz) Large regional communities.

biotechnology (bi-o-tek-nol′uh-je) The science of gene manipulation.

birthrate (burth′rat) The number of individuals entering the population by reproduction per thousand individuals in the population; also called *natality*.

blood (blud) The fluid medium consisting of cells and plasma that assists in the transport of materials and heat.

bloom (bloom) A rapid increase in the number of microorganisms in a body of water.

buffer (bu′fer) A mixture of a weak acid and the salt of a weak acid that operates to maintain a constant pH.

bulimia (bu-lim′e-ah) A nutritional deficiency disease characterized by a binge-and-purge cycle of eating. It is thought to stem from psychological disorders.

calorie (kal′o-rē) The amount of heat energy necessary to raise the temperature of one gram of water one degree Celsius.

cancer (kan′sur) A tumor that is malignant.

capillaries (cap′i-lair-ēz) Tiny blood vessels through which exchange between cells and the blood takes place.

carbohydrate (kar-bo-hi′drāt) One class of organic molecules composed of carbon, hydrogen, and oxygen. The basic building block of a carbohydrate is a simple sugar.

carnivores (kar′ni-vorz) Animals that eat other animals.

carrier (ka′rē-er) Any individual having a hidden, recessive allele.

carrying capacity (ka′rē-ing kuh-pas′i-tē) The optimum population size an area can support over an extended period of time.

cartilage (kar′tl-ij) A flexible, plasticlike supporting tissue in many animals; in humans, it is in the nose, ears, and certain joints.

catalyst (cat′uh-list) A chemical that speeds up a reaction but is not used up in the reaction.

cell (sel) A characteristic of life; the basic structural unit that makes up all living things.

central nervous system (sen′trul ner′vus sis′tem) The portion of the nervous system consisting of the brain and spinal cord.

centromere (sen′tro-mere) The region where two chromatids are joined.

chlorophyll (klo′ro-fil) The green pigment located in the chloroplasts of plant cells that is associated with capturing light energy.

chloroplasts (klo′ro-plasts) Energy-converting organelles in plant cells. Chloroplasts contain the green pigment chlorophyll.

chromatid (kro′mah-tid) One of two identical component parts of a chromosome attached at the centromere.

chromatin (kro′mah-tin) Areas or structures within the nucleus of a cell composed of DNA in association with proteins.

chromatin fibers (kro′mah-tin fi′berz) The double DNA strands with attached proteins; also called *nucleoproteins*.

chromosomes (kro′mo-somz) Complex structures within the nucleus composed of various kinds of proteins and DNA that contain the cell's genetic information; densely coiled chromatin.

class (klas) A group of closely related orders found within a phylum.

classification (kla-se-fe-ka′shen) The assortment of objects into groups based on their similarities and differences.

climax community (kli′maks ko-miu′ni-te) A relatively stable, long-lasting community.

clitoris (kli′to-ris) Small erectile structure of the female genitals containing many nerve endings that cause pleasurable sensations.

clones (klōnz) All of the individuals reproduced asexually that have exactly the same genes.

cochlea (kok′lē-ah) The part of the ear that converts sound into nerve impulses.

codon (ko′don) A sequence of three nucleotides of an mRNA molecule that directs the placement of a particular amino acid during translation.

colonial (ko-lo′ne-al) A term used to describe a collection of cells that cooperate to a small extent.

commensalism (ko-men′sal-izm) A relationship between two organisms in which one organism is helped and the other is not affected.

community (ko-miu′ni-te) A collection of interacting organisms within an ecosystem.

competition (com-pe-ti′shun) A relationship between two organisms in which both organisms are harmed.

conditioned response (kon-di′shund re-spons′) The behavior displayed when the neutral stimulus is given after association has occurred.

conditioning (kon-di′shun-ing) A kind of learning in which a neutral stimulus is associated with a natural stimulus to produce a particular response.

cones (kōnz) Light-sensitive cells in the retina of the eye that respond to different colors of light.

consumers (kon-soom′urs) Organisms that must obtain energy in the form of organic matter.

continental drift (kan″ten-en′tal drift) The movement of the Earth's crustal plates.

control group (kon-trol′ grup) The situation used as the basis for comparison in a controlled experiment.

controlled experiment (kon-trold′ ik-sper′e-ment) An experiment that allows for a comparison of two events that are identical in all but one respect.

convergent evolution (kon-vur′jent ev-o-lu′shun) An evolutionary pattern in which widely different organisms show similar characteristics.

copulation (kop-yu-la'shun) The mating of male and female; the deposition of the male sex cells, or sperm cells, in the reproductive tract of the female.

crossing-over (kro'sing o'ver) The exchange of a part of a chromatid from one chromosome with an equivalent part of a chromatid from a homologous chromosome.

cytokinesis (si-to-ki-ne'sis) Division of the cytoplasm of one cell into two new cells.

cytoplasm (si'to-plazm) The semifluid portion of the cell. Cytoplasm surrounds the nucleus of eukaryotic cells.

cytosine (si'to-sēn) A single-ring nitrogen-containing molecule in DNA and RNA. It is complementary to guanine.

daughter cells (daw'tur sels) Two cells formed by cell division.

death phase (deth fāz) The portion of some population growth curves in which the size of the population declines.

deathrate (deth' rat) The number of individuals leaving the population by death per thousand individuals in the population; also called *mortality*.

decomposers (de-kom-po'zurs) Organisms that use dead organic matter as a source of energy.

denature (de-na'chur) To permanently change the protein structure of an enzyme so that it loses its ability to function.

dendrites (den'drīts) Neuronal fibers that receive information from axons and carry it toward the nerve cell body.

density-dependent factors (den'si-te de-pen'dent fak'torz) Population-limiting factors that become more effective as the size of the population increases.

density-independent factors (den'si-te in"de-pen'dent fak'torz) Population-controlling factors that are not related to the size of the population.

deoxyribonucleic acid (DNA) (dē-ok"sē-rī-bō-nū-klē'ik as'id) A polymer of nucleotides that serves as genetic information. In prokaryotic cells, it is a double-stranded loop. In eukaryotic cells, it is found in strands with attached proteins. When tightly coiled, it is known as a *chromosome*.

depolarized (de-po'la-rīzd) Having lost the electrical difference existing between two points or objects.

development (di-ve'lep-ment) A characteristic of life; change in form over time.

diaphragm (di'uh-fram) A muscle separating the lung cavity from the abdominal cavity that is involved in exchanging the air in the lungs.

diet (di'et) The food and drink consumed by a person from day to day.

differentiation (dif"fur-ent-she-ā'shun) The process of forming specialized cells within a multicellular organism.

diffusion (di-fiu'zhun) Net movement of a kind of molecule from a place where that molecule is in higher concentration to a place where that molecule is in lower concentration.

diploid (dip'loid) A cell that has two sets of chromosomes: one set from the maternal parent and one set from the paternal parent.

divergent evolution (di-vur'jent ev-o-lu'shun) A basic evolutionary pattern in which individual speciation events cause many branches in the evolution of a group of organisms.

DNA See *deoxyribonucleic acid.*

DNA replication (rep"li-ka'shun) The process by which the genetic material (DNA) of the cell reproduces itself prior to its distribution to the next generation of cells.

dominance hierarchy (dom'in-ants hi'ur-ar-ke) A relatively stable, mutually understood order of priority within a group.

dominant allele (dom'in-ant a-lēl) A gene form that expresses itself and hides the effect of other gene forms for that trait.

double-factor cross (dub'l fak'tur kros) A genetic study in which two pairs of alleles are followed from the parental generation to the offspring.

Down syndrome (down sin'drom) A genetic disorder resulting from the presence of an extra chromosome number 21. Symptoms include slightly slanted eyes, flattened facial features, a large tongue, and a tendency toward short stature and fingers. Some individuals also display mental retardation.

ecological isolation (e-ko-loj'i-kal i-so-la'shun) A genetic isolating mechanism that prevents interbreeding between species because they live in different areas; also called *habitat preference.*

ecology (e-kol'o-je) The branch of biology that studies the relationships between organisms and their environment.

ecosystem (e"ko-sis-tum") An interacting collection of organisms and the nonliving factors that affect them.

ectothermic (ek"te-ther'mik) Animals whose body temperature is determined by external environment.

egg (eg) The haploid sex cells produced in the ovary by sexually mature females.

ejaculation (e-jak"u-la'shun) The release of sperm cells and seminal fluid through the penis of a male.

endocrine glands (en'do-krin glandz) Glands that secrete into the circulatory system.

endocrine system (en'do-krin sis'tem) A number of glands that communicate with one another and other tissues through chemical messengers transported throughout the body by the circulatory system.

endospore (en'do-spor") A unique bacterial structure with a low metabolic rate that germinates under favorable conditions to grow into a new cell.

endosymbiotic theory (en"do-sim-be-ot'ik the'o-re) A theory suggesting that present-day eukaryotic cells evolved from the combining of several different types of primitive prokaryotic cells.

endothermic (en"de-ther'mik) Animals that have internal regulating mechanisms to maintain a relatively constant body temperature in spite of wide temperature variations in their environment.

energy (e'ner-je) The ability to do work.

environment (en-vi'ron-ment) Anything that affects an organism during its lifetime.

enzyme (en'zīm) A specific protein that acts as a catalyst to change the rate of a reaction.

epinephrine (e"pi-nef'rin) A hormone produced by the adrenal medulla that increases heart rate, blood pressure, and breathing rate.

epiphyte (ep'e-fīt) A plant that lives on the surface of another plant.

estrogens (es'tro-jens) Female sex hormones that cause the differentiation in the female embryo of the internal and external genital anatomy responsible for the changes in breasts, vagina, uterus, clitoris, and pelvic bone structure at puberty.

ethology (e-thol'uh-je) The scientific study of the nature of behavior and its ecological and evolutionary significance in its natural setting.

eukaryotic cells (yu-ka-re-ah-tik sels) One of the two major types of cells; complex cells with a true nucleus and organelles composed of membranes; in plants, fungi, protists, and animals.

evolution (ev-o-lu'shun) A characteristic of life; the genetic adaptation of a population of organisms to its environment.

exocrine glands (ek'sa-krin glandz) Glands that secrete through ducts to the surface of the body or into hollow organs of the body.

experimental group (ik-sper"e-men'tl grup) The group in a controlled experiment that is identical to the control group in all respects but one.

exponential growth phase (eks-po-nen'shul groth fāz) A period of time during population growth when the population increases at an accelerating rate.

fact (fakt) Something that is true.

family (fam'i-ly) A group of closely related species within an order.

fertilization (fer"ti-li-za'shun) The joining of haploid nuclei, usually from an egg and a sperm cell, resulting in a diploid cell called the zygote.

fiber (fi'ber) Natural (plant) or industrially produced polysaccharides that are resistant to digestion by the human enzymes in the intestinal tract.

flower (flau'er) The plant structure that produces pollen or eggs and usually contains petals.

follicle-stimulating hormone (FSH) (fol'i-kul stim'yu-la-ting hor'mōn) The pituitary secretion that causes the ovaries to begin to produce larger quantities of estrogen and to develop the follicle and prepare the egg for ovulation.

food chain (food chān) A sequence of organisms that feed on one another, resulting in a flow of energy from a producer through a series of consumers.

Food Guide Pyramid (food gīd pi'ra-mid) A tool developed by the U.S. Department of Agriculture to help the general public plan good nutrition. It contains guidelines for required daily intake from each of the five food groups.

food web (food web) A system of interlocking food chains.

fovea centralis (fo've-ah sen-tral'is) The area of sharpest vision on the retina, where light is normally focused.

free-living nitrogen-fixing bacteria (ni'tro-jen fik'sing bak-te're-ah) Soil bacteria that convert nitrogen gas molecules into nitrogen compounds that plants can use.

fruit (frut) A structure that surrounds the seeds in flowering plants.

fungus (fun'gus) The common name for the kingdom Mycetae whose members are nonphotosynthetic, multicellular organisms, have cell walls, and that reproduce by spores.

gamete (gam'ēt) A haploid sex cell.

gametogenesis (ga-me"to-jen'e-sis) The generating of gametes; the meiotic cell-division process that produces sex cells.

gene (jēn) Inheritable units composed of a segment of DNA that determine the physical, chemical, and behavioral characteristics of an organism. They are able to (1) replicate by directing the manufacture of copies of the molecule; (2) mutate, or chemically change, and transmit these changes to future generations; (3) store information that determines the characteristics of cells and organisms; and (4) use this information to direct the synthesis of structural and regulatory proteins.

gene frequency (jēn fre'kwen-se) A measure of the number of times that a gene occurs in a population. The percentage of sex cells that contain a particular gene.

gene pool (jēn pool) All the genes of all the individuals of a species.

genetic engineering (je-net'ik en-je-ner'ing) The science of gene manipulation.

genetic medicine (je-net'ik med'i-sin) The art and science of manipulating genes for the diagnosis and treatment of disease and the maintenance of health.

genetic recombination (je-net'ik re-kom-bi-na'shun) The gene mixing that occurs during sexual reproduction.

genetics (je-net'iks) The study of genes, how genes produce characteristics, and how characteristics are inherited.

genome (je'nōm) A set of all the genes necessary to specify an organism's complete list of characteristics.

genotype (je'no-tīp) The catalog of genes of an organism, whether or not these genes are expressed.

genus (je'nus) (pl. *genera*) A group of closely related species within a family.

geographic barriers (je-o-graf'ik bar'yurz) Geographic features that keep different portions of a species from exchanging genes.

geological time chart (je-a-la'jik-el tīm chart) A chart of the chronological history of living organisms based on the fossil record.

gland (gland) An organ that manufactures and secretes a material either through ducts or directly into the circulatory system.

gonad (go'nad) In animals, the organs in which meiosis occurs.

gradualism (grad'u-al-izm) The theory stating that evolution occurred gradually with an accumulated series of changes over a long period of time.

growth (growth) A characteristic of life; an increase in size or number of cells.

growth-stimulating hormone (grōth sti'mu-la-ting hor'mōn) A hormone produced by the anterior pituitary gland that stimulates tissues to grow.

guanine (gwah'nēn) A double-ring nitrogen-containing molecule in DNA and RNA. It is the complementary base of cytosine.

guess (ges) To form an opinion without supporting evidence.

gymnosperms (jim'no-spurmz) Plants that produce their seeds in cones.

habitat (hab'i-tat) The place or part of an ecosystem occupied by an organism.

haploid (hap'loid) Having a single set of chromosomes resulting from the reduction division of meiosis.

heart (hart) The muscular pump that forces the blood through the blood vessels of the body.

hemoglobin (he'mo-glo-bin) An iron-containing molecule found in red blood cells, to which oxygen molecules bind.

herbivores (her'bi-vorz) Animals that feed directly on plants.

hermaphroditic (her-ma-fre-di'tik) Having both sexes in the same body.

heterozygous (he"ter-o-zi'gus) A diploid organism that has two different allelic forms of a particular gene.

homeotherms (ho'mee-ah-therms") Animals (birds and mammals) that keep a constant body temperature; they are called warm-blooded and use cellular respiration to generate heat.

homologous chromosomes (ho-mol'o-gus kro'mo-somz) A pair of chromosomes in a diploid cell that contain similar genes at corresponding loci throughout their length.

homozygous (ho"mo-zi'gus) A diploid organism that has two identical alleles for a particular characteristic.

hormones (hor'mōnz) Chemicals produced in one part of an organism that alter the activities of a distant part; chemical messengers secreted by endocrine glands.

host (host) An organism that a parasite lives in or on.

hybrid (hy'brid) The offspring of two different genetic lines produced by sexual reproduction.

hypothalamus (hi"po-thal'a-mus) The region of the brain that causes the production of several kinds of hormones involved in sexual development and function.

hypothesis (hy-pa'the-sis) A possible answer to, or explanation of, a question that accounts for all the observed facts and is testable.

immune system (i-mūn' sis'tem) A system of white blood cells specialized to provide the body with resistance to disease. There are two types of protection: antibody-mediated immunity and cell-mediated immunity.

imprinting (im'prin-ting) Learning in which a very young animal is genetically primed to learn a specific behavior in a very short period.

incus (in'kus) The ear bone that is located between the malleus and the stapes.

independent assortment (in"de-pen'dent a-sort'ment) The segregation, or assortment, of one pair of homologous chromosomes, independently of the segregation, or assortment, of any other pair of chromosomes.

inorganic molecules (in-or-gan'ik mol'ah-kuls) Molecules that do not contain carbon atoms in rings or chains.

insecticide (in-sek'ti-sīd) A poison used to kill insects.

insight (in'sīt) Learning in which past experiences are reorganized to solve new problems.

instinctive behavior (in-stink'tiv be-hav'yur) Automatic, preprogrammed, or genetically determined behavior.

insulin (in'se-len) A hormone produced by the pancreas that regulates the amount of glucose in the blood.

interphase (in'tur-faze) The stage between cell divisions in which the cell is engaged in metabolic activities.

interstitial cell-stimulating hormone (ICSH) (in"ter-sti'shal sel stim'yu-la-ting hor'mōn) The chemical messenger molecule released from the pituitary that causes the testes to produce testosterone.

irrelevant (i-re'le-vent) Not related to the situation under discussion.

kilocalorie (kil"o-kal'o-re) A measure of heat energy one thousand times larger than a calorie; also called a *kcalorie*.

kingdom (king'dom) The largest grouping used in the classification of organisms.

lack of dominance (lak uv dom'in-ans) The condition of two unlike alleles both expressing themselves, neither being dominant.

lag phase (lag fāz) A period of time following colonization when the population remains small or increases slowly.

law of dominance (law uv dom'in-ans) When an organism has two different alleles for a trait, the allele that is expressed and overshadows the expression of the other allele is said to be dominant. The allele whose expression is overshadowed is said to be recessive.

law of independent assortment (law uv in"de-pen'dent a-sort'ment) Members of one allelic pair will separate from each other independently of the members of other allele pairs.

law of segregation (law uv seg"re-ga'shun) When gametes are formed by a diploid organism, the alleles that control a trait separate from one another into different gametes, retaining their individuality.

leaf (leef) The thin, flat, green structure of plants where chloroplasts are typically found and photosynthesis takes place.

learning (lur'ning) A change in behavior as a result of experience.

lichen (li'k'n) A mutualistic relation between fungi and algal protists or cyanobacteria.

limiting factors (lim'i-ting fak'torz) Environmental influences that limit population growth.

linkage group (lingk'ij grūp) Genes located on the same chromosome that tend to be inherited together.

lipids (li'pids) Large organic molecules that do not easily dissolve in water; classes include fats, phospholipids, and steroids.

locus (lo'kus) The spot on a chromosome where an allele is located (pl. *loci*).

lung (lung) A respiratory organ in which air and blood are brought close to one another and gas exchange occurs.

lymph (limf) Liquid material that leaves the circulatory system to surround cells.

lymphatic system (lim-fa'tik sis'tem) A collection of thin-walled tubes that collects, filters, and returns lymph from the body to the circulatory system.

malleus (ma'le-us) The ear bone that is attached to the tympanum.

manipulate energy (ma-ni'pu-late e'ner-jee) A characteristic of life; ability of an organism to transform or utilize energy.

manipulate matter (ma-ni'pu-late ma'ter) A characteristic of life; the ability of an organism to rearrange molecular structures.

mass extinction (mas ik-stink'shen) The disappearance of thousands of species of organisms during a short period of time. Mass extinctions are thought to be brought about by dramatic changes in either the physical or living components of the environment.

masturbation (mas"tur-ba'shun) Stimulation of one's own sex organs.

meiosis (mi-o'sis) The specialized pair of cell divisions that reduce the chromosome number from diploid ($2n$) to haploid (n).

Mendelian genetics (men-de'le-an je-net'iks) The pattern of inheriting characteristics that follows the laws formulated by Gregor Mendel.

menopause (men'o-pawz) The period beginning at about age 50 when the ovaries stop producing eggs and estrogen production drops.

menstrual cycle (men'stru-al si'kul) The repeated building up and shedding of the lining of the uterus.

messenger RNA (mRNA) (mes'-en-jer) A molecule composed of ribonucleotides that functions as a copy of the gene and is used in the cytoplasm of the cell during protein synthesis.

metabolism (me-ta'bol-izm) The total of all chemical reactions within an organism; for example, nutrient uptake and processing and waste elimination.

metamorphosis (me"te-mor'fe-sis) The process of changing from one body form to another during the developmental process; for example, from egg to larva to pupa and finally to adult.

metaphase (me'tah-faze) The second stage in mitosis, characterized by alignment of the chromosomes at the equatorial plane.

microorganisms (microbes) (mi"kro-or'guh-niz"mz) Small organisms that cannot be seen without magnification.

minerals (min'er-alz) Inorganic elements essential to metabolism that cannot be manufactured by the body but are required in low concentrations.

mitosis (mi-to'sis) A process that results in equal and identical separation and distribution of chromosomes into two newly formed nuclei.

motile (mo'tile) Able to move about freely; not sessile.

motor neurons (mo'tur noor'onz) Neurons that carry information from the central nervous system to muscles or glands.

motor unit (mo'tur yoo'nit) All of the muscle cells stimulated by a single nerve cell.

multi-regional continuity hypothesis (mul'ti-re'jen-el kan"ten-oo'e-te hi-poth'e-sis) The idea that *Homo sapiens* evolved throughout the Old World from *Homo erectus* populations that had previously migrated from Africa to Europe and Asia.

multiple alleles (mul'ti-pul a-lēlz') A term used to refer to conditions in which there are several different alleles for a particular characteristic, not just two.

mutation (miu-ta'shun) A change in the genetic material.

mutualism (miu'chu-al-izm) A relationship between two organisms in which both organisms benefit.

mycorrhiza (my"ko-rye'zah) A symbiotic relation between fungi and plant roots.

mycotoxin (mi"ko-tok'sin) A deadly poison produced by fungi.

myosin (mi'o-sin) A protein molecule found in the thick filaments of muscle fibers that attaches to and moves along actin molecules.

natural selection (nat'chu-ral se-lek'shun) A broad term used in reference to the various mechanisms that encourage the passage of beneficial genes and discourage the passage of harmful or less valuable genes to future generations.

negative-feedback control (ne'ga-tiv feed'bak con-trōl') A kind of control mechanism in which the product of one activity inhibits an earlier step in the chain of events.

nerve fibers (nerv fi'berz) Cytoplasmic extensions of nerve cells; axons and dendrites.

nerve impulse (nerv im'puls) A series of changes that take place in the neuron, resulting in a wave of depolarization that passes from one end of the neuron to the other.

nerves (nervz) Bundles of neuronal fibers.

nervous system (ner'vus sis'tem) A network of neurons that carry information from sense organs to the central nervous system and from the central nervous system to muscles and glands.

neuron (noor'on) The cellular unit, consisting of a cell body and fibers, that makes up the nervous system; also called *nerve cell*.

neurotransmitter (noor"o-trans'mit-er) A molecule released by the axons of neurons that stimulates other cells.

niche (nitch) The functional role of an organism.

nondisjunction (non"dis-junk'shun) An abnormal meiotic division that results in sex cells with too many or too few chromosomes.

norepinephrine (nor-e"pi-nef'rin) A hormone produced by the adrenal medulla that increases heart rate, blood pressure, and breathing rate.

nuclear membrane (nu'kle-ar mem'bran) The structure surrounding the nucleus that separates the nucleoplasm from the cytoplasm.

nucleic acids (nu'kle-ik as'ids) Complex molecules that store and transfer information within a cell. They are constructed of subunits known as nucleotides.

nucleoli (nu-kle'o-li) Nuclear structures containing information for ribosome construction.

nucleoplasm (nu'kle-o-pla"zem) The fluid of the nucleus composed of water and the molecules used in the construction of the rest of the nuclear structures.

nucleoproteins (nu-kle-o-pro'tēnz) The double DNA strands with attached proteins; also called *chromatin fibers*.

nucleotide (nu'kle-o-tīd) The building block of the nucleic acids. Each is composed of a 5-carbon sugar, a phosphate, and a nitrogen-containing portion.

nucleus (nu'kle-us) The central body that contains the information system for a eukaryotic cell.

nutrients (nu'tre-ents) Molecules required by the body for growth, reproduction, and repair.

nutrition (nu-tri'shun) Collectively, the processes involved in taking in, assimilating, and utilizing nutrients.

obese (o-bēs') A term describing a person who gains a great deal of unnecessary weight and is 15 percent to 20 percent above ideal weight.

olfactory epithelium (ol-fak'to-re e"pi-thē'le-um) The cells of the nasal cavity that respond to chemicals.

omnivores (om'ni-vorz) Animals that eat both plants and other animals.

opinion (e-pin'yin) A thought, idea, or belief about something.

order (or'der) A group of closely related organisms within a class.

orderliness (or'der-lee-nes) A characteristic of life; organization of parts into a coordinated whole.

organ (or'gun) A structure composed of two or more kinds of tissues.

organelles (or-gan-elz') Cellular structures that perform specific functions in the cell.

organic molecules (or-gan'ik mol'uh-kiuls) Complex molecules whose basic building blocks are carbon atoms in chains or rings.

orgasm (or'gaz-um) A complex series of responses to sexual stimulation that result in an intense frenzy of sexual excitement.

osmosis (os-mo'sis) The net movement of water molecules through a semipermeable membrane.

out of Africa hypothesis (owt uv a-fri'ka hi-poth'e-sis) The idea that *Homo sapiens* evolved in Africa from *Homo erectus* and then migrated throughout Europe and Asia, where they out-competed *Homo erectus* populations that had previously migrated from Africa to Europe and Asia.

oval window (o'val win'dō) The membrane-covered opening of the cochlea, to which the stapes is attached.

ovaries (o'var-ez) The female sex organs that produce haploid sex cells (the eggs or ova).

oviduct (o'vi-dukt) The tube that carries the oocyte to the uterus; also called the *fallopian tube*.

ovulation (ov-yu-la'shun) The release of an egg from the surface of the ovary.

oxytocin (ok"si-to'sin) A hormone released from the posterior pituitary that causes contraction of the uterus.

parasite (par'uh-sīt) An organism that lives in or on another living organism and derives nourishment from it.

parasitism (par'uh-sit-izm) A relationship between two organisms in which one organism lives in or on another organism and derives nourishment from it.

parthenogenesis (par"the-no-je'-ne-sis) A form of asexual reproduction in which females produce eggs that are not fertilized and that develop into offspring genetically identical to the parent.

pathogen (path'uh-jen) An agent that causes a specific disease.

penis (pe'nis) The portion of the male reproductive system that deposits sperm in the female reproductive tract.

perception (per-sep'shun) Recognition by the brain that a stimulus has been received.

perennial (pur-en'e-uhl) A plant that requires many years to complete its life cycle.

peripheral nervous system (pu-ri'fe-ral ner'vus sis'tem) The fibers that communicate between the central nervous system and other parts of the body.

pesticide (pes'ti-sid) A poison used to kill pests. This term is often used interchangeably with insecticide.

pH A scale used to indicate the strength of an acid or base.

phagocytosis (fa"jo-si-to'sis) The process by which the cell wraps around a particle and engulfs it.

phenotype (fen'o-tīp) The physical, chemical, and behavioral expression of the genes possessed by an organism.

pheromone (fer'uh-mōn) A chemical produced by an animal and released into the environment to trigger behavioral or developmental processes in some other animal of the same species.

photoperiod (fo"to-pir'e-ud) The length of the light part of the day.

photosynthesis (fo-to-sin'the-sis) The process of combining carbon dioxide and water with the aid of light energy to form sugar and release oxygen.

phylum (fi'lum) A subdivision of a kingdom.

phytoplankton (fye-tuh-plank'tun) Photosynthetic species that form the basis for most aquatic food chains.

pioneer community (pi"o-ner' ko-miu'ni-te) The first community of organisms in the successional process established in a previously uninhabited area.

pioneer organisms (pi"o-ner' or'gun-izms) The first organisms in the successional process.

pistil (pis'til) The sex organ in plants that produces eggs or ova.

pituitary gland (pi-tu'i-ta-re gland) The structure in the brain that controls the functioning of other glands throughout the organism.

placenta (plah-sen'tah) An organ made up of tissues from the embryo and the uterus of the mother that allows for the exchange of materials between the mother's bloodstream and the embryo's bloodstream. It also produces hormones.

plankton (plank'tun) Small floating or weakly swimming organisms.

plasma (plaz'mah) The watery matrix that contains the molecules and cells of the blood.

plasma membrane (plaz'mah mem-brān) The outer boundary membrane of the cell; surrounds all cells.

pleiotropy (pli-ot'ro-pe) The multiple effects that a gene may have on the phenotype of an organism.

poikilotherms (poy-key'le-therms") Animals whose body temperature fluctuates with the temperature of their environment; often called cold-blooded because their body temperature is usually lower than ours.

pollen (pah'len) Tiny particles produced by plants that contain the plant equivalent of sperm cells and are carried from one flower to another by wind or animals.

pollination (pol"i-na'shun) The transfer of pollen in gymnosperms and angiosperms.

polygenic inheritance (pol"e-jen'ik inher'i-tans) The concept that a number of different pairs of alleles may combine their efforts to determine a characteristic.

population (pop"u-la'shun) A group of organisms of the same species located in the same place at the same time.

population density (pop"u-la'shun den'site) The number of organisms of a species per unit area.

population genetics (pop"u-la'shun je-net'iks) The study of population gene frequencies.

population growth curve (pop"u-la'shun groth kurv) A graph of the change in population size over time.

population pressure (pop"u-la'shun prehs'yur) Intense competition that leads to changes in the environment and dispersal of organisms.

predation (pre-da'shun) A relationship between two organisms that involves the capturing, killing, and eating of one by the other.

predator (pred'uh-tor) An organism that captures, kills, and eats other animals.

prey (pra) An organism captured, killed, and eaten by a predator.

primary succession (pri'mar-e sukse'shun) The orderly series of changes that begins in a previously uninhabited area and leads to a climax community.

probability (prob"a-bil-'i-te) The chance that an event will happen, expressed as a percent or fraction.

producers (pro-du'surz) Organisms that produce new organic material from inorganic material with the aid of sunlight.

prokaryotic cells (pro-ka-re-ot-ik sels) One of the two major types of cells. They do not have a typical nucleus bound by a nuclear membrane and lack many of the other membranous cellular organelles; bacteria.

prophase (pro'faze) The first phase of mitosis during which individual chromosomes become visible.

protein (pro'tēn) Macromolecules made up of amino acid subunits attached to each other; groups of polypeptides.

protein-sparing (pro'tēn spe'ring) The conservation of proteins by first using carbohydrates and fats as a source of energy.

protein synthesis (pro'tēn sin'the-sis) The process whereby the tRNA utilizes the mRNA as a guide to arrange the amino acids in their proper sequence according to the genetic information in the chemical code of DNA.

protozoa (pro"to-zo'ah) Heterotrophic, unicellular organisms.

pseudoscience (su-do-si'ens) The use of the appearance of science to mislead. The assertions made are not valid or reliable.

puberty (pu'ber-te) A time in the life of a developing individual characterized by the increasing production of sex hormones, which cause it to reach sexual maturity.

pulmonary artery (pul'muh-na-rē ar'tuh-rē) The major blood vessel that carries blood from the right ventricle to the lungs.

pulmonary circulation (pul'muh-na-re ser-ku-la'shun) The flow of blood through certain chambers of the heart and blood vessels to the lungs and back to the heart.

punctuated equilibrium (pung'chu-a-ted e-kwi-lib're-um) The theory stating that evolution occurs in spurts, between which there are long periods with little evolutionary change.

Punnett square (pun'net sqwar) A method used to determine the probabilities of allele combinations in a zygote.

recessive alleles (re-se'siv a-lēlz) A gene that is hidden by other gene forms.

Recommended Dietary Allowances (RDA) (re-ko-men'ded di'e-te-rē a-lao'an-ses) U.S. dietary guidelines for a healthy person that focus on the six classes of nutrients.

reduction division (re-duk'shun di-vi'zhun) A type of cell division in which daughter cells get only half the chromosomes from the parent cell.

regulator proteins (reg'yu-la-tor pro'tēns) Proteins that influence the activities that occur in an organism; for example, enzymes and some hormones.

relevant (re'le-vent) Pertaining to the matter at hand.

reproduction (re"pro-duk'shun) A characteristic of life; ability of an organism to make copies of itself.

reproductive capacity (re-pro-duk'tiv kuhpas'i-te) The theoretical maximum rate of reproduction.

reproductive isolating mechanism (re-pro-duk'tiv i-so-la'ting me'kan-izm) A mechanism that prevents interbreeding between species; also called **genetic isolating mechanism**.

response (re-spons') A characteristic of life; the reaction of an organism to a stimulus.

retina (re'ti-nah) The light-sensitive region of the eye.

rhodopsin (ro-dop'sin) A light-sensitive, purple-red pigment found in the retinal rods that is important for vision in dim light.

ribonucleic acid (RNA) (ri-bo-nu-kle'ik as'id) A polymer of nucleotides formed on the template surface of DNA by transcription. Three forms that have been identified are mRNA, rRNA, and tRNA.

ribosomal RNA (rRNA) (ri-bo-sōm'al) A globular form of RNA; a part of ribosomes.

ribosomes (ri-bo-somz) Small structures composed of two protein and ribonucleic acid subunits involved in the assembly of proteins from amino acids.

rods (rahdz) Light-sensitive cells in the retina of the eye that respond to low-intensity light but do not respond to different colors of light.

root (rut) The below-ground portion of a plant responsible for anchoring the plant in the soil and absorbing water and other nutrients.

saprophyte (sap'ruh-fīt) An organism that obtains energy by the decomposition of dead organic material.

science (siens) A process of arriving at a solution to a problem or understanding an event in nature that involves testing possible solutions.

scientific law (si-en-ti-fik law) A uniform or constant fact of nature.

scientific method (si-en-ti-fik meth-ud) A way of gaining information about the world by forming possible solutions to questions, followed by rigorous testing to determine if proposed solutions are valid.

scrotum (skrō'tum) The sac in which the testes are located.

seasonal isolation (se'zun-al i-so-la'shun) A genetic isolating mechanism that prevents interbreeding between species because their reproductive periods differ.

secondary sex characteristics (sek'on-dar-e seks kar-ak-te-ris'tiks) Characteristics of the adult male or female, including the typical shape that develops at puberty.

secondary succession (sek'on-dar-e suk-se'shun) The orderly series of changes that begins with the disturbance of an existing community and leads to a climax community.

seed (seed) A structure containing an embryo plant and stored food within a protective coat.

segregation (seg"re-ga'shun) The separation and movement of homologous chromosomes to the poles of the cell.

selecting agent (se-lek'ting a'jent) Any factor that affects the probability that a gene will be passed to the next generation.

semen (se'men) The sperm-carrying fluid produced by the seminal vesicles, prostate gland, and bulbourethral glands of males.

semicircular canals (se-mi-ser'ku-lar ca-nalz') A set of tubular organs associated with the cochlea that senses changes in the movement or position of the head.

sensory neurons (sen'so-re noor'onz) Neurons that send information from sense organs to the central nervous system.

sessile (se'sile) Stationary.

sex chromosomes (seks kro'mo-somz) Chromosomes that contain the genes that control the development of sex characteristics.

sex ratio (seks ra'sho) The number of males in a population compared to the number of females.

sexuality (sek"shoo-al'i-te) A term used in reference to the totality of the aspects—physical, psychological, and cultural—of our sexual nature.

sexual reproduction (sek'shu-al re"pro-duk'shun) The propagation of organisms involving the union of gametes from two parents.

single-factor cross (sing'ul fak'tur kros) A genetic study in which a single characteristic is followed from the parental generation to the offspring.

society (so-si'uh-te) Interacting groups of animals of the same species that show division of labor.

sociobiology (so-sho-bi-ol'o-je) The systematic study of all forms of social behavior, both human and nonhuman.

soil (soyl) A mixture of mineral particles, organic matter, water, air, and organisms.

solar nebula theory (so'ler ne'byu-la the'o-re) A theory proposing that the solar system was formed from a large cloud of gases that developed 10 billion to 20 billion years ago.

soma (so'mah) The cell body of a neuron, which contains the nucleus.

speciation (spe-she-a'shun) The process of generating new species.

species (spe'shez) The scientific name given to a group of organisms that can potentially interbreed naturally to produce fertile offspring.

specific dynamic action (SDA) (spe-si'fik di-na'mik ak'shun) The amount of energy required to digest and assimilate food. SDA is equal to approximately 10 percent of the total daily kcalorie intake.

sperm (spurm) The haploid sex cells produced by sexually mature males.

spindle (spin'dul) An array of microtubules extending from pole to pole; used in the movement of chromosomes.

spontaneous generation (spon-ta'ne-us jen-uh-ra'shun) The theory that living organisms arise from nonliving material.

stable equilibrium phase (sta'bul e-kwi-lib're-um fāz) A period of time during population growth when the number of individuals entering the population and the number leaving the population are equal, resulting in a stable population.

stamen (sta'men) The male reproductive structure of a flower.

stapes (sta'pēz) The ear bone that is attached to the oval window.

stem (stem) The typically upright portion of the plant that supports the leaves and contains vascular tissue that allows for the flow of material between leaves and roots.

stimulus (stim'yu-lus) Some change in the internal or external environment of an organism that causes it to react.

stomate (sto'mate) A tiny opening in the surface of a leaf that allows for the exchange of gases such as carbon dioxide, oxygen, and water vapor.

structural proteins (struk'chu-ral pro'tēns) Proteins that are important for holding cells and organisms together, such as the proteins that make up the cell membrane, muscles, tendons, and blood.

subspecies (sub'spe-shez) A number of more or less separate groups within the same gene pool that differ from one another in gene frequency; also called **races, breeds, strains,** or **varieties.**

substrate (sub'strāt) A reactant molecule with which the enzyme combines.

succession (suk-se'shun) The process of changing one type of community to another.

successional stage (suk-se'shun-al stayg) An intermediate stage in succession.

symbiosis (sim-be-o'sis) A close physical relationship between two kinds of organisms. It usually includes parasitism, commensalism, and mutualism.

symbiotic nitrogen-fixing bacteria (sim-be-ah'tik ni'tro-jen fik'sing bak-te're-ah) Bacteria that live in the roots of certain kinds of plants, where they convert nitrogen gas molecules into compounds that plants can use.

synapse (si'naps) The space between the axon of one neuron and the dendrite of the next, where chemicals are secreted to cause an impulse to be initiated in the second neuron.

synapsis (sin-ap'sis) The condition in which the two members of a pair of homologous chromosomes come to lie close to one another.

systemic circulation (sis-te'mik ser-ku-la'shun) The flow of blood through certain chambers of the heart and blood vessels to the body and back to the heart.

taxonomy (tak-son'uh-me) The science of classifying and naming organisms.

telophase (tel'uh-faze) The last phase in mitosis characterized by the formation of daughter nuclei.

territoriality (ter"i-tor'e-al'i-te) A behavioral process in which an animal protects space for its exclusive use for food, mating, or other purposes.

territory (ter'i-tor-e) A space that an animal defends against others of the same species.

testes (tes'tēz) The male sex organs that produce haploid cells (the sperm).

testosterone (tes-tos'tur-ōn) The male sex hormone produced in the testes that controls the secondary sex characteristics.

theory (the'o-re) A plausible, scientifically acceptable generalization supported by several hypotheses and experimental trials.

theory of natural selection (the'o-re uv nat'chu-ral se-lek'shun) In a species of genetically differing organisms, the organisms with the genes that enable them to survive better in the environment and thus reproduce more offspring than others will transmit more of their genes to the next generation.

thymine (thi'mēn) A single-ring nitrogen-containing molecule in DNA but not in RNA. It is complementary to adenine.

thyroid-stimulating hormone (thi'roid sti'mu-la-ting hor'mōn) A hormone secreted by the pituitary gland that stimulates the thyroid to secrete thyroxine.

thyroxine (thi-rok'sin) A hormone produced by the thyroid gland that speeds up the metabolic rate.

tissue (tish'yu) A group of specialized cells that work together to perform a particular function.

transcription (tran-skrip'shun) The process of manufacturing RNA from the template surface of DNA.

transfer RNA (tRNA) (trans'fur) A molecule composed of ribonucleic acid. It is responsible for transporting a specific amino acid into a ribosome for assembly into a protein.

translation (trans-la'shun) The assembly of individual amino acids into a protein.

triiodothyronine (tri"ī-ō"do-thī-'row-nen) A hormone produced by the thyroid gland that speeds up the metabolic rate; similar to thyroxine but more potent.

trisomy (tris'oh-me) An abnormal number of chromosomes (3) resulting from the nondisjunction of homologous chromosomes during meiosis; for example, as in Down syndrome.

trophic level (tro'fik le"vel) A step in the flow of energy through an ecosystem.

tympanum (tim'pa-num) The eardrum.

uracil (yu'rah-sil) A single-ring nitrogen-containing molecule in RNA but not in DNA. It is complementary to adenine.

uterus (yu'tur-us) The organ in female mammals in which the embryo develops.

vagina (vuh-ji'nah) The passageway between the uterus and outside of the body; the birth canal.

valid (val'id) A term used to describe meaningful data that fit into the framework of scientific knowledge.

vascular tissue (vas'kyu-ler tish'yu) Tubes constructed of many cells attached end to end that allow for the transfer of materials from one part of a plant to another.

veins (vānz) The blood vessels that return blood to the heart.

ventricles (ven'tri-klz) The powerful muscular chambers of the heart whose contractions force blood to flow through the arteries to all parts of the body.

virus (vi'rus) A noncellular parasite composed of hereditary material surrounded by a protein coat.

vitamin deficiency disease (vi'tah-min de-fish'en-se di-zēs') Poor health caused by the lack of a certain vitamin in the diet, such as scurvy for lack of vitamin C.

vitamins (vi'tah-minz) Organic molecules that cannot be manufactured by the body but are required in very low concentrations.

voltage (vōl'tij) A measure of the electrical difference that exists between two different points or objects.

waste elimination (wayst i-li"me-nay'shun) A characteristic of life; mechanism for ridding the organism of unusable or toxic materials produced by the organism.

X-linked gene (eks-lingkt jēn) A gene located on one of the sex-determining X chromosomes.

zygote (zi'gōt) A diploid cell that results from the union of an egg and a sperm.

c r e d i t s

 illustrations

Chapter 1
1.5: Reprinted with permission from Carolina Biological Supply Company.

Chapter 2
2.11: From John W. Hole, Jr., *Human Anatomy and Physiology,* 6th edition. Copyright © 1993 Wm. C. Brown Communications, Inc. Reprinted by permission of Times Mirror Higher Education Group, Inc., Dubuque, Iowa. All Rights Reserved; **Page 48 right, left:** From Kent M. Van De Graaff, R. Ward Rhees, and Christopher H. Creek, *Biology Study Cards,* 1st edition. Copyright © 1991 Wm. C. Brown Communications, Inc. Reprinted by permission of Times Mirror Higher Education Group, Inc., Dubuque, Iowa. All Rights Reserved.

Chapter 3
Pages 55 top middle; 57 bottom; 64 top left, bottom; 75 top; 77 fish; 79: From Cleveland P. Hickman and Larry S. Roberts, *Animal Diversity,* 1st edition. Copyright © 1995 Wm. C. Brown Communications, Inc. Reprinted by permission of Times Mirror Higher Education Group, Inc., Dubuque, Iowa. All Rights Reserved; **Pages 60 bottom left; 61 bottom right; 76 top, bottom left; 77 turtle; 77 bottom:** From Stephen A. Miller and John P. Harley, *Zoology,* 3d edition. Copyright © 1996 Times Mirror Higher Education Group, Inc., Dubuque, Iowa. Reprinted by permission. All Rights Reserved; **Page 65:** From Stephen A. Miller and John P. Harley, *Zoology,* 2d edition. Copyright © 1994 Wm. C. Brown Communications, Inc. Reprinted by permission of Times Mirror Higher Education Group, Inc., Dubuque, Iowa. All Rights Reserved; **3.6:** From George B. Johnson, *Human Biology: Exploring Concepts,* 1st edition. Copyright © 1994 Wm. C. Brown Communications, Inc. Reprinted by permission of Times Mirror Higher Education Group, Inc., Dubuque, Iowa. All Rights Reserved; **Page 75 bottom:** From Kent M. Van De Graaff and Stuart Ira Fox, *Concepts of Human Anatomy & Physiology,* 4th edition. Copyright © 1995 Wm. C. Brown Communications, Inc. Reprinted by permission of Times Mirror Higher Education Group, Inc., Dubuque, Iowa. All Rights Reserved; **Page 76 bottom right:** From Cleveland P. Hickman and Larry S. Roberts, *Biology of Animals,* 6th edition. Copyright

© Wm. C. Brown Communications, Inc. Reprinted by permission of Times Mirror Higher Education Group, Inc., Dubuque, Iowa. All Rights Reserved.

Chapter 4
4.2; 4.4 left; 4.6; 4.7: From Kingsley R. Stern, *Introductory Plant Biology,* 6th edition. Copyright © 1994 Wm. C. Brown Communications, Inc. Reprinted by permission of Times Mirror Higher Education Group, Inc., Dubuque, Iowa. All Rights Reserved; **Page 94:** From Ricki Lewis, *Life,* 2d edition. Copyright © 1995 Wm. C. Brown Communications, Inc. Reprinted by permission of Times Mirror Higher Education Group, Inc., Dubuque, Iowa. All Rights Reserved.

Chapter 5
Figure 5.5; Page 128b: From Leland G. Johnson, *Biology,* 2d edition. Copyright © 1987 Wm. C. Brown Communications, Inc. Reprinted by permission of Times Mirror Higher Education Group, Inc., Dubuque, Iowa. All Rights Reserved.

Chapter 6
Page 152: Source: Data from Rolf O. Peterson, Michigan Technological University.

Chapter 8
8.1: From Joseph A. McFalls, Jr., "Population: A Lively Introduction" in *Population Bulletin,* 46(2), Oct. 1991. Reprinted by permission of Population Reference Bureau; **8.8:** Source: Data from D. A. MacLulich, *Fluctuations in the Numbers of the Varying Hare (Lepus americanus),* University of Toronto Press, 1937, reprinted 1974; **8.9:** From Elaine M. Murphy, *World Population: Toward the Next Century,* August 1994, Population Reference Bureau. Reprinted by permission; **Page 203:** Source: Data from H. Yuan Tien, "China's Demographic Dilemmas" in *Population Bulletin 1992,* Population Reference Bureau, Inc., Washington, DC, and National Family Planning Commission of China.

Chapter 9
9.9: From "Studies on the Flash Communication System in *Photinus* Fireflies" by James E. Lloyd, miscellaneous publication no. 130, November 25, 1966, Museum of Zoology, University of Michigan. Reprinted by permission.

Chapter 10

Page 238: Source: Data from George Georghiou, University of California at Riverside; Page 240: Source: From *Morbidity and Mortality Weekly Report,* May 27, 1994, Centers for Disease Control and Prevention; 10.10: From Stephen A. Miller and John P. Harley, *Zoology,* 2d edition. Copyright © 1994 Wm. C. Brown Communications, Inc. Reprinted by permission of Times Mirror Higher Education Group, Inc., Dubuque, Iowa. All Rights Reserved.

Chapter 12

Page 275: From E. Peter Volpe, *Understanding Evolution,* 5th edition. Copyright © 1985 Wm. C. Brown Communications, Inc. Reprinted by permission of Times Mirror Higher Education Group, Inc., Dubuque, Iowa. All Rights Reserved; Page 280 right: From Cleveland P. Hickman and Larry S. Roberts, *Animal Diversity,* 1st edition. Copyright © 1995 Wm. C. Brown Communications, Inc. Reprinted by permission of Times Mirror Higher Education Group, Inc., Dubuque, Iowa. All Rights Reserved; Page 286: © 1994 Time Inc. Reprinted by permission.

Chapter 15

Page 332: From Robert H. Tamarin, *Principles of Genetics,* 4th edition. Copyright © 1993 Wm. C. Brown Communications, Inc. Reprinted by permission of Times Mirror Higher Education Group, Inc., Dubuque, Iowa. All Rights Reserved.

Chapter 17

Page 386: From Kent M. Van De Graaff, *Human Anatomy,* 4th edition. Copyright © 1995 Wm. C. Brown Communications, Inc. Reprinted by permission of Times Mirror Higher Education Group, Inc., Dubuque, Iowa. All Rights Reserved; 17.2: Used by permission of General Mills, Inc.

Chapter 18

18.3b: From Kent M. Van De Graaff, *Human Anatomy,* 4th edition. Copyright © 1995 Wm. C. Brown Communications, Inc. Reprinted by permission of Times Mirror Higher Education Group, Inc., Dubuque, Iowa. All Rights Reserved; 18.5: From John W. Hole, Jr., *Human Anatomy and Physiology,* 6th edition. Copyright © 1993 Wm. C. Brown Communications, Inc. Reprinted by permission of Times Mirror Higher Education Group, Inc., Dubuque, Iowa. All Rights Reserved; 18.12: From Kent M. Van De Graaff, *Human Anatomy,* 3d edition. Copyright © 1992 Wm. C. Brown Communications, Inc. Reprinted by permission of Times Mirror Higher Education Group, Inc., Dubuque, Iowa. All Rights Reserved; 18.13: From Kent M. Van De Graaff and Stuart Ira Fox, *Concepts of Human Anatomy & Physiology,* 4th edition. Copyright © 1995 Wm. C. Brown Communications, Inc. Reprinted by permission of Times Mirror Higher Education Group, Inc., Dubuque, Iowa. All Rights Reserved; 18.16a: From Ross M. Durham, *Human Physiology.* Copyright © 1989 Wm. C. Brown Communications, Inc. Reprinted by permission of Times Mirror Higher Education Group, Inc., Dubuque, Iowa. All Rights Reserved.

Chapter 19

19.3: From Kent M. Van De Graaff and Stuart Ira Fox, *Concepts of Human Anatomy and Physiology,* 2d edition. Copyright © 1989 Wm. C. Brown Communications, Inc. Reprinted by permission of Times Mirror Higher Education Group, Inc., Dubuque, Iowa. All Rights Reserved; Page 441 right: From Kent M. Van De Graaff and Stuart Ira Fox, *Concepts of Human Anatomy & Physiology,* 3d edition. Copyright © 1992 Wm. C. Brown Communications, Inc. Reprinted by permission of Times Mirror Higher Education Group, Inc., Dubuque, Iowa. All Rights Reserved; 19.4; 19.10: From John W. Hole, Jr., *Human Anatomy and Physiology,* 5th edition. Copyright © 1990 Wm. C. Brown Communications, Inc. Reprinted by permission of Times Mirror Higher Education Group, Inc., Dubuque, Iowa. All Rights Reserved; Pages 447; 454: From Kent M. Van De Graaff and Stuart Ira Fox, *Concepts of Human Anatomy & Physiology,* 4th edition. Copyright © 1995 Wm. C. Brown Communications, Inc. Reprinted by permission of Times Mirror Higher Education Group, Inc., Dubuque, Iowa. All Rights Reserved; Page 451 top, bottom: From John W. Hole, Jr., *Human Anatomy and Physiology,* 6th edition. Copyright © 1993 Wm. C. Brown Communications, Inc. Reprinted by permission of Times Mirror Higher Education Group, Inc., Dubuque, Iowa. All Rights Reserved; 19.7: From Kent M. Van De Graaff, *Human Anatomy,* 4th edition. Copyright © 1995 Wm. C. Brown Communications, Inc. Reprinted by permission of Times Mirror Higher Education Group, Inc., Dubuque, Iowa. All Rights Reserved; 19.9: From Kent M. Van De Graaff, *Human Anatomy,* 3d edition. Copyright © 1992 Wm. C. Brown Communications, Inc. Reprinted by permission of Times Mirror Higher Education Group, Inc., Dubuque, Iowa. All Rights Reserved.

Chapter 20

20.2; 20.5: From John W. Hole, Jr., *Human Anatomy and Physiology,* 6th edition. Copyright © 1993 Wm. C. Brown Communications, Inc. Reprinted by permission of Times Mirror Higher Education Group, Inc., Dubuque, Iowa. All Rights Reserved; 20.3: From Kent M. Van De Graaff and Stuart Ira Fox, *Concepts of Human Anatomy & Physiology,* 4th edition. Copyright © 1995 Wm. C. Brown Communications, Inc. Reprinted by permission of Times Mirror Higher Education Group, Inc., Dubuque, Iowa. All Rights Reserved; 20.4: Source: Data from W. B. Markes, W. H. Dobelle, and E. F. MacNichol, "Visual Pigments of Single Primate Cones" in *Science* 143:1181–1183 (1964); and P. K. Brown and G. Wald, "Visual Pigments in Single Rods and Cones of the Human Retina" in *Science* 144:45–52 (1964); 20.7: From Stuart Ira Fox, *Human Physiology,* 3d edition. Copyright © 1990 Wm. C. Brown Communications, Inc. Reprinted by permission of Times Mirror Higher Education Group, Inc., Dubuque, Iowa. All Rights Reserved; 20.9: From Kent M. Van De Graaff, *Human Anatomy,* 4th edition. Copyright © 1995 Wm. C. Brown Communications, Inc. Reprinted by permission of Times Mirror Higher Education Group, Inc., Dubuque, Iowa. All Rights Reserved.

Appendix A

Page A-14 FYI box: Courtesy Mazola Corn Oil. Best Foods Division, CPC International Inc., Englewood Cliffs, N.J. 07632. Reprinted by permission.

Design Elements

BioFeature boxes: Photos from Digital Stock CDs; Line art clip art: Quality Computer CDs

Part Openers

1: © Lynn Rogers/Peter Arnold, Inc.; 2: © Tony Tietz/Tony Stone Images; 3: © John Cancalosi/Peter Arnold Inc.; 4: © Randy Wells/Tony Stone Images; 5: © Bob Coyle

Chapter 1

Opener: © Inga Spence/Tom Stack & Associates; Page 13: James Gilroy, British, 1757–1815, *The Cowpock,* Engraving, 1802, William McCallin McKee Memorial Collection, 1928, 1497. © 1994 The Art Institute of Chicago. All Rights Reserved; Page 10: © 1995 M. C Escher/Cordon Art, Baarn, Holland. All Rights Reserved; Page 12: © S. Maslowski/Photo Researchers, Inc.; 1.4: Courtesy of Kingsley Stern; Page 16: © Times Mirror Higher Education Group, Inc./Jim Shaffer, photographer

Chapter 2

Opener: © Kevin Schafer/Martha Hill/Tom Stack & Associates; 2.1: Historical Pictures Service/Stock Montage; 2.3a: © Walter H. Hodge/Peter Arnold, Inc.; 2.3b: © David Young-Wolff/Tony Stone Images; 2.3c: © Art Wolfe/Tony Stone Images; Page 27: © James L. Shaffer; Page 29 top, bottom: © SYGMA; 2.4a: © David M. Dennis/Tom Stack & Associates; 2.4b: © Francois Gohier/Photo Researchers, Inc.; 2.4c: © W. B. Saunders, Bryn Mawr College/BPS; 2.4d: © Richar Weiss/Peter Arnold, Inc.; 2.5: © Frank T. Aubrey/Visuals Unlimited; 2.6 top: © Bill Beatty/Visuals Unlimited; 2.6 bottom: © Martin Land /SPL/ Photo Researchers, Inc.; 2.7a: © Tom Stack/Tom Stack & Associates; 2.7b: © Don Valenti/Tom Stack & Associates; 2.8a: © L. Mellinchamp/Visuals Unlimited; 2.8b: © Kenneth M. Fink/Photo Researchers, Inc.; 2.8c: © C. D. Newman/Visuals Unlimited; 2.10b: Courtesy of William Jensen and Roderick B. Park, *Cell Ultrastructure,* p. 57, 1967 Wadsworth Publishing Co.; 2.12 tree, corn: Courtesy of Kingsley Stern; cow: © Will & Demi McIntyre/Photo Researchers, Inc.; humans: © Jon Riley/Tony Stone Images; 2.14a left: © N. Cattlin, Holt Studios/Photo Researchers, Inc.; 2.14a right: © Beverly Factor/Tony Stone Images; 2.14b inset: © Ron Spomer/Visuals Unlimited; 2.14b background: © Jean Paul Nacivet/Tony Stone Images; 2.14c left: © W. H. Hodge/Peter Arnold, Inc.; 2.14c bottom right: © Manfred Kage/Peter Arnold, Inc.; 2.14c: top right: © David M. Phillips/Visuals Unlimited; 2.14d left: © T. E. Adams/Visuals Unlimited; 2.14d right: © Mack Henley/Visuals

Unlimited; **2.14e left:** © David M. Phillips/Visuals Unlimited; **2.14e right:** Biophoto/Science Source/Photo Researchers, Inc.; **2.15d:** Courtesy of Coulter Corporation

Chapter 3

Opener: Photos by Digital Stock; **Page 52 (1):** © Manoj Shah/Tony Stone Images; **Page 52 (2):** © G. C. Kelley/Photo Researchers, Inc.; **Page 52 (3) top:** © Thomas Kitchin/Tom Stack & Associates; **Page 52 (3) bottom:** © Stephen Krasemann/Tony Stone Images; **Page 53 (4) left:** © Joe McDonald/Animals Animals/Earth Scenes; **Page 53 (4) right:** © Bates Lettlehales/Animals Animals/Earth Scenes; **Page 53 (5):** © Science Vu/Visuals Unlimited; **3.3a:** © Stephen Dalton/Animals Animals/Earth Scenes; **3.3b:** © Joan Alonen/Visuals Unlimited; **Page 55 spider:** © John H. Gerard; **centipede:** © Adrian Wenner/Visuals Unlimited; **crab:** © Kevin Schafer/Tony Stone Images; **millipede:** © Simon D. Pollard/Photo Researchers, Inc.; **Page 60a:** © William J. Weber/Visuals Unlimited; **Page 60b:** © Robert A. Ross; **Page 61c:** © Daniel W. Gotshall; **Page 61d:** © Rick M. Harbo; **Page 62a:** © Daniel W. Gotshall/Visuals Unlimited; **Page 62b:** © Peter Parks/Oxford Scientific Films; **Page 62c:** Courtesy of Dr. Larry S. Roberts; **Page 62d:** © Daniel W. Gotshall; **Page 63a:** © John H. Gerard; **Page 63b:** © Ron West/Nature Photography; **Page 63c:** © Leonard Lee Rue, III; **Page 63d:** © Gary Meiburn/Tom Stack & Associates; **Page 67a:** © Cleveland P. Hickman; **Page 67b:** © Bruce Russell/Bio Media Associates; **Page 67c:** © John H. Gerard; **Page 67d:** Courtesy of Herman Zaiman and G. W. Kelley, Jr.; **3.4a:** © Francois Gohier/Photo Researchers, Inc.; **3.4b:** © Leonard Lee Rue/Visuals Unlimited; **Page 72a:** © Rick M. Harbo; **Page 72b:** © Robert L. Dunne/Photo Researchers, Inc.; **Page 72c:** © C. McDaniel/Visuals Unlimited; **Page 72d:** © Daniel W. Gotshall/Visuals Unlimited; **Page 72e:** © Bruce Iverson/Visuals Unlimited; **Page 72f:** © Fred Bavendam/Peter Arnold, Inc.; **Page 73 all:** Courtesy of Kingsley Stern; **3.7 both:** © John D. Cunningham/Visuals Unlimited; **Page 80:** Courtesy of Charles J. Cole, Dept. of Herpetology, American Museum of Natural History

Chapter 4

Opener: © Gunther Ziesler/Peter Arnold, Inc.; **4.1a:** Courtesy of Kingsley Stern; **4.1b:** Courtesy Lani Stemmerman; **4.1c:** Courtesy of Kingsley Stern; **4.1d:** © David Muench/Tony Stone; **4.1e:** © Steve McCutcheon/Visuals Unlimited; **4.1f:** © Chad Slattery/Tony Stone Images; **Page 87 all:** Courtesy of Kingsley Stern; **Page 88 both:** © Times Mirror Higher Education Group, Inc./Bob Coyle photographer; **Page 89 right:** © Dan Guravich/Photo Researchers, Inc.; **Page 89 left:** © Simon Fraser/SPL/Photo Researchers, Inc.; **Page 90:** © Steve Kaufman/Peter Arnold, Inc.; **Page 91 right:** © Robert Finken/Photo Researchers, Inc.; **Page 91 left:** © Carl Purcell/Photo Researchers, Inc.; **Page 92 top left background:** © Jack S. Grove/Tom Stack & Associates; **Page 92 top left inset:** © Times Mirror Higher Education Group, Inc./Bob Coyle, photographer; **Page 92 bottom left:** © Ken Wagner/Visuals Unlimited; **Page 92 bottom right:** © Dennis Kaltrieder/Visuals Unlimited; **Page 92 top right background:** © John Gerlach/Tom Stack & Associates; **Page 92 top right inset:** © Times Mirror Higher Education Group, Inc./Bob Coyle, photographer; **Page 93 left:** © Spencer Grant/Photo Researchers, Inc.; **Page 93 right:** © Jeff Greenberg/Visuals Unlimited; **4.3b:** Courtesy of Kingsley Stern; **4.4b top:** Courtesy of George S. Ellmore; **4.4c bottom:** Triarch, Inc./Visuals Unlimited; **Page 97a:** © Dan Suzio/Photo Researchers, Inc.; **Page 97b:** © Nigel Cattlin/Photo Researchers, Inc.; **Page 97c:** © John D. Cunningham/Visuals Unlimited; **Page 98d:** © Robert E. Daemmrich/Tony Stone Images; **Page 98e:** © Bill Beatty/Visuals Unlimited; **Page 98f:** © Michael Viard/Peter Arnold, Inc.; **Page 98g:** © N. Cattlin/Holt Studios/Photo Researchers, Inc.; **Page 98h:** © Okapia/Photo Researchers, Inc.; **Page 101 all; 4.10a,b; Page 104a:** Courtesy of Kinglsey Stern; **Page 104b:** SEM Courtesy Robert L. Carr; **Page 104c:** © David Scharf/Peter Arnold, Inc.; **4.11a top left:** © Maslowski/Visuals Unlimited; **4.11b top right:** © Gilbert Grant/Photo Researchers, Inc.; **4.11c bottom left:** © Jim W. Grace/Photo Researchers, Inc.; **4.11d bottom right:** © James P. Rowan/Tony Stone Images; **4.12b:** © John D. Cunningham/Visuals Unlimited; **4.13:** Courtesy of Kingsley Stern; **4.14a:** © Stephen J. Krasemann/Peter Arnold, Inc.; **4.14b:** © Fletcher & Bayliss/Photo Researchers, Inc.; **4.15:** © E. F. Anderson/Visuals Unlimited; **4.16:** © Cathlyn Melloan/Tony Stone Images; **Page 110:** © Larry Ulrich/Tony Stone Images; **4.17:** Courtesy of Kingsley Stern; **4.18b top right:** John Gerlach/Tom Stack & Associates; **4.18c bottom left:** Bruce Wilcox/BPS; **4.18d bottom right:** © L. West/Photo Researchers, Inc.; **4.19a:** © John Shaw/Tom Stack & Associates; **4.19b:** Courtesy Robert A. Schlising; **4.21:** © William E. Ferguson

Chapter 5

Opener: © Science Source/Photo Researchers, Inc.; **5.1a:** © Barbara J. Miller/BPS; **5.1b:** © Bill Keogh/Visuals Unlimited; **5.1c:** © Bruce Iverson/Visuals Unlimited; **5.1d:** Courtesy Richard Critchfield; **5.1e:** © H. C. Huang/Visuals Unlimited; **5.2:** © Cabisco/Visuals Unlimited; **Page 119 left:** © Richard Humbert/BPS; **Page 119 right:** © Dr. Jeremy Burgess/SPL/Photo Researchers, Inc.; **Page 120:** © R. Roncardori/Visuals Unlimited; **5.3:** © David M. Dennis/Tom Stack & Associates; **Page 121 right:** © Nancy M. Wells/Visuals Unlimited; **5.4a:** Courtesy of Leland Johnson; **5.4b:** Carolina Biological Supply Co./Phototake; **5.6:** © Cabisco/Visuals Unlimited; **5.7:** © David J. Wrobel/BPS; **5.8a:** © Michael Abbey/Photo Researchers, Inc.; **5.8b:** © Science VU/Visuals Unlimited; **5.10:** © M. Abbey/Visuals Unlimited; **5.11:** © E.S. Ross; **Page 128a:** © Phillip Sze/Visuals Unlimited; **Page 128c:** © Manfred Kage/Peter Arnold, Inc.; **Page 131a:** © R. Kessel-G. Shih/Visuals Unlimited; **Page 131b:** © Moredum Animal/Health/SPL/Photo Researchers, Inc.; **Page 131c:** © David M. Phillips/Visuals Unlimited; **Page 131d:** © CNRI/SPL/Photo Researchers, Inc.; **Page 133 left:** © Science Vu/Miles/Visuals Unlimited; **Page 133 right:** © David M. Phillips/Visuals Unlimited; **5.13:** © Y. Arthus-Bertrand/Peter Arnold; **5.14:** U.S. Department of Agriculture; **5.16:** © T. J. Beveridge, Univ. of Guelph/BPS; **5.17a,b:** © E. S. Ross

Chapter 6

Opener: © Charles A. Mauzy/Tony Stone Images; **6.1a:** © Gregory K. Scott/Photo Researchers, Inc.; **6.1b:** © Kevin Schafer/Tom Stack & Associates; **Page 147 top left:** © Maslowski/Visuals Unlimited; **Page 147 right:** © Ron Austing/Photo Researchers, Inc.; **Page 147 bottom left:** © John S. Dunning/Photo Researchers, Inc.; **6.12:** © Willard Clay; **6.14:** © C. P. Hickman/ Visuals Unlimited; **6.15:** © G. R. Roberts/Nelson, New Zealand; **Page 158 bottom:** © Virginia P. Weinland/ Photo Researchers, Inc.; **Page 158 top:** © Glenn Oliver/ Visuals Unlimited; **Page 160:** © Peter K. Ziminiski/ Visuals Unlimited; **6.17:** © Michael Giannechini/Photo Researchers, Inc.; **6.18:** © Steve McCutchion/Alaska Pictorial Service; **6.19:** © S. L. Pimm/Visuals Unlimited; **6.20:** © Irven DeVore/Anthro-Photo

Chapter 7

Opener: © Mike Bacon/Tom Stack & Associates; **Page 168 left:** © W. Perry Conway/Tom Stack & Associates; **7.2a:** © John D. Cunningham/Visuals Unlimited; **7.2b:** © G. R. Roberts/Nelson, New Zealand; **7.2c:** © Tom McHugh/Photo Researchers, Inc.; **7.3:** © David Waters/Envision; **7.4a:** © J. H. Robinson/Photo Researchers, Inc.; **7.4b:** © Gary Milburn/Tom Stack & Associates; **Page 171a:** © Sinclair Stammers/SPL/Photo Researchers, Inc.; **Page 171b:** © K. Maslowski/Visuals Unlimited; **Page 171c:** © Charles Sykes/Visuals Unlimited; **Page 172 top:** © Phil Harrington/Peter Arnold, Inc.; **7.5:** © Douglas Faulkner/Sally Faulkner Collection; **7.6:** Courtesy Mary Lane Powell; **7.7:** © John D. Cunningham/Visuals Unlimited; **7.8:** © Susanna Pashko/Envision; **7.11 beans:** © Alexander Lowry/Photo Researchers, Inc.; **7.11 fawn:** © Maresa Pryor/Animals Animals/Earth Scenes; **7.11 hunter:** © Leonard Lee Rue, III/Tony Stone; **7.11 dead deer:** © Knolan Benfield/Visuals Unlimited; **7.11f bacteria right:** © Kessel-G. Shih/Visuals Unlimited; **7.11g bacteria left:** © David M. Phillips/Visuals Unlimited; **7.11h bacteria top:** Fred Hossler/Visuals Unlimited; **Page 180:** © Renee Lynn/Tony Stone Images

Chapter 8

Opener: © Art Wolfe/Tony Stone Images; **8.2:** © G. R. Higbee/Photo Researchers, Inc.; **8.3a:** © Betty Derig/Photo Researchers, Inc.; **8.3b:** © Kees Van Den Berg/Photo Researchers, Inc.; **Page 200:** © Julia Sims/Peter Arnold, Inc.

Chapter 9

Opener: © Oliver Strewe/Tony Stone Images; **9.5 both:** © Lincoln P. Brower; **9.6:** Courtesy of Sybille Kalas; **9.10:** © G. R. Roberts/Nelson/New Zealand; **9.11:** © Harry Rogers/Photo Researchers, Inc.

Chapter 10

Opener: © Frans Lanting/Photo Researchers, Inc.; **Page 233:** Bettmann Archive; **10.2:** © M. H. Sharp/Photo Researchers, Inc.; **10.4:** © Joe McDonald/Visuals Unlimited; **10.6:** © John Colwell/Grant Heilman Photography; **10.7:** © John D. Cunningham/Visuals Unlimited

Chapter 11

Opener: © Fred Bruemmer/Peter Arnold, Inc.; **11.1a,b:** © Times Mirror Higher Education Group, Inc./Bob Coyle, photographer; **11.2a top left:** © Reynolds Photography; **11.2b top right:** © Jacana/Photo Researchers, Inc.; **11.2c bottom left:** © Reynolds Photography; **11.2d bottom right:** © Jeanne White/Photo Researchers, Inc.; **11.3:** © Richard Humbert/BPS; **11.4a horse:** © Walt Anderson/Visuals Unlimited; **11.4b donkey:** © John D. Cunningham/Visuals Unlimited; **11.4c mule:** © Williams J. Weber/Visuals Unlimited; **Page 251:** © Gary Milburn/Tom Stack & Associates; **11.6b Africa:** © Linda Bartlett/Photo Researchers, Inc.; **11.6c Europe:** © Henry Bradshaw/Photo Researchers, Inc.; **11.6d Asia:** © Lawrence Migdale/Photo Researchers, Inc.; **11.8a,b:** © Stanley L. Flegler/Visuals Unlimited; **Page 255:** © Brownie Harris/Tony Stone Images; **11.10a,b:** © Tom & Pat Leeson; **Page 257 top:** © Kenneth W. Fink/Photo Researchers, Inc.; **Page 257 bottom:** © Stephen J. Lang/Visuals Unlimited; **11.11a:** © Tom Walker/Tony Stone Images; **11.11b:** © D. Cavagnaro/Visuals Unlimited; **Page 258:** Courtesy of USDA-ARS, photo by Scott Bauer; **Page 259 top:** © Tom McHugh/Photo Researchers, Inc.; **Page 260:** © Andy Sacks/Tony Stone Images; **11.12:** © Gary Milburn/Tom Stack & Associates; **11.13a:** © Anthony Mercer/Photo Researchers, Inc.; **11.13b:** © R & N Bowers/VIREO; **11.14a:** © Joel Arrington/Visuals Unlimited; **11.14b:** © Mero/Jacana/Photo Researchers, Inc.

Chapter 12

Opener: © James Amos/Photo Researchers, Inc.; **12.10:** *The Age of Reptiles,* painted by Rudolph F. Zallinger © Peabody Museum of Natural History, Yale University; **Page 280 left:** Courtesy Department of Library Services, American Museum of Natural History, Neg. #5509; **Page 281:** © James P. Rowan/Tony Stone Images; **12.13a:** © Tom McHugh/Photo Researchers, Inc.; **12.13b:** © J. Alcock/Visuals Unlimited; **12.14a,b:** © John D. Cunningham/Visuals Unlimited; **12.17a left:** © Tony Angermayer/Photo Researchers, Inc.; **12.17b top right:** © Y. Arthus-Bertrand/Peter Arnold, Inc.; **12.17c bottom right:** © Clayton A. Fogle/Tony Stone Images; **12.18:** © The Zoological Society of San Diego; **12.19a gorilla:** © Art Wolfe/Tony Stone Images; **12.19b: gibbon:** © Gerard Lacz/Peter Arnold, Inc.; **12.19c: orangutan:** © Art Wolfe/Tony Stone Images; **12.19d: chimp:** © Tom McHugh/Photo Researchers, Inc.; **12.19e: human:** © Don Smetzer/Tony Stone Images; **12.20:** © Dreyfus and Associates Photography

Chapter 13

Opener: © Spike Walker/Tony Stone Images; **Page 294a:** © Manfred Kage/Peter Arnold, Inc.; **Page 294b,d:** © Ed Reschke; **Page 292:** © Jean-Marc Loubat/Photo Researchers, Inc.; **13.1:** © James Stevenson/Photo Researchers, Inc.; **Page 300:** Reproduced courtesy of Dr. Thomas G. Brewster, Foundation for Blood Research, Scarborough, Maine; **Page 303:** © Times Mirror Higher Education Group, Inc./Bob Coyle, photographer; **13.10 all:** © Times Mirror Higher Education Group, Inc./Kingsley Stern, photographer

Chapter 14

Opener: © David Scharf/Peter Arnold, Inc.; **Page 316:** © Times Mirror Higher Education Group, Inc./Bob Coyle, photographer; **Page 318:** © A. C. Barrington Brown/SPL/Photo Researchers, Inc.; **Page 320 top:** © Times Mirror Higher Education Group, Inc./Bob Coyle, photographer; **14.19:** © Bill Nation/SYGMA; **Page 323d:** © M. Coleman/Visuals Unlimited

Chapter 15

Opener: © Bruno DeHogues/Tony Stone Images; **15.1:** © Times Mirror Higher Education Group, Inc./Bob Coyle, photographer; **Page 330a:** © Renee Lynn/Photo Researchers, Inc.; **15.3a,b:** © Times Mirror Higher Education Group, Inc./Bob Coyle, photographer; **15.5:** © John D. Cunningham/Visuals Unlimited; **Page 339:** © Times Mirror Higher Education Group, Inc./Bob Coyle, photographer; **Page 342:** © Adam Hart-Davis/SPL/Photo Researchers, Inc.; **Page 343:** © Times Mirror Higher Education Group, Inc./Bob Coyle, photographer

Chapter 16

Opener: © Charles Thatcher/Tony Stone Images; **Page 349a:** Courtesy of California Polytechnic University, Doug Allen, photographer; **Page 349b:** Courtesy Department of Library Services, American Museum of Natural History, Neg. #K13079; **Page 353b:** E. J. Dupraw; **Page 364:** © David Hiser/Tony Stone Images; **16.5a both:** © Times Mirror Higher Education Group, Inc./Jim Shaffer, photographer; **Page 368:** Coulter Corporation; **Page 374:** Courtesy of Dr. Oscar L. Miller; **16.10:** Courtesy of Calgene, Inc.

Chapter 17

Opener: © Bob Coyle; **Page 385 both, 17.1:** © Times Mirror Higher Education Group, Inc./Bob Coyle, photographer; **17.4:** © Paul A. Souders; **17.6a,b:** Courtesy National Osteoporosis Foundation; **17.7a,b:** Courtesy Fetal Alcohol and Drug Unit, University of Washington, Seattle

Chapter 18

Opener: © David Hiser/Tony Stone Images; **18.1a:** © Times Mirror Higher Education Group, Inc./Bob Coyle, photographer; **18.1b:** © J. DaCunha/Petit Format/Photo Researchers, Inc.; **18.2:** From Thomas G. Brewster and Park S. Gerald "Chromosome Disorders Associated with Mental Retardation," *Pediatric Annuals 7,* No. 2 (1978). Reproduced courtesy of Dr. Thomas G. Brewster, Foundation for Blood Research, Scarborough, Maine; **Page 414 left:** Courtesy of Dr. Meyer M. Melicow: Reprinted from *Medical Aspects of Human Sexuality.* Copyright 1971 Hospital Publications, Inc.; **Page 414 middle, right:** Courtesy Dr. Kenneth L. Becker, from *Fertility and Sterility,* 23:5668–78, 1972 Williams and Wilkins Co.; **Page 415 top:** © Mike Peres/Custom Medical Stock Photos; **Page 415 bottom:** © Les Stone/Sygma; **Page 417:** © David Phillips/Science Source/Photo Researchers, Inc.; **18.4:** © SIU/Peter Arnold, Inc.; **18.6:** © Phillip Hayson/Science Source/Photo Researchers, Inc.; **Page 425 left:** © Porterfiled-chickering/Photo Researchers, Inc.; **Page 425 right:** © Times Mirror Higher Education Group, Inc./Bob Coyle, photographer; **18.14:** © Erika Stone/Peter Arnold, Inc.; **Page 426 right:** © Biophoto Assoc./Science Source/Photo Researchers, Inc.; **18.15a:** © Times Mirror Higher Education Group, Inc./Bob Coyle, photographer; **18.15b:** © Hank Morgan/Photo Researchers, Inc.; **18.15c:** © Times Mirror Higher Education Group, Inc./Bob Coyle, photographer; **18.15d:** Courtesy of Gyno Pharma, Inc.; **18.15e; 18.15f; 18.15g; 18.15h:** © Times Mirror Higher Education Group, Inc./Bob Coyle, photographer

Chapter 19

Opener: © BioPhoto Assoc./Photo Researchers, Inc.; **Page 441:** Courtesy March of Dimes Birth Defects Foundation; **Page 442:** © W. Ormerod/Visuals Unlimited; **Page 458:** © Times Mirror Higher Education Group, Inc./Bob Coyle, photographer; **Page 459 top and bottom left:** © Martin M. Rotker; **Page 459 top and bottom right:** © Matt Meadows/Peter Arnold, Inc.

Chapter 20

Opener: © Nicholas DeVore/Tony Stone Images; **Page 467a:** © Danny Brass/Science Source/Photo Researchers, Inc.; **Page 467b:** © Biophoto Assoc./Science Source/Photo Researchers, Inc.; **Page 467c:** Courtesy American Academy of Ophthalmology; **Page 474a:** © Alex Bartel/SPL/Photo Researchers, Inc.; **Page 474b:** © SPL/Photo Researchers, Inc.; **Page 475:** © Sheryl McNee/Tony Stone Images; **Page 476:** © A. Tannerbaum/SYGMA; **Page 485:** © Bettina Cirone/Photo Resarchers, Inc.

Appendix B

B.3a; B.3b; B.3c: © David M. Phillips/Visuals Unlimited

index

Arteries, function and structure of, 446, 452, 453
Arterioles, function of, 446
Arthropods, types of, 55
Asexual reproduction
    of animals, 78
    of plants, 106
Asthma, and cigarette smoking, 458
Athletes
    drug misuse by, 442–443
    nutrition for, 405–406
Atmosphere, and molecular oxygen, 272
Autosomes, 324
Autotrophs, 36, 38
Axons, 471

b

Bacteria, 129–135
    beneficial uses of, 130
    characteristics of, 129
    and disease, 132
    endospores of, 135
    growth of, 134–135
    mutualism, 130
    pathogenic bacteria, 130
    saprophytic bacteria, 130
    structure of, 129
    types in human body, 131
Bacterial vaginosis, symptoms of, 432
Baldness, genetic factors, 343
Basal metabolic rate (BMR), 396–398
    estimation of, 397
Base, nature of, 70
Basilar membrane, 469, 471
Basophils, 447
Bees
    navigation/communication of, 221–222
    social behavior of, 224
Behavior
    definition of, 208
    dominance hierarchy, 220, 221
    instinct, 210–211
    instinct coupled with learned behavior,
        216–217
    learned behavior, 211, 213–218
    navigation, 220–221, 224
    and organisms, 208–209
    reproductive behavior, 217–220
    social behavior, 224
    territorial behavior, 220
Behavioral isolation, 262
Benign tumors, 295, 296
Benthic organisms, 122
Biennial plants, 94
Binomial system of nomenclature, 28–29
Biodiversity, meaning of, 161
Biogenesis, 10, 12
Biological amplification, 182
Biological clocks, 223
Biomass
    and carrying capacity, 199
    determination of, 146
    pyramid of, 146

Biomes, 153–161
    coniferous forest, 158, 160
    deserts, 157–158
    prairie, 156
    savanna, 157
    temperate deciduous forest, 153, 155, 156
    tropical rain forest, 160–161
    tundra, 160
Birds
    behavior, and hormones, 484
    flight of, 56–58
    migration of, 222, 224
Birthrate, 194
Black death, 171
Bladder control, and muscles, 441
Blind spot, 468
Blood
    amount in body, 446
    components of, 446–448
    function of, 446
Blood tests
    ABO typing, 448
    complete blood count (CBC), 448
    general biochemistry profile, 448
Blood typing, blood types, 450
Bloom, algal, 124
B-lymphocytes, 449
Bodybuilding supplements, 438
Body-surface area, determination of, 397
Botulism, 135
Brain
    functions of, 478
    right brain/left brain dominance, 480
    specialized areas of, 479
Breast milk, compared to cow's milk, 403
Breathing
    control of breathing rate, 455
    movement of, 457
    process of, 88
Breech birth, 426
Breeds, 255
Bronchial tubes, function of, 455
Bronchioles, function of, 455
Bronchitis, and cigarette smoking, 458
Buffers
    carbonic acid/bicarbonate buffer system,
        456
    functions of, 456
Bulimia, 401
    symptoms of, 401

c

California condor, 168
Calories, definition of, 384
Cancer
    causes of, 296
    and cell division, 295–296
    chemotherapy, 296
    radiation therapy, 297–298
    spread of, 296
    warning signs, 297

Candidiasis, symptoms of, 431
Capillaries, function and structure of, 446, 452
Carbohydrate loading, 405–406
Carbohydrates, 393
    and ATP, 393
    complex carbohydrates, 393
    facts about, 268
    structure of, 393
Carbon cycle, 174–176
Carbon dioxide
    and global warming, 177
    and respiration, 453–454
Carbon-14 dating, 31
Carbon monoxide poisoning, and
        hemoglobin, 446
Cardiac muscle, 445
Carnivores, 83, 85, 142
Carnivorous plants, 87
Carrier, of trait, 330
Carrying capacity, population, 198
Cartilage, 59
Catalysts, function of, 360
Cataracts, 467
Cell body, 471
Cell division
    and cancer, 295–296
    cytokinesis, 282
    and DNA, 299, 300, 301, 354–357
    interphase, 299
    mitosis, 282, 299–303
    in plant cells versus animal cells, 304–305
    regulation of, 293
    stages in cell cycle, 299
Cell-mediated immunity, 449
Cells, 34–47
    energy conversion in, 38
    eukaryotic cells, 40–44
    first cells of Earth, 270–272
    mobility of, 38
    nucleus of, 35
    parts of, 34, 35, 37
    prokaryotic cells, 44
    specialization of, 294–295
Cellular respiration, 88, 442
    process of, 88
Cell walls, 39
Cenozoic Era, 274, 275
Central nervous system, 471
Centrioles, 38
Centromere, 300
Cesarean section, 426
Chancroid, symptoms of, 431
Chemotherapy, 296
Childbirth
    complications of, 426
    process of, 426
Childhood, nutrition in, 403–404
China, and population control, 203
Chlamydia, symptoms of, 432
Chlorophyll, 36, 99
Chloroplasts, 36, 38, 99
Chromatids, 299
Chromatin, 35, 352

and characteristics, 329
function of, 231, 248
genotype, 329
linkage group, 340
location on chromosomes, 300
phenotype, 329
and RNA transcription, 362
sex-linked genes, 340–342
See also Alleles
Genetic engineering, 373–377
genetically altered crops, 374, 377
genetic fingerprinting, 376, 377
genetic medicine, 375
recombinant DNA, 373, 377
Genetic isolation, 261–262
behavioral isolation, 262
ecological isolation, 261
seasonal isolation, 261–262
Genetic recombination, and natural
selection, 234–235
Genetics
alleles in, 248–249
clones, 255
genes in, 248
Mendelian genetics, 329, 331, 332–333
study of, 248, 329
See also Population genetics
Genetic study
double-factor crosses, 335–337
for lack of dominance, 337–338
Mendelian genetics, 329, 331, 332–333
for multiple allele problems, 338
probability and genetics, 331
Punnett square, 334–335
single-factor crosses, 334–335
Genetic variation
crossing-over, 316–318
fertilization, 320
independent assortment, 314, 319–320
mutation, 320
segregation, 314, 318–319
Genetic variety, and natural selection, 234
Genital herpes, symptoms of, 431
Genome, nature of, 329
Genotype, 331, 334, 335, 336, 337, 338, 340
nature of, 329
Geographic barriers, and genetics, 255–256
Geological time chart, 275
George Reserve, wildlife study area, 197
Gigantism, 485
Giraffe, and evolution, 232
Glands
endocrine glands, 478–479, 481–483
exocrine glands, 478
Glandular system, components of, 77
Global warming, and carbon dioxide, 177
Golgi apparatus, 38
Gonads, 312
Gonorrhea
and penicillin, 240, 430
symptoms of, 431
Gould, Stephen Jay, 245
Gradualism, 244
Grana, 38

Growth and development, as characteristic of
life, 14
Growth-stimulating hormone, 481, 485
Guanine, 350
Gymnosperms, 111

Habitats, 166
aquatic habitats, 71
destruction of, 180
nature of, 166
terrestrial habitats, 70–71
Haldane, J.B.S., 267, 270
Haploid number, 310, 311, 312
Hay fever, 104
Hearing, 468–469
ear, structure of, 468–469
Heart
function of, 446, 449, 450
parts of, 450–452
sounds of, 450
Heart disease
and cigarette smoking, 459
and heat attack, 454
Hemoglobin, 446, 453
and carbon monoxide poisoning, 446
Hemophilia type B, 341
Herbivores, 85, 142
Heredity, and DNA, 348–357
Hermaphrodites, 415
Hermaphroditic animals, 78, 193
Heterotrophs, 36, 38
as first cells, 272
Heterozygous organism, 329
HIV
and reverse transcriptase, 368–369
symptoms of, 431
HMS *Beagle,* 233
Homeotherms, 71
Homologous chromosomes, 310, 311, 329,
331
Homosexuality, 415
genetic factors, 416
Homozygous organism, 329
Hormones
and bird behavior, 484
and female sexual development, 419, 421
functions of, 109, 412
and male sexual development, 412, 416
types of, 481–483
Host, and parasite, 170–171
Human Genome Project, 318
Human reproduction, 421–426
childbirth, 426
embryo, development of, 423–425
and fertility drugs, 424
fertilization of egg, 419, 423
Humans, evolution of, 281, 285–286
Hunter-gatherers, 162
Hydrologic cycle, 174

Hypothalamus, 419
Hypothesis
formation of, 8, 18
meaning of, 7

Identical twins, 425
Immune system, 448
Immunity
and antibodies, 449
antibody-mediated, 449
cell-mediated, 449
and white blood cells, 449
Immunogens, 449
Immunoglobulins, 449
functions of, 364
types of, 359
and vaccination, 364
Imprinting, 214–215
characteristics of, 214
and humans, 218
Incus, 468
Independent assortment, 314, 319–320, 335
Infancy, nutrition in, 402–403
Inorganic molecules, nature of, 14, 174
Insects
metamorphosis, 79
pollination by, 102, 103, 108, 173
Insight, 215–216
and animals, 218
Instinct, 210–211
combined with learning, 218–219
examples of, 210, 211
and humans, 212
versus learning, 210
Insulin, 481
Interleukin-2, 375
Internal fertilization, 78, 218
Interphase, 299
Interstitial cell-stimulating hormone (ICSH),
412
Intrauterine device (IUD), 427, 428

Jenner, Edward, 13

Kilocalories, 397, 398
definition of, 384
determining for diet, 398
Kingdoms, 24–25
Klinefelter's syndrome, 414
Kwashiorkor, 402

Lack of dominance, allele combinations, 337–338
Lag phase, population growth curve, 195
Lamarck, Jean Baptiste de, 231
Large intestine, digestive function of, 387
Learned behavior, 211, 213–218
  conditioning, 213–214
  imprinting, 214–215
  insight, 215–216
Leaves
  autumn color changes, 110
  of plants, 98–99
Left atria, 450
Left ventricle, 450
Lice, pubic, symptoms of, 432
Lichens, 135
  characteristics of, 135
  habitats of, 135
Life, characteristics of, 12, 14–15
Limiting factors, population, 198–200, 201–202, 204
Linkage group, 340
Linnaeus, Carolus, 24
Lipids, 395
  in bloodstream, 448
  facts about, 268
  nutritional function of, 395
Liver, digestive function of, 387
Lorenz, Konrad, 214–215
Lung cancer, and cigarette smoking, 458
Lungs, 454–455
  and cigarette smoking, 458
  function of, 454, 455
  structure of, 457
Lyme disease, 170
Lymphatic system
  components of, 451
  lymph, 451
Lymphocytes, 447, 449
Lysosomes, 38

Macrophages, 449
Male sexual development, 412–413, 416, 418–419
  and hormones, 412, 416
  secondary sex characteristics, 416
  sperm cell production, 418
Malignant melanoma, 375
Malignant tumors, 295, 296
Malleus, 468
Malnutrition, 402
Mammals
  egg-laying mammals, 279
  evolutionary patterns, 278–279
Marsupials, reproduction of, 279
Mass extinctions, 278
Masturbation, 418

Meiosis, 311, 313–323
  anaphase I, 314
  and genetic variation, 316–320
  metaphase I, 313
  metaphase II, 315
  compared to mitosis, 325
  nondisjunction, 321–323
  prophase I, 313
  prophase II, 314, 315
  reduction division, 313
  telophase I, 314
  telophase II, 315
Memory
  consolidation in, 476
  short- and long-term, 476
Menarche, 430
Mendel, Gregor, 329, 332–333
Mendelian genetics, 329, 331, 332–333
  law of dominance, 333
  law of independent assortment, 333, 335
  law of segregation, 333
  pea plant study, 333
Menopause, 429–430
  hormonal treatment, 429–430
Menstrual cycle, 421
Mental retardation, genetic factors, 343
Mesozoic Era, 274, 275
Messenger RNA (mRNA), 362, 366–367, 370–372
Metabolism, as characteristic of life, 14
Metamorphosis, process of, 79
Metaphase, mitosis, 302
Metaphase I, meiosis, 313
Metaphase II, meiosis, 315
Metastasis, 296
Microfilaments, 38
Micrometers, 122
Microorganisms
  algae, 122–125
  bacteria, 129–135
  characteristics of, 117
  fungi, 117–122
  lichens, 135
  protozoa, 125–126
  slime molds, 127
  water molds, 127, 129
Microtubules, 38
Microwave ovens, 399
Migration, and genetics, 258
Minerals, nutritional functions of, 391
Mitochondria, 36, 38
Mitosis, 282, 299–303
  anaphase, 302–303
  compared to meiosis, 325
  metaphase, 302
  plant cells versus animal cells, 304–305
  prophase, 300–301
  telophase, 303
Mnemonic devices, 25
Models, use of, 303
Molecular paleontologists, 349
Mollusks, 80–81
Monoculture, 260
Monocytes, 447, 449
Mosses, 112–113
Motile animals, 59
Motor neurons, 471–472

Mouth, digestive function of, 386
Multiple alleles, 338
Multi-regional continuity hypothesis, human evolution, 285, 286
Muscles
  anatomy of, 439
  antagonistic, 439, 440
  cardiac muscle, 445
  contraction of, 437–439
  increasing size of, 437
  muscle pain and exercise, 444
  skeletal muscle, 440–442
  smooth muscle, 442, 445
  strains, 44
Muscular system, components of, 77
Mutation, 292
  and natural selection, 234
  nature of, 320, 367
  occurrence of, 258
Mutualism, 130, 172–173
  examples of, 172–173
Mycorrhizae, 120
Mycotoxins, 122
Myosin, 437

Natural selection, 231
  and differential survival, 236–237
  and excess reproduction, 236
  and gene expression, 235–236
  and genetic recombination, 234–235
  and genetic variety, 234
  giraffe example, 232
  and mutations, 234
  selecting agents in, 236
Navigation, of animals, 220–221, 224
Needham, John T., 10
Negative-feedback control, 481
Nerve fibers, 471
Nerve impulse, nature of, 472
Nerves, 471
  disorders of, 474
Nervous system
  actions/effects in, 485
  brain, 478
  central nervous system, 471
  components of, 76
  drugs, effects on, 477
  function of, 464
  neurons, 470–474
  organization of, 474
  peripheral nervous system, 471
Neurons, 470–474
  motor neurons, 471–472
  parts of, 471
  sensory neurons, 472
  synapse, 472, 474–475
Neurotransmitters, and synapse, 472, 474
Neutrophils, 447
Niches, 166–169
  broad and specialized, 167, 168
  meaning of, 166
Night-blindness, 467

Diversity
of
life

FRED ROSS

Delta College

ELDON ENGER

Delta College

REBECCA OTTO

Delta College

RICHARD KORMELINK

Delta College

WCB

Wm. C. Brown Publishers

Dubuque, IA Bogota Boston Buenos Aires Caracas Chicago
Guilford, CT London Madrid Mexico City Sydney Toronto

Book Team

Editor *Carol J. Mills*
Developmental Editor *Connie Haakinson*
Production Editor *Sue Dillon*
Designer *Jeff Storm/Laurie Janssen*
Photo Editor *Carrie Burger*
Permissions Coordinator *Karen L Storlie*
Art Processor *Jennifer L. Osmanski*

Wm. C. Brown Publishers

President and Chief Executive Officer *Beverly Kolz*
Vice President, Publisher *Kevin Kane*
Vice President, Director of Sales and Marketing *Virginia S. Moffat*
Vice President, Director of Production *Colleen A. Yonda*
National Sales Manager *Douglas J. DiNardo*
Marketing Manager *Julie Joyce Keck*
Advertising Manager *Janelle Keeffer*
Production Editorial Manager *Renée Menne*
Publishing Services Manager *Karen J. Slaght*
Royalty/Permissions Manager *Connie Allendorf*

A Times Mirror Company

Copyedited by Julie Bach

Freelance permissions by Karen Dorman

Cover photos: Toucan: © Tony Galindo/The Image Bank; Hands: © Paul Venning/Stock
Imagery; Cell: © D. M. Phillips/Photo Researchers, Inc.; Beetle: Harters Archive;
all others and title page: Digital Stock Photo CDs

Photo research by Connie Mueller

The credits section for this book begins on page C–1 and
is considered an extension of the copyright page.

Library of Congress Catalog Card Number: 95–79524

ISBN 0–697–27862–X

Printed in the United States of America by Times Mirror Higher Education Group, Inc.,
2460 Kerper Boulevard, Dubuque, IA 52001

10 9 8 7 6 5 4 3 2 1

brief contents

contents

2
interrelationships among
living things

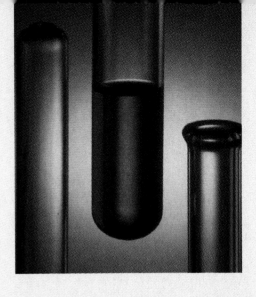

4

cell division and heredity

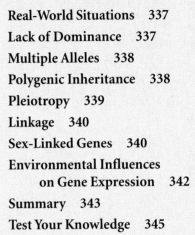

5

the human organism

BIO *feature readings*

📼 Technology—Life Science Animations

The following illustrations in *Diversity of Life* are correlated to the WCB *Life Science Animations* videotapes:

📼 Tape 1   Chemistry, the Cell and Energetics

1 Formation of an Ionic Bond (**figure A.8, p. A–6**)
2 Journey into a Cell (**figure 2.10**)
3 Endocytosis
4 Cellular Secretion
5 Glycolysis
6 Oxidative Respiration (including Krebs cycles)
7 The Electron Transport Chain and the Production of ATP
8 The Photosynthetic Electron Transport Chain and Production of ATP
9 C3 Photosynthesis (Calvin Cycle)
10 C4 Photosynthesis
11 ATP as an Energy Carrier

📼 Tape 2   Cell Division/Heredity/ Genetics/Reproduction and Development

12 Mitosis (**figure 13.10**)
13 Meiosis (**figure 14.11**)
14 Crossing Over (**figure 14.12**)
15 DNA Replication (**figure 16.2**)
16 Transcription of a Gene
17 Protein Synthesis (**figure 16.8**)
18 Regulation of *Lac* Operon
19 Spermatogenesis (**figure 18.8**)
20 Oogenesis (**figure 18.8**)
21 Human Embryonic Development (**figure 18.12**)

📼 Tape 3   Animal Biology I

22 Formation of Myelin Sheath
23 Saltatory Nerve Conduction
24 Signal Integration
25 Reflex Arcs

26 Organ of Static Equilibrium
27 The Organ of Corti
28 Peptide Hormone Action (cAMP)
29 Levels of Muscle Structure (**figure 19.2**)
30 Sliding Filament Model of Muscle Contraction (**figure 19.1**)
31 Regulation of Muscle Contraction (**figure 19.3**)
32 The Cardiac Cycle and Production of Sounds (**figure 19.7**)
33 Peristalsis
34 Digestion of Carbohydrates
35 Digestion of Proteins

📼 Tape 4   Animal Biology II

36 Digestion of Lipids
37 Blood Circulation
38 Production of Electrocardiogram
39 Common Congenital Defects of the Heart
40 A, B, O Blood Types
41 B-Cell Immune Response
42 Structure and Function of Antibodies
43 Types of T Cells
44 Relationship of Helper T Cells and Killer T Cells
45 Life Cycle of Malaria (**figure 5.9**)

📼 Tape 5   Plant Biology/ Evolution/Ecology

46 Journey into a Leaf (**figure 4.6**)
47 How Water Moves Through a Plant (**figure 4.4**)
48 How Food Moves from a Source to a Sink (**figure 4.4**)
49 How Leaves Change Color and Drop in Fall (**p. 110**)
50 Mitosis and Cell Division in Plants (**figure 13.10**)
51 Carbon and Nitrogen Cycles (**figures 7.10 and 7.11**)
52 Energy Flow through an Ecosystem (**figure 6.6**)
53 Continental Drift and Plate Tectonics (**p. 278**)

💿 Explorations CD-ROM Correlations

The following chapters in *Diversity of Life* are correlated to the sixteen topic modules in the WCB *Explorations in Human Biology* CD-ROM by George B. Johnson:

1. Cystic Fibrosis (**chapter 15**)
2. Active Transport (**appendix B**)
3. Life Span and Lifestyle (**chapter 17**)
4. Muscle Contraction (**chapter 19**)
5. Evolution of the Human Heart (**chapter 19**)
6. Smoking and Cancer (**chapter 19**)
7. Diet and Weight Loss (**chapter 17**)
8. Nerve Conduction (**chapter 20**)
9. Synaptic Transmission (**chapter 20**)
10. Drug Addiction (**chapter 20**)
11. Hormone Action (**chapter 20**)
12. Immune Response (**chapter 19**)
13. AIDS (**chapter 18**)
14. Constructing a Genetic Map
15. Heredity in Families (**chapter 15**)
16. Pollution of a Freshwater Lake (**chapter 7**)

Although none of the chapters in *Diversity of Life* are directly correlated to the second CD-ROM by George B. Johnson entitled *Explorations in Cell Biology and Genetics*, the following modules are available to further your learning experience:

1. How Proteins Function: Hemoglobin
2. Cell Size
3. Active Transport
4. Cell-Cell Interactions
5. Mitosis: Regulating the Cell Cycle
6. Cell Chemistry: Thermodynamics
7. Enzymes in Action: Kinetics
8. Oxidative Respiration
9. Photosynthesis
10. Exploring Meiosis: Down Syndrome
11. Constructing a Genetic Map
12. Pedigree Analysis
13. Gene Segregation Within Families
14. DNA Fingerprinting: You Be the Judge
15. Reading DNA
16. Gene Regulation
17. Making a Restriction Map

preface

Purpose

The origin of this book is deeply rooted in our concern for the education of college students in the field of biology. We have observed that many introductory texts are written primarily for the instructor and that large, thick books intimidate introductory-level students who are often already anxious about taking science courses. These large, expensive texts are frequently purchased by the students only to be left on a shelf unread. In contrast, this text was produced to meet students' needs. Its readability and organization were purposely designed to be student-friendly, yet cover all the major biological concepts. Coverage of chemistry has been minimized to make the content more palatable to nonscience majors. In addition, the book's design is similar to a magazine format to enhance student appeal.

Organization

Diversity of Life is written from the familiar to the unfamiliar, from the conceptual to the specific, and from the whole organism to its parts. The book begins with a hypothetical situation designed to lead students through the scientific process. Part One introduces the concept of classification and deals with organisms as they are placed into the five kingdom classification scheme. Part Two centers on organismal interactions with each other and with their environment. Part Three describes the vast diversity of living things and how they came into being. Activities of organisms and the genetic basis of their characteristics are the focus of Part Four. Finally, Part Five describes the organization and function of the human organism.

The Text

Five parts divide the chapters into sections of closely related material. Each part begins with an overview of the material contained in that section. One of the main aspects of the text that we continued to focus on was the appropriateness of the language for an introductory biology course. The informal, easy-to-read writing has been praised by reviewers. A number of pedagogical features enhance student learning. Each chapter contains these elements:

Chapter Outline   As part of the chapter opening, the outline presents the material to be covered in that chapter.

Learning Objectives   At the beginning of each chapter, a list of objectives focuses students' attention on specific goals.

BIO Features   These readings, found in every chapter, expand upon the textual information to provide application or motivation to the student. The articles discuss current events and special interest topics, such as AIDS.

Concept Connections   These short sections highlight vocabulary terms and concepts, and allow students to focus their attention on key points.

CONCEPT CONNECTIONS

Spontaneous generation: The theory that living organisms spontaneously arise from nonliving material.

Biogenesis: The concept that life originates only from preexisting life.

FYI   For Your Information boxes include tidbits of factual information that help students connect the material to their daily lives. FYI boxes are interspersed throughout each chapter.

FYI

The words *organism*, *organ*, *organize*, and *organic* are all related. Organized objects have parts that fit together in a meaningful way. Organisms are separate living things that are organized. Animals have within their organization organs, and the unique kinds of molecules they contain are called organic. Therefore, organisms consist of organized systems of organs containing organic molecules.

Apply Your Knowledge   These boxes challenge students to think critically about science.

apply your **knowledge**

Feeding Starving People

This is a thought puzzle—put it together! Write a paragraph that links all the following bits of information in a way that explains how to solve the problem.

The Problem: How to feed starving people.

Pieces of the Puzzle:

Commercial fertilizer production requires temperatures of 900°C.

Geneticists have developed plants that grow very rapidly and require high amounts of nitrogen to germinate during the normal growing season.

Fossil fuels are stored organic matter.

The rate at which nitrogen can be incorporated into living things depends on the activity of bacteria.

The sun is expected to last for several million years.

Crop rotation is becoming a thing of the past.

The clearing of forests for agriculture changes weather in the area.

Boldface Terms   This highlighting method focuses student attention on key terms when they are first defined in the text. Italic type also emphasizes important terms, phrases, names, and titles.

Graphics   The graphics—often in the form of logical flow diagrams, analogy diagrams, and charts—clarify the textual material.

Appendixes   Two sections at the end of the text provide background information on general and organic chemistry, and energy and molecular movement. This material is placed in appendixes to prevent students from becoming overwhelmed with complex material within the chapters. The placement of this material also allows instructors to be more flexible in designing their courses.

Icons within the text chapters refer students to appendixes A and B for additional information and broader coverage of the particular topic being discussed.

Appendix A

Appendix B

Test Your Knowledge   This section serves as a built-in student study guide and appears at the end of each chapter. Multiple-choice and essay questions test students' comprehension.

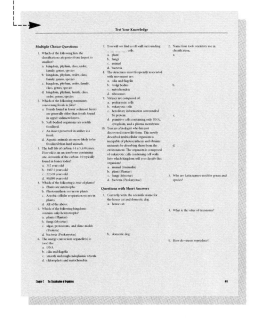

Full-Color Illustrations   More than 800 line drawings and photographs illustrate concepts or associate new concepts with previously mastered information. Every illustration is used to emphasize a point or to help teach a concept.

Glossaries   At the end of each chapter is a glossary that immediately reinforces the terms necessary for student comprehension of concepts.

A master glossary is located at the end of the text and serves as a single resource for essential terminology used throughout the text.

Phonetic Pronunciations   Phonetic spellings follow many glossary entries. They can be helpful if you know how to read the symbols.

An unmarked vowel (a,e,i,o,u) at the end of a syllable has the long sound, as in the word "prey"—PRA. An unmarked vowel followed by a consonant has the short sound, as the phonetic spelling of the word "cell"—SEL.

A vowel in the middle of a syllable may have a mark over it to indicate a short or long sound. A straight bar (ā) indicates the long sound and a small arc (ă), the short sound. The word "acetyl" = ă-sĕt′l shows these two marks plus an accent (′) that tells us to stress the second syllable. Some phonetic spellings may also have a double accent (″). The double-accented syllable is stressed, too, but not as much as a single-accented syllable; for example, res″pĭ-ra′shun.

Software Correlations   Appropriate figures are correlated to the WCB *Life Science Animations* videotapes. The animations bring visual movement to biological processes that are difficult to understand on the text page. The figures correlated to these videotapes are identified with a videotape icon. -----

 Figure 5.9   The Life Cycle of *Plasmodium vivax*   The life cycle of the malaria parasite requires two hosts, the *Anopheles* mosquito and the human. Humans get malaria when they are bitten by a mosquito carrying the larval stage of *Plasmodium*. The larva undergoes asexual reproduction and releases thousands of individuals that invade red blood cells. Their release causes the chills, fever, and headache associated with malaria. Inside the red blood cell, special male and female cells called gametocytes are formed.

When the mosquito bites a person with malaria, it ingests some gametocytes. Fertilization occurs inside the mosquito and zygotes develop. The resulting larvae are housed in the mosquito's salivary glands. Thus, when the mosquito bites another person, some saliva containing the larvae is released into the person's blood and the cycle begins again.

Appropriate chapters are correlated to the WCB *Explorations in Human Biology* and *Explorations in Cell Biology and Genetics* CD-ROMs by George B. Johnson. --------

EXPLORATIONS   Interactive Software

Pollution of a Freshwater Lake
This interactive exercise allows students to explore how addition of certain "harmless" chemicals can pollute a lake by allowing algae to grow. Bacteria feeding on dead algae use up all the dissolved oxygen in the lake water, killing the lake. The exercise presents a map of Lake Washington, indicating the location of sewage treatment plants and the chemical composition of their effluent. Students can investigate the nature of pollution by altering the chemicals in the effluent. They will discover that phosphates in the effluent lead to algal growth, which is followed by bacterial growth (they feed on the dead algae) and a precipitous drop in levels of dissolved oxygen. Students can then attempt to "clean up" the lake by altering the nature and amount of effluent. Their success will depend largely upon how early in the pollution process they initiate their recovery efforts.

1. Why does the growth of algae, which are photosynthetic and make oxygen, lead to oxygen depletion of the lake?

2. Why not simply poison the algae in the lake?

3. How long can you wait before cleaning up the lake and still succeed?

4. Can you envision any way to successfully avoid oxygen depletion of the lake without stopping the discharge of treated sewage into the lake?

Useful Ancillaries

The following supplementary materials have been developed to provide instructors with a complete educational package:

Instructor's Manual/Test Item File   This supplement provides a rationale for the use of the text, with objectives, explanatory information, and answers to the Test Your Knowledge sections of the text.

Microtest   A computerized test bank of the test item file is available in DOS, Windows, and Macintosh formats.

Transparencies   Seventy-five color acetates are available to adopters of *Diversity of Life*. The transparencies are taken from the text and represent important figures that merit extra visual review and discussion.

Laboratory Manual   Laboratory experiments help students experience biology in action. WCB is proud to offer the laboratory manual that accompanies *Concepts in Biology* by Enger et al. with this text.

Additional WCB Supplements

Explorations in Human Biology and Explorations in Cell Biology and Genetics CD-ROMs.

Each of these interactive CDs by Dr. George B. Johnson comprises sixteen modules, featuring fascinating topics in biology. These interactive investigations are correlated to appropriate topical material in *Diversity of Life*.

Life Science Animations Videotapes

Fifty-three animations of key physiological processes are available on videotapes. The animations, correlated to this text by a videotape icon, bring visual movement to biological processes that are difficult to understand on the text page.

Life Science Living Lexicon CD-ROM

This interactive CD-ROM, by Will Marchuk of Red Deer College, contains a comprehensive list of life science terms with definitions of their roots, prefixes, and suffixes as well as audio pronunciations and illustrations. The lexicon is student-interactive, featuring quizzing and note-taking capabilities. It contains 4,500 terms broken down into the following categories: anatomy and physiology, botany, cell and molecular biology, genetics, ecology and evolution, and zoology.

Biology StartUp

This five-disk set of Macintosh tutorials by Myles C. Robinson and Kathleen Hakola Pace, Grays Harbor College, is designed to help non-major students master challenging biological processes like chemistry and cell biology. This set can be a valuable addition to a resource center and is especially helpful to students enrolled in developmental education courses or those who need additional assistance to succeed in an introductory biology course.

You Can Make a Difference
Judith Getis

This short, inexpensive supplement offers students practical guidelines for recycling, conserving energy, disposing of hazardous wastes, and other pollution controls. It can be shrink-wrapped with the text at minimal additional cost. (ISBN 0-697-13923-9)

How to Study Science
Fred Drewes, Suffolk County Community College

This useful workbook offers students helpful suggestions for meeting the challenges of a college science course. It offers tips on how to take notes, how to get the most out of laboratories, and how to overcome science anxiety. The book's unique design helps students develop critical thinking skills while facilitating careful note taking. (ISBN 0-697-14474-7)

Biology Study Cards
Kent Van De Graaff, R. Ward Rhees, and Christopher H. Creek, Brigham Young University

This boxed set of 300 two-sided study cards provides a quick yet thorough visual synopsis of key biological terms and concepts in the general biology curriculum. Each card features a masterful illustration, pronunciation guide, definition, and description in context. (ISBN 0-697-03069-5)

How Scientists Think
George B. Johnson

This paperbound text describes twenty-one experiments that have shaped our understanding of genetics and molecular biology. It fosters critical thinking and reinforces the scientific method.

Acknowledgments

A large number of people have knowingly or unknowingly helped us write this text. Our families continued to give understanding and support as we worked on this project. We acknowledge the thousands of students in our classes who have given us feedback over the years concerning the material and its relevancy. They were the best possible source of criticism.

We gratefully acknowledge the valuable assistance of many reviewers throughout the development and preparation of the manuscript:

Sylvester Allred
Northern Arizona University

Gail F. Baker
La Guardia Community College

Greg Bohm
Florida Institute of Technology

Marilyn P. Boysen
California State University

A. Braganza
Chabot College

Richard D. Brown
Brunswick Community College

Peter Castro
California State Polytechnic University, Pomona

Elizabeth A. Desy
Southwest State University

H. T. Hendrickson
University of North Carolina at Greensboro

Frank M. Mele
Jersey City State College

Gail L. Miller
York College

Robert D. Muckel
Doane College

Paul F. Nicoletto
Shippensburg University of Pennsylvania

James Pawley
University of Wisconsin, Madison

Quentin Reuer
University of Alaska, Anchorage

Lyndon Robinson
Aims Community College

Leba Sarkis
Aims Community College

Edward J. Siden
Felician College

Bradley W. Smith
Sauk Valley Community College

K. Dale Smoak
Piedmont Technical College

Daniel A. Stephens
Wenatchee Valley College

Martin B. Trent
Sinclair Community College

Edmund J. Zimmerer
Murray State University

part

Diversity of Living Things

Science is an important part of a technological society. Without scientific advances we would not enjoy the standard of living we have. Materials science, for example, has developed the computer chip. Engineering science has given us safer buildings, manufacturing processes, and automobiles. Medical science has developed vaccines against measles, smallpox, mumps, whooping cough, chicken pox, and many other "childhood" diseases. These applied sciences are closely tied to the theoretical sciences of chemistry, physics, and biology. Each of these fields of study involves much more than facts and figures. The foundation of all sciences is a way of thinking and a process for finding answers to questions. Biology is the science of living things. All living things share certain common characteristics that distinguish them from nonliving things, yet each species of organism is unique. The similarities and differences that exist among organisms are reflected in a classification scheme. Kingdoms are the largest classification category. The five kingdoms of life are animals, plants, fungi, protists, and bacteria.

Biologist weighing black bear cub.

Diversity of Living Things

▸ **How do scientists solve problems?**

▸ **What qualities do all living things possess?**

▸ **How are life forms classified?**

▸ **What are the common characteristics of animals, plants, and microbes?**